Intermediate Algebra with Applications and Visualization

MyMathLab Edition

Gary K. Rockswold
Minnesota State University, Mankato

Terry A. Krieger
Rochester Community and Technical College

Boston San Francisco New York
London Toronto Sydney Tokyo Singapore Madrid
Mexico City Munich Paris Cape Town Hong Kong Montreal

1, Ingram Publishing/Getty Royalty Free **2,** (top) PhotoDisc Red **2,** (bottom) Blend Images/Getty Royalty Free **13,** Stockbyte Silver/Getty Royalty Free **15,** PhotoDisc **16,** Digital Vision **26,** PhotoDisc Blue **28,** Photographer's Choice/PhotoDisc Red **34,** Corbis Royalty Free **39,** Foodcollection/Getty Royalty Free **47,** Andy Rain/epa/Corbis **48,** Blend Images/Getty Royalty Free **50,** Purestock/Getty Royalty Free **55,** PhotoDisc **59,** PhotoDisc Red **63,** PhotoDisc Red **65,** Photographer's Choice/Getty Royalty Free **67,** Photographer's Choice/Getty Royalty Free **71,** PhotoDisc Red **74,** PhotoDisc Red **89,** NOAA **90,** PhotoDisc **97,** Digital Vision **100,** Stockdisk **104,** Digital Vision **109,** Corbis Royalty Free **117,** PhotoDisc **123,** Beth Anderson **137,** Corbis Royalty Free **138,** Corbis Royalty Free **142,** PhotoDisc Blue **145,** PhotoDisc Red **147,** NASA **154,** PhotoDisc Red **155,** (top) Photographer's Choice/Getty Royalty Free **155,** (bottom) PhotoDisc Red **157,** (top) PhotoDisc **157,** (bottom) PhotoDisc Red **163,** Corbis **166,** Corbis Royalty Free **168,** PhotoDisc **170,** Digital Vision **177,** PhotoDisc Red **184,** Digital Vision **188,** PhotoDisc **197,** Corbis **198,** PhotoDisc **203,** Beth Anderson **205,** Purestock/Getty Royalty Free **207,** Kobal Collection **211,** PhotoDisc Red **214,** PhotoDisc Red **219,** PhotoAlto/Getty Royalty Free **222,** Corbis Royalty Free **227,** PhotoDisc **230,** Digital Vision **234,** PhotoDisc Red **251,** Digital Vision **252,** Digital Vision **267,** PhotoDisc **275,** Image Source/Getty Royalty Free **280,** (top) Corbis Royalty Free **280,** (bottom) PhotoDisc Red **283,** Stockbyte Platinum/Getty Royalty Free **289,** PhotoDisc **291,** PhotoDisc **311,** Stockbyte **312,** Digital Vision **317,** Digital Vision **320,** PhotoDisc Red **334,** Corbis **339,** Stockbyte/Getty Royalty Free **341,** Digital Vision **348,** Sara Anderson **350,** Corbis Royalty Free **363** Corbis Royalty Free **364,** Digital Vision **371,** PhotoDisc **378,** Digital Vision **380,** DigitalStock **384,** Westend61/Getty Royalty Free **388,** Digital Vision **398,** Photographer's Choice Royalty Free/Getty Royalty Free **413,** Purestock/Getty Royalty Free **414,** PhotoDisc **424,** PhotoDisc **429,** Beth Anderson **431,** PhotoDisc **439,** NASA **443,** The Image Bank/Getty Images **447,** iStockphoto **449,** PhotoDisc **455,** PhotoDisc Blue **456,** Space Telescope Science Institute **462,** NASA Headquarters **464,** NASA **466,** NASA **479,** Corbis **480,** Westend61/Getty Royalty Free **485,** PhotoDisc Red **490,** (top) Digital Vision **490,** (bottom) Getty Royalty Free **491,** Getty Royalty Free **493** PhotoDisc **498,** Getty Royalty Free

Taken from *Intermediate Algebra with Applications and Visualization*, Third Edition, by Gary K. Rockswold and Terry A. Krieger.

Copyright © 2009 Pearson Education, Inc.
Publishing as Pearson Addison-Wesley, 75 Arlington Street, Boston, MA 02116.

All rights reserved. No part of this publication may be reproduced, stored in a retrieval system, or transmitted, in any form or by any means, electronic, mechanical, photocopying, recording, or otherwise, without the prior written permission of the publisher. Printed in the United States of America.

ISBN-13: 978-0-321-56680-5
ISBN-10: 0-321-56680-7

1 2 3 4 5 WC 11 10 09 08

CONTENTS

Photo Credits ii
Preface ix

1 REAL NUMBERS AND ALGEBRA   1

1.1 Describing Data with Sets of Numbers   2
Natural and Whole Numbers ▪ Integers and Rational Numbers ▪ Real Numbers
▪ Properties of Real Numbers

1.2 Operations on Real Numbers   8
The Real Number Line ▪ Arithmetic Operations ▪ Data and Number Sense

1.3 Integer Exponents   15
Bases and Positive Exponents ▪ Zero and Negative Exponents ▪ Product, Quotient,
and Power Rules ▪ Order of Operations ▪ Scientific Notation

1.4 Variables, Equations, and Formulas   26
Basic Concepts ▪ Modeling Data ▪ Square Roots and Cube Roots ▪ Tables and Calculators (Optional)

1.5 Introduction to Graphing   31
Relations ▪ The Cartesian Coordinate System ▪ Scatterplots and Line Graphs
▪ The Viewing Rectangle (Optional) ▪ Graphing with Calculators (Optional)

Summary   41
Show Your Work   44

2 LINEAR FUNCTIONS AND MODELS   47

2.1 Functions and Their Representations   48
Basic Concepts ▪ Representations of a Function ▪ Definition of a Function
▪ Identifying a Function ▪ Tables, Graphs, and Calculators (Optional)

2.2 Linear Functions   59
Basic Concepts ▪ Representations of Linear Functions ▪ Modeling Data with Linear Functions

2.3 The Slope of a Line   67
Slope ▪ Slope–Intercept Form of a Line ▪ Interpreting Slope in Applications

	2.4	Equations of Lines and Linear Models	74
		Point–Slope Form ▪ Horizontal and Vertical Lines ▪ Parallel and Perpendicular Lines	

Summary 84
Show Your Work 87

3 LINEAR EQUATIONS AND INEQUALITIES 89

	3.1	Linear Equations	90
		Equations ▪ Symbolic Solutions ▪ Numerical and Graphical Solutions ▪ Identities and Contradictions ▪ Intercepts of a Line	
	3.2	Introduction to Problem Solving	100
		Solving a Formula for a Variable ▪ Steps for Solving a Problem ▪ Percentages	
	3.3	Linear Inequalities	109
		Basic Concepts ▪ Symbolic Solutions ▪ Numerical and Graphical Solutions ▪ An Application	
	3.4	Compound Inequalities	117
		Basic Concepts ▪ Symbolic Solutions and Number Lines ▪ Numerical and Graphical Solutions ▪ Interval Notation	
	3.5	Absolute Value Equations and Inequalities	123
		Basic Concepts ▪ Absolute Value Equations ▪ Absolute Value Inequalities	

Summary 131
Show Your Work 134

4 SYSTEMS OF LINEAR EQUATIONS 137

	4.1	Systems of Linear Equations in Two Variables	138
		Basic Concepts ▪ Graphical and Numerical Solutions ▪ Types of Linear Systems	
	4.2	The Substitution and Elimination Methods	147
		The Substitution Method ▪ The Elimination Method ▪ Models and Applications	
	4.3	Systems of Linear Inequalities	157
		Solving Linear Inequalities in Two Variables ▪ Solving Systems of Linear Inequalities	
	4.4	Introduction to Linear Programming	163
		Basic Concepts ▪ Region of Feasible Solutions ▪ Solving Linear Programming Problems	
	4.5	Systems of Linear Equations in Three Variables	168
		Basic Concepts ▪ Solving Linear Systems with Substitution and Elimination ▪ Modeling Data ▪ Systems of Equations with No Solutions ▪ Systems of Equations with Infinitely Many Solutions	

| 4.6 | Matrix Solutions of Linear Systems | 177 |

Representing Systems of Linear Equations with Matrices ▪ Gauss–Jordan Elimination
▪ Using Technology to Solve Systems of Linear Equations (Optional)

| 4.7 | Determinants | 184 |

Calculation of Determinants ▪ Area of Regions ▪ Cramer's Rule

Summary 190
Show Your Work 194

5 POLYNOMIAL EXPRESSIONS AND FUNCTIONS 197

| 5.1 | Polynomial Functions | 198 |

Monomials and Polynomials ▪ Addition and Subtraction of Polynomials ▪ Polynomial Functions
▪ Evaluating Polynomials ▪ Operations on Functions ▪ Applications and Models

| 5.2 | Multiplication of Polynomials | 207 |

Review of Basic Properties ▪ Multiplying Polynomials ▪ Some Special Products
▪ Multiplying Functions

| 5.3 | Factoring Polynomials | 214 |

Common Factors ▪ Factoring and Equations ▪ Factoring by Grouping

| 5.4 | Factoring Trinomials | 222 |

Factoring $x^2 + bx + c$ ▪ Factoring Trinomials by Grouping ▪ Factoring Trinomials with FOIL
▪ Factoring with Graphs and Tables

| 5.5 | Special Types of Factoring | 230 |

Difference of Two Squares ▪ Perfect Square Trinomials ▪ Sum and Difference of Two Cubes

| 5.6 | Summary of Factoring | 234 |

Guidelines for Factoring Polynomials ▪ Factoring Polynomials

| 5.7 | Polynomial Equations | 238 |

Quadratic Equations ▪ Higher Degree Equations ▪ Equations in Quadratic Form ▪ Applications

Summary 245
Show Your Work 248

6 RATIONAL EXPRESSIONS AND FUNCTIONS 251

| 6.1 | Introduction to Rational Functions and Equations | 252 |

Recognizing and Using Rational Functions ▪ Solving Rational Equations ▪ Operations on Functions

6.2	**Multiplication and Division of Rational Expressions**		260
	Simplifying Rational Expressions ▪ Review of Multiplication and Division of Fractions ▪ Multiplication of Rational Expressions ▪ Division of Rational Expressions		
6.3	**Addition and Subtraction of Rational Expressions**		267
	Least Common Multiples ▪ Review of Addition and Subtraction of Fractions ▪ Addition of Rational Expressions ▪ Subtraction of Rational Expressions		
6.4	**Rational Equations**		275
	Solving Rational Equations ▪ Solving an Equation for a Variable		
6.5	**Complex Fractions**		283
	Basic Concepts ▪ Simplifying Complex Fractions		
6.6	**Modeling with Proportions and Variation**		289
	Proportions ▪ Direct Variation ▪ Inverse Variation ▪ Joint Variation		
6.7	**Division of Polynomials**		297
	Division by a Monomial ▪ Division by a Polynomial ▪ Synthetic Division		
	Summary		303
	Show Your Work		308

7 RADICAL EXPRESSIONS AND FUNCTIONS — 311

7.1	**Radical Expressions and Rational Exponents**		312
	Radical Notation ▪ Rational Exponents ▪ Properties of Rational Exponents		
7.2	**Simplifying Radical Expressions**		320
	Product Rule for Radical Expressions ▪ Quotient Rule for Radical Expressions		
7.3	**Operations on Radical Expressions**		325
	Addition and Subtraction ▪ Multiplication ▪ Rationalizing the Denominator		
7.4	**Radical Functions**		334
	The Square Root Function ▪ The Cube Root Function ▪ Power Functions ▪ Modeling with Power Functions (Optional)		
7.5	**Equations Involving Radical Expressions**		341
	Solving Radical Equations ▪ The Distance Formula ▪ Solving the Equation $x^n = k$		
7.6	**Complex Numbers**		350
	Basic Concepts ▪ Addition, Subtraction, and Multiplication ▪ Powers of i ▪ Complex Conjugates and Division		
	Summary		355
	Show Your Work		360

8 QUADRATIC FUNCTIONS AND EQUATIONS — 363

8.1 Quadratic Functions and Their Graphs — 364
Graphs of Quadratic Functions ▪ Basic Transformations of Graphs ▪ More About Graphing Quadratic Functions (Optional) ▪ Min–Max Applications

8.2 Parabolas and Modeling — 373
Vertical and Horizontal Translations ▪ Vertex Form ▪ Modeling with Quadratic Functions (Optional)

8.3 Quadratic Equations — 380
Basics of Quadratic Equations ▪ The Square Root Property ▪ Completing the Square ▪ Solving an Equation for a Variable ▪ Applications of Quadratic Equations

8.4 The Quadratic Formula — 389
Solving Quadratic Equations ▪ The Discriminant ▪ Quadratic Equations Having Complex Solutions

8.5 Quadratic Inequalities — 398
Basic Concepts ▪ Graphical and Numerical Solutions ▪ Symbolic Solutions

8.6 Equations in Quadratic Form — 404
Higher Degree Polynomial Equations ▪ Equations Having Rational Exponents ▪ Equations Having Complex Solutions

Summary — 407
Show Your Work — 410

9 EXPONENTIAL AND LOGARITHMIC FUNCTIONS — 413

9.1 Composite and Inverse Functions — 414
Composition of Functions ▪ One-to-One Functions ▪ Inverse Functions ▪ Tables and Graphs of Inverse Functions

9.2 Exponential Functions — 424
Basic Concepts ▪ Graphs of Exponential Functions ▪ Models Involving Exponential Functions ▪ The Natural Exponential Function

9.3 Logarithmic Functions — 431
The Common Logarithmic Function ▪ The Inverse of the Common Logarithmic Function ▪ Logarithms with Other Bases

9.4 Properties of Logarithms — 439
Basic Properties ▪ Change of Base Formula

9.5	Exponential and Logarithmic Equations	443
	Exponential Equations and Models ▪ Logarithmic Equations and Models	

Summary	450
Show Your Work	453

10 CONIC SECTIONS — 455

10.1	Parabolas and Circles	456
	Types of Conic Sections ▪ Parabolas with Horizontal Axes of Symmetry ▪ Equations of Circles	
10.2	Ellipses and Hyperbolas	462
	Equations of Ellipses ▪ Equations of Hyperbolas	
10.3	Nonlinear Systems of Equations and Inequalities	468
	Basic Concepts ▪ Solving Nonlinear Systems of Equations ▪ Solving Nonlinear Systems of Inequalities	

Summary	474
Show Your Work	476

11 SEQUENCES AND SERIES — 479

11.1	Sequences	480
	Basic Concepts ▪ Representations of Sequences ▪ Models and Applications	
11.2	Arithmetic and Geometric Sequences	485
	Representations of Arithmetic Sequences ▪ Representations of Geometric Sequences ▪ Applications and Models	
11.3	Series	491
	Basic Concepts ▪ Arithmetic Series ▪ Geometric Series ▪ Summation Notation	
11.4	The Binomial Theorem	498
	Pascal's Triangle ▪ Factorial Notation and Binomial Coefficients ▪ Using the Binomial Theorem	

Summary	502
Show Your Work	504
APPENDIX: Using the Graphing Calculator	AP-1
Glossary	G-1
Bibliography	B-1
Index	I-1

PREFACE

Welcome to the MyMathLab® Edition of *Intermediate Algebra with Applications and Visualization,* **Third Edition.**

MyMathLab is a complete online course available with this text and is perfect for a lecture-based, a self-paced, or an online course. This site offers instructors and students a wide variety of resources from dynamic multimedia—video clips, animations, and more—to course management tools. With MyMathLab, instructors can customize their course and help increase student comprehension and success!

- MyMathLab provides a powerful system for creating and assigning tests and homework, as well as a gradebook for tracking all student performance.
- The entire textbook is available online and is supplemented by multimedia content, such as videos and animations, which is used to explain concepts. With MyMathLab, students can work tutorial exercises tied directly to those in their textbook.
- MyMathLab allows students to do practice work and to complete instructor-assigned tests and homework assignments online. Based on their results, MyMathLab automatically builds individual study plans that students can use to improve their skills.

The MyMathLab Edition is intended for students who are required to use MyMathLab in their mathematics courses. All section exercise material found in the Study Plan of MyMathLab can be assigned as homework by the instructor.

In addition, the MyMathLab Edition correlates examples to tutorial exercises found in the MyMathLab Study Plan.

CHAPTER 1

Real Numbers and Algebra

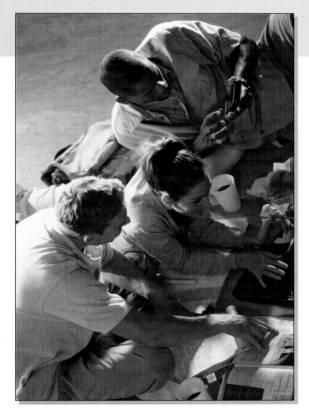

1.1 Describing Data with Sets of Numbers
1.2 Operations on Real Numbers
1.3 Integer Exponents
1.4 Variables, Equations, and Formulas
1.5 Introduction to Graphing

"Why do I need to learn math?" This question is commonly asked by students across the country. A recent cover story in *Business Week* helped explain the answer to this question. In fact, according to James R. Schatz, chief of the mathematics research group at the National Security Agency, "There has never been a better time to be a mathematician."

The world is moving into a new age of numbers. Mathematics is being used to represent all types of information, including digital music and photographs. Mathematics is the *language of technology* and businesses are relying more and more on mathematics. The Internet programs Yahoo! and Google both use a significant amount of mathematics to search millions of items in only tenths of a second. Even consumer behavior is being represented and analyzed mathematically. A company, called Umbria, is breaking down English messages into mathematics and then looking for consumer trends. Mathematicians are in high demand and "top mathematicians are becoming a new global elite." However, the United States needs to produce more mathematics majors—only about 10,000 citizens obtain a math major each year.

Decisions made by students to avoid mathematics can have lifelong ramifications that affect their vocations, incomes, and lifestyles. For example, if you switch majors to avoid mathematics, your decision may prevent you from pursuing your dreams. Like reading and writing, mathematics is an essential component for realizing your full potential.

The human mind has never invented a labor-saving device greater than algebra.

—J. W. GIBBS

Source: Stephen Baker with Bremen Leak, "Math Will Rock Your World," *Business Week*, January 23, 2006.

1.1 DESCRIBING DATA WITH SETS OF NUMBERS

Natural and Whole Numbers ▪ Integers and Rational Numbers ▪ Real Numbers ▪ Properties of Real Numbers

A LOOK INTO MATH ▷

The need for numbers has existed in nearly every society. Numbers first occurred in the measurement of time, currency, goods, and land. As the complexity of a society increased, so did its numbers. One tribe that lived near New Guinea counted only to 6. Any number higher than 6 was referred to as "many." This number system met the needs of that society. However, our highly technical society could not function with so few numbers. For example, cell phones, iPods, and digital cameras have created a need for a wide variety of numbers.
(*Source: Historical Topics for the Mathematics Classroom, Thirty-first Yearbook,* NCTM.)

In this section we discuss numbers that are vital to our technological society. We also show how different types of data can be described with sets of numbers.

Natural and Whole Numbers

One important set of numbers found in most societies is the set of **natural numbers**. These numbers comprise the counting numbers and may be expressed as

$$N = \{1, 2, 3, 4, 5, 6, \ldots\}.$$

Set braces, { }, are used to enclose the elements of a set. Because there are infinitely many natural numbers, three dots show that the list continues in the same pattern without end. A second set of numbers, called the **whole numbers**, is given by

$$W = \{0, 1, 2, 3, 4, 5, \ldots\}.$$

Natural numbers and whole numbers can be used when data are not broken into fractional parts. For example, Table 1.1 lists the number of bachelor's degrees awarded during selected academic years. Note that either natural numbers or whole numbers are appropriate to describe the data because a fraction of a degree cannot be awarded.

TABLE 1.1 Bachelor's Degrees Awarded

Year	1979–1980	1989–1990	1999–2000	2004–2005
Degrees	929,417	1,051,344	1,237,875	1,352,000

Source: Department of Education.

EXAMPLE 1 Describing iPod size

The amount of memory in an iPod is usually measured in gigabytes (GB), where 1 GB is equal to 1,073,741,824 (or approximately 1 billion) bytes of memory.
(a) How many bytes of memory would a 4 GB iPod have?
(b) Would the amount of memory in an iPod be better described by a natural number or a fraction? Explain your answer.

Solution
(a) $4 \times 1{,}073{,}741{,}824 = 4{,}294{,}967{,}296$ bytes.
(b) A natural number is better because a fraction of a byte is not allowed.

Integers and Rational Numbers

The set of **integers** is given by

$$I = \{\ldots, -3, -2, -1, 0, 1, 2, 3, \ldots\}.$$

The integers include both the natural numbers and the whole numbers. During the eighteenth century, negative numbers were not readily accepted by all mathematicians. Such numbers did not seem to have any physical meaning. However, today when a person opens a personal checking account for the first time, negative numbers quickly take on meaning. There is a difference between a positive and a negative balance.

A **rational number** is any number that can be expressed as the ratio of two integers: $\frac{p}{q}$, where q is not equal to 0 because we cannot divide by 0. Rational numbers can be written as fractions and include all integers. Some examples of rational numbers are

$$\frac{8}{1}, \quad \frac{2}{3}, \quad -\frac{3}{5}, \quad -\frac{7}{2}, \quad \frac{22}{7}, \quad 1.2, \quad \text{and} \quad 0.$$

Note that 1.2 and 0 are both rational numbers because they can be written as $\frac{12}{10}$ and $\frac{0}{1}$.

Rational numbers may be expressed in decimal form that either *repeats* or *terminates*. The fraction $\frac{1}{3}$ may be expressed as $0.\overline{3}$, a repeating decimal, and the fraction $\frac{1}{4}$ may be expressed as 0.25, a terminating decimal. The overbar indicates that $0.\overline{3} = 0.3333333\ldots$.

▶ REAL-WORLD CONNECTION Integers and rational numbers are used to describe things such as temperature. Table 1.2 lists equivalent temperatures in both degrees Fahrenheit and degrees Celsius. Note that both positive and negative numbers are used to describe temperature.

TABLE 1.2 Fahrenheit and Celsius Temperature

°F	°C	Observation
−89	$-67.\overline{2}$	Alcohol freezes
−40	−40	Mercury freezes
0	$-17.\overline{7}$	Snow and salt mixture freezes
32	0	Water freezes
100	$37.\overline{7}$	A very warm day
212	100	Water boils

EXAMPLE 2 Classifying numbers

Classify each number as one or more of the following: natural number, whole number, integer, or rational number.

(a) $\frac{6}{3}$ (b) -1 (c) 0 (d) $-\frac{11}{3}$

Solution
(a) Because $\frac{6}{3} = 2$, the number $\frac{6}{3}$ is a natural number, a whole number, an integer, and a rational number.
(b) The number -1 is an integer and a rational number but not a natural or a whole number.
(c) The number 0 is a whole number, an integer, and a rational number but not a natural number.
(d) The fraction $-\frac{11}{3}$ is a rational number as it is the ratio of two integers. However, it is not a natural number, a whole number, or an integer.

Real Numbers

Real numbers can be represented by decimal numbers. Every fraction has a decimal form, so real numbers include rational numbers. However, some real numbers cannot be expressed by fractions. They are called **irrational numbers**. If a square root of a positive integer is not an integer, then it is an irrational number. The numbers $\sqrt{2}$, $\sqrt{15}$, and π are examples of irrational numbers. They can be expressed by decimals but *not* by decimals that either repeat or terminate. Examples of real numbers include

$$2, \quad -10, \quad 151\frac{1}{4}, \quad -131.37, \quad \frac{1}{3}, \quad -\sqrt{5}, \quad \text{and} \quad \sqrt{11}.$$

CALCULATOR HELP
To evaluate π and square roots, see the Appendix (page AP-1).

Any real number may be approximated by a terminating decimal. We use the symbol $\approx$, which means **approximately equal**, to denote an approximation. Each of the following real numbers has been approximated to three *decimal places*.

$$\pi \approx 3.142, \quad \frac{2}{3} \approx 0.667, \quad \sqrt{200} \approx 14.142$$

Figure 1.1 shows the relationships among the different sets of numbers. Note that each real number is either a rational number or an irrational number but not both. The natural numbers, whole numbers, and integers are contained in the set of rational numbers.

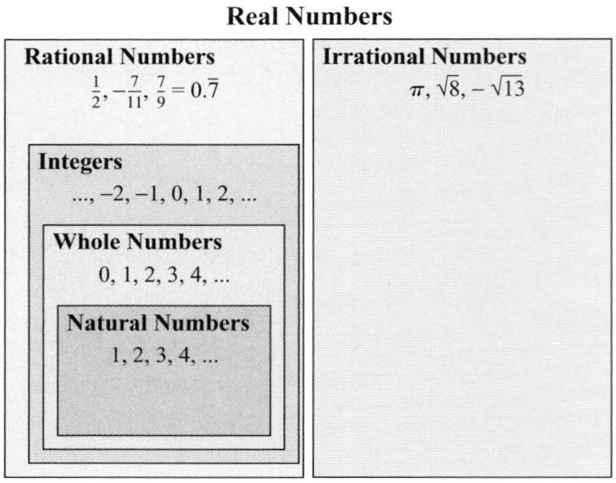

Real Numbers

Rational Numbers	Irrational Numbers
$\frac{1}{2}, -\frac{7}{11}, \frac{7}{9} = 0.\overline{7}$	$\pi, \sqrt{8}, -\sqrt{13}$
Integers ..., −2, −1, 0, 1, 2, ...	
Whole Numbers 0, 1, 2, 3, 4, ...	
Natural Numbers 1, 2, 3, 4, ...	

Figure 1.1 The Set of Real Numbers

EXAMPLE 3 Classifying numbers

Classify each real number as one or more of the following: a natural number, an integer, a rational number, or an irrational number.

$$5, \quad -1.2, \quad \frac{13}{7}, \quad -\sqrt{7}, \quad -12, \quad \sqrt{16}$$

Solution
Natural numbers: 5 and $\sqrt{16} = 4$
Integers: 5, −12, and $\sqrt{16} = 4$
Rational numbers: 5, −1.2, $\frac{13}{7}$, −12, and $\sqrt{16} = 4$
Irrational number: $-\sqrt{7}$

Even though a data set may contain only integers, decimals are often needed to describe it. One common way to do so is to find the **average**. To calculate the average of a set of numbers, we find their sum and divide by the number of numbers in the set.

EXAMPLE 4　Analyzing test scores

A student obtains the following test scores: 81, 96, 79, and 82.
(a) Find the student's average test score.
(b) Is this average a natural, a rational, or a real number?

Solution
(a) To find the average, we find the sum of the four test scores and then divide by 4:

$$\frac{81 + 96 + 79 + 82}{4} = \frac{338}{4} = 84.5.$$

(b) The average of these four test scores is both a rational number *and* a real number but is not a natural number.

CRITICAL THINKING

Is the sum of two irrational numbers ever a rational number? Explain.

Properties of Real Numbers

Several properties of real numbers are used in algebra. They are the identity properties, the commutative properties, the associative properties, and the distributive properties. These properties are essential to understand algebra. If a person does not know "the rules of the game," then it is very difficult to use algebra correctly to solve problems.

IDENTITY PROPERTIES　The **identity property of 0** states that, if 0 is added to any real number a, the result is a. The number 0 is called the **additive identity**. For example,

$$-3 + 0 = -3 \quad \text{and} \quad 0 + 18 = 18.$$

The **identity property of 1** states that, if any number a is multiplied by 1, the result is a. The number 1 is called the **multiplicative identity**. Examples include

$$-7 \cdot 1 = -7 \quad \text{and} \quad 1 \cdot 9 = 9.$$

We can summarize these results as follows.

IDENTITY PROPERTIES

For any real number a,

$$a + 0 = 0 + a = a$$

and

$$a \cdot 1 = 1 \cdot a = a.$$

COMMUTATIVE PROPERTIES　The **commutative property for addition** states that two numbers, a and b, can be added together in any order and the result is the same. That is, $a + b = b + a$. For example, if a person is paid \$5 and then \$7 or paid \$7 and then \$5, the result is the same. Either way the person is paid a total of

$$5 + 7 = 7 + 5 = \$12.$$

There is also a **commutative property for multiplication**. It states that two numbers, a and b, can be multiplied in any order and the result is the same. That is, $a \cdot b = b \cdot a$. For example, 3 groups of 5 people or 5 groups of 3 people both contain

$$3 \cdot 5 = 5 \cdot 3 = 15 \text{ people}.$$

We can summarize these results as follows.

COMMUTATIVE PROPERTIES

For any real numbers a and b,
$$a + b = b + a$$
and
$$a \cdot b = b \cdot a.$$

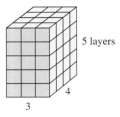

Figure 1.2

ASSOCIATIVE PROPERTIES The commutative properties allow us to interchange the order of two numbers when we add or multiply. The associative properties allow us to change how numbers are grouped. For example, we may add the numbers 3, 4, and 5 as follows.
$$(3 + 4) + 5 = 7 + 5 = 12$$
$$3 + (4 + 5) = 3 + 9 = 12$$

In either case we obtain the same answer, which is the result of the **associative property for addition**. Note that we did not change the order of the numbers; rather, we changed how the numbers were grouped. There is also an **associative property for multiplication**, which is illustrated as follows.
$$(3 \cdot 4) \cdot 5 = 12 \cdot 5 = 60$$
$$3 \cdot (4 \cdot 5) = 3 \cdot 20 = 60$$

We can stack cubes as shown in Figures 1.2 and 1.3 to illustrate the associative property for multiplication. The stacks are shown with either 5 layers of 12 cubes or 3 layers of 20 cubes. In both cases there is a total of 60 cubes in the stack.

We can summarize these results as follows.

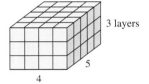

Figure 1.3

ASSOCIATIVE PROPERTIES

For any real numbers a, b, and c,
$$(a + b) + c = a + (b + c)$$
and
$$(a \cdot b) \cdot c = a \cdot (b \cdot c).$$

NOTE: Sometimes we omit the multiplication dot. Thus $a \cdot b = ab$ and $5 \cdot x = 5x$.

EXAMPLE 5 Identifying properties of real numbers

State the property of real numbers that justifies each statement.
(a) $4 \cdot (3x) = (4 \cdot 3)x$ **(b)** $(1 \cdot 5) \cdot 4 = 5 \cdot 4$ **(c)** $5 + ab = ab + 5$

Solution
(a) This equation illustrates the associative property for multiplication, with the grouping of the numbers changed. That is, $4 \cdot (3x) = (4 \cdot 3)x = 12x$.
(b) This equation illustrates the identity property of 1 because $1 \cdot 5 = 5$.
(c) This equation illustrates the commutative property for addition; the order of the terms 5 and ab changed.

1.1 DESCRIBING DATA WITH SETS OF NUMBERS

DISTRIBUTIVE PROPERTIES The **distributive properties** (see Figure 1.4) are used frequently in algebra to simplify expressions. An example of a distributive property is

$$3(6 + 5) = 3 \cdot 6 + 3 \cdot 5.$$

It is important to multiply the 3 by *both* the 6 and 5, not just the 6. This distributive property is valid when addition is replaced by subtraction. For example,

$$3(6 - 5) = 3 \cdot 6 - 3 \cdot 5.$$

We can summarize these results as follows.

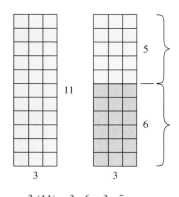

$3(11) = 3 \cdot 6 + 3 \cdot 5$

Figure 1.4

DISTRIBUTIVE PROPERTIES

For any real numbers a, b, and c,

$$a(b + c) = ab + ac$$

and

$$a(b - c) = ab - ac.$$

NOTE: Because multiplication is commutative, the distributive properties may be written as

$$(b + c)a = ba + ca \quad \text{and} \quad (b - c)a = ba - ca.$$

EXAMPLE 6 Applying the distributive properties

Apply a distributive property to each expression.
(a) $5(4 + x)$ **(b)** $10 - (1 + a)$ **(c)** $9x - 5x$ **(d)** $5x + 2x - 3x$

Solution
(a) $5(4 + x) = 5 \cdot 4 + 5 \cdot x = 20 + 5x$
(b) $10 - (1 + a) = 10 - 1(1 + a) = 10 - (1 \cdot 1) - (1 \cdot a) = 9 - a$
(c) $9x - 5x = (9 - 5)x = 4x$
(d) Because each term contains an x, apply the distributive properties to all three terms:

$$5x + 2x - 3x = (5 + 2 - 3)x = 4x.$$

1.1 PUTTING IT ALL TOGETHER

Data and numbers play a central role in a diverse, technological society. Because of the variety of data, it has been necessary to develop different sets of numbers. Without numbers, data could be described qualitatively but not quantitatively. For example, we might say that the day seems hot, but we would not be able to give an actual number for the temperature. The following table summarizes some of the sets of numbers.

Concept	Comments	Examples
Natural Numbers	Are referred to as the *counting numbers*	1, 2, 3, 4, 5, . . .
Whole Numbers	Include the natural numbers	0, 1, 2, 3, 4, . . .
Integers	Include the natural numbers and the whole numbers	. . . , −2, −1, 0, 1, 2, . . .

continued on next page

CHAPTER 1 REAL NUMBERS AND ALGEBRA

continued from previous page

Concept	Comments	Examples
Rational Numbers	Include integers and all fractions $\frac{p}{q}$, where p and q are integers (q is not equal to 0), and all repeating and terminating decimals	$\frac{1}{2}, -3, \frac{128}{6}, -0.335, 0,$ $0.25 = \frac{1}{4},$ and $0.\overline{3} = \frac{1}{3}$
Irrational Numbers	Include real numbers that cannot be expressed *exactly* by a fraction (rational number)	$-\pi, \sqrt{2},$ and $\sqrt{7}$
Real Numbers	Any number that can be expressed in decimal form, including the rational numbers and the irrational numbers	$\pi, \sqrt{3}, -\frac{4}{7}, 0, -10,$ $0.\overline{6} = \frac{2}{3}, 3,$ and $\sqrt{15}$

Real numbers have several important properties, which are summarized in the following table.

Property	Definition	Examples
Identity (0 and 1)	The identity for addition is 0 and the identity for multiplication is 1. For any real number a, $a + 0 = a$ and $a \cdot 1 = a$.	$5 + 0 = 5$ and $5 \cdot 1 = 5$
Commutative	For any real numbers a and b, $a + b = b + a$ and $a \cdot b = b \cdot a$.	$4 + 6 = 6 + 4$ and $4 \cdot 6 = 6 \cdot 4$
Associative	For any real numbers a, b, and c, $(a + b) + c = a + (b + c)$ and $(a \cdot b) \cdot c = a \cdot (b \cdot c)$.	$(3 + 4) + 5 = 3 + (4 + 5)$ and $(3 \cdot 4) \cdot 5 = 3 \cdot (4 \cdot 5)$
Distributive	For any real numbers a, b, and c, $a(b + c) = ab + ac$ and $a(b - c) = ab - ac$.	$5(x + 2) = 5x + 10,$ $5(x - 2) = 5x - 10,$ and $5x + 4x = (5 + 4)x = 9x$

1.2 OPERATIONS ON REAL NUMBERS

The Real Number Line ■ Arithmetic Operations ■ Data and Number Sense

A LOOK INTO MATH ▷ Real numbers are used to describe data. To obtain information from data we frequently perform operations on real numbers. For example, exams are often assigned a score between 0 and 100. This step reduces each exam to a number or *data point*. We might obtain more information about the exams by calculating the average score. To do so we perform the arithmetic operations of addition and division.

In this section we discuss operations on real numbers and provide examples of where these computations occur in real life.

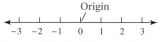

Figure 1.5 The Number Line

The Real Number Line

You can visualize the real number system by using a number line, as shown in Figure 1.5. Each real number corresponds to a point on the number line. The point associated with the real number 0 is called the **origin**.

If a real number a is located to the left of a real number b on the number line, we say that a is **less than** b and write $a < b$. Similarly, if a real number a is located to the right of a real number b, we say that a is **greater than** b and write $a > b$. Thus $-3 < 2$ because -3 is located to the left of 2, and $2 > -3$ because 2 is located to the right of -3. If $a > 0$, then a is a **positive number**, and if $a < 0$, then a is a **negative number**.

The **absolute value** of a real number a, written $|a|$, is equal to its distance from the origin on the number line. Distance may be either a positive number or zero, but it cannot be a negative number. As the points corresponding to 2 and -2 are both 2 units from the origin, $|2| = 2$ and $|-2| = 2$, as shown in Figure 1.6. The absolute value of a real number is *never* negative.

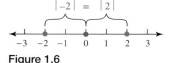
Figure 1.6

EXAMPLE 1 Finding the absolute value of a real number

Find the absolute value of each real number.
(a) 9.12   (b) $-\frac{3}{4}$ (c) $-\pi$
(d) $-a$, if a is a positive number
(e) a, if a is a negative number

Solution
(a) $|9.12| = 9.12$ (b) $\left|-\frac{3}{4}\right| = \frac{3}{4}$
(c) $|-\pi| = \pi$ because $\pi \approx 3.14$ is a positive number.
(d) If a is positive, $-a$ is negative and $|-a| = a$.
(e) If a is negative, $-a$ is positive. Thus $a < 0$ implies that $|a| = -a$. For example, if we let $a = -5$, then $|-5| = -(-5) = 5$.

The absolute value of a number can be defined as follows.

$$|a| = a \quad \text{if } a > 0 \text{ or } a = 0, \text{ and}$$
$$|a| = -a \quad \text{if } a < 0.$$

Arithmetic Operations

The four arithmetic operations are addition, subtraction, multiplication, and division.

ADDITION AND SUBTRACTION OF REAL NUMBERS In an addition problem, the two numbers added are called **addends**, and the answer is called the **sum**. In the addition problem $3 + 5 = 8$, the numbers 3 and 5 are the addends and 8 is the sum.

The **additive inverse** or **opposite** of a real number a is $-a$. For example, the additive inverse of 5 is -5 and the additive inverse of -1.6 is $-(-1.6)$, or 1.6. When we add opposites, the result is 0. That is, $a + (-a) = 0$ for every real number a.

Addends
$\overbrace{3 + 5} = 8$
         Sum

EXAMPLE 2 Finding additive inverses

Find the additive inverse or opposite of each number or expression. Then find the sum of the number or expression and its opposite.

(a) 10,961   (b) π (c) $-\frac{3}{4}$ (d) $6x - 2$

Solution
(a) The opposite of 10,961 is $-10,961$. Their sum is $10,961 + (-10,961) = 0$.
(b) The opposite of π is $-\pi$. Their sum is $\pi + (-\pi) = 0$.
(c) The opposite of $-\frac{3}{4}$ is $\frac{3}{4}$. Their sum is $-\frac{3}{4} + \frac{3}{4} = 0$.
(d) The opposite of $6x - 2$ is $-(6x - 2) = -6x + 2$. Their sum is

$$6x - 2 + (-6x) + 2 = 6x + (-6x) - 2 + 2 = 0.$$

▶ **REAL-WORLD CONNECTION** When you add real numbers, it may be helpful to think of money. A positive number represents being paid an amount of money, whereas a negative number indicates a debt owed. The sum

$$8 + (-6) = 2$$

would represent being paid $8 and owing $6, resulting in $2 being left over. Similarly,

$$-7 + (-5) = -12$$

would represent owing $7 and owing $5, resulting in a debt of $12.

To add two real numbers we may use the following rules.

ADDITION OF REAL NUMBERS

To add two numbers that are either *both positive* or *both negative*, add their absolute values. Their sum has the same sign as the two numbers.

To add two numbers with *opposite signs*, find the absolute value of each number. Subtract the smaller absolute value from the larger. The sum has the same sign as the number with the larger absolute value. The sum of two opposites is 0.

The next example illustrates addition of real numbers.

EXAMPLE 3 Adding real numbers

Evaluate each expression.
(a) $-3 + (-5)$ (b) $-4 + 7$ (c) $8.4 + (-9.5)$

Solution
(a) The addends are both negative, so we add the absolute values $|-3|$ and $|-5|$ to obtain 8. As the signs of the addends are both negative, the sum has the same sign and equals -8. That is, $-3 + (-5) = -8$. If we owe $3 and then owe an additional $5, the total amount owed is $8.
(b) The addends have opposite signs, so we subtract their absolute values to obtain 3. The sum is positive because $|7|$ is greater than $|-4|$. That is, $-4 + 7 = 3$. If we owe $4 and are paid $7, the result is that we have $3 to keep.
(c) $8.4 + (-9.5) = -1.1$ because $|-9.5|$ is 1.1 more than $|8.4|$. If we are paid $8.40 and owe $9.50, we still owe $1.10.

Addition of positive and negative numbers occurs at banks if deposits are represented by positive numbers and withdrawals are represented by negative numbers.

EXAMPLE 4 Balancing a checking account

The initial balance in a checking account is $157. Find the final balance if the following represents a list of withdrawals and deposits: $-55, -19, 123, -98$.

Solution
We find the sum.
$$157 + (-55) + (-19) + 123 + (-98) = 102 + (-19) + 123 + (-98)$$
$$= 83 + 123 + (-98)$$
$$= 206 + (-98)$$
$$= 108$$

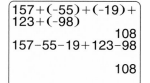

Figure 1.7

$8 - 5 = 3$
Difference

TECHNOLOGY NOTE:
Subtraction and Negation
On graphing calculators, two different keys typically represent subtraction and negation. In the first calculation in Figure 1.7, we used the negation key, and in the second calculation, we used the subtraction key.

The final balance is $108. This result may be supported by evaluating the expression with a calculator. See Figure 1.7, where the sum has been calculated two different ways. Instead of adding the opposite of a number, we can also subtract.

The answer to a subtraction problem is called the **difference**. When you're subtracting two real numbers, it sometimes helps to change the subtraction problem to an addition problem.

SUBTRACTION OF REAL NUMBERS

For any real numbers a and b,
$$a - b = a + (-b).$$
To subtract b from a, add a and the opposite of b.

EXAMPLE 5 Subtracting real numbers

Evaluate each expression.

(a) $-12 - 7$ (b) $-5.1 - (-10.6)$ (c) $\frac{1}{2} - \left(-\frac{2}{3}\right)$ (d) $3t - 7t$

Solution
(a) $-12 - 7 = -12 + (-7) = -19$
(b) $-5.1 - (-10.6) = -5.1 + 10.6 = 5.5$
(c) $\frac{1}{2} - \left(-\frac{2}{3}\right) = \frac{1}{2} + \frac{2}{3} = \frac{3}{6} + \frac{4}{6} = \frac{7}{6}$
(d) $3t - 7t = (3 - 7)t = -4t$

Factors
$3 \cdot 5 = 15$
Product

MULTIPLICATION AND DIVISION OF REAL NUMBERS In a multiplication problem, the two numbers multiplied are called the **factors**, and the answer is called the **product**. In the problem $3 \cdot 5 = 15$, the numbers 3 and 5 are factors and 15 is the product. The **multiplicative inverse** or **reciprocal** of a nonzero number a is $\frac{1}{a}$. The product of a nonzero number a and its reciprocal is $a \cdot \frac{1}{a} = 1$. For example, the reciprocal of -5 is $-\frac{1}{5}$ because $-5 \cdot -\frac{1}{5} = 1$, and the reciprocal of $\frac{2}{3}$ is $\frac{3}{2}$ because $\frac{2}{3} \cdot \frac{3}{2} = 1$. To multiply positive or negative numbers we may use the following rules.

MULTIPLICATION OF REAL NUMBERS

The product of two numbers with *like* signs is positive. The product of two numbers with *unlike* signs is negative.

EXAMPLE 6 Multiplying real numbers

Evaluate each expression.

(a) $-11 \cdot 8$ (b) $\frac{3}{5} \cdot \frac{4}{7}$ (c) $-1.2(-10)$ (d) $(1.2)(5)(-7)$

Solution
(a) The product is negative because the factors -11 and 8 have unlike signs. Thus the product is $-11 \cdot 8 = -88$.
(b) The product is positive because both factors are positive. Thus $\frac{3}{5} \cdot \frac{4}{7} = \frac{3 \cdot 4}{5 \cdot 7} = \frac{12}{35}$.
(c) As both factors are negative, the product is positive. Thus $-1.2(-10) = 12$.
(d) $(1.2)(5)(-7) = (6)(-7) = -42$

Divisor
$20 \div 4 = 5$
Dividend Quotient

In the division problem $20 \div 4 = 5$, the number 20 is the **dividend**, 4 is the **divisor**, and 5 is the **quotient**. This division problem can also be written as $\frac{20}{4} = 5$. Division of real numbers can be defined in terms of multiplication and reciprocals.

DIVISION OF REAL NUMBERS

For real numbers a and b, with $b \neq 0$,

$$\frac{a}{b} = a \cdot \frac{1}{b}.$$

That is, to divide a by b, multiply a by the reciprocal of b.

The expression $b \neq 0$ is read "b **not equal to** 0." Note that division by 0 is always *undefined*. For example, suppose that we try to define $12 \div 0$ to be equal to some number k. Then $\frac{12}{0} = k$ and k must satisfy $0 \cdot k = 12$ because a division problem can be checked by using multiplication. (For example, $\frac{12}{3} = 4$, so $3 \cdot 4 = 12$.) But the product of 0 and any number k is 0, not 12. Thus there is no reasonable value for k, so division by 0 is undefined.

NOTE: The quotient of two numbers with *like* signs is positive, and the quotient of two numbers with *unlike* signs is negative.

EXAMPLE 7 Dividing real numbers

Evaluate each expression.

(a) $-12 \div \frac{1}{2}$ (b) $\frac{\frac{2}{3}}{-7}$ (c) $\frac{-4}{-24}$ (d) $6 \div 0$

Solution

(a) $-12 \div \frac{1}{2} = -12 \cdot \frac{2}{1} = -24$

(b) $\frac{\frac{2}{3}}{-7} = \frac{2}{3} \div (-7) = \frac{2}{3} \cdot \left(-\frac{1}{7}\right) = -\frac{2}{21}$

(c) $\frac{-4}{-24} = -4 \cdot \left(-\frac{1}{24}\right) = \frac{4}{24} = \frac{1}{6}$

(d) $6 \div 0$ is undefined because division by 0 is not possible.

Many calculators have the capability to perform arithmetic on fractions and express the answer as either a decimal or a fraction. The next example illustrates this capability.

EXAMPLE 8 Performing arithmetic operations with technology

Use a calculator to evaluate each expression as a decimal and as a fraction.

(a) $\frac{1}{3} + \frac{2}{5} - \frac{4}{9}$ **(b)** $\left(\frac{4}{9} \cdot \frac{3}{8}\right) \div \frac{2}{3}$

```
(1/3)+(2/5)-(4/9
)
            .2888888889
(1/3)+(2/5)-(4/9
)▶Frac
                  13/45
```

Figure 1.8

Solution

(a) From Figure 1.8,

$$\frac{1}{3} + \frac{2}{5} - \frac{4}{9} = 0.2\overline{8}, \quad \text{or} \quad \frac{13}{45}.$$

In Figure 1.8, the second calculation uses the "Frac" feature. This feature displays the answer as a fraction, rather than a decimal.

NOTE: Generally it is a good idea to put parentheses around fractions when you are using a calculator.

(b) From Figure 1.9, $\left(\frac{4}{9} \cdot \frac{3}{8}\right) \div \frac{2}{3} = 0.25$, or $\frac{1}{4}$.

CALCULATOR HELP

To express answers as fractions, see the Appendix (page AP-2).

```
((4/9)*(3/8))/(2
/3)
                    .25
((4/9)*(3/8))/(2
/3)▶Frac
                    1/4
```

Figure 1.9

Data and Number Sense

In everyday life we commonly make approximations involving a variety of data. To make estimations we often use arithmetic operations on real numbers.

EXAMPLE 9 Developing number sense

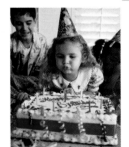

A rectangular birthday cake serves exactly 24 pieces of cake. Suppose that a similar cake is twice as wide and twice as long. How many pieces does the second cake serve?

Solution

Let L be the length and W be the width of the first cake. Then the area of this rectangular cake is $A = LW$. The length of the second cake is $2L$ and the width is $2W$. The area of this cake is $A = 2L \cdot 2W = 4LW$, or four times the area of the first cake. Thus, the second cake can serve $4 \cdot 24 = 96$ pieces of cake.

EXAMPLE 10 Determining a reasonable answer

It is 2823 miles from New York to Los Angeles. Determine mentally which of the following would best estimate the number of hours of driving time required to travel this distance in a car: 50, 100, or 120 hours.

Solution

Speed equals distance divided by time. Dividing by 100 is easy, so start by dividing 100 hours into 2800 miles. The average speed would be $\frac{2800}{100} = 28$ miles per hour, which is too slow for most drivers. A more reasonable choice is 50 hours because then the average speed would be double, or about 56 miles per hour.

EXAMPLE 11 Estimating a numeric value

Table 1.3 lists the number of subscribers of cellular telephones in selected years. Estimate the number of subscribers in 2005.

TABLE 1.3 Cellular Phone Subscribers

Year	2001	2002	2003	2004	2005
Subscribers (millions)	128	141	159	182	?

Source: Cellular Telecommunications Industry Association.

Solution
The data show that the number of subscribers has had annual increases of 13, 18, and 23 million from 2001 to 2004. Note that in each successive year the increase is 5 million more than the previous year. That is, $13 + 5 = 18$ and $18 + 5 = 23$. Following this pattern, we might expect there to be $23 + 5 = 28$ million more subscribers in 2005 than in 2004. Thus, one estimate is that there were $182 + 28 = 210$ million cell phone subscribers in 2005. Other estimates are possible.

1.2 PUTTING IT ALL TOGETHER

The following table summarizes some of the information presented in this section.

Operation	Definition	Examples
Absolute Value of a Real Number	For any real number a, $|a| = a$ if $a > 0$ or $a = 0$, and $|a| = -a$ if $a < 0$.	$|-5| = 5$, $\quad |3.7| = 3.7$, and $|-4 + 4| = 0$
Additive Inverse (Opposite)	The opposite of a is $-a$.	4 and -4 are opposites. -5 and $-(-5) = 5$ are opposites.
Addition of Real Numbers	See the highlighted box: Addition of Real Numbers on page 10.	$3 + (-6) = -3$, $\quad -2 + (-10) = -12$, $-1 + 3 = 2$, and $18 + 11 = 29$
Subtraction of Real Numbers	We can transform a subtraction problem into an addition problem: $a - b = a + (-b)$.	$4 - 6 = 4 + (-6) = -2$, $-7 - (-8) = -7 + 8 = 1$, $-8 - 5 = -8 + (-5) = -13$, and $9 - (-1) = 9 + 1 = 10$
Multiplication of Real Numbers	The product of two numbers with *like* signs is positive. The product of two numbers with *unlike* signs is negative.	$3 \cdot 5 = 15$, $\quad -4(-7) = 28$, $-8 \cdot 7 = -56$, and $5(-11) = -55$
Division of Real Numbers	For real numbers a and b with $b \neq 0$, $\frac{a}{b} = a \cdot \frac{1}{b}.$	$-3 \div \frac{3}{4} = -3 \cdot \frac{4}{3} = -4$ and $\frac{5}{2} \div \left(-\frac{7}{4}\right) = \frac{5}{2}\left(-\frac{4}{7}\right) = -\frac{20}{14} = -\frac{10}{7}$

1.3 INTEGER EXPONENTS

Bases and Positive Exponents ▪ Zero and Negative Exponents ▪ Product, Quotient, and Power Rules ▪ Order of Operations ▪ Scientific Notation

A LOOK INTO MATH ▷ Technology has brought with it the need for both small and large numbers. The size of an average virus is 5 millionths of a centimeter, whereas the distance to the nearest star, Alpha Centauri, is 25 trillion miles. To represent such numbers we often use exponents. In this section we discuss properties of integer exponents and some of their applications.
(**Source:** C. Ronan, *The Natural History of the Universe.*)

Bases and Positive Exponents

The area of a square that is 8 inches on a side is given by the expression

$$\underbrace{8 \cdot 8}_{2 \text{ factors}} = 8^2 = 64 \text{ square inches.}$$

Exponent
8^2
Base

The expression 8^2 is an **exponential expression** with **base** 8 and **exponent** 2. Exponential expressions occur frequently in a variety of applications. For example, suppose that an investment doubles its initial value 3 times. Then its final value is

$$\underbrace{2 \cdot 2 \cdot 2}_{3 \text{ factors}} = 2^3 = 8 \qquad \text{Doubles 3 times}$$

times as large as its original value. In general, if n is a positive integer and a is a real number, then

$$a^n = a \cdot a \cdot a \cdot \cdots \cdot a. \qquad (n \text{ factors of } a)$$

Table 1.4 contains examples of exponential expressions.

TABLE 1.4

Expression	Base	Exponent
2^3	2	3
6^4	6	4
7^1	7	1
0.5^2	0.5	2
x^3	x	3

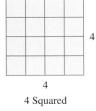

4 Squared
Figure 1.10

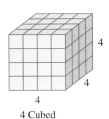

4 Cubed
Figure 1.11

Read 0.5^2 as "0.5 squared," 2^3 as "2 cubed," and 6^4 as "6 to the fourth power." The terms *squared* and *cubed* come from geometry. If the length of a side of a square is 4, then its area is

$$4 \cdot 4 = 4^2 = 16$$

square units, as illustrated in Figure 1.10. Similarly, if the length of an edge (side) of a cube is 4, then its volume is

$$4 \cdot 4 \cdot 4 = 4^3 = 64$$

cubic units, as shown in Figure 1.11.

EXAMPLE 1 Writing numbers in exponential notation

Using the given base, write each number as an exponential expression. Check your results with a calculator.

(a) 10,000 (base 10)   (b) 27 (base 3)   (c) 32 (base 2)

Solution
(a) $10,000 = 10 \cdot 10 \cdot 10 \cdot 10 = 10^4$
(b) $27 = 3 \cdot 3 \cdot 3 = 3^3$
(c) $32 = 2 \cdot 2 \cdot 2 \cdot 2 \cdot 2 = 2^5$

These values are supported in Figure 1.12, where we evaluated exponential expressions with a calculator, using the ^ key.

CALCULATOR HELP
To calculate exponential expressions, see the Appendix (page AP-1).

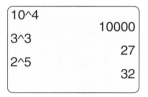

Figure 1.12

▶ REAL-WORLD CONNECTION Computer memory is often measured in bytes. A *byte* is capable of storing one letter of the alphabet. For example, the word "math" requires four bytes to store in a computer. Bytes of computer memory are often manufactured in amounts equal to powers of 2, as illustrated in the next example.

EXAMPLE 2 Using exponents to analyze computer memory

In computer technology, 1 K (kilobyte) of memory is equal to 2^{10} bytes, and 1 MB (megabyte) of memory is equal to 2^{20} bytes. Determine whether 1 K of memory is equal to 1000 bytes and whether 1 MB is equal to 1,000,000 bytes.   (*Source:* D. Horn, *Basic Electronics Theory.*)

Solution
Figure 1.13 shows that $2^{10} = 1024$ and that $2^{20} = 1,048,576$. Thus 1 K represents slightly more than 1000 bytes and 1 MB is more than 1,000,000 bytes.

Figure 1.13

Zero and Negative Exponents

Exponents can be defined for any integer. If a is any nonzero real number, we define
$$a^0 = 1.$$
For example, $3^0 = 1$ and $\left(\frac{1}{7}\right)^0 = 1$. We can also define a^{-n}, where n is a positive integer, as
$$a^{-n} = \frac{1}{a^n}.$$

Thus $5^{-4} = \frac{1}{5^4}$ and $y^{-2} = \frac{1}{y^2}$. Using the previous definition, we obtain

$$\frac{1}{a^{-n}} = \frac{1}{\frac{1}{a^n}} = \frac{a^n}{1} = a^n.$$

Thus $\frac{1}{2^{-5}} = 2^5$ and $\frac{1}{x^{-2}} = x^2$. If a and b are nonzero numbers, then

$$\frac{a^{-n}}{b^{-m}} = \frac{\frac{1}{a^n}}{\frac{1}{b^m}} = \frac{1}{a^n} \cdot \frac{b^m}{1} = \frac{b^m}{a^n}.$$

Thus $\frac{4^{-3}}{z^{-2}} = \frac{z^2}{4^3}$. This discussion leads to the following properties for integer exponents.

INTEGER EXPONENTS

Let a and b be nonzero real numbers and m and n be positive integers. Then

1. $a^n = a \cdot a \cdot a \cdot \cdots \cdot a$ (n factors of a)
2. $a^0 = 1$ (Note: 0^0 is undefined.)
3. $a^{-n} = \frac{1}{a^n}$ and $\frac{1}{a^{-n}} = a^n$
4. $\frac{a^{-n}}{b^{-m}} = \frac{b^m}{a^n}$
5. $\left(\frac{a}{b}\right)^{-n} = \left(\frac{b}{a}\right)^n$

EXAMPLE 3 Evaluating expressions

Evaluate each expression.

(a) 3^{-4} (b) $\frac{1}{2^{-3}}$ (c) $\left(\frac{5}{7}\right)^{-2}$ (d) $\frac{1}{(xy)^{-1}}$ (e) $\frac{2^{-2}}{3t^{-3}}$

Solution

(a) $3^{-4} = \frac{1}{3^4} = \frac{1}{3 \cdot 3 \cdot 3 \cdot 3} = \frac{1}{81}$ (b) $\frac{1}{2^{-3}} = 2^3 = 2 \cdot 2 \cdot 2 = 8$

(c) $\left(\frac{5}{7}\right)^{-2} = \left(\frac{7}{5}\right)^2 = \frac{7}{5} \cdot \frac{7}{5} = \frac{49}{25}$ (d) $\frac{1}{(xy)^{-1}} = (xy)^1 = xy$

 Base is xy.

(e) Note that only t and *not* $3t$ is raised to the power of -3.

$$\frac{2^{-2}}{3t^{-3}} = \frac{t^3}{3(2^2)} = \frac{t^3}{3 \cdot 4} = \frac{t^3}{12}$$

 Base is t.

Powers of 10 are important because they are used in mathematics to express numbers that are either small or large in absolute value. Table 1.5 may be used to simplify powers of 10. Note that, if the power decreases by 1, the result decreases by a factor of $\frac{1}{10}$.

CRITICAL THINKING

Use Table 1.5 to explain why it is reasonable for 10^0 to equal 1.

TABLE 1.5 Powers of Ten

Power of 10	10^3	10^2	10^1	10^0	10^{-1}	10^{-2}	10^{-3}
Value	1000	100	10	1	$\frac{1}{10}$	$\frac{1}{100}$	$\frac{1}{1000}$

Product, Quotient, and Power Rules

We can calculate products and quotients of exponential expressions *provided their bases are the same*. For example,

$$3^2 \cdot 3^3 = \underbrace{(3 \cdot 3)}_{2 \text{ factors}} \cdot \underbrace{(3 \cdot 3 \cdot 3)}_{3 \text{ factors}} = 3^5. \quad 2 + 3 = 5$$

This expression has a total of $2 + 3 = 5$ factors of 3, so the result is 3^5. To multiply exponential expressions with like bases, add exponents. Thus

$$x^3 \cdot x^4 = \underbrace{(x \cdot x \cdot x)}_{3 \text{ factors}} \cdot \underbrace{(x \cdot x \cdot x \cdot x)}_{4 \text{ factors}} = x^7. \quad 3 + 4 = 7$$

THE PRODUCT RULE

For any number a and integers m and n,

$$a^m \cdot a^n = a^{m+n}.$$

Note that the product rule holds for negative exponents. For example,

$$10^5 \cdot 10^{-2} = 10^{5+(-2)} = 10^3.$$

EXAMPLE 4 Using the product rule

Multiply and simplify. Use positive exponents.

(a) $10^2 \cdot 10^4$ **(b)** $7^3 \, 7^{-4}$ **(c)** $x^3 x^{-2} x^4$ **(d)** $3y^2 \cdot 2y^{-4}$

Solution

(a) $10^2 \cdot 10^4 = 10^{2+4} = 10^6 = 1,000,000$ Add exponents.

(b) $7^3 \, 7^{-4} = 7^{3+(-4)} = 7^{-1} = \dfrac{1}{7}$

(c) $x^3 x^{-2} x^4 = x^{3+(-2)+4} = x^5$

(d) $3y^2 \cdot 2y^{-4} = 3 \cdot 2 \cdot y^2 \cdot y^{-4} = 6y^{2+(-4)} = 6y^{-2} = \dfrac{6}{y^2}$

Note that 6 is *not* raised to the power of -2 in the expression $6y^{-2}$.

Consider division of exponential expressions. Here

$$\dfrac{6^5}{6^3} = \dfrac{6 \cdot 6 \cdot 6 \cdot 6 \cdot 6}{6 \cdot 6 \cdot 6} = \dfrac{6}{6} \cdot \dfrac{6}{6} \cdot \dfrac{6}{6} \cdot 6 \cdot 6 = 1 \cdot 1 \cdot 1 \cdot 6 \cdot 6 = 6^2. \quad 5 - 3 = 2$$

Because there are two more 6s in the numerator, the result is $6^{5-3} = 6^2$. Thus, to divide exponential expressions with like bases, subtract exponents.

THE QUOTIENT RULE

For any nonzero number a and integers m and n,

$$\dfrac{a^m}{a^n} = a^{m-n}.$$

1.3 INTEGER EXPONENTS   19

Note that the quotient rule holds for negative exponents. For example,

$$\frac{2^{-6}}{2^{-4}} = 2^{-6-(-4)} = 2^{-2} = \frac{1}{2^2}.$$

However, you may want to evaluate this quotient as

$$\frac{2^{-6}}{2^{-4}} = \frac{2^4}{2^6} = \frac{1}{2^2}.$$

NOTE: The quotient rule can also be used to justify that $a^0 = 1$. For example, $\frac{3^2}{3^2} = \frac{9}{9} = 1$ and by the quotient rule $\frac{3^2}{3^2} = 3^{2-2} = 3^0$. Thus $3^0 = 1$.

EXAMPLE 5   Using the quotient rule

Simplify each expression. Use positive exponents.

(a) $\dfrac{10^4}{10^6}$ (b) $\dfrac{x^5}{x^2}$ (c) $\dfrac{15x^2y^3}{5x^4y}$ (d) $\dfrac{3a^{-2}b^5}{9a^4b^{-3}}$

Solution

(a) $\dfrac{10^4}{10^6} = 10^{4-6} = 10^{-2} = \dfrac{1}{10^2} = \dfrac{1}{100}$ Subtract exponents.

(b) $\dfrac{x^5}{x^2} = x^{5-2} = x^3$

(c) $\dfrac{15x^2y^3}{5x^4y} = \dfrac{15}{5} \cdot \dfrac{x^2}{x^4} \cdot \dfrac{y^3}{y^1} = 3 \cdot x^{2-4}y^{3-1} = 3x^{-2}y^2 = \dfrac{3y^2}{x^2}$

(d) $\dfrac{3a^{-2}b^5}{9a^4b^{-3}} = \dfrac{3b^5b^3}{9a^4a^2} = \dfrac{b^8}{3a^6}$

How should we evaluate $(4^3)^2$? To answer this question consider

$$(4^3)^2 = 4^3 \cdot 4^3 = 4^{3+3} = 4^6. \qquad 3 \cdot 2 = 6$$

Similarly,

$$(x^4)^3 = x^4 \cdot x^4 \cdot x^4 = x^{4+4+4} = x^{12}. \qquad 4 \cdot 3 = 12$$

These results suggest that, to raise a power to a power, multiply the exponents.

RAISING POWERS TO POWERS

For any real number a and integers m and n,

$$(a^m)^n = a^{mn}.$$

EXAMPLE 6 Raising powers to powers

Simplify each expression. Use positive exponents.

(a) $(5^2)^3$ **(b)** $(2^4)^{-2}$ **(c)** $(b^{-7})^5$ **(d)** $\dfrac{(x^3)^{-2}}{(x^{-5})^2}$

Solution

(a) $(5^2)^3 = 5^{2 \cdot 3} = 5^6 = 15{,}625$ Multiply exponents.

(b) $(2^4)^{-2} = 2^{4(-2)} = 2^{-8} = \dfrac{1}{2^8} = \dfrac{1}{256}$

(c) $(b^{-7})^5 = b^{-7 \cdot 5} = b^{-35} = \dfrac{1}{b^{35}}$

(d) $\dfrac{(x^3)^{-2}}{(x^{-5})^2} = \dfrac{x^{-6}}{x^{-10}} = \dfrac{x^{10}}{x^6} = x^4$

How can we simplify the expression $(2x)^3$? Consider
$$(2x)^3 = 2x \cdot 2x \cdot 2x = (2 \cdot 2 \cdot 2) \cdot (x \cdot x \cdot x) = 2^3 x^3.$$

This result suggests that, to cube a product, cube each factor.

RAISING PRODUCTS TO POWERS

For any real numbers a and b and integer n,
$$(ab)^n = a^n b^n.$$

EXAMPLE 7 Raising products to powers

Simplify each expression. Use positive exponents.

(a) $(6y)^2$ **(b)** $(x^2 y)^{-2}$ **(c)** $(2xy^3)^4$ **(d)** $\dfrac{(2a^2 b^{-3})^2}{4(ab^3)^3}$

Solution

(a) $(6y)^2 = 6^2 y^2 = 36 y^2$ Base is $6y$.

(b) $(x^2 y)^{-2} = \dfrac{1}{(x^2 y)^2} = \dfrac{1}{(x^2)^2 y^2} = \dfrac{1}{x^4 y^2}$

(c) $(2xy^3)^4 = 2^4 x^4 (y^3)^4 = 16 x^4 y^{12}$

(d) $\dfrac{(2a^2 b^{-3})^2}{4(ab^3)^3} = \dfrac{2^2 a^4 b^{-6}}{4 a^3 b^9} = \dfrac{4 a^4}{4 a^3 b^9 b^6} = \dfrac{a}{b^{15}}$

The expression $\left(\dfrac{a}{b}\right)^3$ can be simplified as
$$\left(\dfrac{a}{b}\right)^3 = \dfrac{a}{b} \cdot \dfrac{a}{b} \cdot \dfrac{a}{b} = \dfrac{a^3}{b^3}.$$

This result suggests the following rule.

RAISING QUOTIENTS TO POWERS

For nonzero numbers a and b and any integer n,
$$\left(\dfrac{a}{b}\right)^n = \dfrac{a^n}{b^n}.$$

EXAMPLE 8 Raising quotients to powers

Simplify each expression. Use positive exponents.

(a) $\left(\dfrac{3}{x}\right)^3$ (b) $\left(\dfrac{1}{2^3}\right)^{-2}$ (c) $\left(\dfrac{3x^{-3}}{y^2}\right)^4$ (d) $\left(\dfrac{3x^2}{4y^2z}\right)^3$

Solution

(a) $\left(\dfrac{3}{x}\right)^3 = \dfrac{3^3}{x^3} = \dfrac{27}{x^3}$

(b) $\left(\dfrac{1}{2^3}\right)^{-2} = \left(\dfrac{2^3}{1}\right)^2 = \dfrac{(2^3)^2}{1^2} = 2^6 = 64$

(c) $\left(\dfrac{3x^{-3}}{y^2}\right)^4 = \dfrac{3^4(x^{-3})^4}{(y^2)^4} = \dfrac{81x^{-12}}{y^8} = \dfrac{81}{x^{12}y^8}$

(d) $\left(\dfrac{3x^2}{4y^2z}\right)^3 = \dfrac{3^3x^6}{4^3y^6z^3} = \dfrac{27x^6}{64y^6z^3}$

EXAMPLE 9 Simplifying expressions

Write each expression using positive exponents. Simplify the result completely.

(a) $\left(\dfrac{y^{-3}}{3z^{-4}}\right)^{-2}$ (b) $\dfrac{(rt^3)^{-3}}{(r^2t^3)^{-2}}$

Solution

(a) $\left(\dfrac{y^{-3}}{3z^{-4}}\right)^{-2} = \left(\dfrac{3z^{-4}}{y^{-3}}\right)^2 = \left(\dfrac{3y^3}{z^4}\right)^2 = \dfrac{9y^6}{z^8}$

(b) $\dfrac{(rt^3)^{-3}}{(r^2t^3)^{-2}} = \dfrac{(r^2t^3)^2}{(rt^3)^3} = \dfrac{r^4t^6}{r^3t^9} = \dfrac{r}{t^3}$

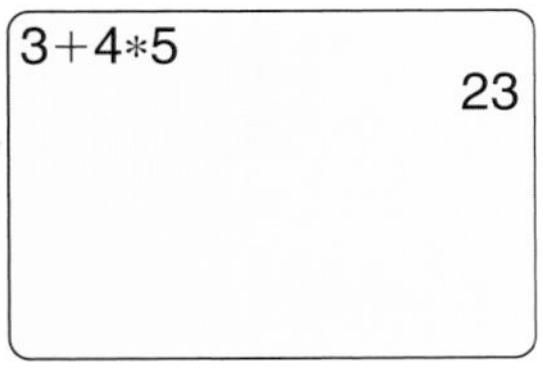

```
3+4*5
            23
```

Figure 1.14

Order of Operations

When we evaluate the expression $3 + 4 \cdot 5$, is the result 35 or 23? Figure 1.14 shows that a calculator gives a result of 23. This is because multiplication is performed before addition.

Because it is important that we evaluate arithmetic expressions consistently, the following rules are used. (Two people should evaluate the same expression the same way.)

> **ORDER OF OPERATIONS**
>
> Using the following order of operations, first perform all calculations within parentheses and absolute values, or above and below the fraction bar. Then use the same order of operations to perform the remaining calculations.
>
> 1. Evaluate all exponential expressions. Do any negations *after* evaluating exponents.
> 2. Do all multiplication and division from *left to right*.
> 3. Do all addition and subtraction from *left to right*.

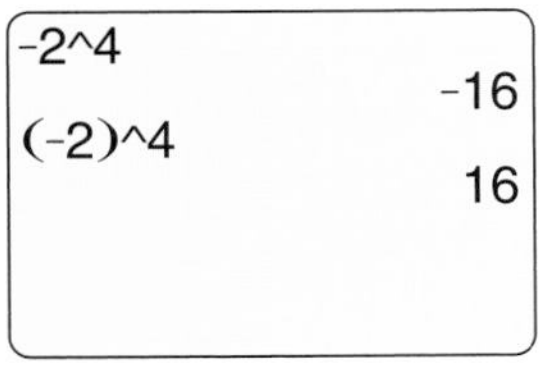

```
-2^4
           -16
(-2)^4
            16
```

Figure 1.15

Be sure to evaluate exponents before performing negation. For example,

$$-2^4 = -(2 \cdot 2 \cdot 2 \cdot 2) = -16, \quad \text{but} \quad (-2)^4 = (-2)(-2)(-2)(-2) = 16.$$

These results are supported by Figure 1.15.

EXAMPLE 10 **Evaluating arithmetic expressions**

Evaluate each expression. Use a calculator to support your results.

(a) $5 - 3 \cdot 2 - (4 + 5)$ **(b)** $-3^2 + \dfrac{5 + 7}{2 + 1}$ **(c)** $4^3 - 5(2 - 6 \cdot 2)$

Solution

(a)
$$\begin{aligned} 5 - 3 \cdot 2 - (4 + 5) &= 5 - 3 \cdot 2 - 9 \\ &= 5 - 6 - 9 \\ &= -1 - 9 \\ &= -10 \end{aligned}$$

(b) Assume that both the numerator and the denominator of a fraction have parentheses around them.

$$\begin{aligned} -3^2 + \frac{5 + 7}{2 + 1} &= -3^2 + \frac{(5 + 7)}{(2 + 1)} \\ &= -3^2 + \frac{12}{3} \\ &= -9 + \frac{12}{3} && \text{Do exponents before division.} \\ &= -9 + 4 \\ &= -5 \end{aligned}$$

(c)
$$\begin{aligned} 4^3 - 5(2 - 6 \cdot 2) &= 4^3 - 5(2 - 12) \\ &= 4^3 - 5(-10) \\ &= 64 - 5(-10) \\ &= 64 + 50 \\ &= 114 \end{aligned}$$

```
5-3*2-(4+5)
                -10
-3²+(5+7)/(2+1)
                 -5
4^3-5(2-6*2)
                114
```

Figure 1.16

These results are supported by Figure 1.16.

Scientific Notation

▶ **REAL-WORLD CONNECTION** Numbers that are large or small in absolute value occur frequently in applications. For simplicity these numbers are often expressed in scientific notation. As mentioned at the beginning of this section, the distance to the nearest star is 25 trillion miles. This number can be expressed in scientific notation as

$$25{,}000{,}000{,}000{,}000 = 2.5 \times 10^{13}.$$

In contrast, a typical virus is about 5 millionths of a centimeter in diameter. In scientific notation this number can be written as

$$0.000005 = 5 \times 10^{-6}.$$

```
25000000000000
             2.5E13
.000005
               5E-6
```

Figure 1.17

CALCULATOR HELP

To display numbers in scientific notation, see the Appendix (page AP-2).

A calculator set in *scientific mode* expresses these numbers in scientific notation, as illustrated in Figure 1.17. The letter E denotes a power of 10. That is, 2.5E13 $= 2.5 \times 10^{13}$ and 5E $-6 = 5 \times 10^{-6}$.

SCIENTIFIC NOTATION

A real number a is in **scientific notation** when a is written as $b \times 10^n$, where $1 \le |b| < 10$ and n is an integer.

Use the following steps to express a positive (rational) number a in scientific notation.

> **WRITING A POSITIVE NUMBER IN SCIENTIFIC NOTATION**
> 1. Move the decimal point in a number a until it represents a number b such that $1 \leq b < 10$.
> 2. Count the number of decimal places that the decimal point was moved. Let this positive integer be n. (If the decimal point is *not* moved, then $a = a \times 10^0$.)
> 3. If the decimal point was moved to the left, then $a = b \times 10^n$.
>   If the decimal point was moved to the right, then $a = b \times 10^{-n}$.

NOTE: The scientific notation for a negative number a is the additive inverse of the scientific notation of $|a|$. For example, $5200 = 5.2 \times 10^3$, so $-5200 = -5.2 \times 10^3$.

Table 1.6 shows the values of some important powers of 10.

TABLE 1.6 Important Powers of 10

Number	10^{-3}	10^{-2}	10^{-1}	10^3	10^6	10^9	10^{12}
Value	Thousandth	Hundredth	Tenth	Thousand	Million	Billion	Trillion

EXAMPLE 11 Writing a number in scientific notation

Express each number in scientific notation.
(a) 360,000   (Dots in 1 square inch of some types of laser print)
(b) 0.00000538   (Time in seconds for light to travel 1 mile)
(c) 10,000,000,000 (Estimated world population in 2050)

Solution
(a) Move the assumed decimal point in 360,000 *five places* to the *left* to obtain 3.6.

$$3.60000.$$

The scientific notation for 360,000 is 3.6×10^5.

(b) Move the decimal point in 0.00000538 *six places* to the *right* to obtain 5.38.

$$0.00000 5.38$$

The scientific notation for 0.00000538 is 5.38×10^{-6}.

(c) Move the decimal point in 10,000,000,000 *ten places* to the *left* to obtain 1.

$$1.0000000000.$$

The scientific notation for 10,000,000,000 is 1×10^{10}. Note that positive powers of 10 indicate a large number, whereas negative powers of 10 indicate a small number.

MAKING CONNECTIONS

Scientific Notation and Moving the Decimal Point

When a *positive* number a is expressed in scientific notation, a negative exponent on 10 indicates that $a < 1$, thus a is relatively small, and a positive exponent on 10 indicates that $a \geq 10$, thus a is relatively large. This can be helpful when converting from scientific notation to standard notation or vice versa. For example, to write the number 3.4×10^7 in standard form, we move the decimal point in 3.4 seven places. Because the exponent on 10 is positive, the resulting number should be relatively large. Moving the decimal point to the right results in 34,000,000. To express the number 0.00087 in scientific notation, we move the decimal point four places to obtain 8.7. Because the number 0.00087 is relatively small, the exponent on 10 will be negative. The resulting scientific notation is 8.7×10^{-4}.

In the next example we convert numbers from scientific notation to **standard form**, or decimal form.

EXAMPLE 12 Writing a number in standard form

Write the number in standard (decimal) form.
(a) 2×10^8 (Number of years for the sun to orbit in the Milky Way)
(b) 9.1×10^{-2} (Fraction of deaths worldwide caused by injuries in 2002)

Solution
(a) Because the exponent of 10 is *positive*, move the assumed decimal point in 2 to the *right* 8 places to obtain 200,000,000.
(b) Because the exponent of 10 is *negative*, move the decimal point in 9.1 to the *left* 2 places to obtain 0.091.

Arithmetic can be performed on expressions in scientific notation. For example, multiplication of the expressions 8×10^4 and 4×10^2 may be done by hand.

$$(8 \times 10^4) \cdot (4 \times 10^2) = (8 \cdot 4) \times (10^4 \cdot 10^2) \quad \text{Properties of real numbers}$$
$$= 32 \times 10^6 \quad \text{Add exponents and simplify.}$$
$$= (3.2 \times 10^1)(10^6) \quad \text{Write in scientific notation.}$$
$$= 3.2 \times 10^7 \quad \text{Add exponents.}$$

CALCULATOR HELP
To enter numbers in scientific notation, see the Appendix (page AP-3).

Division may be performed as follows.

$$\frac{8 \times 10^4}{4 \times 10^2} = \frac{8}{4} \times \frac{10^4}{10^2} \quad \text{Property of fractions}$$
$$= 2 \times 10^2 \quad \text{Subtract exponents.}$$

▶ **REAL-WORLD CONNECTION** The next example illustrates how scientific notation is used in applications.

EXAMPLE 13   Analyzing the federal debt

In 2005, the federal debt held by the public was 4.72 trillion dollars, and the population of the United States was 296 million. Approximate the national debt per person.

Solution
In scientific notation 4.72 trillion equals 4.72×10^{12} and 296 million equals 296×10^6, or 2.96×10^8. The per person debt held by the public is given by

$$\frac{4.72 \times 10^{12}}{2.96 \times 10^8} \approx \$15,946.$$

Figure 1.18 supports this result.

```
(4.72*10^12)/(2.
96*10^8)
          15945.94595
```
Figure 1.18

CRITICAL THINKING
Estimate the number of seconds that you have been alive.

1.3 PUTTING IT ALL TOGETHER

The following table summarizes important properties of exponents, where a and b are nonzero real numbers and m and n are integers.

Property	Definition	Examples
Bases and Exponents	In the expression a^n, a is the base and n is the exponent. *Note:* 0^0 is undefined.	$3^4 = 3 \cdot 3 \cdot 3 \cdot 3 = 81$, $7^0 = 1$, $2^{-3} = \frac{1}{2^3} = \frac{1}{2 \cdot 2 \cdot 2} = \frac{1}{8}$, and $-3^2 = -9$
The Product Rule	$a^m \cdot a^n = a^{m+n}$	$8^4 \cdot 8^2 = 8^{4+2} = 8^6$ and $5^6 \cdot 5^{-3} = 5^{6+(-3)} = 5^3$
The Quotient Rule	$\frac{a^m}{a^n} = a^{m-n}$	$\frac{6^7}{6^4} = 6^{7-4} = 6^3$ and $\frac{7^{-4}}{7^{-2}} = 7^{(-4-(-2))} = 7^{-2} = \frac{1}{7^2}$
The Power Rules	1. $(a^m)^n = a^{mn}$ 2. $(ab)^n = a^n b^n$ 3. $\left(\frac{a}{b}\right)^n = \frac{a^n}{b^n}$	1. $(2^2)^3 = 2^6$ 2. $(3y)^4 = 3^4 y^4$ 3. $\left(\frac{x^2}{y}\right)^4 = \frac{x^8}{y^4}$
Quotients and Negative Exponents	1. $\frac{1}{a^{-n}} = a^n$ 2. $\frac{a^{-n}}{b^{-m}} = \frac{b^m}{a^n}$ 3. $\left(\frac{a}{b}\right)^{-n} = \left(\frac{b}{a}\right)^n$	1. $\frac{1}{x^{-2}} = x^2$ 2. $\frac{z^{-3}}{y^{-5}} = \frac{y^5}{z^3}$ 3. $\left(\frac{4}{t}\right)^{-2} = \left(\frac{t}{4}\right)^2 = \frac{t^2}{4^2}$
Scientific Notation	A positive number a is in scientific notation when a is written as $b \times 10^n$, where $1 \leq b < 10$ and n is an integer.	$52{,}600 = 5.26 \times 10^4$ and $0.0068 = 6.8 \times 10^{-3}$

1.4 VARIABLES, EQUATIONS, AND FORMULAS

Basic Concepts ▪ Modeling Data ▪ Square Roots and Cube Roots ▪ Tables and Calculators (Optional)

A LOOK INTO MATH ▷

Most of the mathematics that people have discovered throughout the ages can be derived by using only pencil and paper, not from science and measured data. However, one of the amazing aspects of mathematics is that it can be used in countless applications that improve our quality of life. In fact, this ability of mathematics to describe the real world is so amazing that Nobel Laureate Eugene Wigner wrote the article "The Unreasonable Effectiveness of Mathematics in the Natural Sciences." Without mathematics, we would not have compact disc players, cars, warm buildings, or accurate weather forecasts. Mathematics can even be used to predict the increase in sea level if the polar ice caps were to melt. In this section we introduce some of the mathematical concepts that are used to model our world.

(**Source:** Eugene Wigner, *Communications of Pure and Applied Mathematics*, February 1960.)

Basic Concepts

▶ **REAL-WORLD CONNECTION** Suppose that we want to calculate the distance traveled by a car moving at a constant speed of 30 miles per hour. One method would be to make a table of values, as shown in Table 1.7.

TABLE 1.7

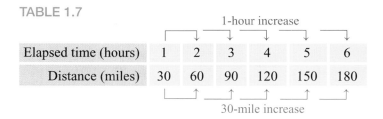

Note that for each 1-hour increase the distance increases by 30 miles. Many times it is not possible to list all relevant values in a table. Instead, we use *variables* to describe data. For example, we might let elapsed time be represented by the variable t and let distance be represented by the variable d. If $t = 2$, then $d = 60$; if $t = 5$, then $d = 150$. In this example, the value of d is always equal to the value of t multiplied by 30. We can *model* this situation by using the *equation* or *formula* $d = 30t$. The values for distance in Table 1.7 can be calculated by letting $t = 1, 2, 3, 4, 5, 6$ in this formula.

A **variable** is a symbol, such as x, y, or t, used to represent any unknown number or quantity. An **algebraic expression** can consist of numbers, variables, arithmetic symbols ($+$, $-$, $\times$, $\div$), exponents, and grouping symbols, such as parentheses, brackets, and square roots. Examples of algebraic expressions include

$$6, \quad x + 2, \quad 4(t - 5) + 1, \quad \sqrt{x + 1}, \quad \text{and} \quad LW. \qquad \text{Expressions}$$

An **equation** is a statement that two algebraic expressions are equal. An equation *always* contains an equals sign. Examples of equations include

$$3 + 6 = 9, \quad x + 1 = 4, \quad d = 30t, \quad \text{and} \quad x + y = 20. \qquad \text{Equations}$$

The first equation contains only constants, the second equation contains one variable, and both the third and fourth equations contain two variables. A **formula** is an equation that can calculate one quantity by using a known value of another quantity. (Formulas can also contain known values of more than one quantity.) Formulas show relationships between variables. The formula $y = \frac{x}{3}$ computes the number of yards in x feet. If $x = 15$, then $y = \frac{15}{3} = 5$. That is, in 15 feet there are 5 yards.

EXAMPLE 1 Writing and using a formula

If a car travels at a constant speed of 70 miles per hour, write a formula that calculates the distance d that the car travels in t hours. Evaluate your formula when $t = 1.5$ and interpret the result.

Solution
Traveling at 70 miles per hour, the car will travel a distance of $d = 70t$ miles in t hours. Evaluating this formula at $t = \mathbf{1.5}$ results in

$$d = 70(\mathbf{1.5}) = 105.$$

After 1.5 hours the car has traveled 105 miles.

EXAMPLE 2 Writing formulas

Write a formula that does the following.
(a) Finds the circumference C of a circle with radius r
(b) Calculates the pay P for working H hours at $9 per hour
(c) Converts Q quarts to C cups

Solution
(a) The circumference of a circle is $C = 2\pi r$, where $\pi \approx 3.14$.
(b) Pay equals the product of the hourly wage and the hours worked. Thus $P = 9H$.
(c) There are 4 cups in each quart, so $C = 4Q$.

EXAMPLE 3 Evaluating formulas from geometry

Evaluate each formula for the given value(s) of the variable(s). See Figure 1.19.
(a) $A = \pi r^2$, $r = 4$ Area of a circle
(b) $P = 2L + 2W$, $L = 8$ and $W = 4$ Perimeter of a rectangle
(c) $V = s^3$, $s = 3$ Volume of a cube

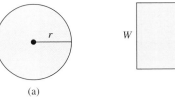
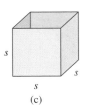

    (a)   (b)   (c)

Figure 1.19

Solution
(a) $A = \pi(4)^2 = \pi(16) = 16\pi$ Let $r = 4$.
(b) $P = 2(8) + 2(4) = 16 + 8 = 24$ Let $L = 8$ and $W = 4$.
(c) $V = 3^3 = 3 \cdot 3 \cdot 3 = 27$ Let $s = 3$.

Modeling Data

▶ **REAL-WORLD CONNECTION** Faster moving automobiles require more distance to stop. For example, at 60 miles per hour it takes about four times the distance to stop as it does at 30 miles per hour. Highway engineers have developed formulas to estimate the braking distance of a car.

EXAMPLE 4 Calculating braking distance

The braking distances in feet for a car traveling on wet, level pavement are shown in Table 1.8. Distances have been rounded to the nearest foot.

TABLE 1.8 Braking Distances

Speed (mph)	10	20	30	40	50	60	70
Distance (feet)	11	44	100	178	278	400	544

Source: L. Haefner, *Introduction to Transportation Systems.*

(a) If a car doubles its speed, what happens to the braking distance?
(b) If the speed is represented by the variable x and the braking distance by the variable d, then the braking distance may be calculated by the formula $d = \frac{x^2}{9}$. Verify the distance values in Table 1.8 for $x = 10, 30, 60$.
(c) Calculate the braking distance for a car traveling at 90 miles per hour. If a football field is 300 feet long, how many football field lengths does this braking distance represent?

Solution

(a) When the speed increases from 10 to 20 miles per hour, the braking distance increases by a factor of $\frac{44}{11} = 4$. Similarly, if the speed doubles from 30 to 60 miles per hour, the distance increases by a factor of $\frac{400}{100} = 4$. Thus it appears that, if the speed of a car doubles, the braking distance quadruples.

(b) Let $x = 10, 30, 60$ in the formula $d = \frac{x^2}{9}$. Then

$$d = \frac{10^2}{9} = \frac{100}{9} \approx 11 \text{ feet,}$$

$$d = \frac{30^2}{9} = \frac{900}{9} = 100 \text{ feet, and}$$

$$d = \frac{60^2}{9} = \frac{3600}{9} = 400 \text{ feet.}$$

These values agree with the values in Table 1.8.

(c) If $x = 90$, then $d = \frac{90^2}{9} = 900$ feet. At 90 miles per hour the braking distance equals three football fields stretched end to end.

In the next example we find a formula that models data in a table.

> **CRITICAL THINKING**
> Write a formula that calculates the time T for a bike rider to travel 100 miles moving at x miles per hour. Test your formula for different values of x.

EXAMPLE 5 Finding a formula

The data in Table 1.9 can be modeled by the formula $y = ax$. Find a.

TABLE 1.9

x	y
1	3
2	6
3	9
4	12
5	15

Solution

Each value of y is 3 times the corresponding value of x, so $a = 3$. We can also find a symbolically. If $x = 1$, then $y = 3$. We can substitute these values into the equation.

$y = ax$	Given equation
$3 = a \cdot 1$	Let $x = 1$ and $y = 3$.
$3 = a$	Identity

Thus $a = 3$. The formula $y = 3x$ models the data in Table 1.9.

Square Roots and Cube Roots

The number b is a **square root** of a number a if $b^2 = a$. For example, one square root of 9 is 3 because $3^2 = 9$. The other square root of 9 is -3 because $(-3)^2 = 9$. We use the symbol $\sqrt{9}$ to denote the *positive* or **principal square root** of 9. That is, $\sqrt{9} = 3$. The following are examples of how to evaluate the square root symbol. A calculator is sometimes needed to approximate square roots.

$$\sqrt{16} = 4, \quad -\sqrt{100} = -10, \quad \sqrt{3} \approx 1.732, \quad \pm\sqrt{4} = \pm 2$$

The symbol "$\pm$" is read "plus or minus." Note that ± 2 represents the numbers 2 or -2.

The number b is a **cube root** of a number a if $b^3 = a$. The cube root of 8 is 2 because $2^3 = 8$, which may be written as $\sqrt[3]{8} = 2$. Similarly, $\sqrt[3]{-27} = -3$ because $(-3)^3 = -27$. Each real number has *exactly one* real cube root.

EXAMPLE 6 Finding square roots and cube roots

Evaluate each expression.
(a) $\sqrt{3^2 + 4^2}$ (b) $\sqrt[3]{64}$ (c) $\sqrt[3]{-2^3 - 19}$

CALCULATOR HELP
To calculate square roots and cube roots, see the Appendix (page AP-1).

Solution
(a) $\sqrt{3^2 + 4^2} = \sqrt{9 + 16} = \sqrt{25} = 5$. Note: $\sqrt{3^2 + 4^2} \neq \sqrt{3^2} + \sqrt{4^2} = 3 + 4 = 7$.
(b) $\sqrt[3]{64} = 4$ because $4 \cdot 4 \cdot 4 = 4^3 = 64$.
(c) $\sqrt[3]{-2^3 - 19} = \sqrt[3]{-8 - 19} = \sqrt[3]{-27} = -3$ because $(-3)(-3)(-3) = -27$.

EXAMPLE 7 Finding lengths of sides

Find the length of a side s for each geometric shape.
(a) A square with area 100 square feet
(b) A cube with volume 125 cubic inches

Solution
(a) The area of a square is $A = s^2$. Thus $s = \sqrt{100} = 10$ feet because $10^2 = 100$.
(b) The volume of a cube is $V = s^3$. Thus $s = \sqrt[3]{125} = 5$ inches because $5^3 = 125$.

MAKING CONNECTIONS

Square Roots and Cube Roots

The square root of a negative number is not a real number. However, the cube root of a negative (or positive) number is a real number. For example, $\sqrt{-8}$ is not a real number, whereas $\sqrt[3]{-8} = -2$ is a real number.

▶ **REAL-WORLD CONNECTION** Roots of numbers often occur in biology, as illustrated in the next example.

EXAMPLE 8 Analyzing the walking speed of animals

When smaller animals walk, they tend to take faster, shorter steps, whereas larger animals tend to take slower, longer steps. For example, a hyena is about 0.8 meter high at the shoulder and takes roughly 1 step per second when walking, whereas an elephant 3 meters high at the shoulder takes 1 step every 2 seconds. If an animal is h meters high at the shoulder, then the frequency F in steps per second while it is walking can be estimated with the formula $F = \frac{0.87}{\sqrt{h}}$. The value of F is referred to as the animal's *stepping frequency*. (**Source:** C. Pennycuick, *Newton Rules Biology*.)

(a) A Thomson's gazelle is 0.6 meter high at the shoulder. Estimate its stepping frequency.
(b) A giraffe is 2.7 meters high at the shoulder. Estimate its stepping frequency.
(c) What happens to the stepping frequency as h increases?

Solution

(a) $F = \frac{0.87}{\sqrt{0.6}} \approx 1.12$. A Thomson's gazelle takes about 1.12 steps per second when walking.

(b) $F = \frac{0.87}{\sqrt{2.7}} \approx 0.53$. A giraffe takes roughly half a step per second when walking, or 1 step every 2 seconds.

(c) As h increases, the denominator of $\frac{0.87}{\sqrt{h}}$ also increases, so the ratio becomes smaller. Thus, as h increases, the stepping frequency decreases.

Tables and Calculators (Optional)

Many calculators are able to generate tables. To generate a table, we specify the formula, the starting x-value (TblStart), and the increment (Δ Tbl) between x-values. The calculator generates the required table automatically, as demonstrated in the next example.

EXAMPLE 9 Using the table feature

Make a table for $y = \frac{x^2}{9}$, starting at $x = 10$ and incrementing by 10. Compare this table to Table 1.8 in Example 4.

Solution

In Figure 1.20 the desired table is generated. Note that, if values are rounded to the nearest foot, the values in Figure 1.20(c) agree with those in Table 1.8.

CALCULATOR HELP
To display a table of values, see the Appendix (page AP-3).

```
Plot1  Plot2  Plot3
\Y1■X^2/9
\Y2=
\Y3=
\Y4=
\Y5=
\Y6=
\Y7=
```
(a)

```
TABLE SETUP
 TblStart=10
 ΔTbl=10
 Indpnt: Auto  Ask
 Depend: Auto  Ask
```
(b)

```
 X    | Y1
 10   | 11.111
 20   | 44.444
 30   | 100
 40   | 177.78
 50   | 277.78
 60   | 400
 70   | 544.44
 Y1■X^2/9
```
(c)

Figure 1.20

1.4 PUTTING IT ALL TOGETHER

The following table summarizes some of the important topics discussed in this section.

Concept	Comments	Examples
Variable	Represents an unknown quantity	x, y, z, A, V
Algebraic Expression	Can consist of numbers, variables, operation symbols, exponents, and grouping symbols but *no* equals sign	$2x - 8$ $3 - (5y + 6)$ s^3 $2L + 2W$
Equation	A statement that two algebraic expressions are equal—*always contains an equals sign*	$5x = 10$ $y = 2x + 1$ $n + 5 = 3 - n$ $z^2 + 1 = 17$
Formula	An equation used to calculate one quantity, using known values of other quantities—shows relationships between variables	$A = \pi r^2$ Area of a circle $C = 2\pi r$ Circumference of a circle $V = s^3$ Volume of a cube $P = 2L + 2W$ Perimeter of a rectangle $A = \frac{1}{2}bh$ Area of a triangle
Square Root	The *positive* or *principal square root* of a is written $\sqrt{a}$. The square root of a negative number is not a real number.	$\sqrt{25} = 5$ $\pm\sqrt{100} = \pm 10$ $-\sqrt{16} = -4$
Cube Root	The cube root of a is written $\sqrt[3]{a}$	$\sqrt[3]{-8} = -2$ because $(-2)^3 = -8$. $\sqrt[3]{64} = 4$ because $4^3 = 64$.

1.5 INTRODUCTION TO GRAPHING

Relations ■ The Cartesian Coordinate System ■ Scatterplots and Line Graphs ■ The Viewing Rectangle (Optional) ■ Graphing with Calculators (Optional)

A LOOK INTO MATH ▷ Computers, the Internet, and other types of electronic communication are creating large amounts of data. The challenge for society is to use these data to solve important problems and create new knowledge. Before conclusions can be drawn, data must be analyzed. A powerful tool in this step is visualization. Pictures and graphs are capable of communicating large quantities of information in short periods of time. A full page of computer graphics typically contains a hundred times more information than a page of text.

The map in Figure 1.21 shows the average date of the first 32°F temperature in autumn. Imagine trying to describe this map by using *only* words. In this section we discuss how graphs are used to visualize data. (**Source:** J. Williams, *The Weather Almanac*.)

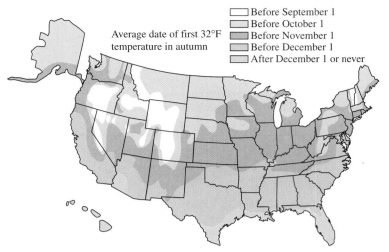

Figure 1.21 First Frost

Relations

▶ **REAL-WORLD CONNECTION** Table 1.10 lists monthly average wind speeds in miles per hour for San Francisco. In this table January corresponds to 1, February to 2, and so on, until December is represented by 12. For example, in April the average wind speed is 12 miles per hour.

TABLE 1.10 Average Wind Speeds in San Francisco

Month	1	2	3	4	5	6	7	8	9	10	11	12
Wind Speed (mph)	7	9	11	12	13	14	14	13	11	9	8	7

Source: J. Williams, *The Weather Almanac.*

If we let x be the month and y be the wind speed, then the **ordered pair** (x, y) represents the average wind speed y during month x. For example, the ordered pair $(2, 9)$ indicates that in February the average wind speed is 9 miles per hour, whereas the ordered pair $(9, 11)$ indicates that the average wind speed in September is 11 miles per hour. *Order is important* in an ordered pair.

The data in Table 1.10 establish a *relation*; that is, each month is associated with a wind speed in an ordered pair (**month, wind speed**). This relation can be represented by a set S, which contains 12 ordered pairs:

$$S = \{(1, 7), (2, 9), (3, 11), (4, 12), (5, 13), (6, 14),$$
$$(7, 14), (8, 13), (9, 11), (10, 9), (11, 8), (12, 7)\}.$$

> **RELATION**
>
> A **relation** is a set of ordered pairs.

If we denote the ordered pairs in a relation (x, y), then the set of all x-values is called the **domain** of the relation and the set of all y-values is called the **range**. In Table 1.10 the domain is

$$D = \{1, 2, 3, 4, 5, 6, 7, 8, 9, 10, 11, 12\},$$

which corresponds to the 12 months. The range is

$$R = \{7, 8, 9, 11, 12, 13, 14\},$$

which corresponds to the monthly average wind speeds. Note that an average wind speed of 14 miles per hour occurs more than once in Table 1.10, but it is listed *only once* in the range set R. The same is true for the values 7, 9, 11, and 13.

EXAMPLE 1 Finding the domain and range of a relation

Find the domain and range for the relation given by

$$S = \{(-1, 5), (0, 1), (2, 4), (4, 2), (5, 1)\}.$$

Solution
The domain D is determined by the first element in each ordered pair, or

$$D = \{-1, 0, 2, 4, 5\}.$$

The range R is determined by the second element in each ordered pair, or

$$R = \{1, 2, 4, 5\}.$$

▶ **REAL-WORLD CONNECTION**   We are all well aware that tuition and fees go up each year. In the next example, we describe this interaction between the year and the cost as a relation.

EXAMPLE 2 Analyzing tuition and fees

Table 1.11 lists the average cost of tuition and fees at *public* colleges from 2001 through 2004. Express this table as a relation S. Identify the domain and range of S.

TABLE 1.11 Tuition and Fees

Year	2001	2002	2003	2004
Cost	$3725	$4081	$4694	$5132

Source: The College Board.

Solution
Let the year be the first element in the ordered pair and the cost of tuition and fees be the second element. Then relation S is given by the following set of ordered pairs.

$$S = \{(2001, 3725), (2002, 4081), (2003, 4694), (2004, 5132)\}$$

The domain of S is

$$D = \{2001, 2002, 2003, 2004\}$$

and the range of S is

$$R = \{3725, 4081, 4694, 5132\}.$$

It is possible for a relation to contain *infinitely many* ordered pairs. For example, let the equation $y = 2x$ define a relation S, where x is any real number. Then S contains infinitely many ordered pairs of the form $(x, 2x)$, such as $(-2, -4)$, $(3, 6)$, and $(0.1, 0.2)$.

EXAMPLE 3 Analyzing iPod memory

The formula $S = 250m$ identifies a relationship between the memory m of an iPod in gigabytes and the number of songs S that can be stored on it. Give four ordered pairs in this relation.

Solution

One way to determine four ordered pairs is to assign the variable m four different values. For example, when $m = 4$, $S = 250(4) = $ **1000**. The ordered pair (**4**, **1000**) means that a 4 GB iPod can hold 1000 songs. Table 1.12 lists this ordered pair along with three additional ordered pairs: (8, 2000), (20, 5000), and (30, 7500).

TABLE 1.12

m	4	8	20	30
S	1000	2000	5000	7500

The Cartesian Coordinate System

We can use the **Cartesian coordinate system**, or **xy-plane**, to visualize or *graph* a relation. The horizontal axis is the **x-axis** and the vertical axis is the **y-axis**. The axes intersect at the **origin** and determine four regions called **quadrants**. They are numbered I, II, III, and IV counterclockwise, as illustrated in Figure 1.22. We can plot the ordered pair (x, y) by using the x-axis and the y-axis. For example, the point (1, 2) is located in quadrant I, 1 unit to the right of the origin and 2 units above the x-axis, as shown in Figure 1.23.

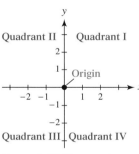

Figure 1.22 The xy-plane

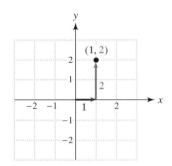

Figure 1.23 Plotting a Point

Similarly, the ordered pair $(-2, 3)$ is located in quadrant II, $(-3, -3)$ is in quadrant III, and $(3, -2)$ is in quadrant IV. See Figure 1.24. A point lying on a coordinate axis does not belong to any quadrant. The point $(-2, 0)$ is located on the x-axis, whereas the point $(0, -2)$ lies on the y-axis.

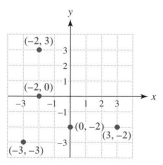

Figure 1.24 Plotting Points

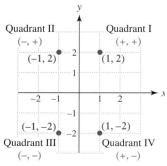

Figure 1.25

CRITICAL THINKING

If a is negative, then what quadrant does the point $(a, -a^2)$ lie in?

For any point (x, y) with x and y nonzero, we can determine the quadrant in which it is located. For example, the point $(1, 2)$ lies in quadrant I because both x and y are positive, whereas $(-1, -2)$ lies in quadrant III because both x and y are negative. The point $(-1, 2)$ lies in quadrant II where x is negative and y is positive, and $(1, -2)$ lies in quadrant IV where x is positive and y is negative. These concepts are illustrated in Figure 1.25, where $(+, +)$ indicates that $x > 0$ and $y > 0$ for any point (x, y) in quadrant I. Other quadrants and ordered pairs can be interpreted similarly.

EXAMPLE 4 Plotting points

Plot the data listed in Table 1.13. State the quadrant containing each point or the axis on which each point lies.

TABLE 1.13

x	-3	0	1	4
y	1	4	-2	3

Solution

We plot the points $(-3, 1)$, $(0, 4)$, $(1, -2)$, and $(4, 3)$ in the xy-plane, as shown in Figure 1.26. The point $(-3, 1)$ is in quadrant II, $(1, -2)$ is in quadrant IV, and $(4, 3)$ is in quadrant I. The point $(0, 4)$ lies on the y-axis and does not belong to any quadrant.

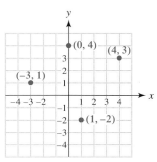

Figure 1.26

EXAMPLE 5 Graphing points given an equation

Evaluate $y = x^2 + 1$ for $x = -2, -1, 0, 1,$ and 2. Plot the resulting ordered pairs.

Solution

Start by evaluating the formula $y = x^2 + 1$ for each x-value.

$x = -2: \quad y = (-2)^2 + 1 = 5$
$x = -1: \quad y = (-1)^2 + 1 = 2$
$x = 0: \quad y = 0^2 + 1 = 1$
$x = 1: \quad y = 1^2 + 1 = 2$
$x = 2: \quad y = 2^2 + 1 = 5$

The points $(-2, 5)$, $(-1, 2)$, $(0, 1)$, $(1, 2)$, and $(2, 5)$ are plotted in Figure 1.27.

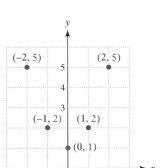

Figure 1.27

EXAMPLE 6 Determining the domain and range

Use the graph in Figure 1.28 to determine the domain and range of the relation.

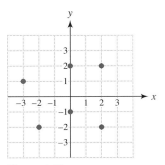

Figure 1.28

Solution
The relation shown in Figure 1.28 includes the points $(-3, 1)$, $(-2, -2)$, $(0, 2)$, $(0, -1)$, $(2, 2)$, and $(2, -2)$. The domain of this relation consists of the x-values of these ordered pairs, or $D = \{-3, -2, 0, 2\}$. The range of this relation consists of the y-values of these ordered pairs, or $R = \{-2, -1, 1, 2\}$.

Scatterplots and Line Graphs

If distinct points are plotted in the xy-plane, the resulting graph is called a **scatterplot**. Figure 1.28 is an example. A scatterplot of a different relation is shown in Figure 1.29, where the points $(1, 2)$, $(2, 4)$, $(3, 5)$, $(4, 6)$, $(5, 4)$, and $(6, 3)$ have been plotted. Its domain is $D = \{1, 2, 3, 4, 5, 6\}$, and its range is $R = \{2, 3, 4, 5, 6\}$.

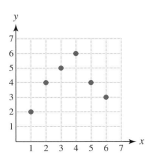

Figure 1.29 A Scatterplot

▶ REAL-WORLD CONNECTION The next example illustrates how to make a scatterplot from real data.

EXAMPLE 7 Making a scatterplot of gasoline prices

Table 1.14 lists the average price of a gallon of gasoline for selected years. Make a scatterplot of these data.

TABLE 1.14 Average Prices of Gasoline

Year	1955	1965	1975	1985	1995	2005
Price (per gallon)	29¢	31¢	57¢	120¢	121¢	227¢

Source: Department of Energy.

Solution

Plot the points (1955, 29), (1965, 31), (1975, 57), (1985, 120), (1995, 121), and (2005, 227). The *x*-values vary from 1955 to 2005, so we label the *x*-axis from 1955 to 2005. The *y*-values vary from 29 to 227, so we label the *y*-axis from 0 to 250. (Note that labels on the *x*- and *y*-axes may vary.) Figure 1.30 shows these points as plotted and labeled. Note that the double hash marks // on the *x*-axis indicate that there is a break in the scale, which starts at 0 and then jumps to 1955.

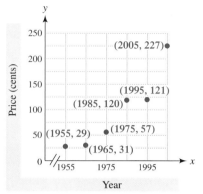

Figure 1.30 Price of Gasoline

▶ **REAL-WORLD CONNECTION** Sometimes it is helpful to connect the data points in a scatterplot with straight line segments. This type of graph emphasizes changes in the data and is called a **line graph**.

EXAMPLE 8 Interpreting a line graph

The line graph shown in Figure 1.31 depicts the total number of all types of college degrees awarded in millions for selected years. (*Source:* Department of Education.)

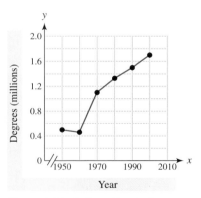

Figure 1.31 College Graduates

(a) Did the number of graduates ever decrease during this time period? Explain.
(b) Approximate the number of college graduates in the year 1970.
(c) Determine the 10-year period when the increase in the number of college graduates was greatest. What was this increase?

Solution

(a) Yes, the number decreased slightly between 1950 and 1960. For this time period, the line segment slopes slightly downward from left to right.
(b) In 1970, about 1.1 million degrees were awarded.
(c) The greatest increase corresponds to the line segment that slopes upward most from left to right. This increase occurred between 1960 and 1970 and was about $1.1 - 0.5 = 0.6$ million graduates.

The Viewing Rectangle (Optional)

Graphing calculators provide several features beyond those found on scientific calculators. Graphing calculators have additional keys that can be used to create tables, scatterplots, and graphs.

▶ **REAL-WORLD CONNECTION** The **viewing rectangle**, or **window**, on a graphing calculator is similar to the viewfinder in a camera. A camera cannot take a picture of an entire scene. The camera must be centered on some object and can photograph only a portion of the available scenery. A camera can capture different views of the same scene by zooming in and out, as can graphing calculators. The xy-plane is infinite, but the calculator screen can show only a finite, rectangular region of the xy-plane. The viewing rectangle must be specified by setting minimum and maximum values for both the x- and y-axes before a graph can be drawn.

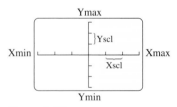

Figure 1.32

We use the following terminology regarding the size of a viewing rectangle. **Xmin** is the minimum x-value along the x-axis, and **Xmax** is the maximum x-value. Similarly, **Ymin** is the minimum y-value along the y-axis, and **Ymax** is the maximum y-value. Most graphs show an x-scale and a y-scale with tick marks on the respective axes. Sometimes the distance between consecutive tick marks is 1 unit, but at other times it might be 5 or 10 units. The distance represented by consecutive tick marks on the x-axis is called **Xscl**, and the distance represented by consecutive tick marks on the y-axis is called **Yscl** (see Figure 1.32).

This information about the viewing rectangle can be written as [Xmin, Xmax, Xscl] by [Ymin, Ymax, Yscl]. For example, $[-10, 10, 1]$ by $[-10, 10, 1]$ means that Xmin $= -10$, Xmax $= 10$, Xscl $= 1$, Ymin $= -10$, Ymax $= 10$, and Yscl $= 1$. This setting is referred to as the **standard viewing rectangle**. The window in Figure 1.32 is $[-3, 3, 1]$ by $[-3, 3, 1]$.

EXAMPLE 9 Setting the viewing rectangle

Show the viewing rectangle $[-2, 3, 0.5]$ by $[-100, 200, 50]$ on your calculator.

Solution
The window setting and viewing rectangle are displayed in Figure 1.33. Note that in Figure 1.33(b) there are 6 tick marks on the positive x-axis because its length is 3 units and the distance between consecutive tick marks is 0.5 unit.

CALCULATOR HELP
To set a viewing rectangle, see the Appendix (page AP-4).

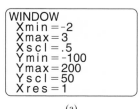

(a)

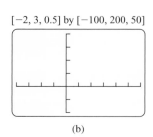

(b)

Figure 1.33

Graphing with Calculators (Optional)

Many graphing calculators have the capability to create scatterplots and line graphs. The next example illustrates how to make a scatterplot with a graphing calculator.

> **EXAMPLE 10** Making a scatterplot with a graphing calculator
>
> Plot the points $(-2, -2)$, $(-1, 3)$, $(1, 2)$, and $(2, -3)$ in $[-4, 4, 1]$ by $[-4, 4, 1]$.
>
> **Solution**
> We entered the points $(-2, -2)$, $(-1, 3)$, $(1, 2)$, and $(2, -3)$ shown in Figure 1.34(a), using the STAT EDIT feature. The variable L1 represents the list of x-values, and the variable L2 represents the list of y-values. In Figure 1.34(b) we set the graphing calculator to make a scatterplot with the STATPLOT feature, and in Figure 1.34(c) the points have been plotted. If you have a different model of calculator you may need to consult your owner's manual.

CALCULATOR HELP
To make a scatterplot, see the Appendix (page AP-4).

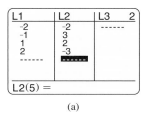
(a)

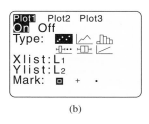
(b)

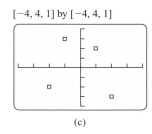
(c)

Figure 1.34

▶ **REAL-WORLD CONNECTION** In the next example, a graphing calculator is used to create a line graph of sales of cordless telephones.

> **EXAMPLE 11** Making a line graph with a graphing calculator
>
> Table 1.15 lists numbers of cordless telephones sold for selected years from 1990 through 2005. Make a line graph of these sales in an appropriate viewing rectangle. Then interpret the line graph.

TABLE 1.15 Cordless Phone Sales

Year	1990	1995	2000	2005
Phones (millions)	10.1	19.5	39.0	38.3

Source: Cellular Telecommunications Industry Association.

Solution
Plot the points (1990, 10.1), (1995, 19.5), (2000, 39.0), and (2005, 38.3). The x-values vary from 1990 to 2005, and the y-values vary between 10.1 and 38.3. We selected the viewing rectangle [1988, 2007, 5] by [0, 50, 10], although other viewing rectangles are possible. The viewing rectangle should be large enough to show all four data points without being too large. A line graph can be created by selecting this option on the graphing calculator.

CALCULATOR HELP

To make a line graph, see the Appendix (page AP-4).

Figures 1.35(a) and 1.35(b) show the data entries and plotting scheme. Figure 1.35(c) shows the resulting graph. It reveals that sales have increased dramatically and then decreased slightly during this time period.

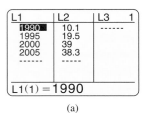
(a)

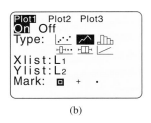
(b)

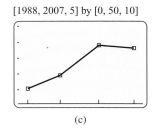
[1988, 2007, 5] by [0, 50, 10]
(c)

Figure 1.35

1.5 PUTTING IT ALL TOGETHER

Graphs are frequently used in mathematics, science, and business as a way to summarize and understand data better. The xy-plane is commonly used to visualize relations.

Concept	Explanation	Example
Relation	A set of ordered pairs	$S = \{(1, 2), (-2, 3), (4, 2)\}$
Domain and Range	If a relation consists of a set of ordered pairs (x, y), then the set of x-values is the domain and the set of y-values is the range.	If $S = \{(1, 2), (-2, 3), (4, 2)\}$, then $D = \{-2, 1, 4\}$ and $R = \{2, 3\}$.
Cartesian Coordinate System or xy-plane	Has four quadrants, two axes, and an origin. Used to plot points and graphs	
Scatterplot	A scatterplot results when individual points are plotted in the xy-plane.	
Line Graph	A line graph is similar to a scatterplot except that line segments are drawn between consecutive points.	

CHAPTER 1 SUMMARY

SECTION 1.1 ■ DESCRIBING DATA WITH SETS OF NUMBERS

Sets of Numbers

Natural Numbers $N = \{1, 2, 3, 4, \ldots\}$

Whole Numbers $W = \{0, 1, 2, 3, \ldots\}$

Integers $I = \{\ldots, -2, -1, 0, 1, 2, \ldots\}$

Rational Numbers Can be written as $\frac{p}{q}$, where p and $q \neq 0$ are integers; includes repeating and terminating decimals

Examples: $\frac{1}{2}, -3, 6, \sqrt{9}, 0.\overline{7}, 0.123$

Irrational Numbers A real number that is not rational

Examples: $\pi, \sqrt{11}, -\sqrt{3}$

Real Numbers Any number that can be written in decimal form

Examples: $-\frac{2}{3}, 0, 12.6, \sqrt{11}, \pi$

Properties of Real Numbers

Identity Properties

$a + 0 = a \qquad a \cdot 1 = a$

Commutative Properties

$a + b = b + a \qquad a \cdot b = b \cdot a$

Associative Properties

$(a + b) + c = a + (b + c) \qquad (a \cdot b) \cdot c = a \cdot (b \cdot c)$

Distributive Properties

$a(b + c) = ab + ac \qquad a(b - c) = ab - ac$

SECTION 1.2 ■ OPERATIONS ON REAL NUMBERS

Real Number Line The number line is used to graph real numbers.

Absolute Value $|a|$ equals a if $a > 0$ or $a = 0$, and $-a$ if $a < 0$.

Examples: $|-4| = 4, |7| = 7$, and $|\pi - 7| = 7 - \pi$ because $7 > \pi$

Arithmetic Operations

Addition **Examples:** $-3 + 4 = 1, 3 + (-4) = -1$, and $-3 + (-4) = -7$

Subtraction Use $a - b = a + (-b)$.

Examples: $-5 - 6 = -5 + (-6) = -11$ and
$4 - (-3) = 4 + 3 = 7$

Multiplication The product of two numbers with like signs is positive. The product of two numbers with unlike signs is negative.

Examples: $3 \cdot (-4) = -12$ and $-5 \cdot (-6) = 30$

Division Use $\dfrac{a}{b} = a \cdot \dfrac{1}{b}$.

Examples: $\dfrac{1}{2} \div -\dfrac{4}{5} = \dfrac{1}{2} \cdot -\dfrac{5}{4} = -\dfrac{5}{8}$ and

$-\dfrac{3}{2} \div 6 = -\dfrac{3}{2} \cdot \dfrac{1}{6} = -\dfrac{3}{12} = -\dfrac{1}{4}$

SECTION 1.3 ■ INTEGER EXPONENTS

Exponential Expression

Base $\rightarrow \mathbf{6^2} \leftarrow$ Exponent

Integer Exponents Let n be a positive integer and a be a nonzero number.

$$a^n = a \cdot a \cdot a \cdot \cdots \cdot a \quad (n \text{ factors of } a)$$
$$a^0 = 1 \quad (\text{Note: } 0^0 \text{ is undefined.})$$
$$a^{-n} = \dfrac{1}{a^n}$$

Examples: $4^3 = 4 \cdot 4 \cdot 4 = 64,\ 5^0 = 1,$ and $2^{-3} = \dfrac{1}{2^3}$

Properties of Exponents

Product Rule $a^m \cdot a^n = a^{m+n}$

Example: $z^3 \cdot z^5 = z^8$

Quotient Rule $\dfrac{a^m}{a^n} = a^{m-n}$

Example: $\dfrac{x^5}{x^7} = x^{-2} = \dfrac{1}{x^2}$

Power Rules $(a^m)^n = a^{mn},\ (ab)^n = a^n b^n,$ and $\left(\dfrac{a}{b}\right)^n = \dfrac{a^n}{b^n}$

Examples: $(5^2)^3 = 5^6,\ (2x)^3 = 8x^3,$ and $\left(\dfrac{2x}{y}\right)^3 = \dfrac{8x^3}{y^3}$

Negative Exponents $\dfrac{1}{a^{-n}} = a^n,\ \dfrac{a^{-n}}{b^{-m}} = \dfrac{b^m}{a^n},$ and $\left(\dfrac{a}{b}\right)^{-n} = \left(\dfrac{b}{a}\right)^n$

Examples: $\dfrac{1}{2^{-3}} = 2^3,\ \dfrac{x^{-4}}{y^{-3}} = \dfrac{y^3}{x^4},$ and $\left(\dfrac{2}{5}\right)^{-4} = \left(\dfrac{5}{2}\right)^4$

Order of Operations

Using the following order of operations, first perform all calculations within parentheses and absolute values, or above and below the fraction bar. Then use the same order of operations to perform the remaining calculations.

1. Evaluate all exponential expressions. Do any negation *after* evaluating exponents.
2. Do all multiplication and division from *left to right*.
3. Do all addition and subtraction from *left to right*.

Example: $-2^4 - 2 \cdot 3 = -16 - 2 \cdot 3 = -16 - 6 = -22$

Scientific Notation A number a written as $b \times 10^n$, where $1 \leq |b| < 10$ and n is an integer.

Examples: $23{,}400 = 2.34 \times 10^4$ and $0.0034 = 3.4 \times 10^{-3}$

SECTION 1.4 ■ VARIABLES, EQUATIONS, AND FORMULAS

Terminology

Variable — Symbol that represents an unknown quantity

  Examples: x, y, z, A, and T

Algebraic Expression — Can consist of numbers, variables, operation symbols, exponents, and grouping symbols but *no* equals sign

  Examples: $3z$, $(x - y)^3$, $4a + 3b$, 5, and $|x - 2|$

Equation — A statement that two algebraic expressions are *equal*—always contains an equals sign

  Examples: $2 + 4 = 6$, $2x = 8$, and $x^2 + 2 = 10$

Formula — An equation used to calculate one quantity by using known values of other quantities

  Examples: $P = 2W + 2L$ and $A = \pi r^2$

Square Root The number b is a square root of a number a if $b^2 = a$.

Example: The square roots of 36 are 6 and -6.

Principal Square Root The positive square root of a number, denoted $\sqrt{a}$

Examples: $\sqrt{4} = 2$, $\sqrt{100} = 10$, and $\sqrt{81} = 9$

Cube Root The number b is a cube root of a number a if $b^3 = a$.

Examples: $\sqrt[3]{8} = 2$, $\sqrt[3]{-27} = -3$, and $\sqrt[3]{64} = 4$

SECTION 1.5 ■ INTRODUCTION TO GRAPHING

Relation A set of ordered pairs

Example: $S = \{(-2, 3), (0, 3), (1, 2)\}$

Domain and Range In a relation consisting of ordered pairs (x, y), the set of x-values is the domain and the set of y-values is the range.

Example: For $S = \{(-2, 3), (0, 3), (1, 2)\}$, $D = \{-2, 0, 1\}$ and $R = \{2, 3\}$.

The Cartesian Coordinate System (*xy*-plane)

Points — Plotted as (x, y) ordered pairs

Four Quadrants — I, II, III, and IV; the axes do not lie in a quadrant.
  NOTE: The point $(1, 0)$ does not lie in a quadrant.

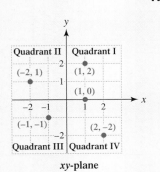
xy-plane

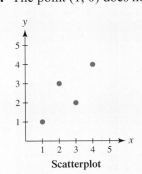
Scatterplot

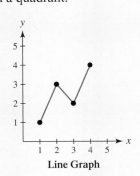
Line Graph

CHAPTER 1 REAL NUMBERS AND ALGEBRA

Assignment Name _____ Name _____ Date _____

Show all work for these items: _____

# _____	# _____
# _____	# _____
# _____	# _____
# _____	# _____

CHAPTER 1   SHOW YOUR WORK   45

Assignment Name _____ Name _____ Date _____
Show all work for these items: _____

# _____	# _____
# _____	**# _____**
# _____	**# _____**
# _____	**# _____**

CHAPTER 2
Linear Functions and Models

Every day our society creates enormous amounts of data, and mathematics is an important tool for summarizing those data and discovering trends. For example, the table shows the number of Toyota vehicles sold in the United States for selected years.

Year	1998	1999	2000	2001	2002
Vehicles (millions)	1.4	1.5	1.6	1.7	1.8

Source: Autodata.

These data contain an obvious pattern: Sales increased by 0.1 million each year. A scatterplot of these data and a line that models this situation are shown in the figure. In this chapter you will learn how to determine the equation of this and other lines. (See Section 2.2, Example 6.)

2.1 Functions and Their Representations
2.2 Linear Functions
2.3 The Slope of a Line
2.4 Equations of Lines and Linear Models

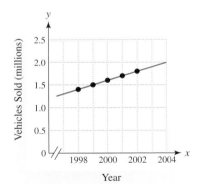

No great thing is created suddenly.
—EPICTETUS

2.1 FUNCTIONS AND THEIR REPRESENTATIONS

Basic Concepts ▪ Representations of a Function ▪ Definition of a Function ▪ Identifying a Function ▪ Tables, Graphs, and Calculators (Optional)

A LOOK INTO MATH ▷

In Chapter 1 we showed how to use numbers to describe data. For example, instead of simply saying that it is *hot* outside, we might use the number 102°F to describe the temperature. We also showed that data can be modeled with formulas and graphs. Formulas and graphs are sometimes used to represent *functions*, which are important in mathematics. In this section we introduce functions and their representations.

Basic Concepts

▶ **REAL-WORLD CONNECTION** Functions are used to calculate many important quantities. For example, suppose that a person works for $7 per hour. Then we could use a function f to calculate the amount of money the person earned after working x hours simply by multiplying the *input* x by 7. The result y is called the *output*. This concept is shown visually in the following diagram.

$$\text{Input } x \longrightarrow \text{Function } f \longrightarrow \text{Output } y = f(x)$$

For each valid input x, a function computes *exactly one* output y, which may be represented by the ordered pair (x, y). If the input is 5 hours, f outputs $7 \cdot 5 = \$35$; if the input is 8 hours, f outputs $7 \cdot 8 = \$56$. These results can be represented by the ordered pairs (5, 35) and (8, 56). Sometimes an input may not be valid. For example, if $x = -3$, there is no reasonable output because a person cannot work -3 hours.

We say that *y is a function of x* because the output y is determined by and *depends* on the input x. As a result, y is called the *dependent variable* and x is the *independent variable*. To emphasize that y is a function of x, we use the notation $y = f(x)$. The symbol $f(x)$ does not represent multiplication of a variable f and a variable x. The notation $y = f(x)$ is called *function notation*, is read "y equals f of x," and means that function f with input x produces output y. For example, if $x = 3$ hours, $y = f(3) = \$21$.

FUNCTION NOTATION

The notation $y = f(x)$ is called **function notation**. The **input** is x, the **output** is y, and the *name* of the function is f.

$$\underset{\text{Output}}{y} = \overset{\text{Name}}{f}(\underset{\text{Input}}{x})$$

The variable y is called the **dependent variable** and the variable x is called the **independent variable**. The expression $f(4) = \mathbf{28}$ is read "f of 4 equals 28" and indicates that f outputs 28 when the input is 4. A function computes *exactly one* output for each valid input. The letters f, g, and h are often used to denote names of functions.

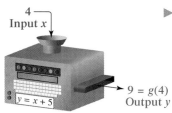

Figure 2.1 Function Machine

▶ **REAL-WORLD CONNECTION** Functions can be used to compute a variety of quantities. For example, suppose that a boy has a sister who is exactly 5 years older than he is. If the age of the boy is x, then a function g can calculate the age of his sister by adding 5 to x. Thus $g(4) = 4 + 5 = 9$, $g(10) = 10 + 5 = 15$, and in general $g(x) = x + 5$. That is, function g adds 5 to every input x to obtain the output $y = g(x)$.

Functions can be represented by an input–output machine, as illustrated in Figure 2.1. This machine represents function g and receives input $x = 4$, adds 5 to this value, and then outputs $g(4) = 4 + 5 = 9$.

Representations of a Function

▶ **REAL-WORLD CONNECTION** A function f forms a relation between inputs x and outputs y that can be represented verbally, numerically, symbolically, and graphically. Functions can also be represented with diagrams. We begin by considering a function f that converts yards to feet.

x (yards)	y (feet)
1	3
2	6
3	9
4	12
5	15
6	18
7	21

VERBAL REPRESENTATION (WORDS) To convert x yards to y feet we must multiply x by 3. Therefore, if function f computes the number of feet in x yards, a **verbal representation** of f is "Multiply the input x in yards by 3 to obtain the output y in feet."

NUMERICAL REPRESENTATION (TABLE OF VALUES) A function f that converts yards to feet is shown in Table 2.1, where $y = f(x)$.

A *table of values* is called a **numerical representation** of a function. Many times it is impossible to list all valid inputs x in a table. On the one hand, if a table does not contain every x-input, it is a *partial* numerical representation. On the other hand, a *complete* numerical representation includes *all* valid inputs. Table 2.1 is a partial numerical representation of f because many valid inputs, such as $x = 10$ or $x = 5.3$, are not shown in it. Note that for each valid input x there is exactly one output y. *For a function, inputs are not listed more than once in a table.*

SYMBOLIC REPRESENTATION (FORMULA) A *formula* provides a **symbolic representation** of a function. The computation performed by f to convert x yards to y feet is expressed by $y = 3x$. A formula for f is $f(x) = 3x$, where $y = f(x)$. We say that function f is *defined by* or *given by* $f(x) = 3x$. Thus $f(2) = 3 \cdot 2 = 6$.

GRAPHICAL REPRESENTATION (GRAPH) A **graphical representation**, or **graph**, visually associates an x-input with a y-output. The ordered pairs

$$(1, 3), (2, 6), (3, 9), (4, 12), (5, 15), (6, 18), \text{ and } (7, 21)$$

from Table 2.1 are plotted in Figure 2.2(a). This scatterplot suggests a line for the graph f. For each real number x there is exactly one real number y determined by $y = 3x$. If we restrict inputs to $x \geq 0$ and plot all ordered pairs $(x, 3x)$, then a line with no breaks will appear, as shown in Figure 2.2(b).

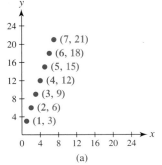

(a)

Figure 2.2

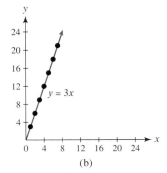

(b)

(Figure 2.2)

(a) Function

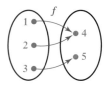
(b) Function

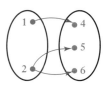
(c) Not a Function

Figure 2.3

MAKING CONNECTIONS
Functions, Points, and Graphs

If $f(a) = b$, then the point (a, b) lies on the graph of f. Conversely, if the point (a, b) lies on the graph of f, then $f(a) = b$. Thus each point on the graph of f can be written in the form $(a, f(a))$.

DIAGRAMMATIC REPRESENTATION (DIAGRAM) Functions may be represented by **diagrams**. Figure 2.3(a) is a diagram of a function, where an arrow is used to identify the output y associated with input x. For example, input **2** results in output **6**, which is written in function notation as $f(2) = 6$. That is, **2** yards are equivalent to **6** feet. Figure 2.3(b) shows a function f even though $f(1) = 4$ and $f(2) = 4$. Although two inputs for f have the same output, each valid input has exactly one output. In contrast, Figure 2.3(c) shows a relation that is not a function because input 2 results in two different outputs, 5 and 6.

MAKING CONNECTIONS
Four Representations of a Function

Symbolic Representation $f(x) = x + 1$

Numerical Representation   *Graphical Representation*

x	y
−2	−1
−1	0
0	1
1	2
2	3

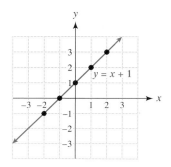

Verbal Representation f adds 1 to an input x to produce an output y.

▶ **REAL-WORLD CONNECTION** In the next example we calculate sales tax by evaluating different representations of a function.

EXAMPLE 1 Calculating sales tax

Let a function f compute a sales tax of 7% on a purchase of x dollars. Use the given representation to evaluate $f(2)$.
(a) *Verbal Representation* Multiply a purchase of x dollars by 0.07 to obtain a sales tax of y dollars.
(b) *Numerical Representation* Shown in Table 2.2
(c) *Symbolic Representation* $f(x) = 0.07x$
(d) *Graphical Representation* Shown in Figure 2.4
(e) *Diagrammatic Representation* Shown in Figure 2.5

2.1 FUNCTIONS AND THEIR REPRESENTATIONS

TABLE 2.2

x	$f(x)$
$1.00	$0.07
$2.00	$0.14
$3.00	$0.21
$4.00	$0.28

Figure 2.4 Sales Tax of 7%

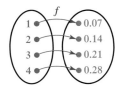

Figure 2.5

Solution

(a) Multiply the input 2 by 0.07 to obtain 0.14. The sales tax on $2.00 is $0.14.
(b) From Table 2.2, $f(2) = \$0.14$.
(c) Because $f(x) = 0.07x$, $f(2) = 0.07(2) = \mathbf{0.14}$, or $0.14.
(d) To evaluate $f(2)$ with a graph, first find 2 on the x-axis. Then move vertically upward until you reach the graph of f. The point on the graph may be estimated as $(\mathbf{2, 0.14})$, meaning that $f(\mathbf{2}) = \mathbf{0.14}$ (see Figure 2.6). Note that it may not be possible to find the exact answer from a graph. For example, one might estimate $f(2)$ to be 0.13 or 0.15 instead of 0.14.

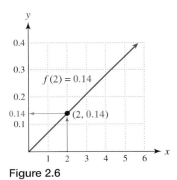

Figure 2.6

(e) In Figure 2.5, follow the arrow from **2** to **0.14**. Thus $f(\mathbf{2}) = \mathbf{0.14}$.

EXAMPLE 2 Evaluating symbolic representations (formulas)

Evaluate each function f at the given value of x.

(a) $f(x) = 3x - 7$ $x = -2$
(b) $f(x) = \dfrac{x}{x + 2}$ $x = 0.5$
(c) $f(x) = \sqrt{x - 1}$ $x = 10$

Solution

(a) $f(-2) = 3(-2) - 7 = -6 - 7 = -13$
(b) $f(0.5) = \dfrac{0.5}{0.5 + 2} = \dfrac{0.5}{2.5} = 0.2$
(c) $f(10) = \sqrt{10 - 1} = \sqrt{9} = 3$

▶ **REAL-WORLD CONNECTION** There are many examples of functions. To give more meaning to a function, sometimes we change both its name and its input variable. For instance, if we know the radius r of a circle, we can calculate its circumference by using $C(r) = 2\pi r$. The next example illustrates how functions are used in physical therapy.

EXAMPLE 3 Computing crutch length

People who sustain leg injuries often require crutches. A proper crutch length can be estimated without using trial and error. The function L, given by $L(t) = 0.72t + 2$, outputs an appropriate crutch length in inches for a person t inches tall. (**Source:** Journal of the American Physical Therapy Association.)

(a) Find $L(60)$ and interpret the result.
(b) If one person is 70 inches tall and another person is 71 inches tall, what should be the difference in their crutch lengths?

Solution
(a) $L(60) = 0.72(60) + 2 = 45.2$. Thus a person 60 inches tall needs crutches that are about 45.2 inches long.
(b) From the formula, $L(t) = 0.72t + 2$, we can see that each 1-inch increase in t results in a 0.72-inch increase in L. For example,
$$L(71) - L(70) = 53.12 - 52.4 = 0.72.$$

In the next example we find a formula and then sketch a graph of a function.

EXAMPLE 4 Finding representations of a function

Let function f square the input x and then subtract 1 to obtain the output y.
(a) Write a formula, or symbolic representation, for f.
(b) Make a table of values, or numerical representation, for f. Use $x = -2, -1, 0, 1, 2$.
(c) Sketch a graph, or graphical representation, of f.

Solution
(a) *Symbolic Representation* If we square x and then subtract 1, we obtain $x^2 - 1$. Thus a formula for f is $f(x) = x^2 - 1$.
(b) *Numerical Representation* Make a table of values for $f(x)$, as shown in Table 2.3. For example,
$$f(-2) = (-2)^2 - 1 = 4 - 1 = 3.$$
(c) *Graphical Representation* To obtain a graph of $f(x) = x^2 - 1$, plot the points from Table 2.3 and then connect them with a smooth curve, as shown in Figure 2.7. Note that we need to plot enough points so that we can determine the overall shape of the graph.

TABLE 2.3

x	$f(x)$
-2	3
-1	0
0	-1
1	0
2	3

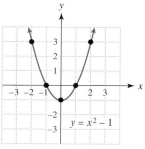

Figure 2.7

Definition of a Function

A function is a fundamental concept in mathematics. Its definition should allow for all representations of a function. *A function receives an input x and produces exactly one output y,* which can be expressed as an ordered pair:

(x, y).
Input   Output

A relation is a set of ordered pairs, and a function is a special type of relation.

FUNCTION

A **function** f is a set of ordered pairs (x, y), where each x-value corresponds to exactly one y-value.

The **domain** of f is the set of all x-values, and the **range** of f is the set of all y-values. For example, a function f that converts 1, 2, 3, and 4 yards to feet could be expressed as

$$f = \{(1, 3), (2, 6), (3, 9), (4, 12)\}.$$

The domain of f is $D = \{1, 2, 3, 4\}$, and the range of f is $R = \{3, 6, 9, 12\}$.

MAKING CONNECTIONS

Relations and Functions

A relation can be thought of as a set of input–output pairs. A function is a special type of relation whereby each input results in exactly one output.

▶ **REAL-WORLD CONNECTION** In the next example, we see how education can improve a person's chances for earning a higher income.

EXAMPLE 5 Computing average income

The function f computes the average 2004 individual income in dollars by educational attainment. This function is defined by $f(N) = 18,900$, $f(H) = 25,900$, $f(B) = 45,400$, and $f(M) = 62,300$, where N denotes no diploma, H a high school diploma, B a bachelor's degree, and M a master's degree. (*Source:* U.S. Census Bureau.)
(a) Write f as a set of ordered pairs.
(b) Give the domain and range of f.
(c) Discuss the relationship between education and income.

Solution
(a) $f = \{(N, 18900), (H, 25900), (B, 45400), (M, 62300)\}$.
(b) The domain of function f is $D = \{N, H, B, M\}$, and the range of function f is $R = \{18900, 25900, 45400, 62300\}$.
(c) Education pays—the greater the educational attainment, the greater are annual earnings.

EXAMPLE 6 Finding the domain and range graphically

Use the graphs of f shown in Figures 2.8 and 2.9 to find each function's domain and range.
(a) 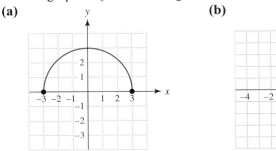 (b)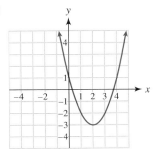

Figure 2.8 Figure 2.9

Solution
(a) The domain is the set of all x-values that correspond to points on the graph of f. Figure 2.10 shows that the domain D includes all x-values satisfying $-3 \leq x \leq 3$.

(Recall that the symbol ≤ is read "*less than or equal to.*") Because the graph is a semicircle with no breaks, the domain includes all real numbers between and including −3 and 3. The range R is the set of y-values that correspond to points on the graph of f. Thus R includes all y-values satisfying $0 \leq y \leq 3$.

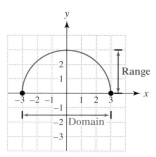

Figure 2.10

(b) The arrows on the ends of the graph in Figure 2.9 indicate that the graph extends indefinitely left and right, as well as upward. Thus D includes all real numbers. The smallest y-value on the graph is $y = -3$, which occurs when $x = 2$. Thus the range is $y \geq -3$. (Recall that the symbol ≥ is read "*greater than or equal to.*")

Symbolic, numerical, and graphical representations of three common functions are shown in Figure 2.11. Note that their graphs are not lines. Use the graphs to find the domain and range of each function.

Absolute value: $f(x) = |x|$

x	−2	−1	0	1	2
$|x|$	2	1	0	1	2

Square: $f(x) = x^2$

x	−2	−1	0	1	2
x^2	4	1	0	1	4

Square root: $f(x) = \sqrt{x}$

x	0	1	4	9
$\sqrt{x}$	0	1	2	3

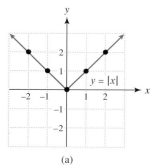

(a)

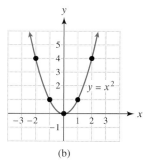

(b)

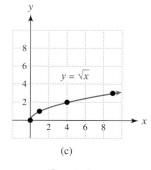

(c)

Figure 2.11
(a) D: all real numbers; R: $y \geq 0$
(b) D: all real numbers; R: $y \geq 0$
(c) D: $x \geq 0$; R: $y \geq 0$

CRITICAL THINKING

Suppose that a car travels at 50 miles per hour to a city that is 250 miles away. Sketch a graph of a function f that gives the distance y traveled after x hours. Identify the domain and range of f.

The domain of a function is the set of all valid inputs. To determine the domain of a function from a formula, we must determine x-values for which the formula is defined. This concept is demonstrated in the next example.

EXAMPLE 7 Finding the domain of a function

Use $f(x)$ to find the domain of f.
(a) $f(x) = 5x$ **(b)** $f(x) = \dfrac{1}{x-2}$ **(c)** $f(x) = \sqrt{x}$

Solution
(a) Because we can always multiply a real number x by 5, $f(x) = 5x$ is defined for all real numbers. Thus the domain of f includes all real numbers.
(b) Because we cannot divide by 0, input $x = 2$ is not valid for $f(x) = \dfrac{1}{x-2}$. The expression for $f(x)$ is defined for all other values of x. Thus the domain of f includes all real numbers except 2, or $x \neq 2$.
(c) Because square roots of negative numbers are not real numbers, the inputs for $f(x) = \sqrt{x}$ cannot be negative. Thus the domain of f includes all nonnegative numbers, or $x \geq 0$.

Identifying a Function

Recall that for a function each valid input x produces exactly one output y. In the next three examples we demonstrate techniques for identifying a function.

EXAMPLE 8 Determining whether a set of ordered pairs is a function

The set S of ordered pairs (x, y) represents the monthly average temperature y in degrees Fahrenheit for the month x in Washington, D.C. Determine whether S is a function.

$S = \{$(January, 33), (February, 37), (March, 45), (April, 53), (May, 66), (June, 73), (July, 77), (August, 77), (September, 70), (October, 51), (November, 48), (December, 37)$\}$

(**Source:** A. Miller and J. Thompson, *Elements of Meteorology*.)

Solution
The input x is the month and the output y is the monthly average temperature. The set S *is* a function because each month x is paired with exactly one monthly average temperature y. Note that, even though an average temperature of $37°F$ occurs in both February and December, S is nonetheless a function.

EXAMPLE 9 Determining whether a table of values represents a function

TABLE 2.4   Determine whether Table 2.4 represents a function.

x	y
1	−4
2	8
3	2
1	5
4	−6

Solution
The table does not represent a function because input $x = 1$ produces two outputs: -4 and 5. That is, the following two ordered pairs both belong to this relation.

Same input x
$(1, -4)$ $(1, 5)$
Different outputs y

VERTICAL LINE TEST To determine whether a graph represents a function, we must be convinced that it is impossible for an input x to have two or more outputs y. If two distinct points have the same x-coordinate on a graph, then the graph cannot represent a function. For example, the ordered pairs $(-1, 1)$ and $(-1, -1)$ could not lie on the graph of a function because input -1 results in *two* outputs: 1 and -1. When the points $(-1, 1)$ and $(-1, -1)$ are plotted, they lie on the same vertical line, as shown in Figure 2.12(a). A graph passing through these points intersects the vertical line twice, as illustrated in Figure 2.12(b).

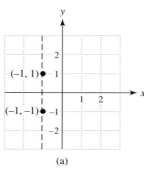
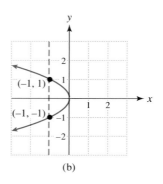

Figure 2.12

To determine whether a graph represents a function, visualize vertical lines moving across the *xy*-plane. If each vertical line intersects the graph *at most once*, then it is a graph of a function. This test is called the **vertical line test**. Note that the graph in Figure 2.12(b) fails the vertical line test and therefore does not represent a function.

VERTICAL LINE TEST

If every vertical line intersects a graph at no more than one point, then the graph represents a function.

EXAMPLE 10 Determining whether a graph represents a function

Determine whether the graphs shown in Figures 2.13 and 2.14 represent functions.

(a)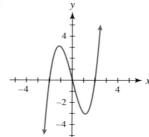

(b)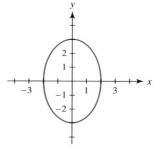

Figure 2.13

Figure 2.14

Solution

(a) Any vertical line will cross the graph at most once, as depicted in Figure 2.15. Therefore the graph *does* represent a function.

(b) The graph *does not* represent a function because there exist vertical lines that can intersect the graph twice. One such line is shown in Figure 2.16.

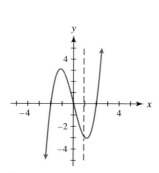

Figure 2.15 Passes Vertical Line Test

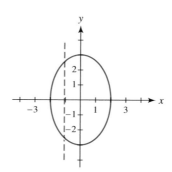

Figure 2.16 Fails Vertical Line Test

2.1 FUNCTIONS AND THEIR REPRESENTATIONS

Tables, Graphs, and Calculators (Optional)

We can use graphing calculators to create graphs and tables, usually more efficiently and reliably than with pencil-and-paper techniques. However, a graphing calculator uses the same techniques that we might use to sketch a graph. For example, one way to sketch a graph of $y = 2x - 1$ is first to make a table of values, as shown in Table 2.5.

We can plot these points in the xy-plane, as shown in Figure 2.17. Next we might connect the points, as shown in Figure 2.18.

TABLE 2.5

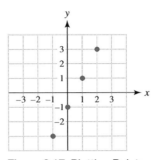

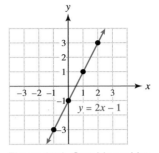

Figure 2.17 Plotting Points   Figure 2.18 Graphing a Line

In a similar manner a graphing calculator plots numerous points and connects them to make a graph. To create a similar graph with a graphing calculator, we enter the formula $Y_1 = 2X - 1$, set an appropriate viewing rectangle, and graph as shown in Figures 2.19(a) and (b). A table of values can also be generated as illustrated in Figure 2.19(c).

CALCULATOR HELP

To make a graph, see the Appendix (page AP-5). To make a table, see the Appendix (page AP-3).

(a)

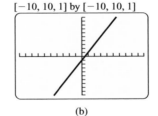
(b)

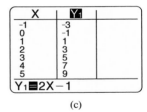
(c)

Figure 2.19

2.1 PUTTING IT ALL TOGETHER

One important concept in mathematics is that of a function. A function calculates exactly one output for each valid input and produces input–output ordered pairs in the form (x, y). A function typically computes something, such as area, speed, or sales tax. The following table summarizes some concepts related to functions.

Concept	Explanation	Examples
Function	A set of ordered pairs (x, y), where each x-value corresponds to exactly one y-value	$f = \{(1, 3), (2, 3), (3, 1)\}$ $f(x) = 2x$ A graph of $y = x + 2$ A table of values for $y = 4x$
Independent Variable	The *input* variable for a function	*Function* *Independent Variable* $f(x) = 2x$ x $A(r) = \pi r^2$ r $V(s) = s^3$ s

continued on next page

continued from previous page

Concept	Explanation	Examples
Dependent Variable	The *output* variable of a function. There is exactly one output for each valid input.	Function — Dependent Variable $y = f(x)$ — y $T = F(r)$ — T $V = g(r)$ — V
Domain and Range of a Function	The domain D is the set of all valid inputs. The range R is the set of all outputs.	For $S = \{(-1, 0), (3, 4), (5, 0)\}$, $D = \{-1, 3, 5\}$ and $R = \{0, 4\}$. For $f(x) = \frac{1}{x}$ the domain includes all real numbers except 0, or $x \neq 0$.
Vertical Line Test	If every vertical line intersects a graph at no more than one point, the graph represents a function.	This graph does *not* pass this test and thus does not represent a function.

A function can be represented verbally, symbolically, numerically, and graphically. The following table summarizes these four representations.

Type of Representation	Explanation	Comments
Verbal	Precise word description of what is computed	May be oral or written Must be stated *precisely*
Symbolic	Mathematical formula	Efficient and concise way of representing a function (e.g., $f(x) = 2x - 3$)
Numerical	List of specific inputs and their outputs	May be in the form of a table or an explicit set of ordered pairs
Graphical, diagrammatic	Shows inputs and outputs visually	No words, formulas, or tables Many types of graphs and diagrams are possible.

2.2 LINEAR FUNCTIONS

Basic Concepts ▪ Representations of Linear Functions ▪ Modeling Data with Linear Functions

A LOOK INTO MATH ▷

In mathematics, functions are used to model real-world phenomena, such as electricity, weather, and the economy. Because there are so many different applications of mathematics, a wide assortment of functions has been created. In fact, new functions are invented every day for use in business, education, and government. In this section we discuss an important type of function called a *linear function*.

Basic Concepts

▶ **REAL-WORLD CONNECTION** Suppose that the air conditioner is turned on when the temperature inside a house is 80°F. The resulting temperatures are listed in Table 2.6 for various elapsed times. Note that for each 1-hour increase in elapsed time, the temperature decreases by 2°F.

TABLE 2.6 **House Temperature**

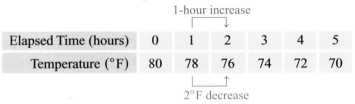

A scatterplot is shown in Figure 2.20, which suggests that a line models these data.

CALCULATOR HELP

To make a scatterplot, see the Appendix (page AP-4).

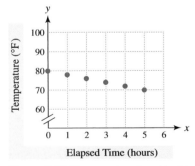

Figure 2.20 Temperature in a Home

Over this 5-hour period, the air conditioner lowers the temperature by 2°F for each hour that it runs. The temperature is found by multiplying the elapsed time x by -2 and adding the initial temperature of 80°F. This situation is modeled by $f(x) = -2x + 80$. For example,

$$f(2.5) = -2(2.5) + 80 = 75$$

means that the temperature is $75°F$ after the air conditioner has run for 2.5 hours. A graph of $f(x) = -2x + 80$, where $0 \leq x \leq 5$, is shown in Figure 2.21. We call f a *linear function* because its graph is a *line*. If a function is *not* a linear function, then it is a **nonlinear function**.

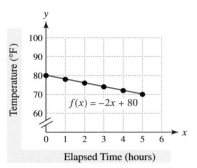

Figure 2.21 A Linear Function

LINEAR FUNCTION

A function f defined by $f(x) = ax + b$, where a and b are constants, is a **linear function**.

For $f(x) = -2x + 80$, we have $a = -2$ and $b = 80$. The constant a represents the rate at which the air conditioner cools the building, and the constant b represents the initial temperature.

▶ **REAL-WORLD CONNECTION** In general, a linear function defined by $f(x) = ax + b$ changes by a units for each unit increase in x. This **rate of change** is an increase if $a > 0$ and a decrease if $a < 0$. For example, if new carpet costs $20 per square yard, then the linear function defined by $C(x) = 20x$ gives the cost of buying x square yards of carpet. The value of $a = 20$ gives the cost (rate of change) for each additional square yard of carpet. For function C, the value of b is 0 because it costs $0 to buy 0 square yards of carpet.

NOTE: If f is a linear function, then $f(0) = a(0) + b = b$. Thus b can be found by evaluating $f(x)$ at $x = 0$.

EXAMPLE 1 Identifying linear functions

Determine whether f is a linear function. If f is a linear function, find values for a and b so that $f(x) = ax + b$.
(a) $f(x) = 4 - 3x$ **(b)** $f(x) = 8$ **(c)** $f(x) = 2x^2 + 8$

Solution
(a) Let $a = -3$ and $b = 4$. Then $f(x) = -3x + 4$, and f is a linear function.
(b) Let $a = 0$ and $b = 8$. Then $f(x) = 0x + 8$, and f is a linear function.
(c) Function f is not linear because its formula contains x^2. The formula for a linear function cannot contain an x with an exponent other than 1.

EXAMPLE 2 Determining linear functions

Use each table of values to determine whether $f(x)$ could represent a linear function. If f could be linear, write a formula for f in the form $f(x) = ax + b$.

(a)
x	0	1	2	3
$f(x)$	10	15	20	25

(b)
x	−2	0	2	4
$f(x)$	4	2	0	−2

(c)
x	0	1	2	3
$f(x)$	1	2	4	7

Solution

(a) For each unit increase in x, $f(x)$ increases by 5 units so $f(x)$ could be linear with $a = 5$. Because $f(0) = 10$, $b = 10$. Thus $f(x) = 5x + 10$.

(b) For each 2-unit increase in x, $f(x)$ decreases by 2 units. Equivalently, each unit increase in x results in a 1-unit decrease in $f(x)$, so $f(x)$ could be linear with $a = -1$. Because $f(0) = 2$, $b = 2$. Thus $f(x) = -x + 2$.

(c) Each unit increase in x does not result in a constant change in $f(x)$. Thus $f(x)$ does not represent a linear function.

Representations of Linear Functions

The graph of a linear function is a line. To graph a linear function f we usually start by making a table of values and then plot three or more points. We can then sketch the graph of f by drawing a line through these points, as demonstrated in the next example.

EXAMPLE 3 Graphing a linear function by hand

Sketch a graph of $f(x) = x - 1$. Use the graph to evaluate $f(-2)$.

Solution
Begin by making a table of values containing at least three points. Pick convenient values of x, such as $x = -1, 0, 1$.

$$f(-1) = -1 - 1 = -2$$
$$f(0) = 0 - 1 = -1$$
$$f(1) = 1 - 1 = 0$$

Display the results, as shown in Table 2.7.

Plot the points $(-1, -2)$, $(0, -1)$, and $(1, 0)$. Then sketch a line through the points to obtain the graph of f. A graph of a line results when *infinitely* many points are plotted, as shown in Figure 2.22.

To evaluate $f(-2)$, first find $x = -2$ on the x-axis. See Figure 2.23. Then move downward to the graph of f. By moving across to the y-axis, we see that the corresponding y-value is -3. Thus $f(-2) = -3$.

TABLE 2.7

x	y
−1	−2
0	−1
1	0

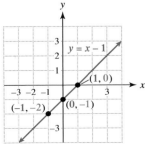

Figure 2.22

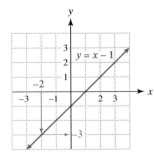

Figure 2.23 $f(-2) = -3$

CRITICAL THINKING

Two points determine a line. Why is it a good idea to plot at least three points when graphing a linear function by hand?

EXAMPLE 4 Representing a linear function

A linear function is given by $f(x) = -3x + 2$.
(a) Give a verbal representation of f.
(b) Make a numerical representation (table) of f by letting $x = -1, 0, 1$.
(c) Plot the points listed in the table from part (b). Then sketch a graph of $y = f(x)$.

Solution

TABLE 2.8

x	$f(x)$
-1	5
0	2
1	-1

(a) *Verbal Representation* Multiply the input x by -3 and then add 2 to obtain the output.

(b) *Numerical Representation* Evaluate the formula $f(x) = -3x + 2$ at $x = -1, 0, 1$, which results in Table 2.8. Note that $f(-1) = 5, f(0) = 2$, and $f(1) = -1$.

(c) *Graphical Representation* To make a graph of f by hand without a graphing calculator, plot the points $(-1, 5), (0, 2)$, and $(1, -1)$ from Table 2.8. Then draw a line passing through these points, as shown in Figure 2.24.

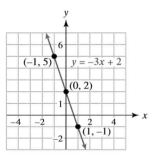

Figure 2.24

In the next example a graphing calculator is used to create a graph and table.

EXAMPLE 5 Using a graphing calculator

Give numerical and graphical representations of $f(x) = \frac{1}{2}x - 2$.

Solution

Numerical Representation To make a numerical representation, construct the table for $Y_1 = .5X - 2$, starting at $x = -3$ and incrementing by 1, as shown in Figure 2.25(a). (Other tables are possible.)

Graphical Representation Graph Y_1 in the standard viewing rectangle, as shown in Figure 2.25(b). (Other viewing rectangles may be used.)

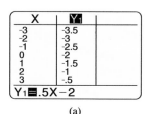

(a)

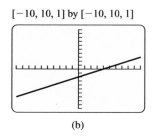

$[-10, 10, 1]$ by $[-10, 10, 1]$

(b)

Figure 2.25

CALCULATOR HELP

To make a table, see the Appendix (page AP-3). To make a graph, see the Appendix (page AP-5).

MAKING CONNECTIONS

Mathematics in Newspapers

Think of the mathematics that you see in newspapers. Often percentages are described *verbally*, numbers are displayed in *tables*, and data are shown in *graphs*. Seldom are *formulas* given, which is an important reason not to study only symbolic representations.

Modeling Data with Linear Functions

▶ **REAL-WORLD CONNECTION** A distinguishing feature of a linear function is that when the input x increases by 1 unit, the output $f(x) = ax + b$ always changes by an amount equal to a. For example, the number of doctors in the private sector from 1970 to 2000 can be modeled by

$$f(x) = 15{,}260x - 29{,}729{,}000,$$

where x is the year. The value $a = 15{,}260$ indicates that the number of doctors has increased, on average, by 15,260 each year. (*Source:* American Medical Association.)

The following are other examples of quantities that are modeled by linear functions. Try to determine the value of the constant a.

- The wages earned by an individual working x hours at $8 per hour
- The distance traveled by light in x seconds if the speed of light is 186,000 miles per second
- The cost of tuition and fees when registering for x credits if each credit costs $80 and the fees are fixed at $50

When we are modeling data with a linear function defined by $f(x) = ax + b$, the following concepts are helpful to determine a and b.

MODELING DATA WITH A LINEAR FUNCTION

The formula $f(x) = ax + b$ may be interpreted as follows.

$$f(x) \quad = \quad ax \quad + \quad b$$

(New amount) = (Change) + (Fixed amount)

When x represents time, *change* equals (rate of change) × (time).

$$f(x) \quad = \quad a \quad \times \quad x \quad + \quad b$$

(Future amount) = (Rate of change) × (Time) + (Initial amount)

▶ **REAL-WORLD CONNECTION** These concepts are applied in the next three examples.

EXAMPLE 6 Modeling car sales

Table 2.9 shows numbers of Toyota vehicles sold in the United States. (Refer to the introduction to this chapter.)

TABLE 2.9 **Toyota Vehicles Sold (millions)**

Year	1998	1999	2000	2001	2002
Vehicles	1.4	1.5	1.6	1.7	1.8

Source: Autodata.

(a) What were the sales in 1998?
(b) What was the annual increase in sales?
(c) Find a linear function f that models these data. Let $x = 0$ correspond to 1998, $x = 1$ to 1999, and so on.
(d) Use f to predict sales in 2004.

Solution
(a) In 1998, 1.4 million vehicles were sold.
(b) Sales have increased by 0.1 million (100,000) vehicles per year. Because this rate of change is the same each year, we can model the data *exactly* with a linear function.
(c) Initial sales in 1998 ($x = 0$) were 1.4 million vehicles, and each year sales increased by 0.1 million vehicles. Thus

$$f(x) \quad = \quad 0.1 \quad \times \quad x \quad + \quad 1.4$$
$$(\text{Future sales}) = (\text{Rate of change in sales}) \times (\text{Time}) + (\text{Initial sales}),$$

or $f(x) = 0.1x + 1.4$.
(d) Because $x = 6$ corresponds to 2004, evaluate $f(6)$.

$$f(6) = 0.1(6) + 1.4 = 2 \text{ million vehicles}$$

Note that the actual sales in 2004 were 2 million vehicles, so the estimate is accurate.

In the next example we model tuition and fees.

EXAMPLE 7 Modeling the cost of tuition and fees

Suppose that tuition costs $80 per credit and that student fees are fixed at $50. Find a formula for a linear function that models tuition and fees.

Solution
Total cost is found by multiplying $80 (rate or cost per credit) by the number of credits x and then adding the fixed fees (fixed amount) of $50. Thus $f(x) = 80x + 50$.

In the next example we consider a simple linear function that models the speed of a car.

EXAMPLE 8 Modeling with a constant function

A car travels on a freeway with its speed recorded at regular intervals, as listed in Table 2.10.

TABLE 2.10 Speed of a Car

Elapsed Time (hours)	0	1	2	3	4
Speed (miles per hour)	70	70	70	70	70

(a) Discuss the speed of the car during this time interval.
(b) Find a formula for a function f that models these data.
(c) Sketch a graph of f together with the data.

Solution
(a) The speed of the car appears to be constant at 70 miles per hour.
(b) Because the speed is constant, the rate of change is 0. Thus

$$f(x) \quad = \quad 0x \quad + \quad 70$$
$$\text{(Future speed)} = \text{(Change in speed)} + \text{(Initial speed)}$$

and $f(x) = 70$.

(c) Because $y = f(x)$, graph $y = 70$ with the data points

$$(0, 70), (1, 70), (2, 70), (3, 70), \text{ and } (4, 70)$$

to obtain Figure 2.26.

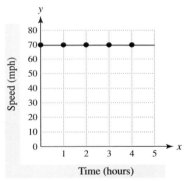

Figure 2.26 Speed of a Car

The function defined by $f(x) = 70$ is an example of a *constant function*. A **constant function** *is a linear function* with $a = 0$ and can be written as $f(x) = b$. Regardless of the input, a constant function always outputs the same value, b. Its graph is a horizontal line.

▶ **REAL-WORLD CONNECTION** The following are two applications of constant functions.

- A thermostat calculates a constant function regardless of the weather outside by maintaining a set temperature.
- A cruise control in a car calculates a constant function by maintaining a fixed speed, regardless of the type of road or terrain.

2.2 PUTTING IT ALL TOGETHER

A linear function is a relatively simple function. A clear understanding of linear functions is essential to the understanding of functions in general.

Concept	Explanation	Examples
Linear Function	Can be represented by $f(x) = ax + b$	$f(x) = 2x - 6$, $a = 2$ and $b = -6$ $f(x) = 10$, $a = 0$ and $b = 10$
Constant Function	Can be represented by $f(x) = b$	$f(x) = -7$, $b = -7$ $f(x) = 22$, $b = 22$
Rate of Change for a Linear Function	The output of a linear function changes by a constant amount for each unit increase in the input.	$f(x) = -3x + 8$ decreases 3 units for each unit increase in x. $f(x) = 5$ neither increases nor decreases. The rate of change is 0.

The following table summarizes symbolic, verbal, numerical, and graphical representations of a linear function.

Type of Representation	Comments	Example
Symbolic	Mathematical formula in the form $f(x) = ax + b$	$f(x) = 2x + 1$, where $a = 2$ and $b = 1$
Verbal	Multiply the input x by a and add b.	Multiply the input x by 2 and then add 1 to obtain the output.
Numerical (table of values)	For each unit increase in x in the table, the output of $f(x) = ax + b$ changes by an amount equal to a.	1-unit increase | x | 0 | 1 | 2 | | $f(x)$ | 1 | 3 | 5 | 2-unit increase
Graphical	The graph of a linear function is a line. Plot at least 3 points and then sketch the line.	$y = 2x + 1$

2.3 THE SLOPE OF A LINE

Slope ■ Slope–Intercept Form of a Line ■ Interpreting Slope in Applications

A LOOK INTO MATH ▷

Figure 2.27 shows some graphs of lines, where the x-axis represents time.

Which graph might represent the amount of gas in your car's tank while you are driving?   c

Which graph might represent the temperature inside a refrigerator?   a

Which graph might represent the amount of water in a pool that is being filled?   b

To answer these questions, you probably used the concept of slope. In mathematics, slope is a real number that measures the "tilt" of a line in the xy-plane. We assume throughout the text that lines are always straight. In this section we discuss how slope relates to the graph of a linear function and how to interpret slope.

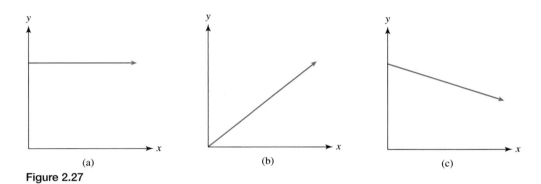

Figure 2.27

Slope

▶ **REAL-WORLD CONNECTION**   The graph shown in Figure 2.28 illustrates the cost of buying x pounds of candy. The graph tilts upward from left to right, which indicates that the cost increases as the number of pounds purchased increases. Note that for every 2 pounds purchased, the cost increases by \$3. We say that the graph *rises* 3 units for every 2 units of *run*. The ratio $\frac{\text{rise}}{\text{run}}$ equals the *slope* of the line. The slope of this line is $\frac{3}{2}$, or 1.5. That is, for every unit of run along the x-axis, the graph rises 1.5 units. A slope of 1.5 indicates that candy costs \$1.50 per pound.

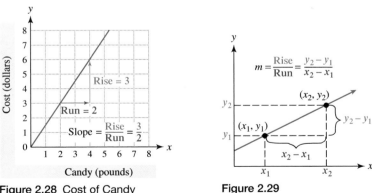

Figure 2.28 Cost of Candy   Figure 2.29

A more general case is shown in Figure 2.29 where a line passes through the points (x_1, y_1) and (x_2, y_2). The **rise** or **change in y** is $y_2 - y_1$, and the **run** or **change in x** is $x_2 - x_1$. The slope m is given by

$$m = \frac{\text{rise}}{\text{run}} = \frac{y_2 - y_1}{x_2 - x_1}.$$

NOTE: The expression x_1 has a **subscript** of 1 and is read "x sub one" or "x one." Thus x_1 and x_2 denote two different x-values. Similar comments can be made about y_1 and y_2.

> ### SLOPE
>
> The **slope** m of the line passing through the points (x_1, y_1) and (x_2, y_2) is
>
> $$m = \frac{y_2 - y_1}{x_2 - x_1},$$
>
> where $x_1 \neq x_2$. That is, slope equals *rise over run*.

NOTE: *Change in x* is sometimes denoted Δx and equals $x_2 - x_1$. The expression Δx is read "delta x." Similarly, *change in y* is sometimes denoted Δy and equals $y_2 - y_1$. Using this notation, we can express slope as $m = \frac{\Delta y}{\Delta x} = \frac{y_2 - y_1}{x_2 - x_1}$.

EXAMPLE 1   Calculating the slope of a line

Find the slope of the line passing through the points $(-4, 1)$ and $(2, 4)$. Plot these points and graph the line. Interpret the slope.

Solution
Begin by letting $(x_1, y_1) = (-4, 1)$ and $(x_2, y_2) = (2, 4)$. The slope is

$$m = \frac{y_2 - y_1}{x_2 - x_1} = \frac{4 - 1}{2 - (-4)} = \frac{3}{6} = \frac{1}{2}.$$

A graph of the line passing through these two points is shown in Figure 2.30. A slope of $\frac{1}{2}$ indicates that the line rises 1 unit for every 2 units of run. This slope is equivalent to 3 units of rise for every 6 units of run.

We would calculate the same slope in Example 1 if we let $(x_1, y_1) = (2, 4)$ and $(x_2, y_2) = (-4, 1)$. In this case the calculation would be

$$m = \frac{y_2 - y_1}{x_2 - x_1} = \frac{1 - 4}{-4 - 2} = \frac{-3}{-6} = \frac{1}{2}.$$

If a line has **positive slope**, the line *rises* from left to right. In Figure 2.31 the rise is 2 units for each unit of run, so the slope is 2. If a line has **negative slope**, the line *falls* from left to right. In Figure 2.32 the line *falls* 1 unit for every 2 units of run, so the slope is $-\frac{1}{2}$. Slope 0 indicates that a line is horizontal, as shown in Figure 2.33. If (x_1, y_1) and (x_2, y_2) are two points on a vertical line, $x_1 = x_2$. In Figure 2.34 the run is $x_2 - x_1 = 0$, so a vertical line has **undefined slope** because division by 0 is undefined.

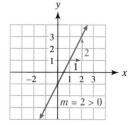

Figure 2.30

Figure 2.31 Positive Slope

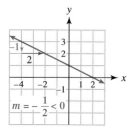

Figure 2.32 Negative Slope

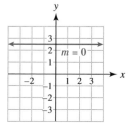

Figure 2.33 Zero Slope

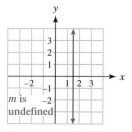

Figure 2.34 Undefined Slope

EXAMPLE 2 Calculating slope

Find the slope of the line passing through each pair of points, if possible
(a) $(-1, 4), (8, 4)$ (b) $\left(-\frac{3}{4}, \frac{1}{4}\right), \left(-\frac{3}{4}, \frac{5}{4}\right)$ (c) $(a, 3b), (3a, 5b)$

Solution

(a) For $(-1, 4)$ and $(8, 4)$, $m = \dfrac{y_2 - y_1}{x_2 - x_1} = \dfrac{4 - 4}{8 - (-1)} = \dfrac{0}{9} = 0$.

(b) For $\left(-\frac{3}{4}, \frac{1}{4}\right)$ and $\left(-\frac{3}{4}, \frac{5}{4}\right)$, $m = \dfrac{y_2 - y_1}{x_2 - x_1} = \dfrac{\frac{5}{4} - \frac{1}{4}}{-\frac{3}{4} - \left(-\frac{3}{4}\right)} = \dfrac{1}{0}$, which is undefined.

(c) For $(a, 3b)$ and $(3a, 5b)$, $m = \dfrac{y_2 - y_1}{x_2 - x_1} = \dfrac{5b - 3b}{3a - a} = \dfrac{2b}{2a} = \dfrac{b}{a}$.

EXAMPLE 3 Sketching a line with a given slope

Sketch a line passing through the point $(0, 4)$ and having slope $-\frac{2}{3}$.

Solution

Start by plotting the point $(0, 4)$. Because $m = \dfrac{\text{change in } y}{\text{change in } x}$, a slope of $-\frac{2}{3}$ indicates that the y-values *decrease* **2** units each time the x-values increase by **3** units. That is, the line *falls* **2** units for every **3**-unit increase in the run. The line passes through $(0, 4)$, so a 2-unit decrease in y and a 3-unit increase in x results in the line passing through the point $(0 + 3, 4 - 2)$ or $(3, 2)$, as shown in Figure 2.35.

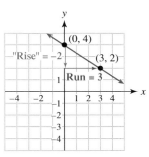

Figure 2.35

Slope–Intercept Form of a Line

Because $f(0) = 3$ and $f(1) = 5$, the graph of $f(x) = 2x + 3$ is a line that passes through $(0, 3)$ and $(1, 5)$, as shown in Figure 2.36. Therefore the slope of this line is

$$m = \dfrac{5 - 3}{1 - 0} = 2.$$

Note that slope 2 equals the coefficient of x in the formula $f(x) = 2x + 3$. In general, if $f(x) = ax + b$, the slope of the graph of f is $m = a$. For example, the graph of $f(x) = 6x - 5$ has slope $m = 6$, and the graph of $f(x) = -\frac{4}{5}x + 1$ has slope $m = -\frac{4}{5}$.

The point $(0, 3)$ lies on the graph of $f(x) = 2x + 3$ and is located on the y-axis. The y-value of 3 is called the *y-intercept*. A **y-intercept** is the y-coordinate of a point where a graph intersects the y-axis. To find a y-intercept let $x = 0$ in $f(x)$. If $f(x) = ax + b$, then

$$f(0) = a(0) + b = b.$$

Thus if $f(x) = -4x + 7$, the y-intercept is 7, and if $f(x) = \frac{1}{2}x - 8$, the y-intercept is -8.

Because $y = f(x)$, any linear function is given by $y = mx + b$, where m is the slope and b is the y-intercept. The form $y = mx + b$ is called the *slope–intercept form* of a line.

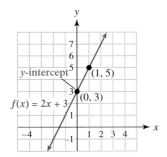

Figure 2.36

> ### SLOPE–INTERCEPT FORM
>
> The line with slope m and y-intercept b is given by
> $$y = mx + b,$$
> the **slope–intercept form** of a line.

EXAMPLE 4   Graphing lines

Identify the slope and y-intercept for the three lines $y = \frac{1}{2}x - 2$, $y = \frac{1}{2}x$, and $y = \frac{1}{2}x + 2$. Graph and compare the lines.

Solution
The graph of $y = \frac{1}{2}x - 2$ has slope $\frac{1}{2}$ and y-intercept -2. This line passes through the point $(0, -2)$ and rises 1 unit for each 2 units of run (see Figure 2.37). The graph of $y = \frac{1}{2}x$ has slope $\frac{1}{2}$ and y-intercept 0, and the graph of $y = \frac{1}{2}x + 2$ has slope $\frac{1}{2}$ and y-intercept 2. These lines are parallel, and the vertical distance between adjacent lines is always 2.

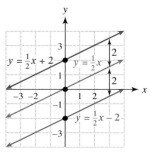

Figure 2.37

EXAMPLE 5   Using a graph to write the slope–intercept form

For each graph shown in Figures 2.38 and 2.39, write the slope–intercept form of the line.

(a)
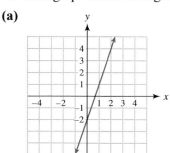

Figure 2.38

(b)
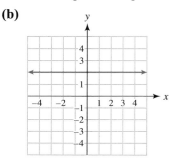

Figure 2.39

Solution
(a) The graph passes through $(0, -2)$, so the y-intercept is -2. Because the graph rises 3 units for each unit increase in x, the slope is 3. The slope–intercept form is $y = 3x - 2$.
(b) The graph passes through $(0, 2)$, so the y-intercept is 2. Because the graph is a horizontal line, its slope is 0. The slope–intercept form is $y = 0x + 2$, or more simply, $y = 2$.

EXAMPLE 6   Finding the slope–intercept form

TABLE 2.11

x	y
−1	1
0	?
2	10

The points listed in Table 2.11 all lie on a line.
(a) Find the missing value in the table.
(b) Write the slope–intercept form of the line.

Solution
(a) The line passes through $(-1, 1)$ and $(2, 10)$ so its slope is

$$m = \frac{10 - 1}{2 - (-1)} = \frac{9}{3} = 3.$$

For each unit increase in x, y increases by 3. When x increases from -1 to 0, y increases from 1 to $1 + 3 = 4$. Therefore the missing value is 4.

(b) Because the line passes through the point $(0, 4)$, its y-intercept is 4. The slope–intercept form is $y = 3x + 4$.

Interpreting Slope in Applications

▶ **REAL-WORLD CONNECTION** When a linear function is used to model physical quantities, the slope of its graph provides certain information. Slope can be interpreted as a **rate of change** of a quantity, which we illustrate in the next four examples.

EXAMPLE 7   Interpreting slope

The distance y in miles that an athlete riding a bicycle is from home after x hours is shown in Figure 2.40.
(a) Find the y-intercept. What does the y-intercept represent?
(b) The graph passes through the point $(2, 6)$. Discuss the meaning of this point.
(c) Find the slope–intercept form of this line. Interpret the slope as a rate of change.

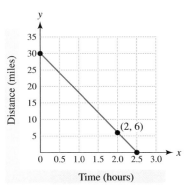

Figure 2.40 Distance from Home

Solution
(a) The y-intercept is 30, which indicates that the athlete is initially 30 miles from home.
(b) The point $(2, 6)$ means that after 2 hours the athlete is 6 miles from home.
(c) The line passes through the points $(0, 30)$ and $(2, 6)$. Thus its slope is

$$m = \frac{6 - 30}{2 - 0} = -12,$$

and the slope–intercept form is $y = -12x + 30$. A slope of -12 indicates that the athlete is traveling at 12 miles per hour *toward* home. The negative sign indicates that the distance between the athlete and home is *decreasing*.

EXAMPLE 8　Interpreting slope

Water is being pumped into a 500-gallon tank during a 10-minute interval. The amount of water W in the tank after t minutes is given by $W(t) = 40t + 100$, where W is in gallons.
(a) Evaluate $W(0)$ and $W(10)$. Interpret each result.
(b) Graph W for $0 \leq t \leq 10$.
(c) What is the slope of the graph of W? Interpret this slope.

Solution
(a) $W(0) = 40(0) + 100 = 100$; $W(10) = 40(10) + 100 = 500$. Initially, there are 100 gallons of water in the tank. After 10 minutes, there are 500 gallons and the tank is full.
(b) Because $W(0) = 100$ and $W(10) = 500$, plot the points $(0, 100)$ and $(10, 500)$. Function W is linear, so connect these points with a line segment. See Figure 2.41.

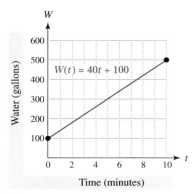

Figure 2.41

(c) The slope is 40, which indicates that water is being pumped into the tank at a rate of 40 gallons per minute.

EXAMPLE 9　Interpreting slope

If a sample of a gas is heated, it will expand. The expression $V(t) = 0.183t + 50$ gives the volume V of a sample of helium in cubic inches at a temperature of t degrees Celsius.
(a) Find the slope of the graph of V.
(b) Interpret the slope as a rate of change.

Solution
(a) The slope of the graph of $V(t) = 0.183t + 50$ is $m = 0.183$.
(b) A slope of $m = 0.183$ means that the sample of helium increases in volume by 0.183 cubic inch for each 1°C increase in temperature.

EXAMPLE 10　Analyzing growth of Wal-Mart

Table 2.12 lists numbers of Wal-Mart employees.
(a) Make a line graph of the data.
(b) Find the slope of each line segment in the graph.
(c) Interpret these slopes as rates of change.

TABLE 2.12　Wal-Mart Employees (millions)

Year	1997	1999	2002	2007
Employees	0.7	1.1	1.4	2.2

Source: Wal-Mart.

2.3 THE SLOPE OF A LINE

Solution

(a) To make a line graph start by plotting the points (1997, 0.7), (1999, 1.1), (2002, 1.4), and (2007, 2.2). Connecting these points with line segments results in Figure 2.42.

CRITICAL THINKING

An athlete runs away from home at 10 miles per hour for 30 minutes and then jogs back home at 5 miles per hour. Sketch a graph that shows the distance between the athlete and home.

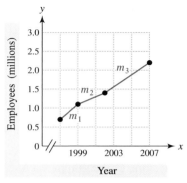

Figure 2.42 Wal-Mart Employees

(b) The slopes of the line segments in Figure 2.42 are

$$m_1 = \frac{1.1 - 0.7}{1999 - 1997} = 0.2, \quad m_2 = \frac{1.4 - 1.1}{2002 - 1999} = 0.1, \quad \text{and}$$

$$m_3 = \frac{2.2 - 1.4}{2007 - 2002} = 0.16.$$

(c) Slope $m_1 = 0.2$ means that, on average, the number of Wal-Mart employees increased by 0.2 million (or 200,000) per year between 1997 and 1999. Slopes m_2 and m_3 can be interpreted similarly.

2.3 PUTTING IT ALL TOGETHER

The graph of a linear function is a line. The "tilt" of a line is called the slope and equals rise over run. A positive slope indicates that the line *rises* from left to right, whereas a negative slope indicates that the line *falls* from left to right. A horizontal line has slope 0 and a vertical line has undefined slope. The following table summarizes some basic concepts about slope and slope–intercept form.

Concept	Explanation	Example
Slope	The slope of the line passing through the points (x_1, y_1) and (x_2, y_2) is given by $$m = \frac{\text{rise}}{\text{run}} = \frac{y_2 - y_1}{x_2 - x_1}, \ (x_1 \neq x_2).$$ Rise = change in y; run = change in x	The slope of the line passing through $(-2, 3)$ and $(1, 5)$ is $$m = \frac{5 - 3}{1 - (-2)} = \frac{2}{3}.$$ If the x-values increase by 3 units, the y-values increase by 2 units.

continued on next page

74 CHAPTER 2 LINEAR FUNCTIONS AND MODELS

continued from previous page

Concept	Explanation	Example
Slope–Intercept Form for a Line	The slope equals m, and the y-intercept equals b. $$y = mx + b$$	If $y = \frac{1}{2}x + 1$, the slope of the graph is $\frac{1}{2}$ and the y-intercept is 1.
Slope as a Rate of Change	The slope of the graph of a linear function indicates the rate at which a quantity is either increasing or decreasing.	From 1981 to 2000, average public college tuition and fees can be modeled by $$f(x) = 136x + 772,$$ where $x = 1$ corresponds to 1981. The slope of the graph of f is $m = 136$ and indicates that, on average, tuition and fees increased by \$136 per year between 1981 and 2000.

2.4 EQUATIONS OF LINES AND LINEAR MODELS

Point–Slope Form ■ Horizontal and Vertical Lines ■ Parallel and Perpendicular Lines

A LOOK INTO MATH ▷

In 1999, there were approximately 100 million Internet users in the United States, and this number grew to about 200 million in 2005. This growth is illustrated in Figure 2.43, where the line passes through the points (1999, 100) and (2005, 200). In this section we discuss how to find the equation of the line that models these data. To do so we start by discussing the *point–slope form* of a line.

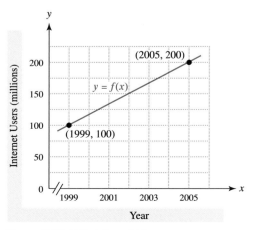

Figure 2.43 U.S. Internet Users

Point–Slope Form

If we know the slope m and y-intercept b of a line, we can write its slope–intercept form, $y = mx + b$. The slope–intercept form is an example of an **equation of a line**. The point–slope form is a different form of the equation of a line.

Suppose that a (nonvertical) line with slope m passes through the point (x_1, y_1). If (x, y) is a different point on this line, then $m = \frac{y - y_1}{x - x_1}$ (see Figure 2.44). We can use this slope formula to find the point–slope form.

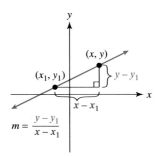

Figure 2.44

$$m = \frac{y - y_1}{x - x_1} \qquad \text{Slope formula}$$

$$m(x - x_1) = y - y_1 \qquad \text{Multiply each side by } (x - x_1).$$

$$y - y_1 = m(x - x_1) \qquad \text{Rewrite the equation.}$$

$$y = m(x - x_1) + y_1 \qquad \text{Add } y_1 \text{ to each side.}$$

The equation $y - y_1 = m(x - x_1)$ is traditionally called the *point–slope form*. We frequently think of y as being a function of x, written $y = f(x)$, so the equivalent form $y = m(x - x_1) + y_1$ is also referred to as the point–slope form.

> **POINT–SLOPE FORM**
>
> The line with slope m passing through the point (x_1, y_1) is given by
> $$y = m(x - x_1) + y_1,$$
> or equivalently,
> $$y - y_1 = m(x - x_1),$$
> the **point–slope form** of a line.

EXAMPLE 1 Using the point–slope form

Find the point–slope form of a line passing through the point $(1, 2)$ with slope -3. Does the point $(2, -1)$ lie on this line?

Solution
Let $m = -3$ and $(x_1, y_1) = (1, 2)$ in the point–slope form.

$$y = m(x - x_1) + y_1 \qquad \text{Point–slope form}$$
$$y = -3(x - 1) + 2 \qquad \text{Substitute.}$$

To determine whether the point $(2, -1)$ lies on the line, substitute 2 for x and -1 for y in the equation.

$$-1 \stackrel{?}{=} -3(2 - 1) + 2 \qquad \text{Let } x = 2 \text{ and } y = -1.$$
$$-1 \stackrel{?}{=} -3 + 2 \qquad \text{Simplify.}$$
$$-1 = -1 \qquad \text{The point satisfies the equation.}$$

The point $(2, -1)$ lies on the line because it satisfies the point–slope form.

We can use the point–slope form to find the equation of a line passing through two points.

CHAPTER 2 LINEAR FUNCTIONS AND MODELS

EXAMPLE 2 Applying the point-slope form

Find an equation of the line passing through $(-2, 3)$ and $(6, -1)$.

Solution
Before we can apply the point–slope form, we must find the slope.

$$m = \frac{y_2 - y_1}{x_2 - x_1} \qquad \text{Slope formula}$$

$$= \frac{-1 - 3}{6 - (-2)} \qquad \text{Substitute.}$$

$$= -\frac{1}{2} \qquad \text{Simplify.}$$

We can use either $(-2, 3)$ or $(6, -1)$ for (x_1, y_1) in the point–slope form. If we choose $(-2, 3)$, the point–slope form becomes the following.

$$y = m(x - x_1) + y_1 \qquad \text{Point–slope form}$$

$$y = -\frac{1}{2}(x - (-2)) + 3 \qquad \text{Let } x_1 = -2 \text{ and } y_1 = 3.$$

$$y = -\frac{1}{2}(x + 2) + 3 \qquad \text{Simplify.}$$

If we choose $(6, -1)$, the point–slope form with $x_1 = 6$ and $y_1 = -1$ becomes

$$y = -\frac{1}{2}(x - 6) - 1.$$

Note that, although the two point–slope forms look different, they are equivalent equations because their graphs are identical.

Example 2 illustrates the fact that the point–slope form *is not* unique for a given line. However, the slope–intercept form *is* unique because each line has a unique slope and a unique *y*-intercept. If we simplify both point–slope forms in Example 2, they reduce to the same slope–intercept form.

$$y = -\frac{1}{2}(x + 2) + 3 \qquad y = -\frac{1}{2}(x - 6) - 1 \qquad \text{Point–slope forms}$$

$$y = -\frac{1}{2}x - 1 + 3 \qquad y = -\frac{1}{2}x + 3 - 1 \qquad \text{Distributive property}$$

$$y = -\frac{1}{2}x + 2 \qquad y = -\frac{1}{2}x + 2 \qquad \text{Identical slope–intercept forms}$$

We can use the *slope–intercept form*, $y = mx + b$, instead of the point–slope form to find the equation of a line, as illustrated in the next example.

EXAMPLE 3 Applying the slope–intercept form

Find the equation of the line that passes through the points $(-3, 9)$ and $(1, 1)$.

Solution
First, find the slope of the line.

$$m = \frac{y_2 - y_1}{x_2 - x_1} = \frac{1 - 9}{1 - (-3)} = -\frac{8}{4} = -2$$

Now substitute -2 for m and $(-3, 9)$ for x and y in the slope–intercept form. The point $(1, 1)$ could be used instead.

$$y = mx + b \qquad \text{Slope–intercept form}$$
$$9 = -2(-3) + b \qquad \text{Let } y = 9, m = -2, \text{ and } x = -3.$$
$$9 = 6 + b \qquad \text{Simplify.}$$
$$3 = b \qquad \text{Solve for } b.$$

Thus the slope–intercept form is $y = -2x + 3$.

▶ **REAL-WORLD CONNECTION** In the next example we model the data presented in the introduction to this section.

EXAMPLE 4 Modeling growth in Internet usage

In 1999, there were approximately 100 million Internet users in the United States, and this number grew to about 200 million in 2005 (see Figure 2.43).
(a) Find values for m, x_1, and y_1, so that $f(x) = m(x - x_1) + y_1$ models these data.
(b) Interpret m as a rate of change.
(c) Use f to estimate Internet usage in 2007.

Solution
(a) The slope of the line passing through (1999, 100) and (2005, 200) is

$$m = \frac{200 - 100}{2005 - 1999} = \frac{100}{6} = \frac{50}{3}.$$

Thus, by choosing the point $(1999, 100)$ for the point–slope form, we can write

$$f(x) = \frac{50}{3}(x - 1999) + 100.$$

(b) Slope $m = \frac{50}{3} \approx 16.7$ indicates that the number of Internet users is increasing by about 16.7 million users per year.
(c) $f(2007) = \frac{50}{3}(2007 - 1999) + 100 \approx 233$ million.

MAKING CONNECTIONS

Modeling and the Dependent Variable x

From Example 4, $f(x) = \frac{50}{3}(x - 1999) + 100$ models the number of Internet users in the United States. In this formula x represents the actual year. We could also model Internet users with $g(x) = \frac{50}{3}x + 100$, where $x = 0$ corresponds to 1999, $x = 1$ to 2000, and so on. Then to estimate the Internet users in 2007, we let $x = 8$.

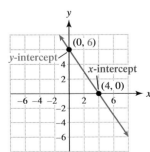

Figure 2.45 x-intercept 4; y-intercept 6

If a line intersects the y-axis at the point $(0, 6)$, the y-intercept is 6. Similarly, if the line intersects the x-axis at the point $(4, 0)$, the x-intercept is 4. This line and its intercepts are illustrated in Figure 2.45. The x-coordinate of a point where a graph intersects the x-axis is called the ***x*-intercept**. The next example interprets intercepts in a physical situation.

EXAMPLE 5 Modeling water in a pool

A small swimming pool containing 6000 gallons of water is emptied by a pump removing 1200 gallons per hour.
(a) How long does it take to empty the pool?
(b) Sketch a linear function f that models the amount of water in the pool after x hours.
(c) Identify the x-intercept and the y-intercept. Interpret each intercept.
(d) Find the slope–intercept form of the line. Interpret the slope as a rate of change.
(e) What are the domain and range of f?

Solution
(a) The time needed to empty the pool is $\frac{6000}{1200} = 5$ hours.
(b) Initially the pool contained 6000 gallons, and after 5 hours the pool was empty. Thus the graph of f is a line passing through (0, 6000) and (5, 0), as shown in Figure 2.46.
(c) The x-intercept is 5, which means that after 5 hours the pool is empty. The y-intercept of 6000 means that initially (when $x = 0$) the pool contained 6000 gallons of water.
(d) To find the equation of the line shown in Figure 2.46, we first find the slope of the line passing through the points (0, 6000) and (5, 0).

$$m = \frac{y_2 - y_1}{x_2 - x_1} \quad \text{Slope formula}$$

$$= \frac{0 - 6000}{5 - 0} \quad \text{Substitute.}$$

$$= -1200 \quad \text{Simplify.}$$

The slope is -1200 and the y-intercept is 6000, so the slope–intercept form is

$$y = -1200x + 6000.$$

The pump *removed* water at the rate of 1200 gallons per hour.
(e) The domain is D: $0 \leq x \leq 5$, and the range is R: $0 \leq y \leq 6000$.

CRITICAL THINKING

Can the graph of a function have more than one y-intercept? Explain.
Can the graph of a function have more than one x-intercept? Explain.

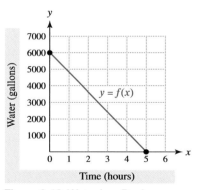

Figure 2.46 Water in a Pool

In the next example we introduce a method for modeling linear data by hand.

EXAMPLE 6 Modeling linear data by hand

Find a line $y = mx + b$ that models the data in Table 2.13.

TABLE 2.13

x	10	20	30	40	50
y	15	24	30	39	45

Solution

STEP 1: *Carefully make a scatterplot of the data* Be sure to label properly the *x*- and *y*-axes. For these data in Table 2.13 we can label each axis from 0 to 60 and have each hash mark represent 10 units. A scatterplot of the data is shown in Figure 2.47.

TABLE 2.13 (Repeated)

x	y
10	15
20	24
30	30
40	39
50	45

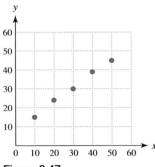

Figure 2.47

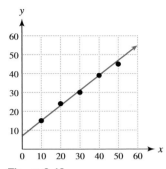

Figure 2.48

STEP 2: *Sketch a line that models the data* You may want to use a ruler for this step. In Figure 2.48 a line is drawn that passes through the first and fourth data points. Your line may be slightly different. Note that the line does not have to pass through any of the data points.

STEP 3: *Choose two points on the line and find the equation of the line* The line in Figure 2.48 passes through (10, 15) and (40, 39). Therefore its slope is

$$m = \frac{39 - 15}{40 - 10} = \frac{24}{30} = \frac{4}{5}.$$

The equation of the line is

$$y = \frac{4}{5}(x - 10) + 15 \quad \text{or} \quad y = \frac{4}{5}x + 7.$$

Note that answers may vary because the data are not exactly linear.

MAKING CONNECTIONS

Modeling, Lines, and Linear Functions

If a set of data is modeled by $y = mx + b$, then the data are also modeled by the linear function defined by $f(x) = mx + b$ because $y = f(x)$. In Example 6 the data set is modeled by $y = \frac{4}{5}x + 7$, so it is also modeled by $f(x) = \frac{4}{5}x + 7$.

Horizontal and Vertical Lines

The graph of a constant function is a horizontal line. For example, the graph of $f(x) = 3$ is a horizontal line with *y*-intercept 3, as shown in Figure 2.49. Its equation may be expressed as $y = 3$, so every point on the line has a *y*-coordinate of 3. In general, the equation $y = b$ represents a horizontal line with *y*-intercept *b*, as shown in Figure 2.50.

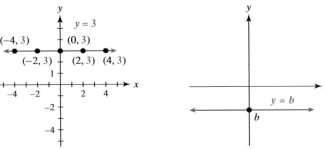

Figure 2.49   Figure 2.50

A vertical line cannot be represented by a function because different points on a vertical line have the same *x*-coordinate. The equation of the vertical line depicted in Figure 2.51 is $x = 3$. Every point on the line $x = 3$ has an *x*-coordinate equal to 3. In general, the equation of a vertical line with *x*-intercept *h* is $x = h$, as shown in Figure 2.52.

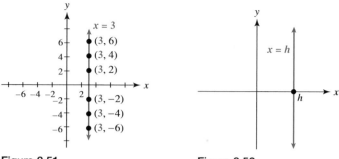

Figure 2.51   Figure 2.52

EQUATIONS OF HORIZONTAL AND VERTICAL LINES

The equation of a horizontal line with *y*-intercept *b* is $y = b$.

The equation of a vertical line with *x*-intercept *h* is $x = h$.

EXAMPLE 7 Finding equations of horizontal and vertical lines

Find equations of the vertical and horizontal lines that pass through the point $(-3, 4)$. Graph these two lines.

Solution

The *x*-coordinate of the point $(-3, 4)$ is -3. The vertical line $x = -3$ passes through *every* point in the *xy*-plane with an *x*-coordinate of -3, including the point $(-3, 4)$.

Similarly, the horizontal line $y = 4$ passes through *every* point with a *y*-coordinate of 4, including the point $(-3, 4)$. The lines $x = -3$ and $y = 4$ are graphed in Figure 2.53.

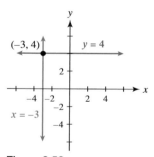

Figure 2.53

TECHNOLOGY NOTE: Graphing Vertical Lines

The equation of a vertical line is $x = h$ and cannot be expressed on a graphing calculator in the form "$Y_1 =$". Some graphing calculators graph a vertical line by accessing the DRAW menu. The accompanying figure shows a calculator graph of Figure 2.53.

CALCULATOR HELP

To graph a vertical line, see the Appendix (page AP-6).

$[-6, 6, 1]$ by $[-6, 6, 1]$

Parallel and Perpendicular Lines

Slope is important when we are determining whether two lines are parallel. For example, the lines $y = 2x$ and $y = 2x + 1$ are parallel because they both have slope 2.

PARALLEL LINES

Two lines with the same slope are parallel.

Two nonvertical parallel lines have the same slope.

EXAMPLE 8 Finding parallel lines

Find the slope–intercept form of a line parallel to $y = -2x + 5$, passing through $(-4, 3)$.

Solution
Because the line $y = -2x + 5$ has slope -2, any parallel line also has slope -2. The line passing through $(-4, 3)$ with slope -2 is determined as follows.

$$y = -2(x - (-4)) + 3 \qquad \text{Point–slope form}$$
$$y = -2x - 8 + 3 \qquad \text{Distributive property}$$
$$y = -2x - 5 \qquad \text{Slope–intercept form}$$

Figure 2.54 shows three pairs of perpendicular lines with their slopes labeled. Note in Figures 2.54(a) and 2.54(b) that the product $m_1 m_2$ equals -1. That is,

$$m_1 m_2 = 1 \cdot (-1) = -1 \quad \text{and} \quad m_1 m_2 = 2 \cdot \left(-\frac{1}{2}\right) = -1.$$

A more general situation for two perpendicular lines is shown in Figure 2.54(c), where

$$m_1 m_2 = m_1 \cdot \left(-\frac{1}{m_1}\right) = -1.$$

That is, if two nonvertical lines are perpendicular, then the product of their slopes is -1.

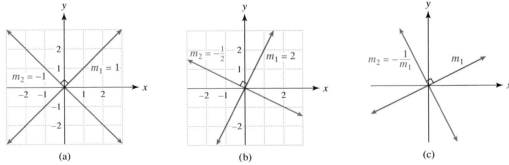

Figure 2.54 Perpendicular Lines

These results are summarized in the following box.

PERPENDICULAR LINES

Two lines with nonzero slopes m_1 and m_2 are perpendicular if $m_1 m_2 = -1$.

If two lines have slopes m_1 and m_2 such that $m_1 \cdot m_2 = -1$, then they are perpendicular.

Table 2.14 shows examples of slopes m_1 and m_2 that result in perpendicular lines because $m_1 m_2 = -1$. Note that m_1 and m_2 are **negative reciprocals** of each other; that is, $m_2 = -\frac{1}{m_1}$ and $m_1 = -\frac{1}{m_2}$.

TABLE 2.14 Slopes of Perpendicular Lines

m_1	1	$-\frac{1}{2}$	-4	$\frac{2}{3}$	$\frac{3}{4}$	$-\frac{5}{4}$
m_2	-1	2	$\frac{1}{4}$	$-\frac{3}{2}$	$-\frac{4}{3}$	$\frac{4}{5}$
$m_1 m_2$	-1	-1	-1	-1	-1	-1

EXAMPLE 9 Finding perpendicular lines

Find the slope–intercept form of the line perpendicular to $y = -\frac{1}{2}x + 1$, passing through the point (3, 2). Graph the lines.

Solution
The line $y = -\frac{1}{2}x + 1$ has slope $m_1 = -\frac{1}{2}$. From Table 2.14 the slope of a perpendicular line is $m_2 = 2$. The slope–intercept form of a line having slope 2 and passing through (3, 2) can be found as follows.

$$y = 2(x - 3) + 2 \quad \text{Point–slope form}$$
$$y = 2x - 6 + 2 \quad \text{Distributive property}$$
$$y = 2x - 4 \quad \text{Slope–intercept form}$$

A graph of these perpendicular lines is shown in Figure 2.55.

Figure 2.55

TECHNOLOGY NOTE: Square Viewing Rectangles

The accompanying figure shows a square viewing rectangle, in which the perpendicular lines from Example 9 intersect at 90°. Try graphing the two perpendicular lines in Example 9 by using the viewing rectangle [−6, 6, 1] by [−10, 10, 1]. Do the lines appear perpendicular? For many graphing calculators a square viewing rectangle results when the distance along the y-axis is about $\frac{2}{3}$ the distance along the x-axis. On some graphing calculators you can create a square viewing rectangle automatically by using the ZOOM menu.

CALCULATOR HELP
To set a square viewing rectangle, see the Appendix (page AP-6).

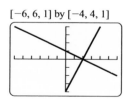
[−6, 6, 1] by [−4, 4, 1]

EXAMPLE 10 Equations of perpendicular lines

Find the slope–intercept form of each line shown in Figure 2.56. Verify that the two lines are perpendicular.

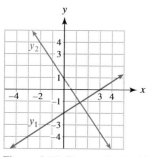

Figure 2.56 Perpendicular Lines

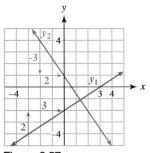

Figure 2.57

Solution
In Figure 2.57, the graph of y_1 has slope $m_1 = \frac{2}{3}$ because the line rises 2 units for every 3 units of run. Its y-intercept is -2, and its slope–intercept form is $y_1 = \frac{2}{3}x - 2$. The graph of y_2 has slope $m_2 = -\frac{3}{2}$ because the line falls 3 units for every 2 units of run. Its y-intercept is 1, and its slope–intercept form is $y = -\frac{3}{2}x + 1$. To be perpendicular, the product of their slopes must equal -1; that is,

$$m_1 \cdot m_2 = \frac{2}{3} \cdot \left(-\frac{3}{2}\right) = -1.$$

2.4 PUTTING IT ALL TOGETHER

The following table shows important forms of an equation of a line.

Concept	Comments	Example
Point–Slope Form $y = m(x - x_1) + y_1$ or $y - y_1 = m(x - x_1)$	Used to find an equation of a line, given two points or one point and the slope	Given two points $(1, 2)$ and $(3, 5)$, first compute $m = \frac{5 - 2}{3 - 1} = \frac{3}{2}.$ An equation of this line is $y = \frac{3}{2}(x - 1) + 2.$
Slope–Intercept Form $y = mx + b$	A unique equation for a line, determined by the slope m and the y-intercept b	An equation of the line with slope $m = 3$ and y-intercept $b = -5$ is $y = 3x - 5$.

The graph of a linear function f is a line. Therefore linear functions can be represented by

$$f(x) = mx + b \quad \text{or} \quad f(x) = m(x - x_1) + y_1.$$

The following table summarizes the important concepts involved with special types of lines.

Concept	Equation(s)	Example
Horizontal Line	$y = b$, where b is a constant	A horizontal line with y-intercept 5 has the equation $y = 5$.
Vertical Line	$x = h$, where h is a constant	A vertical line with x-intercept -3 has the equation $x = -3$.
Parallel Lines	$y = m_1 x + b_1$ and $y = m_2 x + b_2$, where $m_1 = m_2$	The lines $y = 2x - 1$ and $y = 2x + 5$ are parallel because both have slope 2.
Perpendicular Lines	$y = m_1 x + b_1$ and $y = m_2 x + b_2$, where $m_1 m_2 = -1$	The lines $y = 3x - 5$ and $y = -\frac{1}{3}x + 2$ are perpendicular because $m_1 m_2 = 3\left(-\frac{1}{3}\right) = -1$.

CHAPTER 2 SUMMARY

SECTION 2.1 ■ FUNCTIONS AND THEIR REPRESENTATIONS

Function A function is a set of ordered pairs (x, y), where each x-value corresponds to exactly one y-value. A function takes a valid input x and computes exactly one output y, forming the ordered pair (x, y).

Domain and Range of a Function The domain D is the set of all valid inputs, or x-values, and the range R is the set of all outputs, or y-values.

Examples: $f = \{(1, 2), (2, 3), (3, 3)\}$ has $D = \{1, 2, 3\}$ and $R = \{2, 3\}$.

$f(x) = x^2$ has domain all real numbers and range $y \geq 0$. (See the graph below.)

Function Notation $y = f(x)$ and is read "y equals f of x."

Example: $f(x) = \frac{2x}{x-1}$ implies that $f(3) = \frac{2 \cdot 3}{3-1} = \frac{6}{2} = 3$. Thus the point $(3, 3)$ is on the graph of f.

Function Representations A function can be represented symbolically, numerically, graphically, or verbally.

Symbolic Representation (Formula) $f(x) = x^2$

Numerical Representation (Table)

x	y
−2	4
−1	1
0	0
1	1
2	4

Graphical Representation (Graph)

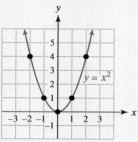

Verbal Representation (Words) f computes the square of the input x.

Vertical Line Test If every vertical line intersects a graph at most once, then the graph represents a function.

SECTION 2.2 ■ LINEAR FUNCTIONS

Linear Function A linear function can be represented by $f(x) = ax + b$. Its graph is a (straight) line. For each unit increase in x, $f(x)$ changes by an amount equal to a.

Example: $f(x) = 2x - 1$ represents a linear function with $a = 2$ and $b = -1$.

Numerical Representation

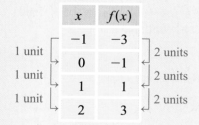

Graphical Representation

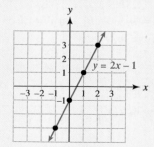

Each 1-unit increase in x results in a 2-unit increase in $f(x)$.

NOTE: A numerical representation is a table of values of $f(x)$.

Modeling Data with Linear Functions When data have a constant rate of change, they can be modeled by $f(x) = ax + b$. The constant a represents the *rate of change*, and the constant b represents the *initial amount* or the value when $x = 0$. That is,

$$f(x) = (\text{Rate of change})x + (\text{Initial amount}).$$

Example: In the following table, the y-values decrease by 3 units for each unit increase in x. When $x = 0$, $y = 4$. Thus the data are modeled by $f(x) = -3x + 4$.

x	-2	-1	0	1	2
y	10	7	4	1	-2

SECTION 2.3 ■ THE SLOPE OF A LINE

Slope The slope m of the line passing through the points (x_1, y_1) and (x_2, y_2) is

$$m = \frac{\text{rise}}{\text{run}} = \frac{y_2 - y_1}{x_2 - x_1}, \quad \text{where } x_1 \neq x_2.$$

Example: The slope of the line connecting $(-2, 3)$ and $(4, 0)$ is

$$m = \frac{0 - 3}{4 - (-2)} = \frac{-3}{6} = -\frac{1}{2}.$$

Slope–Intercept Form The equation $y = mx + b$ gives the slope m and y-intercept b of a line.

Example: The graph of $y = -\frac{1}{2}x + 1$ has slope $-\frac{1}{2}$ and y-intercept 1.

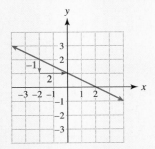

x	-2	0	2	4
y	2	1	0	-1

y decreases by 1 unit for each 2-unit increase in x.

Slope as a Rate of Change The slope of the graph of a linear function indicates how fast the function is increasing or decreasing.

Example: If $V(t) = 10t$ models the volume of water in a tank in gallons after t minutes, water is *entering* the tank at 10 gallons per minute.

SECTION 2.4 ■ EQUATIONS OF LINES AND LINEAR MODELS

Point–Slope Form

$$y = m(x - x_1) + y_1 \quad \text{or} \quad y - y_1 = m(x - x_1),$$

where m is the slope and (x_1, y_1) is a point on the line.

Example: The point–slope form of the line with slope 4 passing through $(2, -3)$ is

$$y = 4(x - 2) - 3.$$

Equations of Horizontal and Vertical Lines

$$y = b \quad \text{(horizontal)}, \quad x = h \quad \text{(vertical)}$$

Example: The equation of the horizontal line passing through $(2, 3)$ is $y = 3$. The equation of the vertical line passing through $(2, 3)$ is $x = 2$.

Parallel Lines

Two lines with the same slope are parallel.

Two nonvertical parallel lines have the same slope.

Example: The lines $y = 2x - 1$ and $y = 2x + 3$ are parallel with slope 2.

Perpendicular Lines Two lines with nonzero slopes m_1 and m_2 are perpendicular if $m_1 m_2 = -1$. If two lines have slopes m_1 and m_2 such that $m_1 \cdot m_2 = -1$, then they are perpendicular.

Example: The lines $y = 2x - 1$ and $y = -\frac{1}{2}x + 3$ are perpendicular because the product of their slopes equals -1. That is, $2\left(-\frac{1}{2}\right) = -1$.

Assignment Name _____ Name _____ Date _____
Show all work for these items: _____

# _____	# _____
# _____	# _____
# _____	# _____
# _____	# _____

Assignment Name _____ Name _____ Date _____

Show all work for these items: _____

# _____	# _____
# _____	# _____
# _____	# _____
# _____	# _____

CHAPTER 3
Linear Equations and Inequalities

3.1 Linear Equations
3.2 Introduction to Problem Solving
3.3 Linear Inequalities
3.4 Compound Inequalities
3.5 Absolute Value Equations and Inequalities

A recent report predicts that 90% of the fish and shellfish species taken from the ocean to feed people may be gone by 2048. As more marine species disappear, the ability of others to survive decreases. This trend is due in part to the increase in people's consumption of fish. The graph shows that the average American's appetite for seafood has increased substantially during past decades.

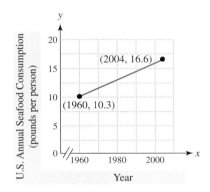

By using the data in the graph and our knowledge of mathematics, we can estimate past consumption of seafood and make predictions about the future. To accomplish this task, we first need to discuss linear equations. An understanding of linear equations is essential to understanding many concepts in mathematics.

> Say it, I'll forget. Demonstrate it, I may recall. But if I do it, I'll understand.
>
> —OLD CHINESE PROVERB

Source: Science Journal, November 3, 2006.

3.1 LINEAR EQUATIONS

Equations ▪ Symbolic Solutions ▪ Numerical and Graphical Solutions ▪ Identities and Contradictions ▪ Intercepts of a Line

A LOOK INTO MATH ▷

A primary objective of mathematics is solving equations. Billions of dollars are spent each year to solve equations that hold the answers for creating better products, such as high-definition televisions, DVD players, iPods, fiber optics, CAT scans, computers, and accurate weather forecasts. In this section we discuss linear equations and their applications.

Equations

▶ **REAL-WORLD CONNECTION** In Chapter 2 we discussed modeling data with *linear* functions. Linear functions can also be used to solve time and distance problems. For example, suppose $f(x) = 50x + 100$ models the distance in miles that a car is from the Texas border after x hours. We could use $f(x)$ to determine when the car is **300** miles from the border by solving the equation

$$50x + 100 = 300.$$

This is an example of a *linear equation* in one variable.

> **LINEAR EQUATION IN ONE VARIABLE**
>
> A **linear equation** in one variable is an equation that can be written in the form
> $$ax + b = 0,$$
> where a and b are constants with $a \neq 0$.

Examples of linear equations include

$$2x - 1 = 0, \quad -5x = 10 + x, \quad \text{and} \quad 3x + 8 = 2.$$

Although the second and third equations do not appear to be in the form $ax + b = 0$, they can be transformed by using properties of algebra, which we discuss later in this section.

To *solve* an equation means to find all values for a variable that make the equation a true statement. Such values are called **solutions**, and the set of all solutions is called the **solution set**. For example, substituting **2** for x in the equation $3x - 1 = 5$ results in $3(2) - 1 = 5$, which is a true statement. The value 2 *satisfies* the equation $3x - 1 = 5$ and is the only solution. The solution set is $\{2\}$. Two equations are *equivalent* if they have the same solution set.

Because every linear equation can be written in the form $ax + b = 0$ with $a \neq 0$, linear equations have *one solution*. To understand this condition visually, consider the graph of $y = ax + b$ shown in Figure 3.1. It is a line that cannot be horizontal. The equation of the x-axis is $y = 0$, so a solution to the linear equation $ax + b = 0$ corresponds to the x-intercept h of the line $y = ax + b$. Because this line intersects the x-axis exactly once, the equation $ax + b = 0$ has one solution.

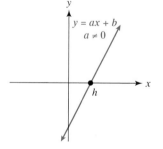

Figure 3.1

> **MAKING CONNECTIONS**
>
> Linear Functions and Equations
>
> A linear function can be written as $f(x) = ax + b$.
>
> A linear equation can be written as $ax + b = 0$ with $a \neq 0$.

Symbolic Solutions

Linear equations can be solved symbolically. *One advantage of a symbolic method is that the solution is always exact.* To solve an equation symbolically, we write a sequence of equivalent equations, using algebraic properties. For example, to solve $3x - 5 = 0$, we might add 5 to each side of the equation and then divide each side by 3 to obtain $x = \frac{5}{3}$.

$$3x - 5 = 0 \quad \text{Given equation}$$
$$3x - 5 + 5 = 0 + 5 \quad \text{Add 5 to each side.}$$
$$3x = 5 \quad \text{Simplify.}$$
$$\frac{3x}{3} = \frac{5}{3} \quad \text{Divide each side by 3.}$$
$$x = \frac{5}{3} \quad \text{Simplify.}$$

The solution is $\frac{5}{3}$. That is, $3\left(\frac{5}{3}\right) - 5 = 0$ is a true statement.

Adding 5 to each side is an example of the *addition property of equality* and dividing each side by 3 is an example of the *multiplication property of equality*. Note that dividing each side by 3 is equivalent to multiplying each side by $\frac{1}{3}$.

PROPERTIES OF EQUALITY

Addition Property of Equality

If a, b, and c are real numbers, then

$$a = b \quad \text{is equivalent to} \quad a + c = b + c.$$

Multiplication Property of Equality

If a, b, and c are real numbers with $c \neq 0$, then

$$a = b \quad \text{is equivalent to} \quad ac = bc.$$

The addition property states that an equivalent equation results if the same number is added to (or subtracted from) each side of an equation. Similarly, the multiplication property states that an equivalent equation results if each side of an equation is multiplied (or divided) by the same *nonzero* number.

EXAMPLE 1 Applying properties of equality

Solve each equation. Check your answer.
(a) $x - 4 = 5$ **(b)** $5x = 13$

Solution
(a) Isolate x in the equation $x - 4 = 5$ by applying the addition property of equality. To do this, we add **4** to each side of the equation.

$$x - 4 = 5 \quad \text{Given equation}$$
$$x - 4 + 4 = 5 + 4 \quad \text{Add 4 to each side.}$$
$$x = 9 \quad \text{Simplify.}$$

Check: To check that **9** is the solution, we substitute **9** into the *given* equation.

$$x - 4 = 5 \quad \text{Given equation}$$
$$9 - 4 \stackrel{?}{=} 5 \quad \text{Let } x = 9.$$
$$5 = 5 \quad \text{It checks.}$$

(b) Isolate x in the equation $5x = 13$ by applying the multiplication property of equality. To do this, we multiply each side of the equation by $\frac{1}{5}$, the reciprocal of 5. Note that this step is equivalent to dividing each side of the equation by **5**.

$$5x = 13 \qquad \text{Given equation}$$
$$\frac{5x}{5} = \frac{13}{5} \qquad \text{Divide each side by 5.}$$
$$x = \frac{13}{5} \qquad \text{Simplify.}$$

Check: To check that $\frac{13}{5}$ is the solution, we substitute $\frac{13}{5}$ into the *given* equation.

$$5x = 13 \qquad \text{Given equation}$$
$$5 \cdot \frac{13}{5} \stackrel{?}{=} 13 \qquad \text{Let } x = \tfrac{13}{5}.$$
$$13 = 13 \qquad \text{It checks.}$$

EXAMPLE 2 Solving a linear equation symbolically

Solve $2 - \frac{1}{2}x = 1$.

Solution
To solve, we write a sequence of equivalent equations by using the properties of equality.

$$2 - \frac{1}{2}x = 1 \qquad \text{Given equation}$$
$$-2 + 2 - \frac{1}{2}x = 1 + (-2) \qquad \text{Add } -2 \text{ to each side.}$$
$$-\frac{1}{2}x = -1 \qquad \text{Simplify.}$$
$$-2\left(-\frac{1}{2}x\right) = -1(-2) \qquad \text{Multiply each side by } -2, \text{ the reciprocal of } -\tfrac{1}{2}.$$
$$x = 2 \qquad \text{Simplify.}$$

The solution is 2.

MAKING CONNECTIONS

Expressions and Equations

Expressions and equations are *different* mathematical concepts. An expression does not contain an equals sign, whereas an equation *always* contains an equals sign. *An equation is a statement that two expressions are equal.* For example,

$$2x - 1 \quad \text{and} \quad x + 1$$

are two different expressions, and

$$2x - 1 = x + 1$$

is an equation. We often *solve an equation* for a value of x that makes the equation a true statement, but we *do not solve an expression*. We sometimes *simplify* expressions.

In the next example we use the distributive property to solve a linear equation.

EXAMPLE 3   Solving a linear equation symbolically

Solve $2(x - 1) = 4 - \frac{1}{2}(4 + x)$. Check your answer.

Solution

We begin by applying the distributive property.

$2(x - 1) = 4 - \frac{1}{2}(4 + x)$ Given equation

$2x - 2 = 4 - 2 - \frac{1}{2}x$ Distributive property

$2x - 2 = 2 - \frac{1}{2}x$ Simplify.

Next, we move the constant terms to the right and the *x*-terms to the left.

$2x - 2 + 2 = 2 - \frac{1}{2}x + 2$ Add 2 to each side.

$2x = 4 - \frac{1}{2}x$ Simplify.

$2x + \frac{1}{2}x = 4 - \frac{1}{2}x + \frac{1}{2}x$ Add $\frac{1}{2}x$ to each side.

$\frac{5}{2}x = 4$ Simplify.

Finally, we multiply by $\frac{2}{5}$, which is the reciprocal of $\frac{5}{2}$.

$\frac{2}{5} \cdot \frac{5}{2}x = 4 \cdot \frac{2}{5}$ Multiply each side by $\frac{2}{5}$.

$x = \frac{8}{5} = 1.6$ Simplify.

The solution is $\frac{8}{5}$.

To check our answer, we substitute $x = \frac{8}{5}$ in the *given* equation.

$2(x - 1) = 4 - \frac{1}{2}(4 + x)$ Given equation

$2\left(\frac{8}{5} - 1\right) \stackrel{?}{=} 4 - \frac{1}{2}\left(4 + \frac{8}{5}\right)$ Let $x = \frac{8}{5}$.

$\frac{16}{5} - 2 \stackrel{?}{=} 4 - 2 - \frac{4}{5}$ Distributive property

$\frac{6}{5} = \frac{6}{5}$ It checks.

CRITICAL THINKING

When you are checking an answer, why is it important to substitute the answer in the *given* equation?

The equation in Example 3 contained a fraction. Sometimes it is easier to avoid working with fractions. To clear fractions we can multiply each side by a common denominator.

EXAMPLE 4 Solving equations involving fractions or decimals

Solve each equation.

(a) $\frac{1}{3}(2z - 3) - \frac{1}{2}z = -2$ (b) $0.4t + 0.3 = 0.75 - 0.05t$

Solution

(a) The least common denominator of $\frac{1}{3}$ and $\frac{1}{2}$ is 6, so multiply each side by 6.

$$\frac{1}{3}(2z - 3) - \frac{1}{2}z = -2 \qquad \text{Given equation}$$

$$6\left(\frac{1}{3}(2z - 3) - \frac{1}{2}z\right) = 6(-2) \qquad \text{Multiply each side by 6.}$$

$$2(2z - 3) - 3z = -12 \qquad \text{Distributive property}$$

$$4z - 6 - 3z = -12 \qquad \text{Distributive property}$$

$$z - 6 = -12 \qquad \text{Combine like terms.}$$

$$z = -6 \qquad \text{Add 6 to each side.}$$

The solution is -6.

(b) The decimals 0.4, 0.75, 0.3, and 0.05 can be written as $\frac{4}{10}, \frac{75}{100}, \frac{3}{10}$, and $\frac{5}{100}$. A common denominator is 100, so multiply each side by 100.

$$0.4t + 0.3 = 0.75 - 0.05t \qquad \text{Given equation}$$

$$100(0.4t + 0.3) = 100(0.75 - 0.05t) \qquad \text{Multiply each side by 100.}$$

$$40t + 30 = 75 - 5t \qquad \text{Distributive property}$$

$$45t + 30 = 75 \qquad \text{Add } 5t \text{ to each side.}$$

$$45t = 45 \qquad \text{Subtract 30 from each side.}$$

$$t = 1 \qquad \text{Divide each side by 45.}$$

The solution is 1.

▶ **REAL-WORLD CONNECTION** In the next example, we model some real data with a linear function and then use this function to make an estimate.

EXAMPLE 5 Modeling numbers of LCD screens

Flat screens or LCD (liquid crystal display) screens are becoming increasingly popular. In 2002 about 30 million LCD screens were manufactured and this number increased to 90 million in 2006.
(a) Find a linear function that models these data.
(b) Estimate the year when 105 million LCD screens were manufactured.

Solution
(a) The graph of this linear function must pass through (2002, 30) and (2006, 90). Its slope is

$$m = \frac{90 - 30}{2006 - 2002} = \frac{60}{4} = 15.$$

Using the point (2002, 30) in the point–slope form gives

$$f(x) = 15(x - 2002) + 30.$$

(b) We must solve the equation $f(x) = 105$.

$$15(x - 2002) + 30 = 105 \quad \text{Equation to be solved}$$
$$15(x - 2002) = 75 \quad \text{Subtract 30 from each side.}$$
$$x - 2002 = 5 \quad \text{Divide each side by 15.}$$
$$x = 2007 \quad \text{Add 2002 to each side.}$$

In 2007, 105 million LCD screens might be manufactured.

Numerical and Graphical Solutions

Linear equations can also be solved numerically (with a table) and graphically. The disadvantage of using a table or graph is that the solution is often an estimate, rather than exact. In the next example these two methods are applied to the equation in Example 2.

EXAMPLE 6 Solving equations numerically and graphically

Solve $2 - \frac{1}{2}x = 1$ numerically and graphically.

Solution

Numerical Solution Begin by constructing a table for the expression $2 - \frac{1}{2}x$, as shown in Table 3.1. This expression equals 1 when $x = 2$. That is, the solution to $2 - \frac{1}{2}x = 1$ is 2.

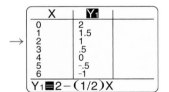

Figure 3.2 A Numerical Solution

TABLE 3.1   A Numerical Solution

x	0	1	2	3	4	5	6
$2 - \frac{1}{2}x$	2	1.5	1	0.5	0	-0.5	-1

In Figure 3.2 a calculator has been used to create the same table.

Graphical Solution One way to find a graphical solution is to let y_1 equal the left side of the equation, to let y_2 equal the right side of the equation, and then to graph $y_1 = 2 - \frac{1}{2}x$ and $y_2 = 1$, as shown in Figure 3.3. The graphs intersect at the point (2, 1). We are seeking an x-value that satisfies $2 - \frac{1}{2}x = 1$, so 2 is the solution.

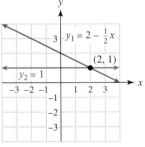

Figure 3.3 Graphical Solution

TECHNOLOGY NOTE: Finding a Numerical Solution

The solution to Example 3 is the decimal number 1.6. It is possible to find it numerically. Let

$$Y_1 = 2(X - 1) \quad \text{and} \quad Y_2 = 4 - (1/2)(4 + X)$$

and start incrementing by 1. There is no value where $y_1 = y_2$, as shown in the left-hand figure. However, note that when $x = 1$, $y_1 < y_2$ and when $x = 2$, $y_1 > y_2$. This change indicates that there is a solution *between* $x = 1$ and $x = 2$. When x is incremented by 0.1, $y_1 = y_2$ when $x = 1.6$, as shown in the right-hand figure.

Many times a graphical solution is an *approximate* solution because it depends on how accurately a graph can be read. To verify that a graphical solution is exact, check it by substituting it in the given equation, as is done in the next two examples.

EXAMPLE 7 Solving a linear equation graphically

Figure 3.4 shows graphs of $y_1 = 2x + 1$ and $y_2 = -x + 4$. Use the graph to solve the equation $2x + 1 = -x + 4$. Check your answer.

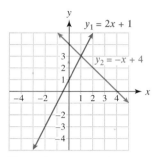

Figure 3.4

Solution
The graphs of y_1 and y_2 intersect at the point $(1, 3)$. Therefore **1** is the solution. We can check this solution by substituting $x = 1$ in the equation.

$$2x + 1 = -x + 4 \quad \text{Given equation}$$
$$2(1) + 1 \stackrel{?}{=} -1 + 4 \quad \text{Let } x = 1.$$
$$3 = 3 \quad \text{The answer checks.}$$

EXAMPLE 8 Solving a linear equation graphically

Solve $3(1 - x) = 2$ graphically.

Solution
We begin by graphing $Y_1 = 3(1 - X)$ and $Y_2 = 2$, as shown in Figure 3.5. Their graphs intersect near the point $(0.3333, 2)$. Because $\frac{1}{3} = 0.\overline{3}$, the solution appears to be $\frac{1}{3}$. Note that this graphical solution is *approximate*. We can verify our result as follows.

$$3(1 - x) = 2 \quad \text{Given equation}$$
$$3\left(1 - \frac{1}{3}\right) \stackrel{?}{=} 2 \quad \text{Substitute } x = \frac{1}{3}.$$
$$2 = 2 \quad \text{It checks.}$$

[−6, 6, 1] by [−4, 4, 1]

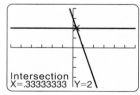

Figure 3.5

CALCULATOR HELP
To find a point of intersection, see the Appendix (page AP-7).

Technology can be helpful when we are solving an equation that is complicated. In the next example, we solve an application graphically.

EXAMPLE 9 Solving an application graphically

From 1987 to 2004, the (combined) percentage of music sales from rap and hip hop music can be calculated by $H(x) = \frac{83}{170}x + 3.8$, where x represents the number of years *after* 1987. Similarly, $R(x) = -\frac{22}{17}x + 46$ can calculate the percentage of music sales from rock and roll music. (*Source:* Recording Industry Association of America.)

(a) Evaluate $R(3)$ and interpret the result.
(b) Estimate graphically the year when rap and hip hop sales might equal rock and roll sales.

Solution
(a) $R(3) = -\frac{22}{17}(3) + 46 = -\frac{66}{17} + 46 \approx 42.1$. Three years after 1987, or in 1990, rock and roll music accounted for about 42.1% of music sales.
(b) Let $Y_1 = (83/170)X + 3.8$ and $Y_2 = -(22/17)X + 46$, as shown in Figure 3.6(a). Their graphs intersect near (23.7, 15.4), as shown in Figure 3.6(b). An x-value of 23.7 corresponds to the year $1987 + 23.7 = 2010.7$, or approximately 2011. Thus, sales of rap and hip hop music could equal rock and roll sales during 2011. At this time each type of music would account for 15.4% of music sales.

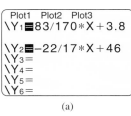
(a)

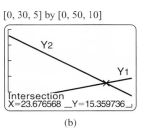
(b)

Figure 3.6

Identities and Contradictions

A linear equation has exactly one solution. However, there are other types of equations, called *identities* and *contradictions*. An **identity** is an equation that is always *true*, regardless of the values of any variables. For example, $x + x = 2x$ is an identity, because it is true for all real numbers x. A **contradiction** is an equation that is always *false*, regardless of the values of any variables. For example, $x = x + 1$ is a contradiction because no real number x can be equal to itself plus 1. If an equation is true for some, but not all, values of any variables, then it is called a **conditional equation**. The equation $x + 1 = 4$ is conditional because 3 is the only solution. Thus far, we have only discussed conditional equations.

EXAMPLE 10 Determining identities and contradictions

Determine whether each equation is an identity, contradiction, or conditional equation.
(a) $0 = 1$ (b) $5 = 5$ (c) $2x = 6$ (d) $2(z - 1) + z = 3z + 2$

Solution
(a) The equation $0 = 1$ is always false. It is a contradiction.
(b) The equation $5 = 5$ is always true. It is an identity.
(c) The equation $2x = 6$ is true only when $x = 3$. It is a conditional equation.
(d) Simplify the equation to obtain the following.

$$2(z - 1) + z = 3z + 2 \quad \text{Given equation}$$
$$2z - 2 + z = 3z + 2 \quad \text{Distributive property}$$
$$3z - 2 = 3z + 2 \quad \text{Combine like terms.}$$
$$-2 = 2 \quad \text{Subtract } 3z \text{ from each side.}$$

Because $-2 = 2$ is false, the given equation is a contradiction.

Intercepts of a Line

Equations of lines can be written in slope–intercept form or point–slope form. A third form for the equation of a line is called *standard form*, which is defined as follows.

STANDARD FORM OF A LINE

An equation for a line is in **standard form** when it is written as

$$ax + by = c,$$

where a, b, and c are constants with a and b not both 0.

Examples of lines in standard form include

$$3x - 7y = -2, \quad -5x - 6y = 0, \quad 2y = 5, \quad \text{and} \quad \tfrac{1}{2}x + y = 4.$$

To graph a line in standard form, we often start by locating the x- and y-intercepts. For example, the graph of $-x + 2y = 2$ is shown in Figure 3.7. All points on the x-axis have a y-coordinate of 0. To find the x-intercept, we let $y = 0$ in the equation $-x + 2y = 2$ and then solve for x.

$$-x + 2(0) = 2 \quad \text{or} \quad x = -2.$$

The x-intercept is -2. Note that the graph intersects the x-axis at $(-2, 0)$. Similarly, all points on the y-axis have an x-coordinate of 0. To find the y-intercept, we let $x = 0$ in the equation $-x + 2y = 2$ and then solve for y to obtain 1. Note that the graph intersects the y-axis at the point $(0, 1)$. This discussion is summarized by the following.

FINDING INTERCEPTS

To find the x-intercept of a line, let $y = 0$ in its equation and solve for x.

To find the y-intercept of a line, let $x = 0$ in its equation and solve for y.

NOTE: In some texts intercepts are defined to be points rather than real numbers. In this case the x-intercept in Figure 3.7 is $(-2, 0)$ and the y-intercept is $(0, 1)$.

EXAMPLE 11 Finding intercepts of a line in standard form

Let the equation of a line be $3x - 2y = 6$.
(a) Find the x- and y-intercepts.
(b) Graph the line.
(c) Solve the equation for y to obtain the slope–intercept form.

Solution
(a) *x-intercept:* Let $y = 0$ in $3x - 2y = 6$ to obtain $3x - 2(0) = 6$, or $x = 2$. The x-intercept is 2.
y-intercept: Let $x = 0$ in $3x - 2y = 6$ to obtain $3(0) - 2y = 6$, or $y = -3$. The y-intercept is -3.
(b) Sketch a line passing through $(2, 0)$ and $(0, -3)$, as shown in Figure 3.8.
(c) To solve for y, start by subtracting $3x$ from each side.

$$3x - 2y = 6 \quad \text{Given equation.}$$
$$-2y = -3x + 6 \quad \text{Subtract } 3x \text{ from each side.}$$
$$-\tfrac{1}{2}(-2y) = -\tfrac{1}{2}(-3x + 6) \quad \text{Multiply each side by } -\tfrac{1}{2}.$$
$$y = \tfrac{3}{2}x - 3 \quad \text{Distributive property}$$

The slope–intercept form is $y = \tfrac{3}{2}x - 3$.

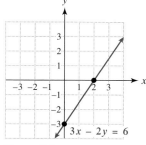

Figure 3.8

3.1 PUTTING IT ALL TOGETHER

The following table summarizes some important topics related to linear equations.

Concept	Explanation	Examples
Linear Equations	Can be written as $ax + b = 0$, where $a \neq 0$; has *one solution*	$3x - 4 = 0$, $2(x + 3) = -2$, and $2x = \frac{2}{3} - 5x$
Solution Set	The set of all solutions	The solution set for $x - 4 = 0$ is $\{4\}$ because 4 is the only solution to the equation.
Addition Property of Equality	$a = b$ is equivalent to $a + c = b + c$. "Equals added to equals are equal."	If 2 is added to each side of $x - 2 = 1$, the equation becomes $x = 3$.
Multiplication Property of Equality	$a = b$ is equivalent to $ac = bc$ for $c \neq 0$. "Equals multiplied by equals are equal."	If each side of $\frac{1}{2}x = 3$ is multiplied by 2, the resulting equation is $x = 6$.
Standard Form for a Line	$ax + by = c$, where a, b, and c are constants with a and b not both zero	$3x + 5y = 10$, $-2x + y = 0$, $3y = 18$, and $x = 4$
Finding Intercepts	To find the x-intercept, let $y = 0$ and solve for x. To find the y-intercept, let $x = 0$ and solve for y.	Let $2x + 4y = 8$. x-intercept: $2x + 4(0) = 8$, or $x = 4$ y-intercept: $2(0) + 4y = 8$, or $y = 2$

Linear equations can be solved symbolically, graphically, and numerically. Symbolic solutions are *always exact*, whereas graphical and numerical solutions may be *approximate*. The following illustrates how to solve the equation $2x - 1 = 3$ with each method.

Symbolic Solution

$2x - 1 = 3$

$2x = 4$

$x = 2$

The solution is 2.

Graphical Solution

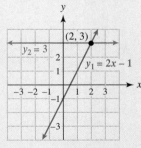

The solution is 2.

Numerical Solution

x	0	1	2	3
$2x - 1$	-1	1	3	5

Because $2x - 1$ equals 3 when $x = 2$, the solution is 2.

3.2 INTRODUCTION TO PROBLEM SOLVING

Solving a Formula for a Variable ■ Steps for Solving a Problem ■ Percentages

A LOOK INTO MATH ▷ Problem-solving skills are essential because every day we must solve problems—both small and large—for survival. A logical approach to a problem is often helpful, which is especially true in mathematics. In this section we discuss steps that can be used to solve mathematical problems. We begin the section by discussing how to solve a formula for a variable.

Solving a Formula for a Variable

The perimeter P of a rectangle with width W and length L is given by $P = 2W + 2L$, as illustrated in Figure 3.9. Suppose that we know the perimeter P and length L of this rectangle. Even though the formula is written to calculate P, we can still use it to find the width W. This technique is demonstrated in the next example.

Figure 3.9 $P = 2W + 2L$

EXAMPLE 1 Finding the width of a rectangle

A rectangle has a perimeter of 80 inches and a length of 23 inches.
(a) Write a formula to find the width W of a rectangle with known perimeter P and length L.
(b) Use the formula to find W.

Solution
(a) We must solve $P = 2W + 2L$ for W.

$$P = 2W + 2L \qquad \text{Given formula}$$
$$P - 2L = 2W \qquad \text{Subtract } 2L.$$
$$\frac{1}{2}(P - 2L) = \frac{1}{2}(2W) \qquad \text{Multiply by } \frac{1}{2}.$$
$$\frac{P}{2} - L = W \qquad \text{Distributive property}$$

The required formula is $W = \frac{P}{2} - L$.

(b) Substitute $P = 80$ and $L = 23$ into this formula. The width is

$$W = \frac{80}{2} - 23 = 17 \text{ inches.}$$

▶ **REAL-WORLD CONNECTION** In the next example we use a formula that converts degrees Fahrenheit to degrees Celsius to find a formula that converts degrees Celsius to degrees Fahrenheit.

EXAMPLE 2 **Converting temperature**

The formula $C = \frac{5}{9}(F - 32)$ is used to convert degrees Fahrenheit to degrees Celsius.
(a) Use this formula to convert 212°F to an equivalent Celsius temperature.
(b) Solve the formula for F to obtain a formula that can convert degrees Celsius to degrees Fahrenheit. Interpret the slope of the graph of F.
(c) Use this new formula to convert 25°C to an equivalent Fahrenheit temperature.

Solution

(a) $C = \frac{5}{9}(212 - 32) = \frac{5}{9}(180) = 100°C$

(b)
$$C = \frac{5}{9}(F - 32) \qquad \text{Given formula}$$

$$\frac{9}{5}C = \frac{9}{5} \cdot \frac{5}{9}(F - 32) \qquad \text{Multiply by } \tfrac{9}{5}.$$

$$\frac{9}{5}C = F - 32 \qquad \text{Simplify.}$$

$$\frac{9}{5}C + 32 = F \qquad \text{Add 32.}$$

Because $F = \frac{9}{5}C + 32$, the slope of the graph of F is $\frac{9}{5}$, which indicates that a 1° increase in Celsius temperature is equivalent to a $\frac{9°}{5}$ increase in Fahrenheit temperature.

(c) Substitute 25 for C in this formula to obtain $F = \frac{9}{5}(25) + 32 = 77°F$.

Sometimes an equation in two variables can be used to determine the formula for a function. To do so, we must let one variable be the dependent variable and the other variable be the independent variable. For example, $2x - 5y = 10$ is the equation of a line written in standard form. Solving for y gives the following result.

$$-5y = -2x + 10 \qquad \text{Subtract } 2x.$$

$$-\frac{1}{5}(-5y) = -\frac{1}{5}(-2x + 10) \qquad \text{Multiply by } -\tfrac{1}{5}.$$

$$y = \frac{2}{5}x - 2 \qquad \text{Distributive property}$$

If y is the dependent variable, x is the independent variable, and $y = f(x)$, then

$$f(x) = \frac{2}{5}x - 2$$

defines a function f whose graph is the line determined by $2x - 5y = 10$.

EXAMPLE 3 Writing a function

Solve each equation for y and write a formula for a function f defined by $y = f(x)$.

(a) $4(x - 2y) = -3x$ **(b)** $\dfrac{x + y}{2} - 5 = 20$

Solution

(a)

$4(x - 2y) = -3x$	Given equation
$4x - 8y = -3x$	Distributive property
$-8y = -7x$	Subtract $4x$.
$y = \dfrac{7}{8}x$	Divide by -8.

Thus $f(x) = \dfrac{7}{8}x$.

(b)

$\dfrac{x + y}{2} - 5 = 20$	Given equation
$\dfrac{x + y}{2} = 25$	Add 5.
$\dfrac{2(x + y)}{2} = 2(25)$	Multiply by 2.
$x + y = 50$	Simplify.
$y = 50 - x$	Subtract x.

Thus $f(x) = 50 - x$.

Steps for Solving a Problem

Solving problems in mathematics can be challenging, especially when formulas and equations are not given to us. In these situations we have to write them, but to do so, we need a strategy. The following steps are often a helpful strategy. They are based on George Polya's (1888–1985) four-step process for problem solving.

STEPS FOR SOLVING A PROBLEM

STEP 1: Read the problem carefully to be sure that you understand it. (You may need to read the problem more than once.) Assign a variable to what you are being asked to find. If necessary, write other quantities in terms of this variable.

STEP 2: Write an equation that relates the quantities described in the problem. You may need to sketch a diagram, make a table, or refer to known formulas.

STEP 3: Solve the equation and determine the solution.

STEP 4: Look back and check your answer. Does it seem reasonable? Did you find the required information?

In the next example we apply these steps to a word problem involving unknown numbers.

EXAMPLE 4 Solving a number problem

The sum of three consecutive *even* integers is 108. Find the three numbers.

Solution
STEP 1: *Start by assigning a variable to an unknown quantity.*

n: smallest of the three integers

Next, write the other two numbers in terms of n.

$n + 2$: next consecutive *even* integer

$n + 4$: largest of the three even integers

STEP 2: *Write an equation that relates these unknown quantities.* As the sum of these three even integers is 108, the needed equation is
$$n + (n + 2) + (n + 4) = 108.$$

STEP 3: *Solve the equation in Step 2.*

$n + (n + 2) + (n + 4) = 108$	Equation to be solved
$(n + n + n) + (2 + 4) = 108$	Commutative and associative properties
$3n + 6 = 108$	Combine like terms.
$3n = 102$	Subtract 6 from each side.
$n = 34$	Divide each side by 3.

The smallest of the three numbers is 34, so the three numbers are 34, 36, and 38.

STEP 4: *Check your answer.* The sum of these three even integers is
$$34 + 36 + 38 = 108.$$
The answer checks.

In the next example we apply this procedure to find the dimensions of a room.

EXAMPLE 5 Solving a geometry problem

The length of a rectangular room is 2 feet more than its width. If the perimeter of the room is 80 feet, find the width and length of the room.

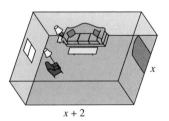

Figure 3.10 Perimeter = 80

Solution
STEP 1: *Start by assigning a variable to an unknown quantity.* See Figure 3.10.

x: width of the room

$x + 2$: length of the room

STEP 2: *Write an equation that relates these unknown quantities.* The perimeter is the distance around the room and equals 80 feet. Thus
$$x + (x + 2) + x + (x + 2) = 80,$$
which simplifies to $4x + 4 = 80$.

STEP 3: *Solve the equation in Step 2.*

$4x + 4 = 80$	Equation to be solved
$4x = 76$	Subtract 4 from each side.
$\dfrac{4x}{4} = \dfrac{76}{4}$	Divide each side by 4.
$x = 19$	Simplify.

The width is 19 feet and the length is $19 + 2 = 21$ feet.

STEP 4: *Check your answer.* The perimeter is $19 + 21 + 19 + 21 = 80$ feet. The answer checks.

EXAMPLE 6　Solving a geometry problem

A wire, 100 inches long, needs to be cut into two pieces so that it can be bent into a circle and a square. The length of a side of the square must equal the diameter of the circle. Approximate where the 100-inch wire should be cut.

Solution

STEP 1: *Assign a variable.* Let x be both the diameter of the circle and the length of a side of the square, as shown in Figure 3.11.

STEP 2: *Write an equation.* Because $C = \pi d$, the circumference of the circle is $C = \pi x$, and the perimeter of the square is $P = 4x$. The wire is 100 inches long, so

$$\pi x + 4x = 100.$$

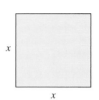

Figure 3.11

STEP 3: *Solve the equation in Step 2.*

$\pi x + 4x = 100$	Equation to be solved
$(\pi + 4)x = 100$	Distributive property
$x = \dfrac{100}{\pi + 4}$	Divide by $\pi + 4$.
$x \approx 14$	Approximate.

The wire should be cut so that the piece for the square is $4 \times 14 = 56$ inches and the piece for the circle is $100 - 56 = 44$ inches.

STEP 4: *Check the answer.* If the square's perimeter is 56 inches, then each of its sides is $\frac{56}{4} = 14$ inches. If the diameter of the circle is 14 inches, then its circumference is $C = \pi(14) \approx 43.98$, or about 44 inches. Because $56 + 44 = 100$, the answer checks.

EXAMPLE 7　Solving a motion problem

Two cars are traveling in opposite lanes on a freeway. Three hours after meeting, they are 420 miles apart. If one car is traveling 10 miles per hour faster than the other car, find the speeds of the two cars.

Solution
STEP 1: *Assign a variable.*

$$x : \text{speed of slower car}$$
$$x + 10 : \text{speed of faster car}$$

STEP 2: *Write an equation.* The information is summarized in Table 3.2.

NOTE: Distance = Rate × Time.

TABLE 3.2

	Rate	Time	Distance
Slower Car	x	3	$3x$
Faster Car	$x + 10$	3	$3(x + 10)$
Total			420

The sum of the distances traveled by the cars is 420 miles. Thus

$$3x + 3(x + 10) = 420.$$

STEP 3: *Solve the equation in Step 2.*

$$3x + 3(x + 10) = 420 \qquad \text{Equation to be solved}$$
$$3x + 3x + 30 = 420 \qquad \text{Distributive property}$$
$$6x + 30 = 420 \qquad \text{Combine like terms.}$$
$$6x = 390 \qquad \text{Subtract 30 from each side.}$$
$$x = 65 \qquad \text{Divide each side by 6.}$$

The speed of the slower car is 65 miles per hour and the speed of the faster car is 10 miles per hour faster, or 75 miles per hour.

STEP 4: *Check the answer.*

$$\text{Distance traveled by slower car: } 65 \times 3 = 195 \text{ miles}$$
$$\text{Distance traveled by faster car: } 75 \times 3 = 225 \text{ miles}$$

The total distance is 195 + 225 = 420 miles, so it checks.

▶ **REAL-WORLD CONNECTION**   Many times we are called upon to solve a mixture problem. Such problems may involve a mixture of acid solutions or a mixture of loan amounts. In the next example an athlete jogs at two different speeds, and mathematics is used to determine how much time is spent jogging at each speed.

EXAMPLE 8   Solving a "mixture" problem

An athlete begins jogging at 8 miles per hour and then jogs at 7 miles per hour, traveling 10.9 miles in 1.5 hours. How long did the athlete jog at each speed?

Solution

STEP 1: *Assign a variable.* Let t represent the time spent jogging at 8 miles per hour. Because total time spent jogging is 1.5 hours, the time spent jogging at 7 miles per hour must be $1.5 - t$.

STEP 2: *Write an equation.* A table is often helpful in solving a mixture problem. Because $d = rt$, Table 3.3 shows that the distance the athlete jogs at 8 miles per hour is $8t$. The distance traveled at 7 miles per hour is $7(1.5 - t)$, and the total distance traveled is 10.9 miles. From the third column, we can write the equation

$$8t + 7(1.5 - t) = 10.9.$$

TABLE 3.3

	Speed	Time	Distance
First Part	8	t	$8t$
Second Part	7	$1.5 - t$	$7(1.5 - t)$
Total		1.5	10.9

STEP 3: *Solve the equation in Step 2.*

$$8t + 7(1.5 - t) = 10.9 \qquad \text{Equation to be solved}$$
$$8t + 10.5 - 7t = 10.9 \qquad \text{Distributive property}$$
$$t + 10.5 = 10.9 \qquad \text{Combine like terms.}$$
$$t = 0.4 \qquad \text{Subtract 10.5 from each side.}$$

The athlete jogged 0.4 hour at 8 miles per hour and $1.5 - 0.4 = 1.1$ hours at 7 miles per hour.

STEP 4: *Check your answer.* Start by calculating the distance traveled at each speed.

$$8 \cdot 0.4 = 3.2 \text{ miles} \quad \text{Distance at 8 mph}$$
$$7 \cdot 1.1 = 7.7 \text{ miles} \quad \text{Distance at 7 mph}$$

The total distance jogged is $3.2 + 7.7 = 10.9$ miles and the total time is $0.4 + 1.1 = 1.5$ hours. The answer checks.

Percentages

▶ **REAL-WORLD CONNECTION** Applications involving percentages often make use of linear equations. Taking P percent of x is given by Px, where P is written in decimal form. For example, to calculate 35% of x, we compute $0.35x$. As a result, 35% of $150 is $0.35(150) = 52.5$, or $52.50.

NOTE: The word *of* often indicates multiplication when working with percentages.

EXAMPLE 9 Analyzing smoking data

In 2005 an estimated 20.9% of Americans aged 18 and older, or 45.1 million people, were smokers. Use these data to estimate the number of Americans aged 18 and older in 2005.
(***Source:*** Department of Health and Human Services.)

Solution
STEP 1: *Assign a variable.* Let x be the number of Americans aged 18 and older.
STEP 2: *Write an equation.* Note that 20.9% of x equals 45.1 million. Thus

$$0.209x = 45.1.$$

STEP 3: *Solve the equation in Step 2.*

$$0.209x = 45.1 \quad \text{Equation to be solved}$$
$$\frac{0.209x}{0.209} = \frac{45.1}{0.209} \quad \text{Divide each side by 0.209.}$$
$$x \approx 216 \quad \text{Approximate with a calculator.}$$

In 2005 there were about 216 million Americans aged 18 and older.

STEP 4: *Check your answer.* If there were 216 million Americans aged 18 and older, then 20.9% of 216 million would equal $0.209 \times 216 \approx 45.1$ million. The answer checks.

EXAMPLE 10 Solving a percent problem

In 2004 an estimated $264 billion were spent on advertisement in the United States. This was an increase of 380% over the amount spent in 1980. Determine how much was spent on advertisement in 1980. (***Source:*** Advertising Age.)

Solution
STEP 1: *Assign a variable.* Let x be the amount spent on advertisement in 1980.
STEP 2: *Write an equation.* Note that the increase was 380% of x, or $3.8x$.

$$x \quad + \quad 3.8x \quad = \quad 264$$
Amount in 1980 + Increase = Amount in 2004

STEP 3: *Solve the equation in Step 2.*

$$x + 3.8x = 264 \quad \text{Equation to be solved}$$
$$4.8x = 264 \quad \text{Add like terms.}$$
$$x = \frac{264}{4.8} \quad \text{Divide each side by 4.8.}$$
$$x = 55 \quad \text{Simplify.}$$

In 1980, $55 billion were spent on advertisement.

STEP 4: *Check your answer.* An increase of 380% of $55 billion is $3.8 \times 55 = \$209$ billion. Thus the amount spent in 2004 would be $55 + 209 = \$264$ billion. The answer checks.

EXAMPLE 11 Solving an interest problem

A college student takes out two unsubsidized loans to pay tuition. The first loan is for 3% annual interest and the second loan is for 7% annual interest. If the total of the two loans is $4000 and the student must pay $170 in interest at the end of the first year, determine the amount borrowed at each interest rate.

Solution

STEP 1: *Assign a variable.*

$$x\text{: the loan amount at 3\%}$$
$$4000 - x\text{: the loan amount at 7\%}$$

STEP 2: *Write an equation.* Note that the total interest is $170.

$$\underbrace{0.03x}_{\text{Interest at 3\%}} + \underbrace{0.07(4000 - x)}_{\text{Interest at 7\%}} = \underbrace{170}_{\text{Total Interest}}$$

STEP 3: *Solve the equation in Step 2.*

$$0.03x + 0.07(4000 - x) = 170 \quad \text{Equation to be solved}$$
$$3x + 7(4000 - x) = 17{,}000 \quad \text{Multiply by 100 to clear decimals.}$$
$$3x + 28{,}000 - 7x = 17{,}000 \quad \text{Distributive property}$$
$$-4x + 28{,}000 = 17{,}000 \quad \text{Combine like terms.}$$
$$-4x = -11{,}000 \quad \text{Subtract 28,000.}$$
$$x = 2750 \quad \text{Divide by } -4.$$

The amount at 3% is $2750 and the amount at 7% is $4000 - 2750 = \$1250$.

STEP 4: *Check the answer.* The sum is $2750 + 1250 = \$4000$. Also, 3% of $2750 is $82.50 and 7% of $1250 is $87.50. The total interest is $\$82.50 + \$87.50 = \$170$. The answer checks.

▶ **REAL-WORLD CONNECTION** In chemistry acids are frequently mixed. In the next example, percentages are used to determine how to mix an acid solution with a prescribed concentration.

EXAMPLE 12 Mixing acid

A chemist mixes 2 liters of 20% sulfuric acid with another sample of 60% sulfuric acid to obtain a sample of 50% sulfuric acid. How much of the 60% sulfuric acid was used? See Figure 3.12.

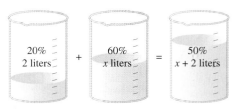

Figure 3.12 Mixing acid

Solution

STEP 1: *Assign a variable.* Let x be as follows.

x: liters of 60% sulfuric acid

$x + 2$: liters of 50% sulfuric acid

STEP 2: *Write an equation.* Table 3.4 can be used to organize our calculations. The total amount of pure sulfuric acid in the 20% and 60% samples must equal the amount of pure sulfuric acid in the final 50% acid solution.

TABLE 3.4 Mixing Acid

Concentration (as a decimal)	Solution Amount (liters)	Pure Acid (liters)
20% = 0.20	2	0.20(2)
60% = 0.60	x	0.60x
50% = 0.50	$x + 2$	0.50($x + 2$)

From the third column, we can write the equation

$$0.20(2) + 0.60x = 0.50(x + 2).$$

Pure acid in 20% solution + Pure acid in 60% solution = Pure acid in 50% solution

STEP 3: *Solve the equation in Step 2.*

$0.20(2) + 0.60x = 0.50(x + 2)$	Equation to be solved
$2(2) + 6x = 5(x + 2)$	Multiply by 10.
$4 + 6x = 5x + 10$	Distributive property
$x = 6$	Subtract $5x$ and 4 from each side.

Six liters of the 60% acid solution was added to the 2 liters of 20% acid solution.

STEP 4: *Check your answer.* If 6 liters of 60% acid solution are added to 2 liters of 20% solution, then there will be 8 liters of acid solution containing

$$0.60(6) + 0.20(2) = 4 \text{ liters}$$

of pure acid. The concentration is $\frac{4}{8} = 0.50$, or 50%. The answer checks.

3.2 PUTTING IT ALL TOGETHER

When solving an application problem, follow the *Steps for Solving a Problem* discussed on page 102. Be sure to read the problem carefully and check your answer. The following table summarizes some other important concepts from this section.

Concept	Explanation	Examples
Writing a Function	If possible, solve an equation for a variable. Express the result in function notation.	$5x + y = 10$ $y = -5x + 10$ If $y = f(x)$, then $f(x) = -5x + 10$.
Changing a Percentage to a Decimal	Move the decimal point two places to the left.	$73\% = 0.73$, $5.3\% = 0.053$, and $125\% = 1.25$
Percent Problems	To find $P\%$ of a quantity Q, change $P\%$ to a decimal and multiply by Q.	To find 45% of $200, calculate $0.45 \times 200 = \$90$.

3.3 LINEAR INEQUALITIES

Basic Concepts ▪ Symbolic Solutions ▪ Numerical and Graphical Solutions ▪ An Application

A LOOK INTO MATH ▷

On a freeway, the speed limit might be 75 miles per hour. A driver traveling x miles per hour is obeying the speed limit if $x \leq 75$ and breaking the speed limit if $x > 75$. A speed of $x = 75$ represents the boundary between obeying the speed limit and breaking it. A posted speed limit, or *boundary*, allows drivers to easily determine whether they are speeding.

Solving linear inequalities is closely related to solving linear equations because equality is the boundary between *greater than* and *less than*. In this section we discuss techniques needed to solve linear inequalities.

Basic Concepts

An **inequality** results whenever the equals sign in an equation is replaced with any one of the symbols $<, \leq, >$, or $\geq$. Examples of linear equations include

$$2x + 1 = 0, \quad 1 - x = 6, \quad \text{and} \quad 5x + 1 = 3 - 2x,$$

and, therefore, examples of linear inequalities include

$$2x + 1 < 0, \quad 1 - x \geq 6, \quad \text{and} \quad 5x + 1 \leq 3 - 2x.$$

A **solution** to an inequality is a value of the variable that makes the statement true. The set of all solutions is called the **solution set**. Two inequalities are *equivalent* if they have the same solution set. Inequalities frequently have *infinitely many solutions*. For example, the solution set to the inequality $x - 5 > 0$ includes all real numbers greater than 5, which can be written as $x > 5$. Using **set-builder notation**, we can write the solution set as $\{x \mid x > 5\}$. This expression is read as "the set of all real numbers x such that x is greater than 5." The vertical line | in set-builder notation is read "such that." To summarize, we have the following.

Inequality	Set-Builder Notation	Meaning
$t \leq 3$	$\{t \mid t \leq 3\}$	The set of all real numbers t such that t is less than or equal to 3
$z > 8$	$\{z \mid z > 8\}$	The set of all real numbers z such that z is greater than 8

Next we more formally define a linear inequality in one variable.

> ### LINEAR INEQUALITY IN ONE VARIABLE
>
> A **linear inequality** in one variable is an inequality that can be written in the form
> $$ax + b > 0,$$
> where $a \neq 0$. (The symbol $>$ may be replaced with $\geq$, $<$, or $\leq$.)

> ### MAKING CONNECTIONS
>
> **Linear Functions, Equations, and Inequalities**
>
> A *linear function* is given by $f(x) = ax + b$, a *linear equation* by $ax + b = 0$, and a *linear inequality* by $ax + b > 0$. A linear equation (with $a \neq 0$) has one solution. A linear inequality has infinitely many solutions.

▶ **REAL-WORLD CONNECTION**   To understand linear inequalities better, suppose that it costs a student $100 to make a large batch of candy. If the student sells bags of this candy for $5 each, then the profit y is $y = 5x - 100$, where x represents the number of bags sold. The graph of y is a line with slope 5 and y-intercept -100, as shown in Figure 3.13.

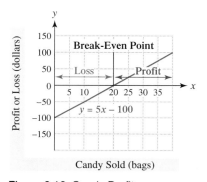

Figure 3.13 Candy Profit

Because the x-intercept is 20, the student will *break even* when $5x - 100 = 0$, or when 20 bags of candy are sold. (The **break-even point** occurs when the *revenue* from selling the candy equals the *cost* of making the candy.) Note that the student incurs a *loss* when the line is below the x-axis, or when $5x - 100 < 0$. This situation corresponds to selling less than 20 bags, or $x < 20$. A profit occurs when the graph is above the x-axis, or when $5x - 100 > 0$. This situation corresponds to selling more than 20 bags, or $x > 20$. Selling $x = 20$ bags of candy represents the *boundary* between loss and profit.

These concepts for the solution set for a linear inequality are summarized as follows.

> ### SOLUTION SET FOR A LINEAR INEQUALITY
>
> The solution set for $ax + b > 0$ with $a \neq 0$ is either $\{x \mid x < k\}$ or $\{x \mid x > k\}$, where k is the solution to $ax + b = 0$ and corresponds to the x-intercept for the graph of $y = ax + b$. Similar statements can be made for the symbols $<$, $\leq$, and $\geq$.

Symbolic Solutions

The inequality $3 < 5$ is equivalent to $3 + 1 < 5 + 1$. That is, we can add the same number to each side of an inequality. This is an example of one property of inequalities.

PROPERTIES OF INEQUALITIES

Let a, b, and c be real numbers.

1. $a < b$ and $a + c < b + c$ are equivalent.
   (The same number may be added to or subtracted from each side of an inequality.)
2. If $c > 0$, then $a < b$ and $ac < bc$ are equivalent.
   (Each side of an inequality may be multiplied or divided by the same positive number.)
3. If $c < 0$, then $a < b$ and $ac > bc$ are equivalent.
   (Each side of an inequality may be multiplied or divided by the same negative number provided the inequality symbol is reversed.)

Similar properties exist for the $\leq$ and $\geq$ symbols.

When applying Property 2 or 3, we need to determine whether the inequality symbol should be reversed. For example, if each side of the inequality $3 < 5$ is multiplied by the *positive* number 2, we obtain

$$2 \cdot 3 < 2 \cdot 5 \quad \text{or} \quad 6 < 10,$$

which is a true statement. However, if each side of the inequality $3 < 5$ is multiplied by the *negative* number -2, we obtain

$$-2 \cdot 3 > -2 \cdot 5 \quad \text{or} \quad -6 > -10,$$

↑ Reverse inequality symbol.

which is a true statement because the inequality symbol is reversed from $<$ to $>$.

To solve an inequality we apply properties of inequalities to find a simpler, equivalent inequality, as illustrated in the next example.

EXAMPLE 1 Solving linear inequalities

Solve each inequality.
(a) $2x - 1 > 4$ (b) $\frac{1}{2}(z - 3) - (2 - z) \leq 1$

Solution
(a) Begin by adding 1 to each side of the inequality.

$2x - 1 > 4$	Given inequality
$2x - 1 + 1 > 4 + 1$	Add 1 to each side.
$2x > 5$	Simplify.
$\dfrac{2x}{2} > \dfrac{5}{2}$	Divide by 2; do *not* reverse inequality symbol because $2 > 0$.
$x > \dfrac{5}{2}$	Simplify.

The solution set is $\{x \mid x > \frac{5}{2}\}$.

(b) To clear fractions multiply each term by 2.

$\frac{1}{2}(z - 3) - (2 - z) \leq 1$	Given inequality
$(z - 3) - 2(2 - z) \leq 2$	Multiply by 2; do *not* reverse inequality symbol because $2 > 0$.
$z - 3 - 4 + 2z \leq 2$	Distributive property
$3z - 7 \leq 2$	Combine like terms.
$3z - 7 + 7 \leq 2 + 7$	Add 7 to each side.
$3z \leq 9$	Simplify.
$z \leq 3$	Divide by 3; do *not* reverse inequality symbol because $3 > 0$.

The solution set is $\{z \mid z \leq 3\}$.

EXAMPLE 2 Solving linear inequalities

Solve each inequality.

(a) $5 - 3x \leq x - 3$ (b) $\dfrac{2t - 3}{5} \geq \dfrac{t + 1}{3} + 3t$

Solution

(a) Begin by subtracting 5 from each side of the inequality.

$$5 - 3x \leq x - 3 \quad \text{Given inequality}$$
$$5 - 3x - 5 \leq x - 3 - 5 \quad \text{Subtract 5 from each side.}$$
$$-3x \leq x - 8 \quad \text{Simplify.}$$
$$-3x - x \leq x - 8 - x \quad \text{Subtract } x \text{ from each side.}$$
$$-4x \leq -8 \quad \text{Simplify.}$$

Next divide each side by -4. As we are dividing by a *negative* number, Property 3 requires reversing the inequality by changing $\leq$ to $\geq$.

$$\dfrac{-4x}{-4} \geq \dfrac{-8}{-4} \quad \text{Divide by } -4 \text{; reverse the inequality because } -4 < 0.$$
$$x \geq 2 \quad \text{Simplify.}$$

The solution set is $\{x \mid x \geq 2\}$.

(b) To clear fractions multiply each term by 15.

$$\dfrac{2t - 3}{5} \geq \dfrac{t + 1}{3} + 3t \quad \text{Given inequality}$$
$$15 \cdot \left(\dfrac{2t - 3}{5}\right) \geq 15 \cdot \left(\dfrac{t + 1}{3}\right) + 15 \cdot 3t \quad \text{Multiply each term by 15; do not reverse the inequality symbol because } 15 > 0.$$
$$3(2t - 3) \geq 5(t + 1) + 45t \quad \text{Simplify: } \tfrac{15}{5} = 3 \text{ and } \tfrac{15}{3} = 5$$
$$6t - 9 \geq 5t + 5 + 45t \quad \text{Distributive property}$$
$$6t - 9 + 9 \geq 50t + 5 + 9 \quad \text{Add 9; combine like terms.}$$
$$6t \geq 50t + 14 \quad \text{Simplify.}$$
$$6t - 50t \geq 50t - 50t + 14 \quad \text{Subtract } 50t \text{ from each side.}$$
$$-44t \geq 14 \quad \text{Simplify.}$$
$$t \leq \dfrac{14}{-44} \quad \text{Divide by } -44 \text{; reverse the inequality symbol because } -44 < 0.$$
$$t \leq -\dfrac{7}{22} \quad \text{Simplify.}$$

The solution set is $\left\{t \mid t \leq -\dfrac{7}{22}\right\}$.

Numerical and Graphical Solutions

In Section 3.1 we solved linear equations with symbolic, numerical, and graphical methods. We can also use these methods to solve linear inequalities.

EXAMPLE 3 Solving linear inequalities numerically

Use Table 3.5 to find the solution set to each equation or inequality.
(a) $-\frac{1}{2}x + 1 = 0$ (b) $-\frac{1}{2}x + 1 > 0$ (c) $-\frac{1}{2}x + 1 < 0$

TABLE 3.5

x	-1	0	1	2	3	4	5
$-\frac{1}{2}x + 1$	$\frac{3}{2}$	1	$\frac{1}{2}$	0	$-\frac{1}{2}$	-1	$-\frac{3}{2}$

Solution
(a) The expression $-\frac{1}{2}x + 1$ equals 0 when $x = 2$. Thus the solution set is $\{x \mid x = 2\}$.
(b) The expression $-\frac{1}{2}x + 1$ is positive when $x < 2$. Thus the solution set is $\{x \mid x < 2\}$.
(c) The expression $-\frac{1}{2}x + 1$ is negative when $x > 2$. Thus the solution set is $\{x \mid x > 2\}$.

EXAMPLE 4 Solving linear inequalities graphically

Use Figure 3.14 to find the solution set to each equation or inequality.
(a) $-\frac{1}{2}x + 1 = 0$ (b) $-\frac{1}{2}x + 1 > 0$ (c) $-\frac{1}{2}x + 1 < 0$

Solution
(a) The graph of $y = -\frac{1}{2}x + 1$ in Figure 3.14 crosses the x-axis at $x = 2$. Thus the solution set is $\{x \mid x = 2\}$.
(b) The graph is *above* the x-axis when $x < 2$. Thus the solution set is $\{x \mid x < 2\}$.
(c) The graph is *below* the x-axis when $x > 2$. Thus the solution set is $\{x \mid x > 2\}$.

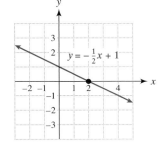

Figure 3.14

NOTE: Numerical and graphical solutions can sometimes be difficult to find if the x-value that determines equality is not an integer. Symbolic methods work well in such situations.

Sometimes linear inequalities are written in the form $y_1 < y_2, y_1 \leq y_2, y_1 > y_2$, or $y_1 \geq y_2$, where both y_1 and y_2 contain variables. These types of inequalities can also be solved graphically or numerically.

▶ **REAL-WORLD CONNECTION** Figure 3.15 shows the distances that two cars are from Chicago, Illinois, after x hours while traveling in the same direction on a freeway. The distance for Car 1 is denoted y_1, and the distance for Car 2 is denoted y_2.

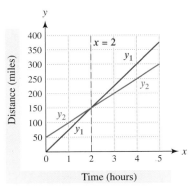

Figure 3.15 Distances of Two Cars

When $x = 2$ hours, $y_1 = y_2$ and both cars are 150 miles from Chicago. To the left of the dashed vertical line $x = 2$, the graph of y_1 is below the graph of y_2, so Car 1 is closer to Chicago than Car 2. Thus

$$y_1 < y_2 \quad \text{when} \quad x < 2.$$

To the right of the dashed vertical line $x = 2$, the graph of y_1 is above the graph of y_2, so Car 1 is farther from Chicago than Car 2. Thus

$$y_1 > y_2 \quad \text{when} \quad x > 2.$$

CRITICAL THINKING

A linear equation has one solution that can be easily checked. A linear inequality has infinitely many solutions. Discuss ways that the solution set to a linear inequality could be checked.

EXAMPLE 5 Solving an inequality graphically

Solve $5 - 3x \leq x - 3$.

Solution
The graphs of $y_1 = 5 - 3x$ and $y_2 = x - 3$ intersect at the point $(2, -1)$, as shown in Figure 3.16(a). Equality, or $y_1 = y_2$, occurs when $x = 2$ and the graph of y_1 is below the graph of y_2 when $x > 2$. Thus $5 - 3x \leq x - 3$ is satisfied when $x \geq 2$. The solution set is $\{x \mid x \geq 2\}$. Figure 3.16(b) shows the same graph generated with a graphing calculator.

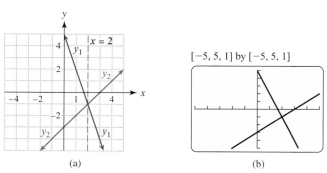

(a)   (b)

Figure 3.16

CRITICAL THINKING

Use the results from Example 5 to write the solution set for $5 - 3x \geq x - 3$.

An Application

▶ **REAL-WORLD CONNECTION** In the lower atmosphere, the air generally becomes colder as the altitude increases. See Figure 3.17. One mile above Earth's surface the temperature is about 19°F colder than the ground-level temperature. As the air temperature cools, the chance of clouds forming increases. In the next example we estimate the altitudes at which clouds will not form.

Figure 3.17

EXAMPLE 6 Finding the altitude of clouds

If ground-level temperature is 90°F, the temperature T above Earth's surface is modeled by $T(x) = 90 - 19x$, where x is the altitude in miles. Suppose that clouds will form only if the temperature is 53°F or colder. (*Source:* A. Miller and R. Anthes, *Meteorology*.)
(a) Determine symbolically the altitudes at which there are no clouds.
(b) Give graphical support for your answer.
(c) Give numerical support for your answer.

Solution
(a) **Symbolic Solution** Clouds will not form at altitudes at which the temperature is greater than 53°F. Thus we must solve the inequality $T(x) > 53$.

$$90 - 19x > 53 \qquad \text{Inequality to be solved}$$
$$90 - 19x - 90 > 53 - 90 \qquad \text{Subtract 90 from each side.}$$
$$-19x > -37 \qquad \text{Simplify.}$$
$$\frac{-19x}{-19} < \frac{-37}{-19} \qquad \text{Divide by } -19; \text{ reverse inequality.}$$
$$x < \frac{37}{19} \qquad \text{Simplify.}$$

The result, $\frac{37}{19} \approx 1.95$, indicates that clouds will not form below about 1.95 miles. Note that models are usually not exact, so rounding values is appropriate.

(b) **Graphical Solution** Graph $Y_1 = 90 - 19X$ and $Y_2 = 53$. In Figure 3.18(a) their graphs intersect near the point (1.95, 53). The graph of y_1 is above the graph of y_2 when $x < 1.95$.

(c) **Numerical Solution** Make a table of y_1 and y_2, as shown in Figure 3.18(b). When $x = 2$, the air temperature is 52°F. Thus the air temperature would be 53°F when the altitude is slightly less than 2 miles, which supports our symbolic solution of $x \approx 1.95$.

CALCULATOR HELP
To find a point of intersection, see the Appendix (page AP-7).

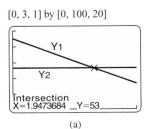
(a)

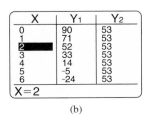
(b)

Figure 3.18

3.3 PUTTING IT ALL TOGETHER

The following table summarizes some important topics related to linear inequalities.

Concept	Explanation	Examples
Linear Inequalities in One Variable	Can be written as $$ax + b > 0,$$ where $a \neq 0$ and $>$ may be replaced by $<$, $\leq$, or $\geq$; has infinitely many solutions	$3x - 4 > 0$, $-(x - 5) < 2$, $2x \geq 5 - x$, and $4 - 2(x + 1) \leq 7x - 1$
Set-Builder Notation	Used to express sets of real numbers	$\{x \mid x > 4\}$ represents the set of real numbers x such that x is greater than 4.
Properties of Inequalities	See the box on page 111. Be sure to note Property 3: When multiplying or dividing by a negative number, *reverse* the inequality symbol.	*Property 1*: $x - 3 \geq 2$ is equivalent to $x \geq 5$. *Property 2*: $2x \leq 6$ is equivalent to $x \leq 3$. *Property 3*: $-3x < 6$ is equivalent to $x > -2$.

Linear inequalities can be solved symbolically, graphically, or numerically. Each method is used as follows to solve the inequality $2x - 4 \geq 0$.

Symbolic Solution

$2x - 4 \geq 0$

$2x \geq 4$

$x \geq 2$

Graphical Solution

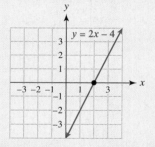

The graph of $y = 2x - 4$ is above or intersects the x-axis when $x \geq 2$.

Numerical Solution

x	$2x - 4$
0	-4
1	-2
2	0
3	2
4	4

The values of $2x - 4$ are greater than or equal to 0 when $x \geq 2$.

3.4 COMPOUND INEQUALITIES

Basic Concepts ▪ Symbolic Solutions and Number Lines ▪ Numerical and Graphical Solutions ▪ Interval Notation

A LOOK INTO MATH ▷ A person weighing 143 pounds and needing to purchase a life vest for white-water rafting is not likely to find one designed exactly for this weight. Life vests are manufactured to support a range of body weights. A vest approved for weights between 100 and 160 pounds might be appropriate for this person. In other words, if a person's weight is w, this life vest is safe if $w \geq 100$ *and* $w \leq 160$. This example illustrates the concept of a *compound inequality*.

Basic Concepts

A **compound inequality** consists of two inequalities joined by the words *and* or *or*. The following are two examples of compound inequalities.

$$2x \geq -3 \quad \text{and} \quad 2x < 5 \quad (1)$$
$$x + 2 \geq 3 \quad \text{or} \quad x - 1 < -5 \quad (2)$$

If a compound inequality contains the word *and*, a solution must satisfy *both* inequalities. For example, 1 is a solution to the first compound inequality because

$$\underset{\text{True}}{2(1) \geq -3} \quad \text{and} \quad \underset{\text{True}}{2(1) < 5}$$

are both true statements.

If a compound inequality contains the word *or*, a solution must satisfy *at least one* of the two inequalities. Thus 5 is a solution to the second compound inequality, because the first statement is true.

$$\underset{\text{True}}{5 + 2 \geq 3} \quad \text{or} \quad \underset{\text{False}}{5 - 1 < -5}$$

Note that 5 does not need to satisfy both statements for this compound inequality to be true.

EXAMPLE 1 Determining solutions to compound inequalities

Determine whether the given x-values are solutions to the compound inequalities.
(a) $x + 1 < 9$ and $2x - 1 > 8$ $\quad x = 5, -5$
(b) $5 - 2x \leq -4$ or $5 - 2x \geq 4$ $\quad x = 2, -3$

Solution
(a) Substitute $x = 5$ in the given compound inequality.

$$\underset{\text{True}}{5 + 1 < 9} \quad \text{and} \quad \underset{\text{True}}{2(5) - 1 > 8}$$

Both inequalities are true, so 5 is a solution. Now substitute $x = -5$.

$$\underset{\text{True}}{-5 + 1 < 9} \quad \text{and} \quad \underset{\text{False}}{2(-5) - 1 > 8}$$

To be a solution both inequalities must be true, so -5 is not a solution.

(b) Substitute $x = 2$ into the given compound inequality.

$$\underset{\text{False}}{5 - 2(2) \leq -4} \quad \text{or} \quad \underset{\text{False}}{5 - 2(2) \geq 4}$$

Neither inequality is true, so 2 is not a solution. Now substitute $x = -3$.

$$\underset{\text{False}}{5 - 2(-3) \leq -4} \quad \text{or} \quad \underset{\text{True}}{5 - 2(-3) \geq 4}$$

At least one of the two inequalities is true, so -3 is a solution.

CRITICAL THINKING

Graph the following inequalities and discuss your results.
1. $x < 2$ and $x > 5$
2. $x > 2$ or $x < 5$

Symbolic Solutions and Number Lines

We can use a number line to graph solutions to compound inequalities, such as

$$x \leq 6 \quad \text{and} \quad x > -4.$$

The solution set for $x \leq 6$ is shaded to the left of 6, with a bracket placed at $x = 6$, as shown in Figure 3.19. The solution set for $x > -4$ can be shown by shading a different number line to the right of -4 and placing a left parenthesis at -4. Because the inequalities are connected by *and*, the solution set consists of all numbers that are shaded on *both* number lines. The final number line represents the *intersection* of the two solution sets. That is, the solution set includes real numbers where the graphs "overlap." For any two sets A and B, the **intersection** of A and B, denoted $A \cap B$, is defined by

$$A \cap B = \{x \mid x \text{ is an element of } A \text{ and an element of } B\}.$$

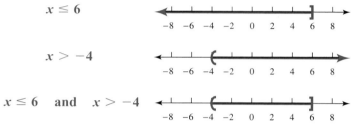

Figure 3.19

NOTE: A bracket, either [or], is used when an inequality contains $\leq$ or $\geq$. A parenthesis, either (or), is used when an inequality contains $<$ or $>$. This notation makes clear whether an endpoint is included in the inequality.

EXAMPLE 2 Solving a compound inequality containing "and"

Solve $2x + 4 > 8$ and $5 - x < 9$. Graph the solution.

Solution
First solve each linear inequality separately.

$$2x + 4 > 8 \quad \text{and} \quad 5 - x < 9$$
$$2x > 4 \quad \text{and} \quad -x < 4$$
$$x > 2 \quad \text{and} \quad x > -4$$

Graph the two inequalities on two different number lines. On a third number line, shade solutions that appear on both of the first two number lines. As shown in Figure 3.20, the solution set is $\{x \mid x > 2\}$.

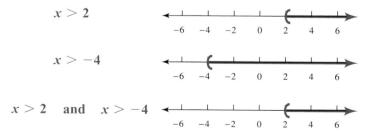

Figure 3.20

Sometimes a compound inequality containing the word *and* can be combined into a three-part inequality. For example, rather than writing

$$x > 5 \quad \text{and} \quad x \leq 10,$$

we could write the **three-part inequality**
$$5 < x \leq 10.$$
This three-part inequality is represented by the number line shown in Figure 3.21.

$5 < x \leq 10$

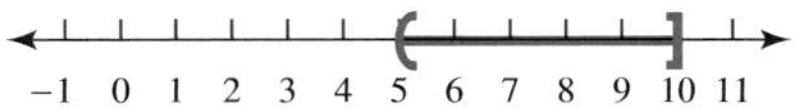

Figure 3.21

EXAMPLE 3 Solving three-part inequalities

Solve each inequality.
(a) $4 < t + 2 \leq 8$ (b) $-3 \leq 3z \leq 6$ (c) $-\dfrac{5}{2} < \dfrac{1-m}{2} < 4$

Solution

(a) To solve a three-part inequality, isolate the variable by applying properties of inequalities to each part of the inequality.

$4 < t + 2 \leq 8$	Given three-part inequality
$4 - 2 < t + 2 - 2 \leq 8 - 2$	Subtract 2 from each part.
$2 < t \leq 6$	Simplify each part.

The solution set is $\{t \mid 2 < t \leq 6\}$.

(b) To simplify, divide each part by 3.

$-3 \leq 3z \leq 6$	Given three-part inequality
$\dfrac{-3}{3} \leq \dfrac{3z}{3} \leq \dfrac{6}{3}$	Divide each part by 3.
$-1 \leq z \leq 2$	Simplify each part.

The solution set is $\{z \mid -1 \leq z \leq 2\}$.

(c) Multiply each part by 2 to clear fractions.

$-\dfrac{5}{2} < \dfrac{1-m}{2} < 4$	Given three-part inequality
$2 \cdot \left(-\dfrac{5}{2}\right) < 2 \cdot \left(\dfrac{1-m}{2}\right) < 2 \cdot 4$	Multiply each part by 2.
$-5 < 1 - m < 8$	Simplify each part.
$-5 - 1 < 1 - m - 1 < 8 - 1$	Subtract 1 from each part.
$-6 < -m < 7$	Simplify each part.
$-1 \cdot (-6) > -1 \cdot (-m) > -1 \cdot 7$	Multiply each part by -1; *reverse* inequality symbols.
$6 > m > -7$	Simplify each part.
$-7 < m < 6$	Rewrite inequality.

The solution set is $\{m \mid -7 < m < 6\}$.

NOTE: Either $6 > m > -7$ or $-7 < m < 6$ is a correct way to write a three-part inequality. *However*, we usually write the smaller number on the left side and the larger number on the right side.

▶ **REAL-WORLD CONNECTION** Three-part inequalities occur frequently in applications. In the next example we find altitudes at which the air temperature is within a certain range.

EXAMPLE 4 Solving a three-part inequality

If the ground-level temperature is 80°F, the air temperature x miles above Earth's surface is cooler and can be modeled by $T(x) = 80 - 19x$. Find the altitudes at which the air temperature ranges from 42°F down to 23°F. (**Source:** A. Miller and R. Anthes, *Meteorology*.)

Solution
We write and solve the three-part inequality $23 \leq T(x) \leq 42$.

$$23 \leq 80 - 19x \leq 42 \qquad \text{Substitute for } T(x).$$
$$-57 \leq -19x \leq -38 \qquad \text{Subtract 80 from each part.}$$
$$\frac{-57}{-19} \geq x \geq \frac{-38}{-19} \qquad \text{Divide by } -19; \textit{reverse} \text{ inequality symbols.}$$
$$3 \geq x \geq 2 \qquad \text{Simplify.}$$
$$2 \leq x \leq 3 \qquad \text{Rewrite inequality.}$$

The air temperature ranges from 42°F to 23°F for altitudes between 2 and 3 miles.

MAKING CONNECTIONS
Writing Three-Part Inequalities

The inequality $-2 < x < 1$ means that $x > -2$ *and* $x < 1$. A three-part inequality should *not* be used when *or* connects a compound inequality. Writing $x < -2$ or $x > 1$ as $1 < x < -2$ is incorrect because it states that x must be both greater than 1 *and* less than -2. It is impossible for any value of x to satisfy this statement.

We can also solve compound inequalities containing the word *or*. To write the solution to such an inequality we sometimes use *union* notation. For any two sets A and B, the **union** of A and B, denoted $A \cup B$, is defined by

$$A \cup B = \{x \mid x \text{ is an element of } A \text{ or an element of } B\}.$$

If the solution to an inequality is $\{x \mid x < 1\}$ or $\{x \mid x \geq 3\}$, then it can also be written as

$$\{x \mid x < 1\} \cup \{x \mid x \geq 3\}.$$

That is, we can replace the word *or* with the $\cup$ symbol.

EXAMPLE 5 Solving a compound inequality containing "or"

Solve $x + 2 < -1$ or $x + 2 > 1$.

Solution
We first solve each linear inequality.

$$x + 2 < -1 \quad \text{or} \quad x + 2 > 1 \qquad \text{Given compound inequality}$$
$$x < -3 \quad \text{or} \quad x > -1 \qquad \text{Subtract 2.}$$

We can graph the simplified inequalities on different number lines, as shown in Figure 3.22. A solution must satisfy at least one of the two inequalities. Thus the solution set for the compound inequality results from taking the *union* of the first two number lines. We can write the solution, using set-builder notation, as $\{x \mid x < -3\} \cup \{x \mid x > -1\}$ or as $\{x \mid x < -3 \text{ or } x > -1\}$.

$x < -3$

$x > -1$

$x < -3$ or $x > -1$

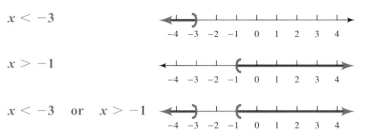

Figure 3.22

CRITICAL THINKING

Carbon dioxide is emitted when human beings breathe. In one study of college students, the amount of carbon dioxide exhaled in grams per hour was measured during both lectures and exams. The average amount exhaled during lectures L satisfied $25.33 \leq L \leq 28.17$, whereas the average amount exhaled during exams E satisfied $36.58 \leq E \leq 40.92$. What do these results indicate? Explain. (*Source:* T. Wang, *ASHRAE Trans.*)

Numerical and Graphical Solutions

Compound inequalities can also be solved graphically and numerically, as illustrated in the next example.

EXAMPLE 6 Solving a compound inequality numerically and graphically

Tuition and fees at private colleges and universities from 1980 to 2000 can be modeled by $f(x) = 575(x - 1980) + 3600$. Estimate when the average tuition and fees ranged from $8200 to $10,500.

Solution

Numerical Solution Let $Y_1 = 575(X - 1980) + 3600$. Make a table of values, as shown in Figure 3.23(a). In 1988, the average tuition and fees were $8200 and in 1992 they were $10,500. Thus from 1988 to 1992 the average tuition and fees ranged from $8200 to $10,500.

Graphical Solution Graph $Y_1 = 575(X - 1980) + 3600$, $Y_2 = 8200$, and $Y_3 = 10,500$. We must find x-values so that $y_2 \leq y_1 \leq y_3$. Figures 3.23(b) and (c) show that y_1 is between y_2 and y_3 when $1988 \leq x \leq 1992$.

CALCULATOR HELP

To find a point of intersection, see the Appendix (page AP-7).

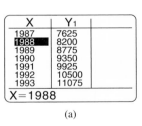
(a)

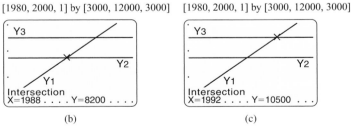
(b)   (c)

Figure 3.23

Interval Notation

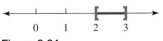

Figure 3.24

The solution set in Example 4 was $\{x \mid 2 \leq x \leq 3\}$. This solution set can be graphed on a number line, as shown in Figure 3.24.

A convenient notation for number line graphs is called **interval notation**. Instead of drawing the entire number line as in Figure 3.24, the solution set can be expressed as $[2, 3]$ in interval notation. Because the solution set includes the endpoints 2 and 3, brackets are used. A solution set that includes all real numbers satisfying $-2 < x < 3$ can be expressed as $(-2, 3)$. Parentheses indicate that the endpoints are *not* included. The interval $0 \leq x < 4$ is represented by $[0, 4)$.

Table 3.6 provides some examples of interval notation. The symbol ∞ refers to **infinity**, and it does not represent a real number. The interval $(5, \infty)$ represents $x > 5$, which has no maximum x-value, so ∞ is used for the right endpoint. The symbol $-\infty$ may be used similarly and denotes **negative infinity**.

TABLE 3.6 Interval Notation

Inequality	Interval Notation	Number Line Graph
$-1 < x < 3$	$(-1, 3)$	
$-3 < x \leq 2$	$(-3, 2]$	
$-2 \leq x \leq 2$	$[-2, 2]$	
$x < -1$ or $x > 2$	$(-\infty, -1) \cup (2, \infty)$ ($\cup$ is the union symbol.)	
$x > -1$	$(-1, \infty)$	
$x \leq 2$	$(-\infty, 2]$	

MAKING CONNECTIONS

Points and Intervals

The expression $(1, 2)$ may represent a point in the xy-plane or the interval $1 < x < 2$. To alleviate confusion, phrases such as "the point $(1, 2)$" or "the interval $(1, 2)$" are used.

EXAMPLE 7 Writing inequalities in interval notation

Write each expression in interval notation.
(a) $-2 \leq x < 5$ (b) $x \geq 3$ (c) $x < -5$ or $x \geq 2$
(d) $\{x \mid x > 0 \text{ and } x \leq 3\}$ (e) $\{x \mid x \leq 1 \text{ or } x \geq 3\}$

Solution
(a) $[-2, 5)$ (b) $[3, \infty)$ (c) $(-\infty, -5) \cup [2, \infty)$
(d) $(0, 3]$ (e) $(-\infty, 1] \cup [3, \infty)$

EXAMPLE 8 Solving an inequality

Solve $2x + 1 \leq -1$ or $2x + 1 \geq 3$. Write the solution in interval notation.

Solution
First solve each inequality.

$2x + 1 \leq -1$	or	$2x + 1 \geq 3$	Given compound inequality
$2x \leq -2$	or	$2x \geq 2$	Subtract 1.
$x \leq -1$	or	$x \geq 1$	Divide by 2.

The solution set may be written as $(-\infty, -1] \cup [1, \infty)$.

3.4 PUTTING IT ALL TOGETHER

The following table summarizes some important topics related to compound inequalities.

Concept	Explanation	Examples
Compound Inequality	Two inequalities joined by *and* or *or*	$x < -1$ or $x > 2$; $2x \geq 10$ and $x + 2 < 6$
Three-Part Inequality	Can be used to write some types of compound inequalities involving *and*	$x > -2$ and $x \leq 3$ is equivalent to $-2 < x \leq 3$.
Interval Notation	Notation used to write sets of real numbers rather than using number lines or inequalities	$-2 \leq z \leq 4$ is equivalent to $[-2, 4]$. $x < 4$ is equivalent to $(-\infty, 4)$. $x \leq -2$ or $x > 0$ is equivalent to $(-\infty, -2] \cup (0, \infty)$.

The following table lists basic strategies for solving compound inequalities.

Type of Inequality	Method to Solve Inequality
Solving a Compound Inequality Containing *and*	**STEP 1:** First solve each inequality individually. **STEP 2:** The solution set includes values that satisfy *both* inequalities from Step 1.
Solving a Compound Inequality Containing *or*	**STEP 1:** First solve each inequality individually. **STEP 2:** The solution set includes values that satisfy *at least one* of the inequalities from Step 1.
Solving a Three-Part Inequality	Work on all three parts at the same time. Be sure to perform the same step on each part. Continue until the variable is isolated in the middle part.

3.5 ABSOLUTE VALUE EQUATIONS AND INEQUALITIES

Basic Concepts ▪ Absolute Value Equations ▪ Absolute Value Inequalities

A LOOK INTO MATH ▷

Monthly average temperatures can vary greatly from one month to another, whereas yearly average temperatures remain fairly constant from one year to the next. In Boston, Massachusetts, the yearly average temperature is 50°F, but monthly average temperatures can vary from 28°F to 72°F. Because 50°F − 28°F = 22°F and 72°F − 50°F = 22°F, the monthly average temperatures are always within 22°F of the yearly average temperature. If T represents a monthly average temperature, we can model this situation by using the absolute value inequality

$$|T - 50| \leq 22.$$

The absolute value is necessary because a monthly average temperature T can be either greater than or less than 50°F by as much as 22°F. In this section we discuss absolute value equations and inequalities. (*Source:* A. Miller and J. Thompson, *Elements of Meteorology.*)

Basic Concepts

In Chapter 1 we discussed the absolute value of a number. We can define the **absolute value function** by $f(x) = |x|$. To graph $y = |x|$, we begin by making a table of values, as shown in Table 3.7. Next we plot these points and then sketch the graph shown in Figure 3.25. Note that the graph is V-shaped.

TABLE 3.7

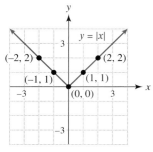

Figure 3.25 Absolute Value

Because the input for $f(x) = |x|$ is any real number, the domain of f is all real numbers, or $(-\infty, \infty)$. The graph of the absolute value function shows that the output y (range) is any real number greater than or equal to 0. That is, the range is $[0, \infty)$.

Absolute Value Equations

An equation that contains an absolute value is an **absolute value equation**. Examples are

$$|x| = 2, \quad |2x - 1| = 5, \quad \text{and} \quad |5 - 3x| - 3 = 1.$$

Consider the absolute value equation $|x| = 2$. This equation has *two* solutions: 2 and -2 because $|2| = 2$ and $|-2| = 2$. We can also demonstrate this result with a table of values or a graph. Refer back to Table 3.7: $|x| = 2$ when $x = -2$ or $x = 2$. In Figure 3.26 the graph of $y_1 = |x|$ intersects the graph of $y_2 = 2$ at the points $(-2, 2)$ and $(2, 2)$. The x-values at these points of intersection correspond to the solutions -2 and 2.

We generalize this discussion in the following manner.

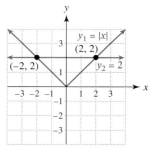

Figure 3.26

> **SOLVING $|x| = k$**
>
> 1. If $k > 0$, then $|x| = k$ is equivalent to $x = k$ or $x = -k$.
> 2. If $k = 0$, then $|x| = k$ is equivalent to $x = 0$.
> 3. If $k < 0$, then $|x| = k$ has no solutions.

EXAMPLE 1 Solving absolute value equations

Solve each equation.
(a) $|x| = 20$ (b) $|x| = -5$

Solution
(a) The solutions are -20 and 20.
(b) There are no solutions because $|x|$ is never negative.

EXAMPLE 2 Solving an absolute value equation

Solve $|2x - 5| = 3$ symbolically.

Solution
If $|2x - 5| = 3$, then either $2x - 5 = 3$ or $2x - 5 = -3$. Solve each equation separately.

$$2x - 5 = 3 \quad \text{or} \quad 2x - 5 = -3 \quad \text{Equations to be solved}$$
$$2x = 8 \quad \text{or} \quad 2x = 2 \quad \text{Add 5.}$$
$$x = 4 \quad \text{or} \quad x = 1 \quad \text{Divide by 2.}$$

The solutions are 1 and 4.

A table of values can be used to solve the equation $|2x - 5| = 3$ from Example 2. Table 3.8 shows that $|2x - 5| = 3$ when $x = 1$ or $x = 4$.

Numerical Solution
TABLE 3.8

x	0	1	2	3	4	5	6
$|2x - 5|$	5	3	1	1	3	5	7

Graphical Solution

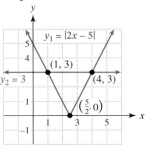

Figure 3.27

This equation can also be solved by graphing $y_1 = |2x - 5|$ and $y_2 = 3$. To graph y_1, first plot some of the points from Table 3.8. Its graph is V-shaped, as shown in Figure 3.27. Note that the x-coordinate of the "point" or vertex of the V can be found by solving the equation $2x - 5 = 0$ to obtain $\frac{5}{2}$. The graph of y_1 intersects the graph of y_2 at the points $(1, 3)$ and $(4, 3)$, giving the solutions **1** and **4**, so the graphical solutions agree with the numerical and symbolic solutions.

This discussion leads to the following result.

ABSOLUTE VALUE EQUATIONS

If $k > 0$, then

$$|ax + b| = k$$

is equivalent to

$$ax + b = k \quad \text{or} \quad ax + b = -k.$$

EXAMPLE 3 Solving absolute value equations

Solve.
(a) $|5 - x| - 2 = 8$ (b) $\left|\frac{1}{2}(x - 6)\right| = \frac{3}{4}$

Solution

(a) Start by adding 2 to each side to obtain $|5 - x| = 10$. This new equation is satisfied by the solution from either of the following equations.

$$5 - x = 10 \quad \text{or} \quad 5 - x = -10 \quad \text{Equations to be solved}$$
$$-x = 5 \quad \text{or} \quad -x = -15 \quad \text{Subtract 5.}$$
$$x = -5 \quad \text{or} \quad x = 15 \quad \text{Multiply by } -1.$$

The solutions are -5 and 15.

(b) This equation is satisfied by the solution from either of the following equations.

$$\frac{1}{2}(x - 6) = \frac{3}{4} \quad \text{or} \quad \frac{1}{2}(x - 6) = -\frac{3}{4} \quad \text{Equations to be solved}$$
$$2(x - 6) = 3 \quad \text{or} \quad 2(x - 6) = -3 \quad \text{Multiply by 4 to clear fractions.}$$
$$2x - 12 = 3 \quad \text{or} \quad 2x - 12 = -3 \quad \text{Distributive property}$$
$$2x = 15 \quad \text{or} \quad 2x = 9 \quad \text{Add 12.}$$
$$x = \frac{15}{2} \quad \text{or} \quad x = \frac{9}{2} \quad \text{Divide by 2.}$$

The solutions are $\frac{9}{2}$ and $\frac{15}{2}$.

The next example illustrates absolute value equations that have either no solutions or one solution.

EXAMPLE 4 Solving absolute value equations

Solve.
(a) $|2x - 1| = -2$ (b) $|4 - 2x| = 0$

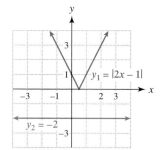

Figure 3.28

Solution

(a) Because an absolute value is never negative, there are no solutions. Figure 3.28 shows that the graph of $y_1 = |2x - 1|$ never intersects the graph of $y_2 = -2$.
(b) If $|y| = 0$, then $y = 0$. Thus $|4 - 2x| = 0$ when $4 - 2x = 0$ or when $x = 2$. The solution is 2.

Sometimes an equation can have an absolute value on each side. An example would be $|2x| = |x - 3|$. In this situation either $2x = x - 3$ (the two expressions are equal), or $2x = -(x - 3)$ (the two expressions are opposites).

These concepts are summarized as follows.

> **SOLVING $|ax + b| = |cx + d|$**
>
> Let a, b, c, and d be constants. Then $|ax + b| = |cx + d|$ is equivalent to
> $$ax + b = cx + d \quad \text{or} \quad ax + b = -(cx + d).$$

3.5 ABSOLUTE VALUE EQUATIONS AND INEQUALITIES

EXAMPLE 5   Solving an absolute value equation

Solve $|2x| = |x - 3|$.

Solution
Solve the following compound equality.

$$2x = x - 3 \quad \text{or} \quad 2x = -(x - 3)$$
$$x = -3 \quad \text{or} \quad 2x = -x + 3$$
$$3x = 3$$
$$x = 1$$

The solutions are -3 and 1.

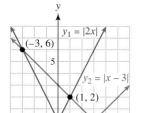

Figure 3.29

Example 5 is solved graphically in Figure 3.29. The graphs of $y_1 = |2x|$ and $y_2 = |x - 3|$ are V-shaped and intersect at $(-3, 6)$ and $(1, 2)$. The solutions are -3 and 1.

Absolute Value Inequalities

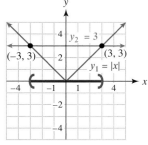

Figure 3.30

We can solve absolute value inequalities graphically. For example, to solve $|x| < 3$, let $y_1 = |x|$ and $y_2 = 3$ (see Figure 3.30). Their graphs intersect at $(-3, 3)$ and $(3, 3)$. The graph of y_1 is *below* the graph of y_2 for x-values *between*, but not including, $x = -3$ and $x = 3$. The solution set is $\{x \mid -3 < x < 3\}$ and is shaded on the x-axis.

Other absolute value inequalities can be solved graphically in a similar way. In Figure 3.31 the solutions to $|2x - 1| = 3$ are -1 and 2. The V-shaped graph of $y_1 = |2x - 1|$ is below the horizontal line $y_2 = 3$ when $-1 < x < 2$. Thus $|2x - 1| < 3$ whenever $-1 < x < 2$. The solution set is shaded on the x-axis.

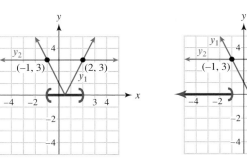

Figure 3.31 $y_1 < y_2$ Figure 3.32 $y_1 > y_2$

In Figure 3.32 the V-shaped graph of $y_1 = |2x - 1|$ is above the horizontal line $y_2 = 3$ both to the left of -1 and to the right of 2. That is, $|2x - 1| > 3$ whenever $x < -1$ or $x > 2$. The solution set is shaded on the x-axis.

This discussion is summarized as follows.

ABSOLUTE VALUE INEQUALITIES

Let the solutions to $|ax + b| = k$ be c and d, where $c < d$ and $k > 0$.

1. $|ax + b| < k$ is equivalent to $c < x < d$.
2. $|ax + b| > k$ is equivalent to $x < c$ or $x > d$.

Similar statements can be made for inequalities involving $\leq$ or $\geq$.

EXAMPLE 6 Solving absolute value equations and inequalities

Solve each absolute value equation and inequality.
(a) $|2 - 3x| = 4$ (b) $|2 - 3x| < 4$ (c) $|2 - 3x| > 4$

Solution
(a) The given equation is equivalent to the following equations.

$2 - 3x = 4$ or $2 - 3x = -4$ Equations to be solved
$-3x = 2$ or $-3x = -6$ Subtract 2.
$x = -\dfrac{2}{3}$ or $x = 2$ Divide by -3.

The solutions are $-\frac{2}{3}$ and 2.

(b) Solutions to $|2 - 3x| < 4$ include x-values **between**, but not including, $-\frac{2}{3}$ and 2. Thus the solution set is $\{x \mid -\frac{2}{3} < x < 2\}$, or in interval notation, $\left(-\frac{2}{3}, 2\right)$.

(c) Solutions to $|2 - 3x| > 4$ include x-values to the **left** of $x = -\frac{2}{3}$ or to the **right** of $x = 2$. Thus the solution set is $\{x \mid x < -\frac{2}{3} \text{ or } x > 2\}$, or in interval notation, $\left(-\infty, -\frac{2}{3}\right) \cup (2, \infty)$.

EXAMPLE 7 Solving an absolute value inequality

Solve $\left|\dfrac{2x - 5}{3}\right| > 3$. Write the solution set in interval notation.

Solution
Start by solving $\left|\dfrac{2x - 5}{3}\right| = 3$ as follows.

$\dfrac{2x - 5}{3} = 3$ or $\dfrac{2x - 5}{3} = -3$ Equations to be solved
$2x - 5 = 9$ or $2x - 5 = -9$ Multiply by 3.
$2x = 14$ or $2x = -4$ Add 5.
$x = 7$ or $x = -2$ Divide by 2.

Because the inequality symbol is $>$, the solution set is $x < -2$ or $x > 7$, or in interval notation, $(-\infty, -2) \cup (7, \infty)$.

The results from Examples 6 and 7 can be generalized as follows.

INEQUALITIES AND ABSOLUTE VALUES

If $k > 0$ and $y = f(x)$, then

$|y| < k$ is equivalent to $-k < y < k$ and
$|y| > k$ is equivalent to $y < -k$ or $y > k$.

Similar statements can be made for inequalities involving $\leq$ and $\geq$.

In the next example, we use the fact that $-k \leq y \leq k$ is equivalent to $|y| \leq k$.

EXAMPLE 8 Analyzing error

An engineer is designing a circular cover for a container. The diameter d of the cover is to be 4.25 inches and must be accurate to within 0.01 inch. Write an absolute value inequality that gives acceptable values for d.

Solution
The diameter d must satisfy $4.24 \leq d \leq 4.26$. Subtracting 4.25 from each part gives
$$-0.01 \leq d - 4.25 \leq 0.01,$$
which is equivalent to $|d - 4.25| \leq 0.01$.

EXAMPLE 9 Modeling temperature in Boston

CALCULATOR HELP
To graph an absolute value, see the Appendix (page AP-7).

In the introduction to this section we discussed how the inequality $|T - 50| \leq 22$ models the range for the monthly average temperatures T in Boston, Massachusetts.
(a) Solve this inequality and interpret the result.
(b) Give graphical support for part (a).

Solution
(a) *Symbolic Solution* Start by solving $|T - 50| = 22$.

$$T - 50 = 22 \quad \text{or} \quad T - 50 = -22 \qquad \text{Equations to be solved}$$
$$T = 72 \quad \text{or} \quad T = 28 \qquad \text{Add 50 to each side.}$$

Thus the solution set to $|T - 50| \leq 22$ is $\{T \mid 28 \leq T \leq 72\}$. Monthly average temperatures in Boston vary from 28°F to 72°F.

[0, 100, 10] by [0, 70, 10]

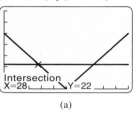

(a)

[0, 100, 10] by [0, 70, 10]

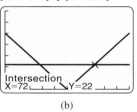

(b)

Figure 3.33

(b) *Graphical Solution* The graphs of $y_1 = |x - 50|$ and $y_2 = 22$ intersect at $(28, 22)$ and $(72, 22)$, as shown in Figures 3.33(a) and (b). The V-shaped graph of y_1 intersects the horizontal graph of y_2, or is below it, when $28 \leq x \leq 72$. Thus the solution set is $\{T \mid 28 \leq T \leq 72\}$. This result agrees with the symbolic result.

Sometimes the solution set to an absolute value inequality can be either the empty set or the set of all real numbers. These two situations are illustrated in the next example.

EXAMPLE 10 Solving absolute value inequalities

Solve if possible.
(a) $|2x - 5| > -1$ (b) $|5x - 1| + 3 \leq 2$

Solution
(a) Because the absolute value of an expression cannot be negative, $|2x - 5|$ is greater than -1 for every x-value. The solution set is all real numbers, or $(-\infty, \infty)$.
(b) Subtracting 3 from each side results in $|5x - 1| \leq -1$. Because the absolute value is always greater than or equal to 0, no x-values satisfy this inequality. There are no solutions.

3.5 PUTTING IT ALL TOGETHER

The absolute value function is given by $f(x) = |x|$, and its graph is V-shaped. Its domain (set of valid inputs) includes all real numbers, and its range (outputs) includes all nonnegative real numbers. The following table summarizes methods for solving absolute value equations and inequalities involving $<$ and $>$ symbols. Inequalities containing $\leq$ and $\geq$ symbols are solved similarly.

Problem	Symbolic Solution	Graphical Solution
$\|ax + b\| = k, k > 0$	Solve the equations $$ax + b = k$$ and $$ax + b = -k.$$	Graph $y_1 = \|ax + b\|$ and $y_2 = k$. Find the x-values of the two points of intersection.
$\|ax + b\| < k, k > 0$	If the solutions to $$\|ax + b\| = k$$ are c and d, $c < d$, then the solutions to $$\|ax + b\| < k$$ satisfy $$c < x < d.$$	Graph $y_1 = \|ax + b\|$ and $y_2 = k$. Find the x-values of the two points of intersection. The solutions are between these x-values on the number line, where the graph of y_1 lies below the graph of y_2.
$\|ax + b\| > k, k > 0$	If the solutions to $$\|ax + b\| = k$$ are c and d, $c < d$, then the solutions to $$\|ax + b\| > k$$ satisfy $$x < c \quad \text{or} \quad x > d.$$	Graph $y_1 = \|ax + b\|$ and $y_2 = k$. Find the x-values of the two points of intersection. The solutions are outside these x-values on the number line, where the graph of y_1 is above the graph of y_2.

CHAPTER 3 SUMMARY

SECTION 3.1 ■ LINEAR EQUATIONS

Linear Equations in One Variable
Can be written as $ax + b = 0$, where $a \neq 0$. Linear equations have *one* solution.

Examples: $3x - 5 = 0$ and $x + 2 = 1 - 3x$

Symbolic, Graphical, and Numerical Solutions

Example: Solve $3x - 1 = 2$.

Symbolic Solution

$3x - 1 = 2$
$3x = 3$
$x = 1$
The solution is 1.

Numerical Solution

x	$3x - 1$
0	-1
1	2
2	5
3	8

$3x - 1 = 2$ when $x = 1$.

Graphical Solution

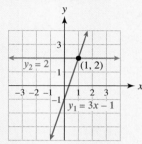

The solution is 1.

Lines

Standard Form $ax + by = c$, where a, b, and c are constants with a and b not both 0.

Finding x-intercepts   Let $y = 0$ and solve for x.

Finding y-intercepts   Let $x = 0$ and solve for y.

Example: $3x - 4y = 24$

$3x - 4(0) = 24$ implies $x = 8$; x-intercept is 8.

$3(0) - 4y = 24$ implies $y = -6$; y-intercept is -6.

Identities and Contradictions
An *identity* is an equation that is always true, regardless of the values of any variables. A *contradiction* is an equation that is always false, regardless of the values of any variables.

Examples: $x + 3x = 4x$ (identity); $4x + 5 = 4x + 1$ (contradiction)

SECTION 3.2 ■ INTRODUCTION TO PROBLEM SOLVING

Solving a Formula for a Variable   Use properties of algebra to solve for a variable.

Example: Solve $N = \frac{a + b}{2}$ for b.

$2N = a + b$ Multiply by 2.

$2N - a = b$ Subtract a.

Steps for Solving a Problem

STEP 1: Read the problem carefully and be sure that you understand it. (You may need to read the problem more than once.) Assign a variable to what you are being asked to find. If necessary, write other quantities in terms of this variable.

STEP 2: Write an equation that relates the quantities described in the problem. You may need to sketch a diagram, make a table, or refer to known formulas.

STEP 3: Solve the equation and determine the solution.

STEP 4: Look back and check your answer. Does it seem reasonable? Did you find the required information?

SECTION 3.3 ■ LINEAR INEQUALITIES

Linear Inequality in One Variable Can be written as $ax + b > 0$, where $a \neq 0$. (The symbol $>$ can be replaced with $<$, $\leq$, or $\geq$.) Linear inequalities have *infinitely many* solutions.

Examples: $3x - 5 < 0$, $x + 2 \geq 1 - 3x$

Symbolic, Numerical, and Graphical Solutions

Example: Solve $4 - 2x \geq 0$.

Symbolic Solution

$4 - 2x \geq 0$

$-2x \geq -4$

$x \leq 2$

Reverse inequality.

Numerical Solution

x	$4 - 2x$
0	4
1	2
2	0
3	-2
4	-4

$4 - 2x \geq 0$ when $x \leq 2$.

Graphical Solution

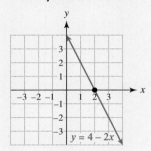

Graph is above or intersects the x-axis for $x \leq 2$.

SECTION 3.4 ■ COMPOUND INEQUALITIES

Compound Inequality Two inequalities connected by *and* or *or*.

Examples: For $x + 1 < 3$ or $x + 1 > 6$, a solution satisfies *at least* one of the inequalities.
For $2x + 1 < 3$ and $1 - x > 6$, a solution satisfies *both* inequalities.

Three-Part Inequality A compound inequality in the form $x > a$ and $x < b$ can be written as the three-part inequality $a < x < b$.

Example: $1 \leq x < 7$ means $x \geq 1$ and $x < 7$.

Interval Notation Can be used to identify intervals on the real number line.

Examples: $-2 < x \leq 3$ is equivalent to $(-2, 3]$.

$x < 5$ is equivalent to $(-\infty, 5)$.

Real numbers are denoted $(-\infty, \infty)$.

SECTION 3.5 ■ ABSOLUTE VALUE EQUATIONS AND INEQUALITIES

Absolute Value Equations The graph of $y = |ax + b|$, $a \neq 0$, is V-shaped and intersects the horizontal line $y = k$ twice if $k > 0$. In this case there are two solutions to the equation $|ax + b| = k$ determined by $ax + b = k$ or $ax + b = -k$.

Example: The equation $|2x - 1| = 5$ has two solutions.

Symbolic Solution

$2x - 1 = 5$ or $2x - 1 = -5$

$\quad 2x = 6$ or $\quad 2x = -4$ Add 1.

$\quad\quad x = 3$ or $\quad\quad x = -2$ Divide by 2.

Graphical Solution

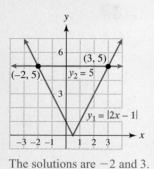

The solutions are -2 and 3.

Numerical Solution

x	-3	-2	-1	0	1	2	3
$\|2x - 1\|$	7	5	3	1	1	3	5

The solutions are -2 and 3.

Absolute Value Inequalities If the solutions to $|ax + b| = k$ are c and d with $c < d$, then the solution set for $|ax + b| < k$ is $\{x \mid c < x < d\}$, and the solution set for $|ax + b| > k$ is $\{x \mid x < c \text{ or } x > d\}$.

Example: The solutions to the equation $|2x - 1| = 5$ are -2 and 3, so the solution set for $|2x - 1| < 5$ is $\{x \mid -2 < x < 3\}$, and the solution set for $|2x - 1| > 5$ is $\{x \mid x < -2 \text{ or } x > 3\}$.

If $k > 0$ and $y = f(x)$, then

$\quad\quad |y| < k$ is equivalent to $-k < y < k$ and

$\quad\quad |y| > k$ is equivalent to $y < -k$ or $y > k$.

Examples: $|3 - x| < 5$ is equivalent to $-5 < 3 - x < 5$ and

$\quad\quad\quad\quad |3 - x| > 5$ is equivalent to $3 - x < -5$ or $3 - x > 5$.

CHAPTER 3 LINEAR EQUATIONS AND INEQUALITIES

Assignment Name _____ Name _____ Date _____
Show all work for these items: _____

# _____	# _____
# _____	# _____
# _____	# _____
# _____	# _____

CHAPTER 3 SUMMARY 135

Assignment Name _____ Name _____ Date _____
Show all work for these items: _____

# _____	# _____
# _____	# _____
# _____	# _____
# _____	# _____

CHAPTER 4
Systems of Linear Equations

In 1940, a physicist named John Atanasoff at Iowa State University needed to solve 29 equations with 29 variables simultaneously. This task was too difficult to do by hand, so he and a graduate student invented the first fully electronic digital computer. Thus the desire to solve a mathematical problem led to one of the most important inventions of the twentieth century. Today people can solve thousands of equations with thousands of variables. Solutions to such equations have resulted in better airplanes, cars, electronic devices, weather forecasts, and medical equipment.

Equations are also widely used in biology. The following table contains the weight W, neck size N, and chest size C for three black bears. Suppose that park rangers find a bear with a neck size of 22 inches and a chest size of 38 inches. Can they use the data in the table to estimate the bear's weight? Using systems of linear equations, they can answer this question. See Example 7 in Section 4.6.

W (pounds)	N (inches)	C (inches)
80	16	26
344	28	45
416	31	54
?	22	38

4.1 Systems of Linear Equations in Two Variables
4.2 The Substitution and Elimination Methods
4.3 Systems of Linear Inequalities
4.4 Introduction to Linear Programming
4.5 Systems of Linear Equations in Three Variables
4.6 Matrix Solutions of Linear Systems
4.7 Determinants

Education is what survives when what has been learned has been forgotten.

—B. F. SKINNER

Sources: A. Tucker, *Fundamentals of Computing*; M. Triola, *Elementary Statistics*; Minitab, Inc.

4.1 SYSTEMS OF LINEAR EQUATIONS IN TWO VARIABLES

Basic Concepts ▪ Graphical and Numerical Solutions ▪ Types of Linear Systems

A LOOK INTO MATH ▷

Many formulas involve more than one variable. For example, to calculate the heat index we need to know both the air temperature and the humidity. To calculate monthly car payments we need the loan amount, interest rate, and duration of the loan. Other applications involve large numbers of variables. To design new aircraft it is necessary to solve equations containing thousands of variables. In this section we consider systems of equations containing only two linear equations in two variables. However, the concepts discussed in this section are used to solve larger systems of equations.

Basic Concepts

▶ **REAL-WORLD CONNECTION** Each year, more and more people buy music subscriptions through the Internet. These subscriptions provide unlimited access to a large assortment of music through a person's computer. A combined total of $400 million was spent in 2004 and 2005 on music subscriptions, with a $200 million increase from 2004 to 2005. (*Source:* Jupiter Research.)

To determine the amount spent each year, we can let x be the amount spent in 2005 and let y be the amount spent in 2004 where both amounts are in millions of dollars. Then the given information is described by the following *system of equations*.

$$x + y = 400 \quad \text{The total is \$400 million.}$$
$$x - y = 200 \quad \text{The difference is \$200 million.}$$

Each equation contains two variables, x and y. These two equations form a **system of two linear equations in two variables**. An ordered pair (x, y) is a **solution** to a system of equations if the values for x and y satisfy *both* equations. Any system of two linear equations in two variables can be written in **standard form** as

$$ax + by = c$$
$$dx + ey = k,$$

where a, b, c, d, e, and k are constants.

EXAMPLE 1 Testing for solutions

Determine which ordered pair is a solution to the system of equations: $(0, 3)$ or $(-1, 2)$.

$$-x + 4y = 9$$
$$3x - 3y = -9$$

Solution
For $(0, 3)$ to be a solution, the values of $x = 0$ and $y = 3$ must satisfy *both* equations.

$$-0 + 4(3) \stackrel{?}{=} 9 \quad \text{False}$$
$$3(0) - 3(3) \stackrel{?}{=} -9 \quad \text{True}$$

Because $(0, 3)$ does not satisfy *both* equations, $(0, 3)$ is *not* a solution. To test $(-1, 2)$, substitute $x = -1$ and $y = 2$ in each equation.

$$-(-1) + 4(2) \stackrel{?}{=} 9 \quad \text{True}$$
$$3(-1) - 3(2) \stackrel{?}{=} -9 \quad \text{True}$$

Both equations are true, so $(-1, 2)$ is a solution.

Graphical and Numerical Solutions

Graphical, numerical, and symbolic techniques can be used to solve systems of equations. In this section we focus on graphical and numerical techniques and delay discussion of symbolic techniques until the next section.

In the next example we solve a system of linear equations graphically. Note that sometimes it may be difficult to read a graph precisely, so it is important to check our solutions by using the technique described in Example 1.

EXAMPLE 2 Solving a system of equations graphically

Solve the system of equations

$$y = x - 2$$
$$y = 4 - x$$

graphically. Check your answer.

Solution
Both equations are in slope–intercept form, so we can graph

$$y = x - 2 \text{ and } y = 4 - x$$

immediately, as shown in Figure 4.1. Their graphs intersect at the point (3, 1). To check that (3, 1) is the solution, substitute $x = 3$ and $y = 1$ into each equation.

$$1 \stackrel{?}{=} 3 - 2 \quad \text{True}$$
$$1 \stackrel{?}{=} 4 - 3 \quad \text{True}$$

The answer checks.

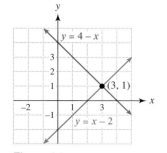
Figure 4.1

In the next example we demonstrate two methods for graphing linear equations that are written in the standard form: $ax + by = c$. The first method uses the slope–intercept form to graph each line and the second method uses the x- and y-intercepts to graph each line.

EXAMPLE 3 Solving a system of equations graphically

Solve the system of equations graphically.

$$x + 2y = 4$$
$$2x - y = 3$$

Solution
Method I: Finding Slope–Intercept Form Solve each equation for y and then use the slope–intercept form to graph the line.

$x + 2y = 4$	First equation		$2x - y = 3$	Second equation
$2y = -x + 4$	Subtract x.		$-y = -2x + 3$	Subtract $2x$.
$y = -\frac{1}{2}x + 2$	Multiply by $\frac{1}{2}$.		$y = 2x - 3$	Multiply by -1.

Now graph $y = -\frac{1}{2}x + 2$ and $y = 2x - 3$. See Figure 4.2. Their graphs intersect at (2, 1).

Method II: Finding x- and y-Intercepts To find the x-intercepts let $y = 0$ in each equation.

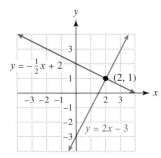
Figure 4.2

$x + 2y = 4$	First equation		$2x - y = 3$	Second equation
$x + 2(0) = 4$	Let $y = 0$.		$2x - 0 = 3$	Let $y = 0$.
$x = 4$	Solve for x.		$x = \frac{3}{2}$	Solve for x.

The x-intercept is 4. The x-intercept is $\frac{3}{2}$.

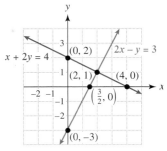

Figure 4.3

To find the *y*-intercepts let $x = 0$ in each equation.

$x + 2y = 4$	First equation		$2x - y = 3$	Second equation
$0 + 2y = 4$	Let $x = 0$.		$2(0) - y = 3$	Let $x = 0$.
$y = 2$	Solve for *y*.		$y = -3$	Solve for *y*.

The *y*-intercept is **2**.   The *y*-intercept is **−3**.

Graph $x + 2y = 4$ by drawing a line that passes through (**4**, 0) and (0, **2**). The graph of $2x - y = 3$ passes through $\left(\frac{3}{2}, 0\right)$ and $(0, -3)$. See Figure 4.3. Their graphs intersect at (2, 1).

In the next example we solve the system of equations presented earlier, which models spending on music subscriptions.

EXAMPLE 4   Solving a system of equations graphically and numerically

Solve the system of equations

$$x + y = 400$$
$$x - y = 200$$

graphically and numerically. Interpret the solution.

Solution
Start by solving each equation for *y*.

$x + y = 400$	First equation		$x - y = 200$	Second equation
$y = -x + 400$	Subtract *x*.		$-y = -x + 200$	Subtract *x*.
			$y = x - 200$	Multiply by -1.

Graphical Solution   The graphs of $y_1 = -x + 400$ and $y_2 = x - 200$ are shown in Figure 4.4. Because *x* and *y* represent sales, which are never negative, we graph these lines only in the first quadrant. Their graphs intersect at the point (300, 100). Thus $300 million was spent in 2005 on music subscriptions, and $100 million was spent in 2004. Check this solution.

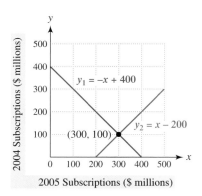

Figure 4.4 Graphical Solution

Numerical Solution   Make a table of values for $y_1 = -x + 400$ and $y_2 = x - 200$. Table 4.1 shows that, when $x = 300$, both of the expressions equal 100. Thus the solution is (300, 100). This type of numerical solution is based on trial and error. If the solution is not a "nice" integer, finding the solution numerically may be difficult or even *impossible*. However, the *important mathematical concept* to remember is that you are looking for an *x*-value where $y_1 = y_2$.

4.1 SYSTEMS OF LINEAR EQUATIONS IN TWO VARIABLES

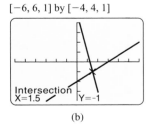

Figure 4.5

TABLE 4.1 A Numerical Solution

x	100	200	300	400
$y_1 = -x + 400$	300	200	100	0
$y_2 = x - 200$	-100	0	100	200

In Figure 4.5 a calculator was used to create a table similar to Table 4.1.

EXAMPLE 5 Solving a system of equations graphically

Solve the system of equations

$$2x - 3y = 6$$
$$4x + y = 5$$

graphically. Check your answer.

Solution
Solve the first equation for y.

$$2x - 3y = 6 \quad \text{First equation}$$
$$-3y = -2x + 6 \quad \text{Subtract } 2x \text{ from each side.}$$
$$\frac{-3y}{-3} = \frac{-2x}{-3} + \frac{6}{-3} \quad \text{Divide each term by } -3.$$
$$y = \frac{2}{3}x - 2 \quad \text{Simplify.}$$

Next, solve the second equation for y.

$$4x + y = 5 \quad \text{Second equation}$$
$$y = -4x + 5 \quad \text{Subtract } 4x \text{ from each side.}$$

Graph $y_1 = \frac{2}{3}x - 2$ and $y_2 = -4x + 5$, as shown in Figure 4.6(a). (A similar calculator graph is shown in Figure 4.6(b).) Their graphs appear to intersect at $(1.5, -1)$. To be *certain*, check this answer in the *given* equations.

$$2x - 3y = 6 \quad \text{First equation}$$
$$2(1.5) - 3(-1) \stackrel{?}{=} 6 \quad \text{Let } x = 1.5 \text{ and } y = -1.$$
$$6 = 6 \quad \text{The answer checks.}$$

Now substitute these values in the second equation. (It is essential to check *both* equations.)

$$4x + y = 5 \quad \text{Second equation}$$
$$4(1.5) + (-1) \stackrel{?}{=} 5 \quad \text{Let } x = 1.5 \text{ and } y = -1.$$
$$5 = 5 \quad \text{The answer checks.}$$

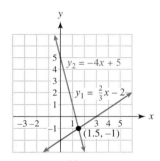

(a)

$[-6, 6, 1]$ by $[-4, 4, 1]$

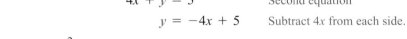

(b)

Figure 4.6

Types of Linear Systems

A system of linear equations that has at least one solution is a **consistent system**; otherwise, it is an **inconsistent system**. A system of linear equations in two variables can be represented graphically by two lines in the xy-plane. Three different situations involving two lines are illustrated in Figure 4.7. In Figure 4.7(a) the lines intersect at a single point, which represents a *unique solution*. In this case the equations of the lines are called **independent equations**. In Figure 4.7(b) the two lines are identical, which occurs when the two equations are *equivalent*. For example, the equations $x + y = 1$ and $2x + 2y = 2$ are equivalent. If we divide each side of the second equation by 2 we obtain the first equation. As a result, their graphs are identical and every point on the line represents a solution to the system of linear equations. Thus there are infinitely many solutions, and the equations are called **dependent equations**. Finally, in Figure 4.7(c) the lines are parallel and do not intersect. There are no solutions, so the system is inconsistent.

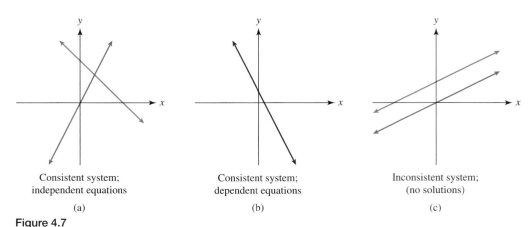

Consistent system; independent equations
(a)

Consistent system; dependent equations
(b)

Inconsistent system; (no solutions)
(c)

Figure 4.7

▶ REAL-WORLD CONNECTION   In the next three examples we illustrate each of these situations with a real-world application.

EXAMPLE 6 Solving a linear system with a unique solution

Suppose that two groups of students go to a football game. The first group buys 3 tickets and 3 soft drinks for $15, and the second group buys 4 tickets and 2 soft drinks for $16. Find the price of a ticket and the price of a soft drink graphically.

Solution
Let x be the cost of a ticket and y be the cost of a soft drink. Then $3x + 3y = 15$ represents 3 tickets and 3 soft drinks costing $15 and $4x + 2y = 16$ represents 4 tickets and 2 soft drinks costing $16. This information can be written as a system of equations.

$$3x + 3y = 15$$
$$4x + 2y = 16$$

To graph each equation, we will use the x- and y-intercepts (Method II), as discussed in Example 3. The x- and y-intercepts for $3x + 3y = 15$ are both 5. The x- and y-intercepts for $4x + 2y = 16$ are 4 and 8, respectively. Plotting the points $(5, 0)$, $(0, 5)$, $(4, 0)$, and $(0, 8)$ and sketching the corresponding lines in the first quadrant results in Figure 4.8. These graphs intersect at $(3, 2)$. Thus the cost of a ticket is $3 and the cost of a soft drink is $2. Note that 3 tickets and 3 soft drinks cost $3(3) + 3(2) = \$15$ and that 4 tickets and 2 soft drinks cost $4(3) + 2(2) = \$16$, so our solution is correct.

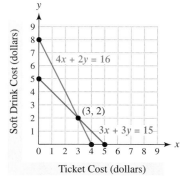

Figure 4.8

EXAMPLE 7 Solving a dependent linear system

Now suppose that two groups of students go to a different football game. The first group buys 4 tickets and 2 soft drinks for $16, and the second group buys 2 tickets and 1 soft drink for $8. If possible, find the price of a ticket and the price of a soft drink graphically.

Solution

Let x be the cost of a ticket and y be the cost of a soft drink. Then this situation can be modeled by the following system of equations.

$$4x + 2y = 16$$
$$2x + y = 8$$

Solve each equation for y.

$$4x + 2y = 16 \qquad\qquad 2x + y = 8$$
$$2y = -4x + 16 \qquad\qquad y = -2x + 8$$
$$y = -2x + 8$$

Both equations simplify to the *same* slope–intercept form, $y = -2x + 8$, and thus the lines are identical, as shown in Figure 4.9.

The system is consistent because there is at least one solution. The equations are dependent because not enough information is available to determine a unique solution. Note that the second group bought half what the first group bought and paid half as much. As a result, the two equations contain essentially the *same information* and are *equivalent*. Thus there are infinitely many solutions because every point on the line is a solution. For example, a ticket could cost $3 and a soft drink could cost $2, or a ticket could cost $2 and a soft drink could cost $4. The solution set can be expressed in set-builder notation as $\{(x, y) \mid 2x + y = 8\}$.

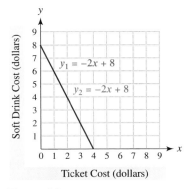

Figure 4.9

CRITICAL THINKING

Suppose that a system of two linear equations with two variables is dependent. If you try to solve this system numerically by using a table, how could you recognize that the equations are indeed dependent? Explain your answer.

EXAMPLE 8   Recognizing an inconsistent linear system

Now suppose that two groups of students go to a concert. The first group buys 4 tickets and 2 soft drinks for $20, and the second group buys 2 tickets and 1 soft drink for $12. If possible, find the price of a ticket and the price of a soft drink graphically.

Solution
Let x be the cost of a ticket and y be the cost of a soft drink. Then the following system models the data.

$$4x + 2y = 20$$
$$2x + y = 12$$

Solving for y, we can write each equation in slope–intercept form.

$$y = -2x + 10$$
$$y = -2x + 12$$

Their graphs are parallel lines with slope -2 and different y-intercepts. Thus they do not intersect (see Figure 4.10). The linear system is *inconsistent* because there are no solutions. Note that the second group purchased half what the first group purchased. If pricing had been *consistent*, the second group would have paid half, or $10, instead of $12. *Inconsistent pricing* resulted in an *inconsistent linear system*.

CRITICAL THINKING
How would a table of values appear when you are solving an inconsistent system of equations?

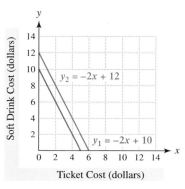

Figure 4.10

EXAMPLE 9   Classifying systems of equations

Classify each system as consistent or inconsistent. If the system is consistent, state whether the equations are dependent or independent.

(a) $x + y = 1$
$x + y = -1$

(b) $x + 2y = 4$
$2x + 4y = 8$

(c) $x + y = 4$
$x - y = 2$

Solution
(a) The slope–intercept forms for these equations are $y = -x + 1$ and $y = -x - 1$. Their graphs are parallel lines with slope -1 and different y-intercepts, so they do not intersect. See Figure 4.11(a). There are no solutions and the system is inconsistent.

NOTE: The sum of two numbers, x and y, cannot equal both 1 and -1 at the same time, so it is reasonable that there are no solutions.

(b) Because the equations both have slope–intercept form $y = -\frac{1}{2}x + 2$, their graphs are identical. The system is consistent and the equations are dependent. See Figure 4.11(b). There are infinitely many solutions of the form $\{(x, y) \,|\, x + 2y = 4\}$.

NOTE: The second equation is exactly double the first equation. When one equation is a *nonzero* multiple of the other, the equations are dependent.

(c) The first equation has slope–intercept form $y = -x + 4$ and the second equation has slope–intercept form $y = x - 2$. These lines have different slopes and intersect at one point: (3, 1). The system is consistent and the equations are independent. See Figure 4.11(c).

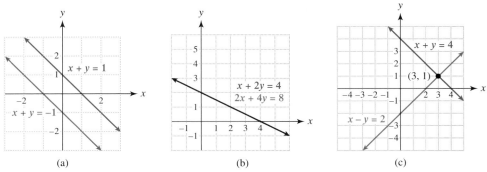

Figure 4.11

A SYSTEM OF TWO LINEAR EQUATIONS IN TWO VARIABLES

A system involving two linear equations in two variables can have no solutions, one solution, or infinitely many solutions. Its graph consists of two lines.

1. If the lines *are parallel*, the system is inconsistent and there are no solutions.
2. If the lines *intersect at a single point*, there is one solution. The system is consistent and the equations are independent.
3. If the lines *are identical*, the equations are dependent and there are infinitely many solutions. The system is consistent.

EXAMPLE 10 Finding an athlete's running speeds

An athlete jogs at a faster pace for 30 minutes and then jogs at a slower pace for 90 minutes. The first pace is 4 miles per hour faster than the second pace, and the athlete covers a total distance of 15 miles.
(a) Write a linear system whose solution gives the athlete's running speeds.
(b) Solve the resulting system graphically and numerically.

Solution
(a) Let x be the faster speed of the runner and y be the slower speed. The athlete runs $\frac{1}{2}$ hour at x miles per hour and $\frac{3}{2}$ hours at y miles per hour. *Rate times time equals distance* and the total distance traveled is 15 miles, so

$$\frac{1}{2}x + \frac{3}{2}y = 15.$$

Because the first pace is 4 miles per hour faster than the second pace,

$$x - y = 4.$$

Thus we need to solve the following linear system of equations.

$$\frac{1}{2}x + \frac{3}{2}y = 15 \qquad \text{First equation}$$

$$x - y = 4 \qquad \text{Second equation}$$

[0, 20, 5] by [0, 20, 5]

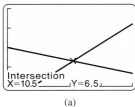

(a)

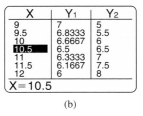

(b)

Figure 4.12

(b) Multiply the first equation by 2 to clear fractions and then solve for y.

$$x + 3y = 30 \quad \text{Multiply first equation by 2.}$$
$$3y = -x + 30 \quad \text{Subtract } x.$$
$$y = -\frac{x}{3} + 10 \quad \text{Divide by 3.}$$

Solving the second equation for y gives

$$y = x - 4.$$

Graphical Solution We graph $Y_1 = -X/3 + 10$ and $Y_2 = X - 4$, as shown in Figure 4.12(a). The graphs intersect at the point (10.5, 6.5). Thus the athlete ran at 10.5 miles per hour and then slowed to 6.5 miles per hour.

Numerical Solution A numerical solution is shown in Figure 4.12(b). To find this solution, we set the increment for the x-values to 0.5.

4.1 PUTTING IT ALL TOGETHER

A system of two linear equations in two variables may be written in standard form as

$$ax + by = c$$
$$dx + ey = k,$$

where a, b, c, d, e, and k are constants (fixed numbers). Linear systems in two variables may be solved graphically or numerically. The following table summarizes the types of systems of equations and the number of solutions that may be encountered.

System	Solution	Graph
Consistent, with a Unique Solution $x + y = 3$ $x - y = 1$ Equations Are Independent	There is one solution: $x = 2$ and $y = 1$. The solution is the ordered pair (2, 1). Check: $2 + 1 = 3$ True $2 - 1 = 1$ True	Graph $y_1 = 3 - x$ and $y_2 = x - 1$. Their graphs intersect at (2, 1).
Consistent, with Infinitely Many Solutions $x + y = 1$ $3x + 3y = 3$ Equations Are Dependent	There are infinitely many solutions, such as (2, −1) and (0, 1). Solution Set: $\{(x, y) \mid x + y = 1\}$ Note that multiplying the first equation by 3 results in the second equation. Also, both equations have the same x- and y-intercepts: 1 and 1.	Graph $y_1 = 1 - x$ and $y_2 = (-3x + 3)/3$. The graphs are identical.

continued on next page

continued from previous page

System	Solution	Graph
Inconsistent $x + y = 1$ $x + y = 2$	There are no solutions. The sum of two variables, x and y, cannot equal both 1 and 2 at the same time.	Graph $y_1 = 1 - x$ and $y_2 = 2 - x$. The lines are parallel with slope -1 and do not intersect.

4.2 THE SUBSTITUTION AND ELIMINATION METHODS

The Substitution Method ■ The Elimination Method ■ Models and Applications

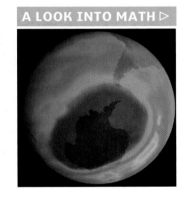

A LOOK INTO MATH ▷ The ability to solve systems of equations has resulted in the development of CAT scans, satellites, DVDs, and accurate weather forecasts. In the preceding section we showed how to solve a system of two linear equations in two variables, using graphical and numerical methods. In this section we demonstrate how to solve these systems symbolically.

The Substitution Method

To apply the **substitution method**, we begin by solving an equation for one of its variables. Then we substitute the result into the other equation. For example, consider the following system of equations.

$$2x + y = 5$$
$$3x - 2y = 4$$

It is convenient to solve the first equation for y to obtain $y = 5 - 2x$. Now substitute $(5 - 2x)$ for y in the second equation,

$$3x - 2(y) = 4 \quad \text{Second equation}$$
$$3x - 2(5 - 2x) = 4, \quad \text{Substitute.}$$

to obtain a linear equation in one variable.

$$3x - 2(5 - 2x) = 4$$
$$3x - 10 + 4x = 4 \quad \text{Distributive property}$$
$$7x - 10 = 4 \quad \text{Combine like terms.}$$
$$7x = 14 \quad \text{Add 10 to each side.}$$
$$x = 2 \quad \text{Divide each side by 7.}$$

To determine y substitute 2 for x in $y = 5 - 2x$ to obtain

$$y = 5 - 2(2) = 1.$$

The solution is $(2, 1)$. To check this solution let $x = 2$ and $y = 1$ in each equation.

$2x + y = 5$ First equation $3x - 2y = 4$ Second equation
$2(2) + 1 \stackrel{?}{=} 5$ Substitute. $3(2) - 2(1) \stackrel{?}{=} 4$ Substitute.
$5 = 5$ It checks. $4 = 4$ It checks.

Because $(2, 1)$ satisfies *both* equations, it is the solution to the system.

Do not stop after solving for the first variable. You must also solve for the second variable. Remember that a solution to a system of equations in two variables consists of an *ordered pair*, not a single number.

NOTE: When using substitution, you may begin by solving for *either variable in either equation*. The same result is obtained regardless of which equation is used. However, it is often simpler to solve for a variable with a coefficient of 1 because there is less likelihood of encountering fractions.

EXAMPLE 1   Applying the substitution method

Solve each system of equations.

(a) $\quad y = 2x$
$\quad\;\; x + y = 21$

(b) $\;2x + y = -1$
$\quad\;\; 2x - y = -3$

(c) $\;-3x + 2y = 3$
$\quad\;\;\; 2x - 4y = -6$

Solution
(a) The first equation is $y = 2x$, so we substitute $(2x)$ for y in the second equation.

$$\begin{aligned} x + y &= 21 & &\text{Second equation} \\ x + (2x) &= 21 & &\text{Let } y = 2x. \\ 3x &= 21 & &\text{Add like terms.} \\ x &= 7 & &\text{Divide by 3.} \end{aligned}$$

Substituting 7 for x in $y = 2x$ gives $y = 14$. The solution is $(7, 14)$.

(b) We start by solving for y in the first equation because its coefficient is 1.

$$\begin{aligned} 2x + y &= -1 & &\text{First equation} \\ y &= -2x - 1 & &\text{Subtract } 2x. \end{aligned}$$

Now we substitute $(-2x - 1)$ for y in the second equation.

$$\begin{aligned} 2x - y &= -3 & &\text{Second equation} \\ 2x - (-2x - 1) &= -3 & &\text{Let } y = (-2x - 1). \\ 2x + 2x + 1 &= -3 & &\text{Distributive property} \\ 4x &= -4 & &\text{Subtract 1; combine like terms.} \\ x &= -1 & &\text{Divide by 4.} \end{aligned}$$

Substituting -1 for x in $y = -2x - 1$ gives $y = 1$. The solution is $(-1, 1)$.

(c) We start by solving for x in the second equation, but we could solve for y.

$$\begin{aligned} 2x - 4y &= -6 & &\text{Second equation} \\ 2x &= 4y - 6 & &\text{Add } 4y. \\ x &= 2y - 3 & &\text{Divide by 2.} \end{aligned}$$

Substitute $(2y - 3)$ for x in the first equation.

$$\begin{aligned} -3x + 2y &= 3 & &\text{First equation} \\ -3(2y - 3) + 2y &= 3 & &\text{Let } x = (2y - 3). \\ -6y + 9 + 2y &= 3 & &\text{Distributive property} \\ -4y + 9 &= 3 & &\text{Combine like terms.} \\ -4y &= -6 & &\text{Subtract 9.} \\ y &= \tfrac{3}{2} & &\text{Divide by } -4. \end{aligned}$$

Substituting $\tfrac{3}{2}$ for y in $x = 2y - 3$ gives $x = 2(\tfrac{3}{2}) - 3 = 0$. The solution is $(0, \tfrac{3}{2})$.

EXAMPLE 2 Finding per capita income

In 2004, the average of the per capita (or per person) incomes for Massachusetts and Maine was $36,000. The per capita income in Massachusetts exceeded the per capita income in Maine by $11,000. Find the 2004 per capita income for each state.

Solution

Let x be the per capita income in Massachusetts and y be the per capita income in Maine. The following system of equations models the data.

$$\frac{x + y}{2} = 36{,}000 \qquad \text{Their average is \$36,000.}$$

$$x - y = 11{,}000 \qquad \text{Their difference is \$11,000.}$$

Begin by solving the second equation for x.

$$x = y + 11{,}000$$

Substitute $(y + 11{,}000)$ for x in the first equation and solve for y.

$$\frac{(y + 11{,}000) + y}{2} = 36{,}000$$

$$(y + 11{,}000) + y = 72{,}000 \qquad \text{Multiply each side by 2.}$$

$$2y + 11{,}000 = 72{,}000 \qquad \text{Combine like terms.}$$

$$2y = 61{,}000 \qquad \text{Subtract 11,000 from each side.}$$

$$y = 30{,}500 \qquad \text{Divide each side by 2.}$$

Substituting for y in $x = y + 11{,}000$ yields $x = 41{,}500$. In 2004, the per capita income in Massachusetts was $41,500, and in Maine it was $30,500.

NOTE: A step-by-step procedure for the substitution method is given in Putting It All Together for this section.

The Elimination Method

The **elimination** (or addition) **method** is a second way to solve linear systems symbolically. This method is based on the property that *equals added to equals are equal*. That is, if

$$a = b \quad \text{and} \quad c = d, \text{ then}$$
$$a + c = b + d.$$

The goal of this method is to obtain an equation from which one of the two variables has been eliminated. This task is sometimes accomplished by adding two equations. This method is demonstrated in the next example.

EXAMPLE 3 Applying the elimination method

Solve each system of equations.

(a) $x + y = 3$ (b) $4x + 3y = 0$
$x - y = 1$ $4x - 2y = -20$

Solution

(a) If we add the two equations, the *y*-variable will be eliminated.

$$x + y = 3$$
$$x - y = 1$$
$$\overline{2x + 0y = 4} \quad \text{or} \quad x = 2 \quad \text{Add equations and solve.}$$

To find the value of *y*, we substitute $x = 2$ in either equation.

$$x + y = 3 \quad \text{First equation}$$
$$2 + y = 3 \quad \text{Let } x = 2.$$
$$y = 1 \quad \text{Subtract 2.}$$

The solution is $(2, 1)$.

(b) If we multiply the first equation by -1 and then add the two equations, the *x*-variable will be eliminated.

$$-4x - 3y = 0 \quad \text{First equation times } -1$$
$$4x - 2y = -20 \quad \text{Second equation}$$
$$\overline{0x - 5y = -20} \quad \text{or} \quad y = 4 \quad \text{Add equations and solve.}$$

To find the value of *x*, we substitute $y = 4$ in either equation.

$$4x + 3y = 0 \quad \text{First equation}$$
$$4x + 3(4) = 0 \quad \text{Let } y = 4.$$
$$4x = -12 \quad \text{Subtract 12.}$$
$$x = -3 \quad \text{Divide by 4.}$$

The solution is $(-3, 4)$.

Solutions to systems of equations can be *supported graphically*. For example, if we graph the equations in Example 3(a), they intersect at the point (2, 1), as shown in Figure 4.13.

Figure 4.13

EXAMPLE 4 Applying the elimination method

Solve the following system by using elimination.

$$2x - y = 4$$
$$x + y = 1$$

Solution

Adding the two equations eliminates the variable *y*.

$$2x - y = 4$$
$$x + y = 1$$
$$\overline{3x + 0y = 5} \quad \text{or} \quad x = \frac{5}{3} \quad \text{Add the two equations and solve for } x.$$

Substituting $x = \frac{5}{3}$ in the second equation gives

$$\frac{5}{3} + y = 1 \quad \text{or} \quad y = -\frac{2}{3}.$$

The solution is $\left(\frac{5}{3}, -\frac{2}{3}\right)$.

NOTE: Example 4 illustrates that the *x*- and *y*-values for a solution can be fractions and *not* integers.

In the next example, we use multiplication before we add the two equations.

EXAMPLE 5 Multiplying before applying elimination

Solve the system of equations.

$$x + \frac{1}{2}y = 1$$
$$-3x + 2y = 11$$

Solution
Neither variable can be eliminated by simply adding the given equations. However, if we multiply each side of the first equation by -4, we eliminate fractions and then addition of the two equations eliminates the *y*-variable.

$$-4x - 2y = -4 \qquad \text{Multiply first equation by } -4.$$
$$\underline{-3x + 2y = 11}$$
$$-7x + 0y = 7 \quad \text{or} \quad x = -1 \qquad \text{Add equations and solve.}$$

To find the value of *y*, we substitute $x = -1$ in the second equation.

$$-3x + 2y = 11 \qquad \text{Second equation}$$
$$-3(-1) + 2y = 11 \qquad \text{Let } x = -1.$$
$$2y = 8 \qquad \text{Subtract 3 from each side.}$$
$$y = 4 \qquad \text{Divide each side by 2.}$$

The solution is $(-1, 4)$.

MAKING CONNECTIONS

Substitution and Elimination

Substitution and elimination are two symbolic methods that accomplish the *same* task: solving a system of linear equations. Be aware that one method may be easier to perform than the other, depending on the system of equations to be solved.

EXAMPLE 6 Multiplying before applying elimination

Solve the following system by using elimination. Support your answer graphically and numerically.

$$3x - 2y = 11$$
$$2x + 3y = 3$$

Solution
Symbolic Solution If we add (or subtract) these equations, neither variable will be eliminated. However, if we multiply the first equation by 3 and multiply the second equation by 2, we can eliminate *y*.

$$9x - 6y = 33 \qquad \text{Multiply first equation by 3.}$$
$$\underline{4x + 6y = 6} \qquad \text{Multiply second equation by 2.}$$
$$13x + 0y = 39 \quad \text{or} \quad x = 3 \qquad \text{Add the equations and solve.}$$

Substituting $x = 3$ in the first equation, $3x - 2y = 11$, gives
$$3(3) - 2y = 11 \quad \text{or} \quad y = -1.$$
The solution is $(3, -1)$.

NOTE: We could have multiplied the first equation by 2 and the second equation by -3. Adding the resulting equations would have eliminated the variable x.

Graphical Solution If we use a calculator, we solve each equation for y.

$$3x - 2y = 11 \quad \text{First equation} \qquad 2x + 3y = 3 \quad \text{Second equation}$$
$$-2y = 11 - 3x \quad \text{Subtract } 3x. \qquad 3y = 3 - 2x \quad \text{Subtract } 2x.$$
$$y = \frac{11 - 3x}{-2} \quad \text{Divide by } -2. \qquad y = \frac{3 - 2x}{3} \quad \text{Divide by } 3.$$

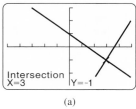

(a)

(b)

Figure 4.14

In Figure 4.14(a), the graphs of $Y_1 = (11 - 3X)/(-2)$ and $Y_2 = (3 - 2X)/3$ intersect at the point $(3, -1)$.

Numerical Solution In Figure 4.14(b), $y_1 = y_2 = -1$, when $x = 3$.

EXAMPLE 7   Recognizing an inconsistent system

Use elimination to solve the following system.
$$3x - 4y = 5$$
$$-6x + 8y = 9$$

Solution
If we multiply the first equation by 2 and add, we obtain the following result.

$$\begin{aligned} 6x - 8y &= 10 \quad \text{Multiply first equation by 2.} \\ \underline{-6x + 8y} &= \underline{9} \\ 0 &= 19 \quad \text{Adding the two equations gives a false result.} \end{aligned}$$

The statement $0 = 19$ is a *contradiction*, which tells us that the system has no solutions. If we solve each equation for y and graph, we obtain two parallel lines with slope $\frac{3}{4}$ that never intersect. This result is shown in Figure 4.15, where the equations are graphed as $Y_1 = (5 - 3X)/(-4)$ and $Y_2 = (6X + 9)/8$.

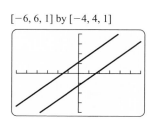

Figure 4.15

EXAMPLE 8 Recognizing dependent equations

Use elimination to solve the following system.
$$3x - 6y = 3$$
$$x - 2y = 1$$

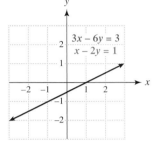

Figure 4.16

Solution
If we multiply the second equation by -3 and add, we obtain the following.

$$\begin{array}{rl} 3x - 6y = & 3 \quad \text{First equation} \\ -3x + 6y = & -3 \quad \text{Multiply second equation by } -3. \\ \hline 0 = & 0 \quad \text{Adding the two equations gives a true result.} \end{array}$$

The statement $0 = 0$ is an *identity*, so the two equations are equivalent. If an ordered pair (x, y) satisfies the first equation, it also satisfies the second equation. Thus the solution set may be expressed as $\{(x, y) \mid x - 2y = 1\}$. For example, $(1, 0)$ and $(5, 2)$ are both solutions because they both satisfy $x - 2y = 1$.

The graphs of these equations are identical lines, as shown in Figure 4.16.

MAKING CONNECTIONS

Graphical Solutions, Substitution, and Elimination

1. **Unique Solution** A graphical solution results in two lines intersecting at a unique point. Substitution and elimination give unique values for x and y.
2. **Inconsistent Linear System** A graphical solution results in two parallel lines. Substitution and elimination result in a contradiction, such as $0 = 1$.
3. **Dependent Linear System** A graphical solution results in two identical lines. Substitution and elimination result in an identity, such as $0 = 0$.

   NOTE: Conditions 2 and 3 occur when *both* variables are eliminated, and the resulting equation is either an identity or a contradiction.

Models and Applications

In Section 3.2, we learned a four-step process for solving problems. We will apply this process to solving applications that involve two variables. A *brief summary* of these four steps is as follows. (See page 102.)

STEPS FOR SOLVING A PROBLEM

STEP 1: Read the problem. Assign two variables to what you are being asked to find.
STEP 2: Write two equations involving these variables.
STEP 3: Solve the systems of equations.
STEP 4: Check your solution.

EXAMPLE 9 Modeling tuition

A student is attempting to graduate on schedule by taking 14 credits of day classes at one college and 4 credits of night classes at a different college. A credit for day classes costs $20 more than a credit for night classes. If the student's total tuition is $2440, how much does each type of credit cost?

Solution

STEP 1: Let x be the cost of a credit for day classes and y be the cost of a credit for night classes.

STEP 2: Then the data can be represented by the following system of equations.

$$x - y = 20$$
$$14x + 4y = 2440$$

Day credits cost $20 more than night credits. The total cost is $2440 for 14 credits during the day and 4 credits at night.

STEP 3: To solve this system we multiply the first equation by 4 and then add the equations.

$$4x - 4y = 80 \quad \text{Multiply by 4.}$$
$$14x + 4y = 2440$$
$$\overline{18x + 0y = 2520} \quad \text{or} \quad x = 140 \quad \text{Add and solve for } x.$$

Because $x = 140$ and $x - y = 20$, $y = 120$. Thus a credit for day classes costs $140 and a credit for night classes costs $120.

STEP 4: Check the solution. The difference is $140 - $120 = $20 and

$$14 \times \$140 = \$1960 \quad \text{Cost of day classes}$$
$$4 \times \$120 = \$480 \quad \text{Cost of night classes}$$
$$\overline{\text{Total cost} = \$2440.} \quad \text{The answer checks.}$$

In the next example we make use of the formula $d = rt$. (*Distance equals rate times time*.) If we solve this formula for r to obtain $r = \frac{d}{t}$, we can find the average speed (rate) of a boat traveling on a river. This method can also be used to determine the speed of an airplane when there is a wind.

EXAMPLE 10 Modeling river travel

A boat travels 150 miles upstream in 10 hours, and the return trip takes 6 hours. Find the speed of the boat in still water and the speed of the current.

Solution

STEP 1: Let x be the speed of the boat and y be the speed of the river current.

STEP 2: The boat travels 150 miles upstream in 10 hours. Thus the speed of the boat *against* the current is $\frac{150}{10} = 15$ miles per hour, or $x - y = 15$. Similarly the boat travels 150 miles downstream in 6 hours. Thus the speed of the boat *with* the current is $\frac{150}{6} = 25$ miles per hour, or $x + y = 25$.

STEP 3: We represent these data with the following equations.

$$x + y = 25$$
$$x - y = 15$$
$$\overline{2x + 0y = 40} \quad \text{or} \quad x = 20 \quad \text{Add equations and solve.}$$

Substituting $x = 20$ in $x + y = 25$ gives $y = 5$. The boat can travel 20 miles per hour in still water and the speed of the current is 5 miles per hour.

STEP 4: Check the solution. The speed of the boat *against* the current is $20 - 5 = 15$ miles per hour and the speed of the boat *with* the current is $20 + 5 = 25$ miles per hour.

$$t = \frac{d}{r} = \frac{150}{15} = 10 \text{ hours} \quad \text{Time to travel upstream}$$

$$t = \frac{d}{r} = \frac{150}{25} = 6 \text{ hours} \quad \text{Time to travel downstream}$$

The answer checks.

4.2 THE SUBSTITUTION AND ELIMINATION METHODS

EXAMPLE 11 Burning calories while exercising

During strenuous exercise, an athlete can burn 12 calories per minute running and 10 calories per minute on a bicycle. In a 60-minute workout, an athlete burns 644 calories. How long did the athlete spend running and bicycling?

Solution

STEP 1: Let x be the time spent running and y be the time spent bicycling.

STEP 2: Because the workout is 60 minutes long, we write $x + y = 60$. The number of calories burned running equals $12x$, and the number of calories burned bicycling equals $10y$. The total calories burned equals 644 so the system is as follows.

$$x + y = 60 \qquad \text{Total workout is 60 minutes.}$$
$$12x + 10y = 644 \qquad \text{Total calories equal 644.}$$

STEP 3: To solve this system we can multiply the first equation by -10 and add the equations to eliminate the y-variable.

$$\begin{array}{r} -10x - 10y = -600 \\ 12x + 10y = 644 \\ \hline 2x + 0y = 44 \quad \text{or} \quad x = 22 \end{array}$$

Multiply first equation by -10.

Add equations and solve.

Substituting $x = 22$ in $x + y = 60$ gives $y = 38$. The athlete spent 22 minutes running and 38 minutes bicycling.

STEP 4: Check the solution. The total time is $22 + 38 = 60$ minutes and

$$\begin{array}{r} 22 \text{ minutes} \times 12 \text{ calories/minute} = 264 \text{ calories} \\ 38 \text{ minutes} \times 10 \text{ calories/minute} = 380 \text{ calories} \\ \hline \text{Total calories burned} = 644 \text{ calories.} \end{array}$$

Calories burned running
Calories burned bicycling
The answer checks.

EXAMPLE 12 Mixing antifreeze

A mixture of water and antifreeze in a car is currently 20% antifreeze. If the radiator holds 5 gallons of fluid and this mixture should be 40% antifreeze, how many gallons of radiator fluid should be drained and replaced with a mixture containing 90% antifreeze?

Solution

STEP 1: Let x be the gallons of 20% antifreeze that should remain in the radiator and y be the gallons of 90% antifreeze that should be added to the radiator.

STEP 2: The radiator holds 5 gallons of fluid, so $x + y = 5$. The final solution in the radiator should contain 5 gallons of 40% solution or $5(0.4) = 2$ gallons of (pure) antifreeze. Thus the antifreeze in the 20% solution plus the antifreeze in the 90% solution must equal 2 gallons. That is, $0.2x + 0.9y = 2$. Table 4.2 summarizes this situation.

TABLE 4.2 Mixing Antifreeze

	20% Solution	90% Solution	40% Solution
Radiator Fluid (gallons)	x	y	5
Pure Antifreeze (gallons)	$0.2x$	$0.9y$	2

The resulting system to be solved is

$$x + y = 5 \qquad \text{First row in table}$$
$$0.2x + 0.9y = 2. \qquad \text{Second row in table}$$

STEP 3: Multiply the first equation by 2 and the second equation by -10.

$$2x + 2y = 10 \quad \text{Multiply first equation by 2.}$$
$$-2x - 9y = -20 \quad \text{Multiply second equation by } -10.$$
$$\overline{0x - 7y = -10} \quad \text{or} \quad y = \tfrac{10}{7} \quad \text{Add and solve.}$$

Thus $\tfrac{10}{7}$ gallons of 90% antifreeze solution should be added.

STEP 4: Check the solution. If $\tfrac{10}{7}$ gallons are drained from the 5-gallon radiator, then there are $5 - \tfrac{10}{7} = \tfrac{35}{7} - \tfrac{10}{7} = \tfrac{25}{7}$ gallons of 20% antifreeze remaining and $\tfrac{10}{7}$ gallons of 90% antifreeze added.

$$20\% \text{ of } \frac{25}{7} = \frac{2}{10} \times \frac{25}{7} = \frac{5}{7} \text{ gallons of pure antifreeze}$$

$$90\% \text{ of } \frac{10}{7} \text{ gallons} = \frac{9}{10} \times \frac{10}{7} = \frac{9}{7} \text{ gallons of pure antifreeze}$$

There are $\tfrac{5}{7} + \tfrac{9}{7} = \tfrac{14}{7} = 2$ gallons of pure antifreeze in the 5-gallon radiator, so the new concentration is $\tfrac{2}{5} = 0.4$ or 40%. The answer checks.

4.2 PUTTING IT ALL TOGETHER

Two symbolic methods for solving a system of linear equations are substitution and elimination. Symbolic methods give exact answers, whereas graphical and numerical methods may give approximate answers. The following table presents a summary.

Concept	Explanation
Substitution Method	1. Solve for a convenient variable such as y in the first equation. $$x + y = 3 \quad \text{or} \quad y = 3 - x$$ $$2x - y = 0$$ 2. Substitute $(3 - x)$ in the second equation for y. Solve the equation for x. $$2x - (3 - x) = 0 \quad \text{or} \quad x = 1$$ 3. Substitute $x = 1$ in one of the given equations and find y. $$1 + y = 3 \quad \text{or} \quad y = 2$$ 4. The solution is $(1, 2)$. Check your answer.
Elimination Method	1. Multiply the first equation by -2 so that the coefficients of x in the two equations are additive inverses. $$x + 2y = 1 \quad \text{or} \quad -2x - 4y = -2$$ $$2x - 3y = 9 \qquad\qquad\quad 2x - 3y = 9$$ 2. Eliminate x by adding the two equations. $$-2x - 4y = -2$$ $$2x - 3y = 9$$ $$\overline{0x - 7y = 7} \quad \text{or} \quad y = -1$$ 3. Substitute $y = -1$ in one of the given equations and solve for x. $$x + 2(-1) = 1 \quad \text{or} \quad x = 3$$ 4. The solution is $(3, -1)$. Check your answer.

4.3 SYSTEMS OF LINEAR INEQUALITIES

Solving Linear Inequalities in Two Variables ■ Solving Systems of Linear Inequalities

A LOOK INTO MATH ▷

People often walk or jog in an effort to increase their heart rates and get in better shape. During strenuous exercise, older people should maintain lower heart rates than younger people. A person cannot maintain precisely one heart rate, so a range of heart rates is recommended by health professionals. For aerobic fitness, a 50-year-old's heart rate might be between 120 and 140 beats per minute, whereas a 20-year-old's heart rate might be between 140 and 160. Systems of linear inequalities can be used to model these situations.

Solving Linear Inequalities in Two Variables

▶ **REAL-WORLD CONNECTION** Suppose that candy costs $2 per pound and that peanuts cost $1 per pound. The total cost C of buying x pounds of candy and y pounds of peanuts is given by

$$C = 2x + y.$$

If we have at most $5 to spend, the inequality

$$2x + y \leq 5$$

must be satisfied. To determine the different weight combinations of candy and peanuts that could be bought, we could use the graph shown in Figure 4.17. Points located on the line $2x + y = 5$ represent weight combinations resulting in a $5 purchase. For example, the point (2, 1) satisfies the equation $2x + y = 5$ and represents buying 2 pounds of candy and 1 pound of peanuts for $5. Ordered pairs (x, y) located in the shaded region below the line represent purchases of less than $5. The point (1, 2) lies in the shaded region and represents buying 1 pound of candy and 2 pounds of peanuts for $4. Note that if we substitute $x = 1$ and $y = 2$ in the inequality, we obtain

$$2(1) + (2) \leq 5,$$

which is a true statement.

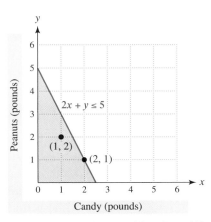

Figure 4.17 Purchases of Candy and Peanuts

When the equals sign in a linear equation in two variables is replaced with one of the symbols $<$, $\leq$, $>$, or $\geq$, a **linear inequality in two variables** results. Examples include

$$2x + y \leq 5, \quad y \geq x - 5, \quad \text{and} \quad \frac{1}{2}x - \frac{3}{5}y < 8.$$

EXAMPLE 1 Solving linear inequalities

Shade the solution set for each inequality.
(a) $x > 1$ (b) $y \leq 2x - 1$ (c) $x - 2y < 4$

Solution

(a) Begin by graphing the vertical line $x = 1$ with a *dashed* line because equality is *not* included. The solution set includes all points with *x*-values greater than 1, so shade the region to the right of this line, as shown in Figure 4.18(a).

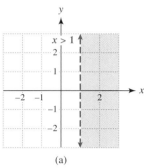
(a)

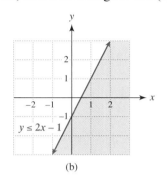
(b)

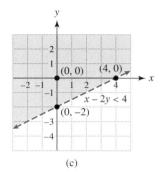
(c)

Figure 4.18

(b) Start by graphing $y = 2x - 1$ with a *solid* line because equality *is* included. The inequality sign is $\leq$, so the solution set includes all points on or below this line, as shown in Figure 4.18(b).

(c) We start by finding the intercepts for the line $x - 2y = 4$.

x-intercept: $y = 0$ implies $x - 2(0) = 4$ or $x = \mathbf{4}$

y-intercept: $x = 0$ implies $0 - 2y = 4$ or $y = \mathbf{-2}$

Plot the points $(4, \mathbf{0})$ and $(\mathbf{0}, -2)$ and sketch a dashed line, as shown in Figure 4.18(c). To determine whether to shade above or below this line, substitute a *test point* in the inequality. For example, if we pick the test point $(0, 0)$ and then substitute $x = \mathbf{0}$ and $y = \mathbf{0}$ in the given inequality, we find the following result.

$$
\begin{aligned}
x - 2y &< 4 &&\text{Given inequality} \\
0 - 2(0) &< 4 &&\text{Let } x = 0 \text{ and } y = 0. \\
0 &< 4 &&\text{A true statement}
\end{aligned}
$$

Because this substitution results in a true statement, shade the region *containing* $(0, 0)$, which is located above the dashed line.

The following steps can be used to graph a linear inequality in two variables.

SOLVING A LINEAR INEQUALITY GRAPHICALLY

1. Replace the inequality symbol with an equals sign.
2. Graph the resulting line. Use a solid line if the inequality symbol is $\leq$ or $\geq$ and a dashed line if it is $<$ or $>$.
3. (a) If the inequality is in the form $x \leq k$ or $x < k$ (where k is a constant) shade to the *left* of the vertical line. If the inequality is in the form $x \geq k$ or $x > k$, shade to the *right* of the vertical line.
   (b) If the inequality is in the form $y \leq mx + b$ or $y < mx + b$, shade *below* this line. If the inequality is in the form $y \geq mx + b$ or $y > mx + b$, shade *above* this line.
   (c) If you are uncertain as to which region to shade, choose a **test point** that is *not* on the line. Substitute it in the given inequality. If the test point makes the inequality true, then shade the region containing the test point. Otherwise, shade on the other side of the line.

EXAMPLE 2 Solving a linear inequality graphically

Shade the solution set for the linear inequality $4x - 3y < 12$.

Solution
Start graphing the line $4x - 3y = 12$ by finding its intercepts.

x-intercept: $y = 0$ implies $4x - 3(0) = 12$ or $x = 3$.

y-intercept: $x = 0$ implies $4(0) - 3y = 12$ or $y = -4$.

Plot the points $(3, 0)$ and $(0, -4)$ and sketch a dashed line. Any point not on this line can be used for a test point. We use the test point $(2, 2)$.

$$4x - 3y < 12 \qquad \text{Given inequality}$$
$$4(2) - 3(2) < 12 \qquad \text{Let } x = 2 \text{ and } y = 2.$$
$$2 < 12 \qquad \text{A true statement}$$

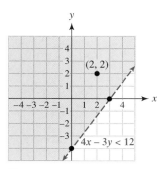

Figure 4.19

Thus shade the region containing the point $(2, 2)$, as shown in Figure 4.19.

Solving Systems of Linear Inequalities

A **system of linear inequalities** results when the equals signs in a system of linear equations are replaced with $<$, $\leq$, $>$, or $\geq$. The system of linear equations

$$x + y = 4$$
$$y = x$$

becomes a system of linear inequalities when it is written as

$$x + y \leq 4$$
$$y \geq x.$$

A solution to a system of inequalities must satisfy *both* inequalities. For example, the ordered pair $(1, 2)$ is a solution to this system because substituting $x = 1$ and $y = 2$ makes both inequalities true.

$$1 + 2 \leq 4 \qquad \text{True}$$
$$2 \geq 1 \qquad \text{True}$$

To see graphically that $(1, 2)$ is a solution, consider the following. The solution set for $x + y \leq 4$ consists of points lying on the line $x + y = 4$ and all points below the line. This region is shaded in Figure 4.20(a). Similarly, the solutions to $y \geq x$ include the line $y = x$ and all points above it. This region is shaded in Figure 4.20(b).

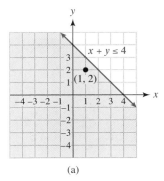
(a)

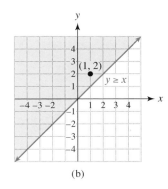
(b)

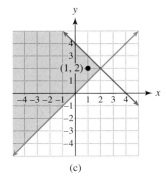
(c)

Figure 4.20

For a point (x, y) to be a solution to the *system* of linear inequalities, it must be located in both shaded regions shown in Figures 4.20(a) and 4.20(b). The *intersection* of the shaded regions is shown in Figure 4.20(c). Note that the point $(1, 2)$ is located in each shaded region shown in Figures 4.20(a) and 4.20(b). Therefore the point $(1, 2)$ is located in the shaded region shown in Figure 4.20(c) and is a solution of the system of linear inequalities.

EXAMPLE 3 Solving a system of linear inequalities

Shade the solution set for each system of inequalities.

(a) $x \leq -1$ $\quad$ (b) $\quad y < 2x$ $\quad$ (c) $2x - y < 2$
$y \geq 2$ $x + y > 3$ $x + 2y \geq 6$

Solution

(a) Graph the vertical line $x = -1$ and the horizontal line $y = 2$ as solid lines. The solution set is to the left of the line $x = -1$ and above the line $y = 2$. See Figure 4.21. The test point $(-2, 3)$ satisfies both inequalities and lies in the shaded region.

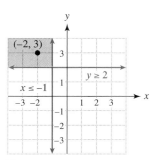

Figure 4.21

CRITICAL THINKING

Is the point where the two lines intersect in Figure 4.20(c) part of the solution set? Why or why not? Repeat this question for Figure 4.22(b).

(b) Graph $y = 2x$ and $x + y = 3$ as dashed lines. These two lines divide the xy-plane into four regions. The test point $(-1, -1)$ does not satisfy the inequalities, so do not shade the region containing it. However, the point $(2, 2)$ does satisfy *both* inequalities. To verify this fact substitute $x = 2$ and $y = 2$ in each inequality.

$$2 < 2(2) \quad \text{True}$$
$$2 + 2 > 3 \quad \text{True}$$

Thus shade the region shown in Figure 4.22(a).

(c) Graph $2x - y = 2$ and $x + 2y = 6$ as dashed and solid lines, respectively. The test point $(2, 3)$ satisfies *both* inequalities, so shade the region containing it, as shown in Figure 4.22(b).

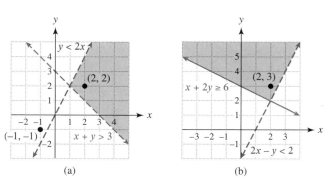

Figure 4.22

NOTE: Finding a test point that satisfies a system of linear inequalities may require trial and error. You may need to pick one test point from each of the four regions determined by the two lines. One of the four points must satisfy the system of inequalities.

▶ **REAL-WORLD CONNECTION** In the next example we demonstrate how systems of inequalities can model the situation discussed in the introduction to this section.

EXAMPLE 4 Modeling target heart rates

When exercising, people often try to maintain target heart rates that are percentages of their maximum heart rates. A person's maximum heart rate (MHR) is MHR = 220 − A, where A represents age and the MHR is in beats per minute. The shaded region shown in Figure 4.23 represents target heart rates for aerobic fitness for various ages. (*Source:* Hebb Industries, Inc.)

(a) Estimate the range R of heart rates that are acceptable for someone 40 years old.

(b) By choosing two points on each line in Figure 4.23 and applying the point–slope form, we can show that the equations for these lines are approximately

$$T = -0.8A + 196 \quad \text{Upper line}$$

and

$$T = -0.7A + 154, \quad \text{Lower line}$$

where A represents age and T represents the target heart rate. Write a system of inequalities whose solution set is the shaded region, including the two lines.

(c) Use Figure 4.23 to determine whether (30, 150) is a solution. Then verify your answer by using the system of inequalities.

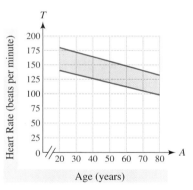

Figure 4.23 Target Heart Rates

Solution

(a) Figure 4.23 reveals that, when the age is 40, $A = 40$, the target heart rates T are *approximately* $125 \leq T \leq 165$ beats per minute.

(b) The region lies **below** the upper line, **above** the lower line, and includes both lines. Therefore this region is modeled by

$$T \leq -0.8A + 196$$
$$T \geq -0.7A + 154.$$

(c) Figure 4.23 shows that the point representing a 30-year-old person with a heart rate of 150 beats per minute lies in the shaded region, so (30, 150) is a solution. This result can be verified by substituting $A = 30$ and $T = \mathbf{150}$ in the system.

$$\mathbf{150} \leq -0.8(\mathbf{30}) + 196 = 172 \quad \text{True}$$
$$\mathbf{150} \geq -0.7(\mathbf{30}) + 154 = 133 \quad \text{True}$$

EXAMPLE 5 Solving a system of linear inequalities with technology

Shade the solution set for the system of inequalities, using a graphing calculator.

$$2x + y \leq 5$$
$$-2x + y \geq 1$$

Solution

Begin by solving each inequality for y to obtain $y \leq 5 - 2x$ and $y \geq 2x + 1$. The graphs of $Y_1 = 5 - 2X$ and $Y_2 = 2X + 1$ are shown in Figure 4.24(a). The solution set lies below y_1 and above y_2. Figure 4.24(b) shows how to shade this region. The solution set is the region comprised of small squares in Figure 4.24(c).

CALCULATOR HELP

To shade a graph, see the Appendix (page AP-7).

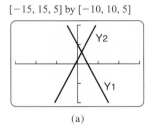
(a)

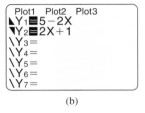
(b)

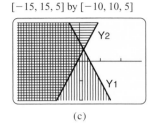
(c)

Figure 4.24

TECHNOLOGY NOTE: Shading of Linear Inequalities

When shading the solution set for a linear inequality, graphing calculators often show solid lines even if a line should be dashed. For example, the graphs for $y < 5 - 2x$ and $y \leq 5 - 2x$ are *identical* on some graphing calculators.

CRITICAL THINKING

Graph the solution set to the following system and discuss your results.

$$3x + y \geq 6$$
$$3x + y \leq 2$$

4.3 PUTTING IT ALL TOGETHER

In this section we presented systems of linear inequalities in two variables. These systems usually have infinitely many solutions and can be represented by a shaded region in the xy-plane. The following table summarizes these concepts.

Concept	Explanation	Examples
Linear Inequality in Two Variables	$ax + by > c$, where a, b, and c are constants. ($>$ may be replaced with $<$, $\leq$, or $\geq$.)	$y \geq 5$, $x - y < -10$, and $6x - 7y \leq 22$
System of Linear Inequalities in Two Variables	Two linear inequalities where a solution must satisfy *both* inequalities	$x + y < 11$ $x - y \geq 4$ $(8, 2)$ is a solution because $x = 8$ and $y = 2$ make *both* inequalities true.

continued on next page

continued from previous page

Concept	Explanation	Examples
Solution Set to a System of Linear Inequalities	Set of all solutions; typically a region in the *xy*-plane. Two intersecting lines divide the *xy*-plane into four regions. To determine which region should be shaded, choose one test point from each region. The region that contains the test point satisfying both inequalities should be shaded.	$y \leq 5 - x$ (below $y = 5 - x$) and $y < 2x - 3$ (below $y = 2x - 3$) Use a dashed line for $<$ or $>$. Use a solid line for $\leq$ or $\geq$. The test point $(3, -2)$ satisfies both inequalities.

4.4 INTRODUCTION TO LINEAR PROGRAMMING

Basic Concepts ■ Region of Feasible Solutions ■ Solving Linear Programming Problems

A LOOK INTO MATH ▷ During World War II large numbers of troops were at the front. Keeping these soldiers supplied with equipment and food was an essential but complex task. To solve this logistics problem, a new type of mathematics called *linear programming* was invented. Today, linear programming is important to business and the social sciences because it is a procedure that can be used to optimize quantities such as cost and profit. Linear programming applications frequently contain thousands of variables. In this section, we focus on problems involving only two variables. However, the concepts discussed in this section are important to your understanding of larger problems.

Basic Concepts

▶ **REAL-WORLD CONNECTION** Suppose that a small business sells candy for $3 per pound and fresh ground coffee for $5 per pound. All inventory is sold by the end of the day. The revenue *R* collected in dollars is given by

$$R = 3x + 5y,$$

where *x* is the pounds of candy sold and *y* is the pounds of coffee sold. For example, if the business sells **80** pounds of candy and **40** pounds of coffee during a day, then its revenue is

$$R = 3(80) + 5(40) = \$440.$$

The function $R = 3x + 5y$ is called an **objective function**.

Suppose also that the company cannot package more than 150 pounds of candy and coffee per day. Then the inequality

$$x + y \leq 150$$

represents a **constraint** on the objective function, which limits the company's revenue for any one day. A goal of this business might be to maximize

$$R = 3x + 5y,$$

subject to the constraints

$$x + y \leq 150$$
$$x \geq 0, \, y \geq 0.$$

Note that the constraints $x \geq 0$ and $y \geq 0$ are included because the number of pounds of candy or coffee cannot be negative. The problem that we have described is called a *linear programming problem*. Before learning how to solve a linear programming problem, we need to discuss the set of *feasible solutions*.

Region of Feasible Solutions

The constraints for a linear programming problem consist of linear inequalities. These inequalities are satisfied by some points in the xy-plane but not by others. The set of solutions to these constraints is called the **feasible solutions**. For example, the region of feasible solutions to the constraints for the business just described is shaded in Figure 4.25.

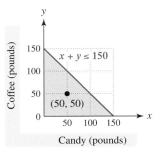

Figure 4.25 Constraints on Sales

The point $(50, 50)$ lies in the shaded region and represents the business selling 50 pounds of candy and 50 pounds of coffee. In the next example, we shade the region of feasible solutions to a set of constraints.

EXAMPLE 1 Finding the region of feasible solutions

Shade the region of feasible solutions to the following constraints.

$$x + 2y \leq 30$$
$$2x + y \leq 30$$
$$x \geq 0, \, y \geq 0$$

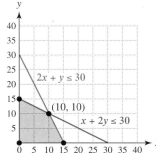

Figure 4.26

Solution

The feasible solutions are the ordered pairs (x, y) that *satisfy all four* inequalities. They lie below the lines $x + 2y = 30$ and $2x + y = 30$, and above the line $y = 0$ and to the right of $x = 0$, as shown in Figure 4.26. Note that the inequalities $x \geq 0$ and $y \geq 0$ restrict the feasible solutions to quadrant I.

Solving Linear Programming Problems

A **linear programming problem** consists of an *objective function* and a system of linear inequalities called *constraints*. The solution set for the system of linear inequalities is called the *region of feasible solutions*. The objective function describes a quantity that is to be optimized. The **optimal value** to a linear programming problem often results in maximum revenue or minimum cost.

When the system of constraints has only two variables, the boundary of the region of feasible solutions often consists of line segments intersecting at points called *vertices* (plural of **vertex**.) To solve a linear programming problem we use the *fundamental theorem of linear programming*.

FUNDAMENTAL THEOREM OF LINEAR PROGRAMMING

If the optimal value for a linear programming problem exists, then it occurs at a vertex of the region of feasible solutions.

The fundamental theorem of linear programming is used to solve the following linear programming problem.

EXAMPLE 2 Maximizing an objective function

Maximize the objective function $R = 2x + 3y$ subject to

$$x + 2y \leq 30$$
$$2x + y \leq 30$$
$$x \geq 0, \ y \geq 0.$$

Solution
The region of feasible solutions is shaded in Figure 4.26. Note that the vertices on the boundary of feasible solutions are $(0, 0)$, $(15, 0)$, $(10, 10)$, and $(0, 15)$. To find the maximum value of R, substitute each vertex in the formula for R, as shown in Table 4.3. The maximum value of R is 50 when $x = 10$ and $y = 10$.

TABLE 4.3

Vertex	$R = 2x + 3y$	
(0, 0)	$2(0) + 3(0) = 0$	
(15, 0)	$2(15) + 3(0) = 30$	
(10, 10)	$2(10) + 3(10) = 50$	← Maximum R
(0, 15)	$2(0) + 3(15) = 45$	

The following steps are helpful in solving linear programming word problems.

STEPS FOR SOLVING A LINEAR PROGRAMMING WORD PROBLEM

STEP 1: Read the problem carefully. Consider making a table.
STEP 2: Write the objective function and all the constraints.
STEP 3: Sketch a graph of the region of feasible solutions. Identify all vertices.
STEP 4: Evaluate the objective function at each vertex. A maximum (or a minimum) occurs at a vertex.

NOTE: If the region is unbounded, a maximum (or minimum) may not exist.

EXAMPLE 3 Minimizing the cost of vitamins

A breeder is mixing two different vitamins, Brand X and Brand Y, into pet food. Each serving of pet food should contain at least 60 units of vitamin A and 30 units of vitamin C. Brand X costs 80 cents per ounce and Brand Y costs 50 cents per ounce. Each ounce of Brand X contains 15 units of vitamin A and 10 units of vitamin C, whereas each ounce of Brand Y contains 20 units of vitamin A and 5 units of vitamin C. Determine how much of each brand of vitamin should be mixed to produce a minimum cost per serving.

Solution

STEP 1: Begin by listing the information, as illustrated in Table 4.4.

TABLE 4.4

Brand	Amount	Vitamin A	Vitamin C	Cost
X	x	15	10	80 cents
Y	y	20	5	50 cents
Minimum		60	30	

STEP 2: If x ounces of Brand X are purchased at 80 cents per ounce and if y ounces of Brand Y are purchased at 50 cents per ounce, then the total cost C is given by $C = 80x + 50y$. Because each ounce of Brand X contains 15 units of vitamin A and each ounce of Brand Y contains 20 units of vitamin A, the total number of units of vitamin A is $15x + 20y$. If each serving of pet food must contain at least 60 units of vitamin A, the constraint is $15x + 20y \geq 60$. Similarly, because each serving requires at least 30 units of vitamin C, $10x + 5y \geq 30$. The linear programming problem then becomes the following.

Minimize: $C = 80x + 50y$ Cost (in cents)
Subject to: $15x + 20y \geq 60$ Vitamin A
$10x + 5y \geq 30$ Vitamin C
$x \geq 0, \; y \geq 0$

STEP 3: The region containing the feasible solutions is shown in Figure 4.27.

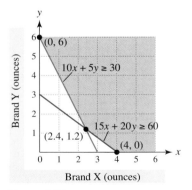

Figure 4.27

NOTE: To determine the vertex (2.4, 1.2) solve the system of equations

$$15x + 20y = 60$$
$$10x + 5y = 30$$

by using elimination.

STEP 4: The vertices for this region are (0, 6), (2.4, 1.2), and (4, 0). Evaluate the objective function C at each vertex, as shown in Table 4.5.

TABLE 4.5

Vertex	$C = 80x + 50y$	
(0, 6)	$80(0) + 50(6) = 300$	
(2.4, 1.2)	$80(2.4) + 50(1.2) = 252$	← Minimum cost (cents)
(4, 0)	$80(4) + 50(0) = 320$	

The minimum cost occurs when 2.4 ounces of Brand X and 1.2 ounces of Brand Y are mixed, at a cost of $2.52 per serving.

4.4 PUTTING IT ALL TOGETHER

The following table summarizes the basic concept of linear programming.

Concept	Comments	Example
Linear Programming	In a linear programming problem, the maximum or minimum of an objective function is found, subject to a set of constraints. If a solution exists, it occurs at a vertex in the region of feasible solutions.	Maximize $R = x + 2y$, subject to $3x + 2y \leq 15$, $2x + 3y \leq 15$, $x \geq 0$, and $y \geq 0$. The vertices are (0, 0), (5, 0), (3, 3), and (0, 5). The maximum, $R = 10$, occurs at vertex (0, 5), and the minimum, $R = 0$, occurs at (0, 0).

4.5 SYSTEMS OF LINEAR EQUATIONS IN THREE VARIABLES

Basic Concepts ▪ Solving Linear Systems with Substitution and Elimination ▪ Modeling Data ▪ Systems of Equations with No Solutions ▪ Systems of Equations with Infinitely Many Solutions

A LOOK INTO MATH ▷

In Sections 4.1 and 4.2, we described how to solve systems of linear equations in two variables. In applications, systems commonly have many variables. Large linear systems are used in the design of electrical circuits, bridges, and ships. They also are used in business, economics, and psychology. Because of the enormous amount of work needed to solve large systems, technology is usually used to find approximate solutions. In this section we discuss symbolic methods for finding solutions of linear systems in three variables. These methods provide the basis for understanding how technology can solve large linear systems.

Basic Concepts

When we solve a linear system in two variables, we can express a solution as an ordered pair (x, y). A linear equation in two variables can be represented graphically by a line. A system of two linear equations with a unique solution can be represented graphically by two lines intersecting at a point, as shown in Figure 4.28.

When solving linear systems in three variables, we often use the variables x, y, and z. A solution is expressed as an **ordered triple** (x, y, z), rather than an ordered pair (x, y). For example, if the ordered triple $(1, 2, 3)$ is a solution, $x = 1, y = 2$, and $z = 3$ satisfy each equation. A linear equation in three variables can be represented by a flat plane in space. If the solution is unique, we can represent a linear system of three equations in three variables graphically by three planes intersecting at a single point, as illustrated in Figure 4.29.

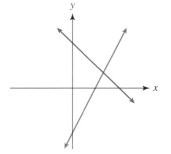

Figure 4.28

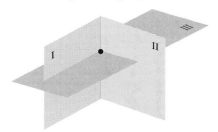

Figure 4.29

NOTE: The three planes in Figure 4.29 all intersect at right angles. In general, three planes can intersect at a point even if they are not at right angles to each other.

CRITICAL THINKING

The following figures of three planes in space represent a system of three linear equations in three variables. How many solutions are there in each case? Explain.

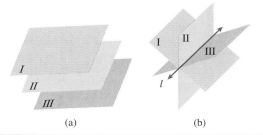

(a)   (b)

The next example shows how to check whether an ordered triple is a solution to a system of linear equations in three variables.

EXAMPLE 1 Checking for solutions to a system of three equations

Determine whether $(4, 2, -1)$ or $(-1, 0, 3)$ is a solution to the system.

$$2x - 3y + z = 1$$
$$x - 2y + 2z = 5$$
$$2y + z = 3$$

Solution
To check $(4, 2, -1)$, substitute $x = 4, y = 2$, and $z = -1$ in each equation.

$$2(4) - 3(2) + (-1) \stackrel{?}{=} 1 \quad \text{True}$$
$$4 - 2(2) + 2(-1) \stackrel{?}{=} 5 \quad \text{False}$$
$$2(2) + (-1) \stackrel{?}{=} 3 \quad \text{True}$$

The ordered triple $(4, 2, -1)$ *does not satisfy all three equations, so it is not a solution.* Next, substitute $x = -1, y = 0$, and $z = 3$ in each equation.

$$2(-1) - 3(0) + 3 \stackrel{?}{=} 1 \quad \text{True}$$
$$-1 - 2(0) + 2(3) \stackrel{?}{=} 5 \quad \text{True}$$
$$2(0) + 3 \stackrel{?}{=} 3 \quad \text{True}$$

The ordered triple $(-1, 0, 3)$ *satisfies all three equations, so it is a solution.*

In the next example we set up a system of three equations in three variables that involves the angles of a triangle.

EXAMPLE 2 Setting up a system of equations

The measure of the largest angle in a triangle is 40° greater than the sum of the two smaller angles and 90° more than the smallest angle. Set up a system of three linear equations in three variables whose solution gives the measure of each angle.

Solution
Let $x, y,$ and z be the measures of the three angles from largest to smallest. Because the sum of the measures of the angles in a triangle equals 180°, we have

$$x + y + z = 180.$$

The measure of the largest angle x is 40° greater than the sum of the measures of the two smaller angles $y + z$, so

$$x - (y + z) = 40 \quad \text{or} \quad x - y - z = 40.$$

The measure of the largest angle x is 90° more than the measure of the smallest angle z, so

$$x - z = 90.$$

Thus the required system of equations can be written as follows.

$$x + y + z = 180$$
$$x - y - z = 40$$
$$x \quad\quad - z = 90$$

▶ **REAL-WORLD CONNECTION** In the next example we show how a linear system involving three equations and three variables can be used to model a real-world situation. We solve this system of equations in Example 7.

EXAMPLE 3 Modeling real data with a linear system

The Bureau of Land Management studies antelope populations in Wyoming. It monitors the number of adult antelope, the number of fawns each spring, and the severity of the winter. The first two columns of Table 4.6 contain counts of fawns and adults for three representative winters. The third column shows the severity of each winter. The severity of the winter is measured from 1 to 5, with 1 being mild and 5 being severe.

TABLE 4.6

Fawns (F)	Adults (A)	Winter (W)
405	870	3
414	848	2
272	684	5
?	750	4

We want to use the data in the first three rows of the table to estimate the number of fawns F in the fourth row when the number of adults is 750 and the severity of the winter is 4. To do so, we use the formula

$$F = a + bA + cW,$$

where a, b, and c are constants. Write a system of linear equations whose solution gives appropriate values for a, b, and c.

Solution
From the first row in the table, when $F = 405$, $A = 870$, and $W = 3$, the formula

$$F = a + bA + cW$$

becomes

$$405 = a + b(870) + c(3).$$

Similarly, $F = 414$, $A = 848$, and $W = 2$ gives

$$414 = a + b(848) + c(2),$$

and $F = 272$, $A = 684$, and $W = 5$ yields

$$272 = a + b(684) + c(5).$$

To find values for a, b, and c we can solve the following system of linear equations.

$$405 = a + 870b + 3c$$
$$414 = a + 848b + 2c$$
$$272 = a + 684b + 5c$$

We can also write these equations as a linear system in the following form.

$$a + 870b + 3c = 405$$
$$a + 848b + 2c = 414$$
$$a + 684b + 5c = 272$$

Finding values for a, b, and c will allow us to use the formula $F = a + bA + cW$ to predict the number of fawns F when the number of adults A is 750 and the severity of the winter W is 4 (see Example 7).

NOTE: Linear systems of two equations can have no solutions, one solution, or infinitely many solutions. The same is true for larger linear systems. In the following subsection we focus on linear systems having one solution.

Solving Linear Systems with Substitution and Elimination

When solving systems of linear equations with more than two variables, we usually use both substitution and elimination. However, in the next example we use only substitution to solve a particular type of linear system in three variables.

EXAMPLE 4 Using substitution to solve a linear system of equations

Solve the following system.

$$2x - y + z = 7$$
$$3y - z = 1$$
$$z = 2$$

Solution

The last equation gives us the value of z immediately. We can substitute $z = 2$ in the second equation and determine y.

$3y - z = 1$	Second equation
$3y - 2 = 1$	Substitute $z = 2$.
$3y = 3$	Add 2 to each side.
$y = 1$	Divide each side by 3.

Knowing that $y = 1$ and $z = 2$ allows us to find x by using the first equation.

$2x - y + z = 7$	First equation
$2x - 1 + 2 = 7$	Let $y = 1$ and $z = 2$.
$2x = 6$	Simplify and subtract 1.
$x = 3$	Divide each side by 2.

Thus $x = 3$, $y = 1$, and $z = 2$ and the solution is $(3, 1, 2)$.

In the next example we use elimination and substitution to solve a system of linear equations. This four-step method is summarized by the following.

SOLVING A LINEAR SYSTEM IN THREE VARIABLES

STEP 1: Eliminate one variable, such as x, from two of the equations.

STEP 2: Use the two resulting equations in two variables to eliminate one of the variables, such as y. Solve for the remaining variable, z.

STEP 3: Substitute z in one of the two equations from Step 2. Solve for the unknown variable y.

STEP 4: Substitute values for y and z in one of the given equations and find x. The solution is (x, y, z).

CHAPTER 4 SYSTEMS OF LINEAR EQUATIONS

EXAMPLE 5 Solving a linear system in three variables

Solve the following system.

$$x - y + 2z = 6$$
$$2x + y - 2z = -3$$
$$-x - 2y + 3z = 7$$

Solution

STEP 1: We begin by eliminating the variable x from the second and third equations. To eliminate x from the second equation we multiply the first equation by -2 and then add it to the second equation. To eliminate x from the third equation we add the first and third equations.

$-2x + 2y - 4z = -12$	First equation times -2		$x - y + 2z = 6$	First equation
$2x + y - 2z = -3$	Second equation		$-x - 2y + 3z = 7$	Third equation
$3y - 6z = -15$	Add.		$-3y + 5z = 13$	Add.

STEP 2: Take the two resulting equations from Step 1 and eliminate either variable. Here we add the two equations to eliminate the variable y.

$$3y - 6z = -15$$
$$-3y + 5z = 13$$
$$-z = -2 \quad \text{Add the equations.}$$
$$z = 2 \quad \text{Multiply by } -1.$$

STEP 3: Now we can use substitution to find the values of x and y. We let $z = 2$ in either equation used in Step 2 to find y.

$$3y - 6z = -15$$
$$3y - 6(2) = -15 \quad \text{Substitute } z = 2.$$
$$3y - 12 = -15 \quad \text{Multiply.}$$
$$3y = -3 \quad \text{Add 12.}$$
$$y = -1 \quad \text{Divide by 3.}$$

STEP 4: Finally, we substitute $y = -1$ and $z = 2$ in one of the given equations to find x.

$$x - y + 2z = 6 \quad \text{First given equation}$$
$$x - (-1) + 2(2) = 6 \quad \text{Let } y = -1 \text{ and } z = 2.$$
$$x + 1 + 4 = 6 \quad \text{Simplify.}$$
$$x = 1 \quad \text{Subtract 5.}$$

The solution is $(1, -1, 2)$. Check this solution.

▶ **REAL-WORLD CONNECTION** In the next example we solve a system of linear equations to determine the number of tickets sold at a play.

EXAMPLE 6 Finding the number of tickets sold

One thousand tickets were sold for a play, which generated $3800 in revenue. The prices of the tickets were $3 for children, $4 for students, and $5 for adults. There were 100 fewer student tickets sold than adult tickets. Find the number of each type of ticket sold.

Solution
Let x be the number of tickets sold to children, y be the number of tickets sold to students, and z be the number of tickets sold to adults. The total number of tickets sold was 1000, so

$$x + y + z = 1000.$$

Each child's ticket cost $3, so the revenue generated from selling x tickets was $3x$. Similarly, the revenue generated from students was $4y$, and the revenue from adults was $5z$. Total ticket sales were $3800, so

$$3x + 4y + 5z = 3800.$$

The equation $z - y = 100$, or $y - z = -100$, must also be satisfied, as 100 fewer tickets were sold to students than adults.

To find the price of a ticket we need to solve the following system of linear equations.

$$\begin{aligned} x + y + z &= 1000 \\ 3x + 4y + 5z &= 3800 \\ y - z &= -100 \end{aligned}$$

STEP 1: We begin by eliminating the variable x from the second equation. To do so, we multiply the first equation by -3 and add the second equation.

$$\begin{aligned} -3x - 3y - 3z &= -3000 \quad &\text{First given equation times } -3 \\ \underline{3x + 4y + 5z} &= \underline{3800} \quad &\text{Second equation} \\ y + 2z &= 800 \quad &\text{Add.} \end{aligned}$$

STEP 2: We then multiply the resulting equation from Step 1 by -1 and add the third equation to eliminate y.

$$\begin{aligned} -y - 2z &= -800 \quad &\text{Equation from Step 1 times } -1 \\ \underline{y - z} &= \underline{-100} \quad &\text{Third equation} \\ -3z &= -900 \quad &\text{Add the equations.} \\ z &= 300 \quad &\text{Divide by } -3. \end{aligned}$$

STEP 3: To find y we can substitute $z = 300$ in the third equation.

$$\begin{aligned} y - z &= -100 \quad &\text{Third equation} \\ y - 300 &= -100 \quad &\text{Let } z = 300. \\ y &= 200 \quad &\text{Add 300.} \end{aligned}$$

STEP 4: Finally, we substitute $y = 200$ and $z = 300$ in the first equation.

$$\begin{aligned} x + y + z &= 1000 \quad &\text{First equation} \\ x + 200 + 300 &= 1000 \quad &\text{Let } y = 200 \text{ and } z = 300. \\ x &= 500 \quad &\text{Subtract 500.} \end{aligned}$$

Thus **500** tickets were sold to children, **200** to students, and **300** to adults.

Modeling Data

In the next example we solve the system of equations that we discussed in Example 3.

EXAMPLE 7 Predicting fawns in the spring

Solve the following linear system for a, b, and c.

$$a + 870b + 3c = 405$$
$$a + 848b + 2c = 414$$
$$a + 684b + 5c = 272$$

Then use $F = a + bA + cW$ to predict the number of fawns when there are 750 adults and the severity of the winter is 4.

Solution

STEP 1: We begin by eliminating the variable a from the second and third equations. To do so, we add the second and third equations to the first equation times -1.

$$\begin{array}{ll} -a - 870b - 3c = -405 & \text{First times } -1 \\ \underline{a + 848b + 2c = 414} & \text{Second/third equation} \\ -22b - c = 9 & \text{Add.} \end{array}$$

$$\begin{array}{ll} -a - 870b - 3c = -405 & \text{First times } -1 \\ \underline{a + 684b + 5c = 272} & \text{Second/third equation} \\ -186b + 2c = -133 & \text{Add.} \end{array}$$

STEP 2: We use the two resulting equations from Step 1 to eliminate c. To do so we multiply $-22b - c = 9$ by 2 and add it to the other equation.

$$\begin{array}{ll} -44b - 2c = 18 & (-22b - c = 9) \text{ times 2} \\ \underline{-186b + 2c = -133} & \\ -230b = -115 & \text{Add the equations.} \\ b = 0.5 & \text{Divide by } -230. \end{array}$$

STEP 3: To find c we substitute $b = 0.5$ in either equation used in Step 2.

$$\begin{array}{ll} -44b - 2c = 18 & \\ -44(0.5) - 2c = 18 & \text{Let } b = 0.5. \\ -22 - 2c = 18 & \text{Multiply.} \\ -2c = 40 & \text{Add 22.} \\ c = -20 & \text{Divide by } -2. \end{array}$$

STEP 4: Finally, we substitute $b = 0.5$ and $c = -20$ in one of the given equations to find a.

$$\begin{array}{ll} a + 870b + 3c = 405 & \text{First given equation} \\ a + 870(0.5) + 3(-20) = 405 & \text{Let } b = 0.5 \text{ and } c = -20. \\ a + 435 - 60 = 405 & \text{Multiply.} \\ a = 30 & \text{Solve for } a. \end{array}$$

The solution is $a = 30$, $b = 0.5$, and $c = -20$. Thus we may write

$$F = a + bA + cW$$
$$= 30 + 0.5A - 20W.$$

If there are 750 adults and the winter has a severity of 4, this model predicts

$$F = 30 + 0.5(750) - 20(4)$$
$$= 325 \text{ fawns.}$$

CRITICAL THINKING

Give reasons why the coefficient for A is positive and the coefficient for W is negative in the formula

$$F = 30 + 0.5A - 20W.$$

Systems of Equations with No Solutions

It is possible for a system of three linear equations in three variables to be inconsistent and have no solutions. If we apply substitution and elimination to this type of system, we arrive at a contradiction. This case is demonstrated in the next example.

EXAMPLE 8 Recognizing an inconsistent system

Solve the system, if possible.

$$x + y + z = 4$$
$$-x + y + z = 2$$
$$y + z = 1$$

Solution
STEP 1: If we add the first two equations, we can eliminate x. The variable x is already eliminated from the third equation.

$$x + y + z = 4 \quad \text{First equation}$$
$$-x + y + z = 2 \quad \text{Second equation}$$
$$\overline{2y + 2z = 6} \quad \text{Add.}$$

STEP 2: If we multiply the third *given* equation by -2 and add it to the resulting equation in Step 1, we arrive at a contradiction.

$$-2y - 2z = -2 \quad \text{Third equation times } -2$$
$$2y + 2z = 6 \quad \text{Equation from Step 1}$$
$$\overline{0 = 4} \quad \text{Add equations.}$$

Because $0 = 4$ is a contradiction, there are no solutions to the given system of equations.

Systems of Equations with Infinitely Many Solutions

It is possible for a system of linear equations in three variables to have infinitely many solutions. If we apply substitution and elimination to this type of system, we arrive at an identity. This case is demonstrated in the next example.

EXAMPLE 9 Solving a system with infinitely many solutions

Solve the system.

$$x + y + z = 2$$
$$x - y + z = 4$$
$$3x - y + 3z = 10$$

Solution
STEP 1: To eliminate y from the second equation, add the first equation to the second. To eliminate y from the third equation, add the first equation to the third equation.

$$x + y + z = 2 \quad \text{First equation}$$
$$x - y + z = 4 \quad \text{Second equation}$$
$$\overline{2x + 2z = 6} \quad \text{Add.}$$

$$x + y + z = 2 \quad \text{First equation}$$
$$3x - y + 3z = 10 \quad \text{Third equation}$$
$$\overline{4x + 4z = 12} \quad \text{Add.}$$

STEP 2: If we multiply the first resulting equation in Step 1 by -2 and add it to the second resulting equation in Step 1, we arrive at an identity.

$$\begin{aligned} -4x - 4z &= -12 \quad &&(2x + 2z = 6) \text{ times } -2 \\ 4x + 4z &= 12 \quad &&\text{Second equation from Step 1} \\ \hline 0 &= 0 \quad &&\text{Add. (Identity)} \end{aligned}$$

The variable x can be written in terms of z by solving $2x + 2z = 6$ for x.

$$\begin{aligned} 2x + 2z &= 6 \quad &&\text{Equation from Step 1} \\ 2x &= 6 - 2z \quad &&\text{Subtract } 2z. \\ x &= 3 - z \quad &&\text{Divide by 2.} \end{aligned}$$

STEP 3: To find y in terms of z, substitute $3 - z$ for x in the first *given* equation.

$$\begin{aligned} x + y + z &= 2 \quad &&\text{First equation} \\ (3 - z) + y + z &= 2 \quad &&\text{Let } x = 3 - z. \\ y &= \mathbf{-1} \quad &&\text{Solve for } y. \end{aligned}$$

All solutions have the form $(3 - z, -1, z)$, where z can be any real number. For example, if $z = 1$, then $(2, -1, 1)$ is one of infinitely many solutions to the system of equations.

4.5 PUTTING IT ALL TOGETHER

In this section we discussed how to solve a system of three linear equations in three variables. Systems of linear equations can have no solutions, one solution, or infinitely many solutions. The following table summarizes some of the important concepts presented in this section.

Concept	Explanation
System of Linear Equations in Three Variables	The following is a system of three linear equations in three variables. $$\begin{aligned} x - 2y + z &= 0 \\ -x + y + z &= 4 \\ -y + 4z &= 10 \end{aligned}$$
Solution to a Linear System in Three Variables	The solution to a linear system in three variables is an ordered triple, expressed as (x, y, z). The solution to the preceding system is $(1, 2, 3)$ because substituting $x = 1$, $y = 2$, and $z = 3$ in each equation results in a true statement. We can check solutions this way. $$\begin{aligned} (1) - 2(2) + (3) &= 0 \quad &&\text{True} \\ -(1) + (2) + (3) &= 4 \quad &&\text{True} \\ -(2) + 4(3) &= 10 \quad &&\text{True} \end{aligned}$$
Solving a Linear System with Substitution and Elimination	**STEP 1:** Eliminate one variable, such as x, from two of the equations. **STEP 2:** Use the two resulting equations in two variables to eliminate one of the variables, such as y. Solve for the remaining variable z. **STEP 3:** Substitute z in one of the two equations from Step 2. Solve for the unknown variable y. **STEP 4:** Substitute values for y and z in one of the given equations and find x. The solution is (x, y, z).

4.6 MATRIX SOLUTIONS OF LINEAR SYSTEMS

Representing Systems of Linear Equations with Matrices ■ Gauss–Jordan Elimination ■ Using Technology to Solve Systems of Linear Equations (Optional)

A LOOK INTO MATH ▷ Suppose that the size of a bear's head and its overall length are known. Can its weight be estimated from these variables? Can a bear's weight be estimated if its neck size and chest size are known? In this section we show that systems of linear equations can be used to make such estimates.

In the previous section we solved systems of three linear equations in three variables by using elimination and substitution. In real life, systems of equations often contain thousands of variables. To solve a large system of equations, we need an efficient method. Long before the invention of the computer, Carl Fredrich Gauss (1777–1855) developed a method called *Gaussian elimination* to solve systems of linear equations. Even though it was developed more than 150 years ago, it is still used today in modern computers and calculators. In this section we introduce the Gauss–Jordan method, which is based on Gaussian elimination.

Representing Systems of Linear Equations with Matrices

Arrays of numbers are used frequently in many different real-world situations. Spreadsheets often make use of arrays. A **matrix** is a rectangular array of numbers. Each number in a matrix is called an **element**. The following are examples of *matrices* (plural of matrix), with their dimensions written below them.

$$\begin{bmatrix} 2 & 0 \\ 3 & 1 \end{bmatrix} \quad \begin{bmatrix} -1.2 & 5 & 0 \\ 1 & 0 & 1 \\ 4 & -5 & 7 \end{bmatrix} \quad \begin{bmatrix} 3 & -6 & 0 & \sqrt{3} \\ 1 & 4 & 0 & 9 \\ -3 & 1 & 1 & 18 \\ -10 & -4 & 5 & -1 \end{bmatrix} \quad \begin{bmatrix} 4 & 2 \\ 0 & 1 \\ 1 & 0 \end{bmatrix} \quad \begin{bmatrix} 1 & 5 & -1 \\ 3 & 4 & 2 \end{bmatrix}$$

2×2 3×3 4×4 3×2 2×3

rows × columns

▶ **REAL-WORLD CONNECTION** The dimension of a matrix is stated much like the dimensions of a rectangular room. We might say that a room is m feet long and n feet wide. Similarly, the **dimension of a matrix** is $m \times n$ (m by n), if it has m rows and n columns. For example, the last matrix in the preceding group has a dimension of 2×3 because it has 2 rows and 3 columns. If the number of rows equals the number of columns, the matrix is a **square matrix**. The first three matrices in that group are square matrices.

Matrices can be used to represent a system of linear equations. For example, if we have the system of equations

$$3x - y + 2z = 7$$
$$x - 2y + z = 0$$
$$2x + 5y - 7z = -9,$$

we can represent the system with the following **augmented matrix**. Note how the coefficients of the variables were placed in the matrix. A vertical line is positioned in the matrix where the equals signs occur in the system. The rows and columns are labeled, and the elements of the **main diagonal** of the augmented matrix are circled. The matrix has dimension 3 × 4.

$$\begin{bmatrix} 3 & -1 & 2 & | & 7 \\ 1 & -2 & 1 & | & 0 \\ 2 & 5 & -7 & | & -9 \end{bmatrix} \begin{matrix} \leftarrow \text{Row 1} \\ \leftarrow \text{Row 2} \\ \leftarrow \text{Row 3} \end{matrix}$$

Main diagonal (circled: 3, −2, −7). Columns 1, 2, 3, 4.

EXAMPLE 1 Representing a linear system

Represent each linear system with an augmented matrix. State the dimension of the matrix.

(a) $\begin{aligned} x - 2y &= 9 \\ 6x + 7y &= 16 \end{aligned}$ (b) $\begin{aligned} x - 3y + 7z &= 4 \\ 2x + 5y - z &= 15 \\ 2x + y &= 8 \end{aligned}$

Solution

(a) This system can be represented by the following 2 × 3 augmented matrix.

$$\begin{bmatrix} 1 & -2 & | & 9 \\ 6 & 7 & | & 16 \end{bmatrix}$$

(b) This system can be represented by the following 3 × 4 augmented matrix.

$$\begin{bmatrix} 1 & -3 & 7 & | & 4 \\ 2 & 5 & -1 & | & 15 \\ 2 & 1 & 0 & | & 8 \end{bmatrix}$$

Gauss–Jordan Elimination

A convenient matrix form for representing a system of linear equations is **reduced row–echelon form**. The following matrices are examples of reduced row–echelon form. Note that there are 1s on the main diagonal with 0s above and below the 1s.

$$\begin{bmatrix} 1 & 0 & | & 3 \\ 0 & 1 & | & -2 \end{bmatrix} \quad \begin{bmatrix} 1 & 0 & 0 & | & 3 \\ 0 & 1 & 0 & | & 1 \\ 0 & 0 & 1 & | & -1 \end{bmatrix} \quad \begin{bmatrix} 1 & 0 & 0 & | & 8 \\ 0 & 1 & 0 & | & 2 \\ 0 & 0 & 1 & | & 3 \end{bmatrix}$$

If an augmented matrix representing a linear system is in reduced row–echelon form, we can usually determine the solution easily.

EXAMPLE 2 Determining a solution from reduced row–echelon form

Each matrix represents a system of linear equations. Find the solution.

(a) $\begin{bmatrix} 1 & 0 & 0 & | & 2 \\ 0 & 1 & 0 & | & -3 \\ 0 & 0 & 1 & | & 5 \end{bmatrix}$ (b) $\begin{bmatrix} 1 & 0 & | & 10 \\ 0 & 1 & | & -4 \end{bmatrix}$

Solution

(a) The top row represents $1x + 0y + 0z = 2$ or $x = 2$. The second and third rows tell us that $y = -3$ and $z = 5$. The solution is $(2, -3, 5)$.

(b) The system involves two equations in two variables. The solution is $(10, -4)$.

4.6 MATRIX SOLUTIONS OF LINEAR SYSTEMS

We can use a numerical method called **Gauss–Jordan elimination** to solve a linear system. It makes use of the following matrix row transformations.

MATRIX ROW TRANSFORMATIONS

For any augmented matrix representing a system of linear equations, the following row transformations result in an equivalent system of linear equations.

1. Any two rows may be interchanged.
2. The elements of any row may be multiplied by a nonzero constant.
3. Any row may be changed by adding to (or subtracting from) its elements a nonzero multiple of the corresponding elements of another row.

Gauss–Jordan elimination can be used to transform an augmented matrix into reduced row–echelon form. Its objective is to use these matrix row transformations to obtain a matrix that has the following reduced row–echelon form, where (a, b) represents the solution.

$$\begin{bmatrix} 1 & 0 & | & a \\ 0 & 1 & | & b \end{bmatrix}$$

This method is illustrated in the next example.

EXAMPLE 3 Transforming a matrix into reduced row–echelon form

Use Gauss–Jordan elimination to transform the augmented matrix of the linear system into reduced row–echelon form. Find the solution.

$$x + y = 5$$
$$-x + y = 1$$

Solution
Both the linear system and the augmented matrix are shown.

Linear System   *Augmented Matrix*
$$x + y = 5 \qquad \begin{bmatrix} 1 & 1 & | & 5 \\ -1 & 1 & | & 1 \end{bmatrix}$$
$$-x + y = 1$$

First, we want to obtain a 0 in the second row, where the -1 is highlighted. To do so, we add row 1 to row 2 and place the result in row 2. This step is denoted $R_2 + R_1$ and eliminates the x-variable from the second equation.

$$x + y = 5$$
$$2y = 6 \qquad R_2 + R_1 \rightarrow \begin{bmatrix} 1 & 1 & | & 5 \\ 0 & 2 & | & 6 \end{bmatrix}$$

To obtain a 1 where the 2 in the second row is located, we multiply the second row by $\frac{1}{2}$, denoted $\frac{1}{2}R_2$.

$$x + y = 5$$
$$y = 3 \qquad \frac{1}{2}R_2 \rightarrow \begin{bmatrix} 1 & 1 & | & 5 \\ 0 & 1 & | & 3 \end{bmatrix}$$

Next, we need to obtain a 0 where the 1 is highlighted. We do so by subtracting row 2 from row 1 and placing the result in row 1, denoted $R_1 - R_2$.

$$x = 2 \qquad R_1 - R_2 \rightarrow \begin{bmatrix} 1 & 0 & | & 2 \\ 0 & 1 & | & 3 \end{bmatrix}$$
$$y = 3$$

This matrix is in reduced row–echelon form. The solution is $(2, 3)$.

In the next example, we use Gauss–Jordan elimination to solve a system with three linear equations and three variables. To do so we transform the matrix into the following reduced row–echelon form, where (a, b, c) represents the solution.

$$\begin{bmatrix} 1 & 0 & 0 & | & a \\ 0 & 1 & 0 & | & b \\ 0 & 0 & 1 & | & c \end{bmatrix}$$

EXAMPLE 4 Transforming a matrix to reduced row–echelon form

Use Gauss–Jordan elimination to transform the augmented matrix of the linear system into reduced row–echelon form. Find the solution.

$$\begin{aligned} x + y + 2z &= 1 \\ -x + z &= -2 \\ 2x + y + 5z &= -1 \end{aligned}$$

Solution
The linear system and the augmented matrix are both shown.

Linear System

$$\begin{aligned} x + y + 2z &= 1 \\ -x + z &= -2 \\ 2x + y + 5z &= -1 \end{aligned}$$

Augmented Matrix

$$\begin{bmatrix} 1 & 1 & 2 & | & 1 \\ -1 & 0 & 1 & | & -2 \\ 2 & 1 & 5 & | & -1 \end{bmatrix}$$

First, we want to put 0s in the second and third rows, where the -1 and 2 are highlighted. To obtain a 0 in the first position of the second row we add row 1 to row 2 and place the result in row 2, denoted $R_2 + R_1$. To obtain a 0 in the first position of the third row we subtract 2 times row 1 from row 3 and place the result in row 3, denoted $R_3 - 2R_1$. Row 1 does not change. These steps eliminate the x-variable from the second and third equations.

$$\begin{aligned} x + y + 2z &= 1 \\ y + 3z &= -1 \\ -y + z &= -3 \end{aligned} \qquad \begin{aligned} R_2 + R_1 \to \\ R_3 - 2R_1 \to \end{aligned} \begin{bmatrix} 1 & 1 & 2 & | & 1 \\ 0 & 1 & 3 & | & -1 \\ 0 & -1 & 1 & | & -3 \end{bmatrix}$$

To eliminate the y-variable in row 1, we subtract row 2 from row 1. To eliminate the y-variable from row 3, we add row 2 to row 3.

$$\begin{aligned} x - z &= 2 \\ y + 3z &= -1 \\ 4z &= -4 \end{aligned} \qquad \begin{aligned} R_1 - R_2 \to \\ \\ R_3 + R_2 \to \end{aligned} \begin{bmatrix} 1 & 0 & -1 & | & 2 \\ 0 & 1 & 3 & | & -1 \\ 0 & 0 & 4 & | & -4 \end{bmatrix}$$

To obtain a 1 in row 3, where the highlighted 4 is located, we multiply row 3 by $\frac{1}{4}$.

$$\begin{aligned} x - z &= 2 \\ y + 3z &= -1 \\ z &= -1 \end{aligned} \qquad \frac{1}{4} R_3 \to \begin{bmatrix} 1 & 0 & -1 & | & 2 \\ 0 & 1 & 3 & | & -1 \\ 0 & 0 & 1 & | & -1 \end{bmatrix}$$

For the matrix to be in reduced row–echelon form, we need 0s in the highlighted locations. We first add row 3 to row 1 and then subtract 3 times row 3 from row 2.

$$\begin{aligned} x &= 1 \\ y &= 2 \\ z &= -1 \end{aligned} \qquad \begin{aligned} R_1 + R_3 \to \\ R_2 - 3R_3 \to \end{aligned} \begin{bmatrix} 1 & 0 & 0 & | & 1 \\ 0 & 1 & 0 & | & 2 \\ 0 & 0 & 1 & | & -1 \end{bmatrix}$$

This matrix is now in reduced row–echelon form. The solution is $(1, 2, -1)$.

CRITICAL THINKING

An *inconsistent* system of linear equations has no solutions, and a system of *dependent* linear equations has infinitely many solutions. Suppose that an augmented matrix row reduces to either of the following matrices. Explain what each matrix indicates about the given system of linear equations.

$$\begin{bmatrix} 1 & 0 & 0 & | & 2 \\ 0 & 1 & 0 & | & 3 \\ 0 & 0 & 0 & | & 1 \end{bmatrix} \quad \begin{bmatrix} 1 & 0 & 1 & | & 2 \\ 0 & 1 & 2 & | & 3 \\ 0 & 0 & 0 & | & 0 \end{bmatrix}$$

In the next example we find the amounts invested in three mutual funds.

EXAMPLE 5 Determining investment amounts in mutual funds

A total of $8000 was invested in three funds that grew at a rate of 5%, 10%, and 20% over 1 year. After 1 year, the combined value of the three funds had grown by $1200. Five times as much money was invested at 20% as at 10%. Find the amount invested in each fund.

Solution
Let x be the amount invested at 5%, y be the amount invested at 10%, and z be the amount invested at 20%. The total amount invested was $8000, so

$$x + y + z = 8000.$$

The growth in the first mutual fund, paying 5% of x, is given by $0.05x$. Similarly, the growths in the other mutual funds are given by $0.10y$ and $0.20z$. As the total growth was $1200, we can write

$$0.05x + 0.10y + 0.20z = 1200.$$

Multiplying each side of this equation by 20 to eliminate decimals results in

$$x + 2y + 4z = 24{,}000.$$

Five times as much was invested at 20% as at 10%, so $z = 5y$, or $5y - z = 0$.

These three equations can be written as a system of linear equations and as an augmented matrix.

Linear System

$$\begin{aligned} x + y + z &= 8{,}000 \\ x + 2y + 4z &= 24{,}000 \\ 5y - z &= 0 \end{aligned}$$

Augmented Matrix

$$\begin{bmatrix} 1 & 1 & 1 & | & 8{,}000 \\ 1 & 2 & 4 & | & 24{,}000 \\ 0 & 5 & -1 & | & 0 \end{bmatrix}$$

A 0 can be obtained in the highlighted position by subtracting row 1 from row 2.

$$\begin{aligned} x + y + z &= 8{,}000 \\ y + 3z &= 16{,}000 \\ 5y - z &= 0 \end{aligned} \quad R_2 - R_1 \rightarrow \quad \begin{bmatrix} 1 & 1 & 1 & | & 8{,}000 \\ 0 & 1 & 3 & | & 16{,}000 \\ 0 & 5 & -1 & | & 0 \end{bmatrix}$$

Zeros can be obtained in the highlighted positions by subtracting row 2 from row 1 and by subtracting 5 times row 2 from row 3.

$$\begin{aligned} x \quad\quad - 2z &= -8{,}000 \\ y + 3z &= 16{,}000 \\ -16z &= -80{,}000 \end{aligned} \quad \begin{matrix} R_1 - R_2 \rightarrow \\ \\ R_3 - 5R_2 \rightarrow \end{matrix} \quad \begin{bmatrix} 1 & 0 & -2 & | & -8{,}000 \\ 0 & 1 & 3 & | & 16{,}000 \\ 0 & 0 & -16 & | & -80{,}000 \end{bmatrix}$$

To obtain a 1 in the highlighted position, multiply row 3 by $-\frac{1}{16}$.

$$\begin{array}{rl} x - 2z = & -8{,}000 \\ y + 3z = & 16{,}000 \\ z = & 5{,}000 \end{array} \qquad -\frac{1}{16}R_3 \to \qquad \begin{bmatrix} 1 & 0 & -2 & -8{,}000 \\ 0 & 1 & 3 & 16{,}000 \\ 0 & 0 & 1 & 5{,}000 \end{bmatrix}$$

To obtain a 0 in each of the highlighted positions, add twice row 3 to row 1 and subtract three times row 3 from row 2.

$$\begin{array}{rl} x = & 2{,}000 \\ y = & 1{,}000 \\ z = & 5{,}000 \end{array} \qquad \begin{array}{l} R_1 + 2R_3 \to \\ R_2 - 3R_3 \to \end{array} \qquad \begin{bmatrix} 1 & 0 & 0 & 2{,}000 \\ 0 & 1 & 0 & 1{,}000 \\ 0 & 0 & 1 & 5{,}000 \end{bmatrix}$$

Thus $2000 was invested at 5%, $1000 at 10%, and $5000 at 20%.

Using Technology to Solve Systems of Linear Equations (Optional)

▶ **REAL-WORLD CONNECTION** Examples 4 and 5 involve a lot of arithmetic. Trying to solve a large system of equations by hand is an enormous—if not impossible—task. In the real world, people use technology to solve large systems. In the next example we solve the linear systems from Examples 3 and 4 with a graphing calculator.

EXAMPLE 6 Using technology

Use a graphing calculator to solve the following systems of equations.

(a) $\begin{array}{l} x + y = 5 \\ -x + y = 1 \end{array}$ (b) $\begin{array}{l} x + y + 2z = 1 \\ -x + z = -2 \\ 2x + y + 5z = -1 \end{array}$

Solution

(a) Enter the 2 × 3 augmented matrix from Example 3 in a graphing calculator, as shown in Figure 4.30(a). Then transform the matrix into reduced row–echelon form (rref), as shown in Figure 4.30(b). The solution is (2, 3).

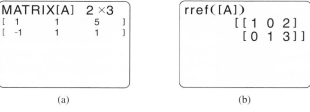

(a) (b)

Figure 4.30

(b) Enter the 3 × 4 augmented matrix from Example 4, as shown in Figure 4.31(a). (The fourth column of A can be seen by scrolling right.) Then transform the matrix into reduced row–echelon form (rref), as shown in Figure 4.31(b). The solution is $(1, 2, -1)$.

CALCULATOR HELP

To enter a matrix and put it in reduced row–echelon form, see the Appendix (pages AP-8 and AP-9).

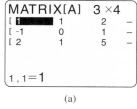

(a) (b)

Figure 4.31

▶ **REAL-WORLD CONNECTION** In the next example we use technology to solve an application.

EXAMPLE 7 Modeling the weight of male bears

The data shown in Table 4.7 give the weight W, head length H, and overall length L of three bears. These data can be modeled with the equation $W = a + bH + cL$, where a, b, and c are constants that we need to determine. (*Sources:* M. Triola, *Elementary Statistics*; Minitab, Inc.)

TABLE 4.7

W (pound)	H (inches)	L (inches)
362	16	72
300	14	68
147	11	52

(a) Set up a system of equations whose solution gives values for constants a, b, and c.
(b) Solve the system.
(c) Predict the weight of a bear with $H = 13$ inches and $L = 65$ inches.

Solution

(a) Substitute each row of Table 4.7 in the equation $W = a + bH + cL$.

$$362 = a + b(16) + c(72)$$
$$300 = a + b(14) + c(68)$$
$$147 = a + b(11) + c(52)$$

Rewrite this system as

$$a + 16b + 72c = 362$$
$$a + 14b + 68c = 300$$
$$a + 11b + 52c = 147$$

and represent it as the augmented matrix

$$A = \begin{bmatrix} 1 & 16 & 72 & | & 362 \\ 1 & 14 & 68 & | & 300 \\ 1 & 11 & 52 & | & 147 \end{bmatrix}.$$

(a)

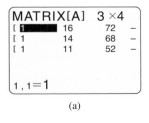

(b)

Figure 4.32

(b) Enter A and put it in reduced row–echelon form, as shown in Figures 4.32(a) and (b), respectively. The solution is $a = -374$, $b = 19$, and $c = 6$.

(c) For $W = a + bH + cL$, use

$$W = -374 + 19H + 6L$$

to predict the weight of a bear with head length $H = 13$ and overall length $L = 65$.

$$W = -374 + 19(13) + 6(65) = 263 \text{ pounds}$$

4.6 PUTTING IT ALL TOGETHER

A matrix is a rectangular array of numbers. An augmented matrix may be used to represent any system of linear equations. One common method for solving a system of linear equations is Gauss–Jordan elimination. Matrix row operations may be used to transform an augmented matrix to reduced row–echelon form. Technology can be used to solve systems of linear equations efficiently. The following table summarizes augmented matrices and reduced row–echelon form.

Concept	Explanation
Augmented Matrix	A linear system can be represented by an augmented matrix. The following matrix has dimension 3×4. **Linear System** $$\begin{aligned} x + 2y - z &= 6 \\ -2x + y - z &= 7 \\ 2x + 3z &= -11 \end{aligned}$$ **Augmented Matrix** $$\begin{bmatrix} 1 & 2 & -1 & 6 \\ -2 & 1 & -1 & 7 \\ 2 & 0 & 3 & -11 \end{bmatrix}$$
Reduced Row–Echelon Form	The following augmented matrix is in reduced row–echelon form, which results from transforming the preceding system to reduced row–echelon form. There are 1s along the main diagonal and 0s elsewhere in the first three columns. The solution to the linear system is $(-1, 2, -3)$. $$\begin{bmatrix} 1 & 0 & 0 & -1 \\ 0 & 1 & 0 & 2 \\ 0 & 0 & 1 & -3 \end{bmatrix}$$

4.7 DETERMINANTS

Calculation of Determinants ■ Area of Regions ■ Cramer's Rule

A LOOK INTO MATH ▷ Surveyors commonly calculate the areas of parcels of land. To do so, they frequently divide the land into triangular regions. When the coordinates of the vertices of a triangle are known, determinants may be used to find the area of the triangle. *A determinant is a real number that can be calculated for any square matrix.* In this section we use determinants to find areas and to solve systems of linear equations.

Calculation of Determinants

▶ **REAL-WORLD CONNECTION** The concept of determinants originated with the Japanese mathematician Seki Kowa (1642–1708), who used them to solve systems of linear equations. Later, Gottfried Leibniz (1646–1716) formally described determinants and also used them to solve systems of linear equations. (**Source:** *Historical Topics for the Mathematical Classroom*, NCTM.)

We begin by defining a determinant of a 2×2 matrix.

> **DETERMINANT OF A 2 × 2 MATRIX**
>
> The **determinant** of
> $$A = \begin{bmatrix} a & b \\ c & d \end{bmatrix}$$
> is a *real number* defined by
> $$\det A = ad - cb.$$

4.7 DETERMINANTS

EXAMPLE 1 Calculating determinants

Find det A for each 2×2 matrix.

(a) $A = \begin{bmatrix} 1 & 2 \\ 3 & 4 \end{bmatrix}$ (b) $A = \begin{bmatrix} -1 & -3 \\ 2 & -8 \end{bmatrix}$

Solution

(a) The determinant is calculated as follows.
$$\det A = \det \begin{bmatrix} 1 & 2 \\ 3 & 4 \end{bmatrix} = (1)(4) - (3)(2) = -2$$

(b) Similarly,
$$\det A = \det \begin{bmatrix} -1 & -3 \\ 2 & -8 \end{bmatrix} = (-1)(-8) - (2)(-3) = 14.$$

We can use determinants of 2×2 matrices to find determinants of 3×3 matrices. This method is called **expansion of a determinant by minors**.

DETERMINANT OF A 3 × 3 MATRIX

$$\det A = \det \begin{bmatrix} a_1 & b_1 & c_1 \\ a_2 & b_2 & c_2 \\ a_3 & b_3 & c_3 \end{bmatrix}$$

$$= a_1 \cdot \det \begin{bmatrix} b_2 & c_2 \\ b_3 & c_3 \end{bmatrix} - a_2 \cdot \det \begin{bmatrix} b_1 & c_1 \\ b_3 & c_3 \end{bmatrix} + a_3 \cdot \det \begin{bmatrix} b_1 & c_1 \\ b_2 & c_2 \end{bmatrix}$$

The 2×2 matrices in this equation are called **minors**.

EXAMPLE 2 Calculating 3×3 determinants

Evaluate det A.

(a) $A = \begin{bmatrix} 2 & 1 & -1 \\ -1 & 3 & 2 \\ 4 & -3 & -5 \end{bmatrix}$ (b) $A = \begin{bmatrix} 5 & -2 & 4 \\ 0 & 2 & 1 \\ -1 & 4 & -4 \end{bmatrix}$

Solution

(a) We evaluate the determinant as follows.
$$\det \begin{bmatrix} 2 & 1 & -1 \\ -1 & 3 & 2 \\ 4 & -3 & -5 \end{bmatrix} = 2 \cdot \det \begin{bmatrix} 3 & 2 \\ -3 & -5 \end{bmatrix} - (-1) \cdot \det \begin{bmatrix} 1 & -1 \\ -3 & -5 \end{bmatrix}$$
$$+ 4 \cdot \det \begin{bmatrix} 1 & -1 \\ 3 & 2 \end{bmatrix}$$
$$= 2(-9) + 1(-8) + 4(5)$$
$$= -6$$

(b) We evaluate the determinant as follows.
$$\det \begin{bmatrix} 5 & -2 & 4 \\ 0 & 2 & 1 \\ -1 & 4 & -4 \end{bmatrix} = 5 \cdot \det \begin{bmatrix} 2 & 1 \\ 4 & -4 \end{bmatrix} - (0) \cdot \det \begin{bmatrix} -2 & 4 \\ 4 & -4 \end{bmatrix}$$
$$+ (-1) \cdot \det \begin{bmatrix} -2 & 4 \\ 2 & 1 \end{bmatrix}$$
$$= 5(-12) - 0(-8) + (-1)(-10)$$
$$= -50$$

Many graphing calculators can evaluate the determinant of a matrix, as illustrated in the next example, where we evaluate the determinants from Example 2.

EXAMPLE 3   Using technology to find determinants

Find each determinant of A, using a graphing calculator.

(a) $A = \begin{bmatrix} 2 & 1 & -1 \\ -1 & 3 & 2 \\ 4 & -3 & -5 \end{bmatrix}$ (b) $A = \begin{bmatrix} 5 & -2 & 4 \\ 0 & 2 & 1 \\ -1 & 4 & -4 \end{bmatrix}$

Solution
(a) Begin by entering the matrix and then evaluate the determinant, as shown in Figure 4.33. The result is $\det A = -6$, which agrees with our earlier calculation.

CALCULATOR HELP
To find a determinant, see the Appendix (page AP-9).

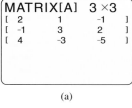
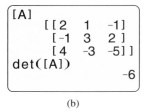

(a)   (b)

Figure 4.33

(b) The determinant of A evaluates to -50 (see Figure 4.34).

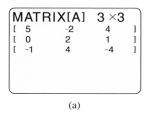

(a)   (b)

Figure 4.34

Area of Regions

▶ REAL-WORLD CONNECTION   A determinant may be used to find the area of a triangle. For example, if a triangle has vertices (a_1, a_2), (b_1, b_2), and (c_1, c_2), its area equals the absolute value of D, where

$$D = \frac{1}{2} \det \begin{bmatrix} a_1 & b_1 & c_1 \\ a_2 & b_2 & c_2 \\ 1 & 1 & 1 \end{bmatrix}.$$

If the vertices are entered in the columns of D counterclockwise as they appear in the xy-plane, D will be positive.   (*Source:* W. Taylor, *The Geometry of Computer Graphics.*)

EXAMPLE 4 Computing the area of a triangular parcel of land

A triangular parcel of land is shown in Figure 4.35. If all units are miles, find the area of the parcel of land by using a determinant.

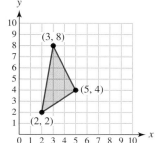

Figure 4.35

Solution
The vertices of the triangular parcel of land are $(2, 2)$, $(5, 4)$, and $(3, 8)$. The area of the triangle is

$$D = \frac{1}{2}\det\begin{bmatrix} 2 & 5 & 3 \\ 2 & 4 & 8 \\ 1 & 1 & 1 \end{bmatrix} = \frac{1}{2} \cdot 16 = 8.$$

The area of the triangle is 8 square miles.

CRITICAL THINKING
Suppose that you are given three distinct points in the xy-plane and $D = 0$. What must be true about the three points?   The points are collinear; they all lie on the same line.

Cramer's Rule

▶ **REAL-WORLD CONNECTION** Determinants were developed independently by Gabriel Cramer (1704–1752). His work, published in 1750, provided a method called **Cramer's rule** for solving systems of linear equations.

CRAMER'S RULE FOR LINEAR SYSTEMS IN TWO VARIABLES

The solution to the system of linear equations

$$a_1 x + b_1 y = c_1$$
$$a_2 x + b_2 y = c_2$$

is given by $x = \frac{E}{D}$ and $y = \frac{F}{D}$, where

$$E = \det\begin{bmatrix} c_1 & b_1 \\ c_2 & b_2 \end{bmatrix}, \quad F = \det\begin{bmatrix} a_1 & c_1 \\ a_2 & c_2 \end{bmatrix}, \quad \text{and} \quad D = \det\begin{bmatrix} a_1 & b_1 \\ a_2 & b_2 \end{bmatrix} \neq 0.$$

NOTE: If $D = 0$, the system has either no solutions or infinitely many solutions.

EXAMPLE 5 Using Cramer's rule

Use Cramer's rule to solve the following linear systems.
(a) $3x - 4y = 18$
    $7x + 5y = -1$
(b) $-4x + 9y = -24$
    $6x + 17y = -25$

Solution
(a) $E = \det\begin{bmatrix} c_1 & b_1 \\ c_2 & b_2 \end{bmatrix} = \det\begin{bmatrix} 18 & -4 \\ -1 & 5 \end{bmatrix} = (18)(5) - (-1)(-4) = \mathbf{86}$

$F = \det\begin{bmatrix} a_1 & c_1 \\ a_2 & c_2 \end{bmatrix} = \det\begin{bmatrix} 3 & 18 \\ 7 & -1 \end{bmatrix} = (3)(-1) - (7)(18) = \mathbf{-129}$

$D = \det\begin{bmatrix} a_1 & b_1 \\ a_2 & b_2 \end{bmatrix} = \det\begin{bmatrix} 3 & -4 \\ 7 & 5 \end{bmatrix} = (3)(5) - (7)(-4) = \mathbf{43}$

Because $x = \frac{E}{D} = \frac{86}{43} = 2$ and $y = \frac{F}{D} = \frac{-129}{43} = -3$, the solution is $(2, -3)$.

(b) $E = \det\begin{bmatrix} c_1 & b_1 \\ c_2 & b_2 \end{bmatrix} = \det\begin{bmatrix} -24 & 9 \\ -25 & 17 \end{bmatrix} = (-24)(17) - (-25)(9) = -183$

$F = \det\begin{bmatrix} a_1 & c_1 \\ a_2 & c_2 \end{bmatrix} = \det\begin{bmatrix} -4 & -24 \\ 6 & -25 \end{bmatrix} = (-4)(-25) - (6)(-24) = 244$

$D = \det\begin{bmatrix} a_1 & b_1 \\ a_2 & b_2 \end{bmatrix} = \det\begin{bmatrix} -4 & 9 \\ 6 & 17 \end{bmatrix} = (-4)(17) - (6)(9) = -122$

Because $x = \frac{E}{D} = \frac{-183}{-122} = 1.5$ and $y = \frac{F}{D} = \frac{244}{-122} = -2$, the solution is $(1.5, -2)$.

Cramer's rule can be applied to systems that have any number of linear equations.

▶ **REAL-WORLD CONNECTION** In applications, equations with hundreds of variables are routinely solved. Such systems of equations could be solved with Cramer's rule. However, using Cramer's rule and the expansion of a determinant with minors to solve a linear system of equations with n variables requires at least

$$1 \cdot 2 \cdot 3 \cdot 4 \cdot \cdots \cdot n \cdot (n + 1)$$

multiplication operations. To solve a linear system involving only 25 variables would require about

$$1 \cdot 2 \cdot 3 \cdot 4 \cdot \cdots \cdot 25 \cdot 26 \approx 4 \times 10^{26}$$

multiplication operations. Supercomputers can perform about 1 trillion (1×10^{12}) multiplication operations per second. This would take about

$$\frac{4 \times 10^{26}}{1 \times 10^{12}} = 4 \times 10^{14} \text{ seconds.}$$

With $60 \times 60 \times 24 \times 365 = 31{,}536{,}000$ seconds in a year, 4×10^{14} seconds equals

$$\frac{4 \times 10^{14}}{31{,}536{,}000} \approx 12{,}700{,}000 \text{ years!}$$

Modern software packages do *not* use Cramer's rule for three or more variables.

4.7 PUTTING IT ALL TOGETHER

The determinant of a square matrix A is a real number, denoted det A. Cramer's rule is a method that uses determinants to solve systems of linear equations. The following table summarizes important topics from this section.

Concept	Explanation
Determinant of a 2×2 Matrix	The determinant of a 2×2 matrix A is given by $$\det A = \det \begin{bmatrix} a & b \\ c & d \end{bmatrix} = ad - cb.$$
Determinant of a 3×3 Matrix	The determinant of a 3×3 matrix A is given by $$\det A = \det \begin{bmatrix} a_1 & b_1 & c_1 \\ a_2 & b_2 & c_2 \\ a_3 & b_3 & c_3 \end{bmatrix}$$ $$= a_1 \cdot \det \begin{bmatrix} b_2 & c_2 \\ b_3 & c_3 \end{bmatrix} - a_2 \cdot \det \begin{bmatrix} b_1 & c_1 \\ b_3 & c_3 \end{bmatrix}$$ $$+ a_3 \cdot \det \begin{bmatrix} b_1 & c_1 \\ b_2 & c_2 \end{bmatrix}$$
Area of a Triangle	If a triangle has vertices (a_1, a_2), (b_1, b_2), and (c_1, c_2), its area equals the absolute value of D, where $$D = \frac{1}{2} \det \begin{bmatrix} a_1 & b_1 & c_1 \\ a_2 & b_2 & c_2 \\ 1 & 1 & 1 \end{bmatrix}.$$
Cramer's Rule for Linear Systems in Two Variables	The solution to the linear system $$a_1 x + b_1 y = c_1$$ $$a_2 x + b_2 y = c_2$$ is given by $x = \frac{E}{D}$ and $y = \frac{F}{D}$, where $$E = \det \begin{bmatrix} c_1 & b_1 \\ c_2 & b_2 \end{bmatrix}, \quad F = \det \begin{bmatrix} a_1 & c_1 \\ a_2 & c_2 \end{bmatrix}, \text{ and}$$ $$D = \det \begin{bmatrix} a_1 & b_1 \\ a_2 & b_2 \end{bmatrix} \neq 0.$$ **NOTE:** If $D = 0$, then the system has either no solutions or infinitely many solutions.

CHAPTER 4 SUMMARY

SECTION 4.1 ■ SYSTEMS OF LINEAR EQUATIONS IN TWO VARIABLES

Systems of Linear Equations in Two Variables

$$ax + by = c$$
$$dx + ey = k$$

Systems of linear equations in two variables can have no solutions, one solution, or infinitely many solutions. A solution is an ordered pair (x, y).

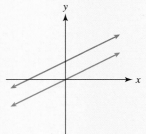 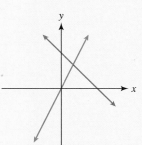 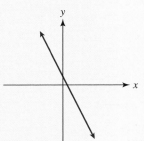

Inconsistent System; (No Solutions) | Consistent System; Independent Equations | Consistent System; Dependent Equations

Examples:
$x + y = 4$ $x + y = 4$ $x + y = 1$
$x + y = 2$ $x - y = 2$ $2x + 2y = 2$

No solutions   The solution is (3, 1). $\{(x, y) \mid x + y = 1\}$

Graphical and Numerical Solutions   These methods use graphs and tables to solve systems.

Example: $2x + y = 4$
           $x + y = 3$

Graph and make a table for $y_1 = 4 - 2x$ and $y_2 = 3 - x$. The solution is (1, 2).

Graphical Solution

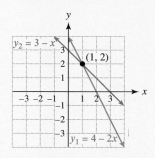

Numerical Solution

x	$4 - 2x$	$3 - x$
0	4	3
1	2	2
2	0	1
3	-2	0

SECTION 4.2 ■ THE SUBSTITUTION AND ELIMINATION METHODS

Two symbolic methods for solving systems of linear equations are *substitution* and *elimination*.

Substitution

$x + y = 5$ or $y = 5 - x$
$x - y = -3$

Substituting in the second equation, $x - (5 - x) = -3$, yields $x = 1$ and $y = 4$. Solution is (1, 4).

Elimination

$x + y = 5$
$x - y = -3$
$\overline{}$
$2x + 0y = 2$ Add the equations.

Thus $x = 1$ and $y = 4$. Solution is (1, 4).

CHAPTER 4 SUMMARY

SECTION 4.3 ■ SYSTEMS OF LINEAR INEQUALITIES

Linear Inequalities in Two Variables

$$ax + by > c,$$

where $>$ can be replaced by $<$, $\leq$, or $\geq$. The solution set is typically a region in the xy-plane.

Example: $x + y \leq 4$

Any point in the shaded region must satisfy the given inequality. The *test point*, $(0, 0)$, lies in the shaded region and satisfies the inequality $x + y \leq 4$ because $0 + 0 \leq 4$.

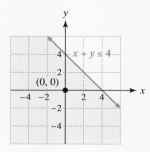

Systems of Linear Inequalities in Two Variables

$$ax + by > c$$
$$dx + ey > k,$$

where $>$ can be replaced by $<$, $\leq$, or $\geq$.

Example: $x + y \leq 2$
$\phantom{\text{Example:}}\ y \geq x$

The test point $(-2, 1)$ satisfies both inequalities so shade the region containing $(-2, 1)$.

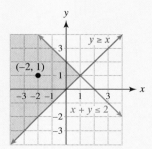

SECTION 4.4 ■ INTRODUCTION TO LINEAR PROGRAMMING

Linear Programming Problem A linear programming problem consists of an *objective function* to be minimized or maximized (optimized) and a system of linear inequalities called *constraints*. The solution set to the constraints is called the *region of feasible solutions*.

Fundamental Theorem of Linear Programming If the optimal value for a linear programming problem exists, then it occurs at a *vertex* of the region of feasible solutions.

Example: The maximum of $R = 2x + 3y$ must occur at one of the vertices $(0, 0)$, $(0, 4.5)$, $(3, 3)$, and $(4.5, 0)$.

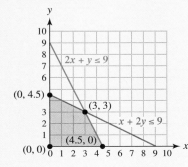

Vertex	$R = 2x + 3y$
$(0, 0)$	$2(0) + 3(0) = 0$
$(0, 4.5)$	$2(0) + 3(4.5) = 13.5$
$(3, 3)$	$2(3) + 3(3) = 15$
$(4.5, 0)$	$2(4.5) + 3(0) = 9$

Maximum of $R = 15$ occurs at vertex $(3, 3)$.

SECTION 4.5 ■ SYSTEMS OF LINEAR EQUATIONS IN THREE VARIABLES

Solution to a System of Linear Equations in Three Variables An ordered triple (x, y, z) that satisfies *every* equation

Example:
$x - y + 2z = 3$
$2x - y + z = 5$
$x + y + z = 6$

The solution is $(3, 2, 1)$ because these values for (x, y, z) satisfy all three equations.

$3 - 2 + 2(1) = 3$ True
$2(3) - 2 + 1 = 5$ True
$3 + 2 + 1 = 6$ True

Elimination and Substitution Systems of linear equations in *three* variables can be solved by elimination and substitution, using the following steps.

STEP 1: Eliminate one variable, such as x, from two of the given equations.

STEP 2: Use the two resulting equations in two variables to eliminate one of the variables, such as y. Solve for the remaining variable z.

STEP 3: Substitute z in one of the two equations from Step 2. Solve for the unknown variable y.

STEP 4: Substitute values for y and z in one of the given equations. Then find x. The solution is (x, y, z).

SECTION 4.6 ■ MATRIX SOLUTIONS OF LINEAR SYSTEMS

Matrix A rectangular array of numbers is a matrix. If a matrix has m rows and n columns, it has dimension $m \times n$.

Example: Matrix $A = \begin{bmatrix} 3 & -1 & 7 \\ 0 & 6 & -2 \end{bmatrix}$ has dimension 2×3.

Augmented Matrix Any linear system can be represented with an augmented matrix.

Linear System
$4x - 3y = 5$
$x + 2y = 4$

Augmented Matrix
$\begin{bmatrix} 4 & -3 & | & 5 \\ 1 & 2 & | & 4 \end{bmatrix}$

Gauss–Jordan Elimination A numerical method that uses matrix row transformations to tranform a matrix into reduced row–echelon form

Example: The matrix $\begin{bmatrix} 4 & -3 & | & 5 \\ 1 & 2 & | & 4 \end{bmatrix}$ reduces to $\begin{bmatrix} 1 & 0 & | & 2 \\ 0 & 1 & | & 1 \end{bmatrix}$.

The solution to the system is $(2, 1)$.

SECTION 4.7 ■ DETERMINANTS

Determinant for a 2 × 2 Matrix A determinant is a *real number*. The determinant of a 2×2 matrix is

$$\det A = \det \begin{bmatrix} a & b \\ c & d \end{bmatrix} = ad - cb.$$

Example: $\det \begin{bmatrix} 2 & 3 \\ 4 & 5 \end{bmatrix} = (2)(5) - (4)(3) = -2$

Determinant for a 3 × 3 Matrix

$$\det A = \det \begin{bmatrix} a_1 & b_1 & c_1 \\ a_2 & b_2 & c_2 \\ a_3 & b_3 & c_3 \end{bmatrix}$$

$$= a_1 \cdot \det \begin{bmatrix} b_2 & c_2 \\ b_3 & c_3 \end{bmatrix} - a_2 \cdot \det \begin{bmatrix} b_1 & c_1 \\ b_3 & c_3 \end{bmatrix} + a_3 \cdot \det \begin{bmatrix} b_1 & c_1 \\ b_2 & c_2 \end{bmatrix}$$

Example: $\det \begin{bmatrix} 2 & 3 & 2 \\ 3 & 7 & -3 \\ 0 & 0 & -1 \end{bmatrix} = 2 \det \begin{bmatrix} 7 & -3 \\ 0 & -1 \end{bmatrix} - 3 \det \begin{bmatrix} 3 & 2 \\ 0 & -1 \end{bmatrix} + 0$

$$= 2(-7) - 3(-3) = -5$$

Cramer's rule uses determinants to solve linear systems of equations. Determinants can also be used to find areas of triangles. See Putting It All Together for Section 4.7.

194 CHAPTER 4 SYSTEMS OF LINEAR EQUATIONS

Assignment Name _____ Name _____ Date _____
Show all work for these items: _____

# _____	# _____
# _____	# _____
# _____	# _____
# _____	# _____

Assignment Name _____ Name _____ Date _____
Show all work for these items: _____

# _____	# _____
# _____	# _____
# _____	# _____
# _____	# _____

CHAPTER 5

Polynomial Expressions and Functions

5.1 Polynomial Functions
5.2 Multiplication of Polynomials
5.3 Factoring Polynomials
5.4 Factoring Trinomials
5.5 Special Types of Factoring
5.6 Summary of Factoring
5.7 Polynomial Equations

Video games have been available to consumers for many years. However, most of these older games lacked the graphic capability to make the action appear real on a screen. During the past three decades the computing power of processors has *doubled every 2 years*. Now video games systems, such as the PlayStation 3, Wii, and Xbox 360, are taking advantage of the new generation of powerful processors to make action scenes look incredibly realistic. Thousands of objects can be tracked and manipulated simultaneously, making the experience more lifelike to the user.

Video games are mathematically intensive because separate equations must be formed and solved in real time to determine the position of each person and object on the screen. Whether it is an athlete, a football, or an exploding object, *mathematics is essential* to making the movements look real.

Because flying objects do not follow straight-line paths, we cannot use linear functions and equations to model their paths. Instead we need functions whose graphs are curves. Polynomial functions are frequently used in modeling *nonlinear motion* of objects and are excellent at describing flying objects. In this chapter you will begin to learn about polynomials and how they can be used to model the motion of objects like golf balls in flight.

The struggle is what teaches us.

—SUE GRAFTON

Source: Microsoft.

5.1 POLYNOMIAL FUNCTIONS

Monomials and Polynomials ■ Addition and Subtraction of Polynomials ■ Polynomial Functions ■ Evaluating Polynomials ■ Operations on Functions ■ Applications and Models

A LOOK INTO MATH ▷

Many quantities in applications cannot be modeled with linear functions and equations. If data points do not lie on a line, we say that the data are *nonlinear*. For example, a scatterplot of the *cumulative* number of AIDS deaths from 1981 through 2004 is shown in Figure 5.1. Monomials and polynomials are often used to model nonlinear data such as these. (*Source:* U.S. Department of Health and Human Services.)

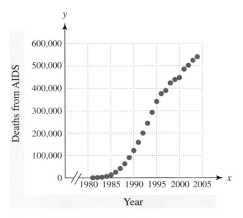

Figure 5.1 U.S. AIDS Deaths

Monomials and Polynomials

A **term** is a number, a variable, or a *product* of numbers and variables raised to powers. Examples of terms include

$$-15, \quad y, \quad x^4, \quad 3x^3z, \quad x^{-1/2}y^{-2}, \quad \text{and} \quad 6x^{-1}y^3.$$

If the variables in a term have only *nonnegative integer* exponents, the term is called a **monomial**. Examples of monomials include

$$-4, \quad 5y, \quad x^2, \quad 5x^2z^6, \quad -xy^7, \quad \text{and} \quad 6xy^3.$$

NOTE: Although a monomial can have a negative sign, it cannot contain any addition or subtraction signs. Also, a monomial cannot have division by a variable.

EXAMPLE 1 Identifying monomials

Determine whether the expression is a monomial.

(a) $-8x^3y^5$ **(b)** xy^{-1} **(c)** $9 + x^2$ **(d)** $\dfrac{2}{x}$

Solution
(a) The expression $-8x^3y^5$ represents a monomial because it is a product of the number -8 and the variables x and y, which have the nonnegative integer exponents 3 and 5.
(b) The expression xy^{-1} is not a monomial because y has a negative exponent.
(c) The expression $9 + x^2$ is not a monomial; it is the sum of two monomials, 9 and x^2.
(d) This expression is not a monomial because it involves division by a variable. Note also that $\dfrac{2}{x} = 2x^{-1}$, which has a negative exponent.

Monomials occur in applications involving geometry, as illustrated in the next example.

EXAMPLE 2 Writing monomials

Write a monomial that represents the volume of a cube with sides of length x.

Solution
The volume of a rectangular box equals the product of its length, width, and height. For a cube with sides of length x, its volume is given by $x \cdot x \cdot x = x^3$. See Figure 5.2.

Figure 5.2 Volume: x^3

The **degree of a monomial** equals the sum of the exponents of the variables. A constant term has degree 0, unless the term is 0 (which has an undefined degree). The numeric constant in a monomial is called its **coefficient**. Table 5.1 shows the degree and coefficient of several monomials.

CRITICAL THINKING

Write a monomial that represents the volume of four identical cubes with sides of length y.

TABLE 5.1

Monomial	64	$4x^2y^3$	$-5x^2$	xy^3
Degree	0	5	2	4
Coefficient	64	4	-5	1

A **polynomial** is either a monomial or a sum of monomials. Examples include

$$5x^4z^2, \quad 9x^4 - 5, \quad 4x^2 + 5xy - y^2, \quad \text{and} \quad 4 - y^2 + 5y^4 + y^5.$$
$\quad$ 1 term $\quad\quad$ 2 terms $\quad\quad\quad$ 3 terms $\quad\quad\quad\quad\quad\quad$ 4 terms

Polynomials containing one variable are called **polynomials of one variable**. The second and fourth polynomials shown are examples of polynomials of one variable. The **leading coefficient** of a polynomial of one variable is the coefficient of the monomial with highest degree. The **degree of a polynomial** equals the degree of the monomial with highest degree. Table 5.2 shows several polynomials of one variable along with their degrees and leading coefficients. A polynomial of degree 1 is a **linear polynomial**, a polynomial of degree 2 is a **quadratic polynomial**, and a polynomial of degree 3 is a **cubic polynomial**.

TABLE 5.2

Polynomial	Degree	Leading Coefficient	Type
-98	0	-98	Constant
$2x - 7$	1	2	Linear
$-5z + 9z^2 + 7$	2	9	Quadratic
$-2x^3 + 4x^2 + x - 1$	3	-2	Cubic
$7 - x + 4x^2 + x^5$	5	1	Fifth degree

Addition and Subtraction of Polynomials

Suppose that we have 3 rectangles of the same dimension having length x and width y, as shown in Figure 5.3. The total area is given by

$$xy + xy + xy.$$

This area is equivalent to 3 times xy, which can be expressed as $3xy$. In symbols we write

$$xy + xy + xy = 3xy.$$

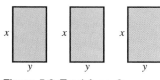

Figure 5.3 Total Area: $3xy$

We can add these three terms because they are *like terms*. If two terms contain the *same variables raised to the same powers*, we call them **like terms**. We can add or subtract *like* terms but not *unlike* terms. For example, if one cube has sides of length x and another cube has sides of length y, their respective volumes are x^3 and y^3. The total volume equals

$$x^3 + y^3,$$

but we cannot combine these terms into one term because they are unlike terms. However, the distributive property can be used to simplify the expression $2x^3 + 3x^3$ as

$$2x^3 + 3x^3 = (2 + 3)x^3 = 5x^3$$

because $2x^3$ and $3x^3$ are like terms. Table 5.3 lists examples of like terms and unlike terms.

TABLE 5.3

Like terms	$-x^4, 2x^4$	$-2x^2y, 4x^2y$	$a^2b^4, \frac{1}{2}a^2b^4$
Unlike terms	$-x^3, 2x^4$	$-2xy^2, 4x^2y$	$ab^5, \frac{1}{2}a^2b^4$

To add or subtract monomials we simply apply the distributive property.

EXAMPLE 3 Adding and subtracting monomials

Simplify each expression by combining like terms.
(a) $8x^2 - 4x^2 + x^3$ (b) $9x - 6xy^2 + 2xy^2 + 4x$ (c) $5ab^2 - 4a^2 - ab^2 + a^2$

Solution
(a) The terms $8x^2$ and $-4x^2$ are like terms, so they can be combined.

$$8x^2 - 4x^2 + x^3 = (8 - 4)x^2 + x^3 \quad \text{Combine like terms.}$$
$$= 4x^2 + x^3 \quad \text{Simplify.}$$

However, $4x^2$ and x^3 are unlike terms and cannot be combined.

(b) The terms $9x$ and $4x$ can be combined, as can $-6xy^2$ and $2xy^2$.

$$9x - 6xy^2 + 2xy^2 + 4x = 9x + 4x - 6xy^2 + 2xy^2 \quad \text{Commutative property}$$
$$= (9 + 4)x + (-6 + 2)xy^2 \quad \text{Combine like terms.}$$
$$= 13x - 4xy^2 \quad \text{Simplify.}$$

(c) The terms $5ab^2$ and $-ab^2$ are like terms, as are $-4a^2$ and a^2, so they can be combined.

$$5ab^2 - 4a^2 - ab^2 + a^2 = 5ab^2 - 1ab^2 - 4a^2 + 1a^2 \quad \text{Commutative property}$$
$$= (5 - 1)ab^2 + (-4 + 1)a^2 \quad \text{Combine like terms.}$$
$$= 4ab^2 - 3a^2 \quad \text{Simplify.}$$

To add two polynomials we combine like terms, as in the next example.

EXAMPLE 4 Adding polynomials

Simplify each expression.
(a) $(2x^2 - 3x + 7) + (3x^2 + 4x - 2)$ (b) $(z^3 + 4z + 8) + (4z^2 - z + 6)$

Solution
(a) $(2x^2 - 3x + 7) + (3x^2 + 4x - 2) = 2x^2 + 3x^2 - 3x + 4x + 7 - 2$
$$= (2 + 3)x^2 + (-3 + 4)x + (7 - 2)$$
$$= 5x^2 + x + 5$$

(b) $(z^3 + 4z + 8) + (4z^2 - z + 6) = z^3 + 4z^2 + 4z - z + 8 + 6$
$$= z^3 + 4z^2 + (4 - 1)z + (8 + 6)$$
$$= z^3 + 4z^2 + 3z + 14$$

We can also add polynomials vertically, as in the next example.

EXAMPLE 5 Adding polynomials vertically

Find the sum.
(a) $(3x^2 - 5xy - 7y^2) + (xy + 4y^2 - x^2)$ (b) $(3x^3 - 2x + 7) + (x^3 + 5x^2 - 9)$

Solution
(a) Polynomials can be added vertically by placing like terms in the same columns and then adding each column.

$$\begin{array}{r} 3x^2 - 5xy - 7y^2 \\ -x^2 + xy + 4y^2 \\ \hline 2x^2 - 4xy - 3y^2 \end{array} \quad \text{Add each column.}$$

(b) Note that the first polynomial does not contain an x^2-term and that the second polynomial does not contain an x-term. When you are adding vertically, leave a blank for a missing term.

$$\begin{array}{r} 3x^3 \qquad\quad -2x + 7 \\ x^3 + 5x^2 \qquad\quad - 9 \\ \hline 4x^3 + 5x^2 - 2x - 2 \end{array} \quad \text{Add each column.}$$

Recall that to subtract integers we add the first integer and the *additive inverse* or *opposite* of the second integer. For example, to evaluate $3 - 5$ we perform the following operations.

$$3 - 5 = 3 + (-5) \quad \text{Add the opposite.}$$
$$= -2 \quad \text{Simplify.}$$

Similarly, to subtract two polynomials we add the first polynomial and the opposite of the second polynomial. To find the **opposite of a polynomial**, we negate each term. Table 5.4 shows three polynomials and their opposites.

TABLE 5.4

Polynomial	Opposite
$9 - x$	$-9 + x$
$5x^2 + 4x - 1$	$-5x^2 - 4x + 1$
$-x^4 + 5x^3 - x^2 + 5x - 1$	$x^4 - 5x^3 + x^2 - 5x + 1$

EXAMPLE 6 Subtracting polynomials

Simplify.
(a) $(y^5 + 3y^3) - (-y^4 + 2y^3)$ (b) $(5x^3 + 9x^2 - 6) - (5x^3 - 4x^2 - 7)$

Solution
(a) The opposite of $(-y^4 + 2y^3)$ is $(y^4 - 2y^3)$.

$$(y^5 + 3y^3) - (-y^4 + 2y^3) = (y^5 + 3y^3) + (y^4 - 2y^3)$$
$$= y^5 + y^4 + (3 - 2)y^3$$
$$= y^5 + y^4 + y^3$$

(b) The opposite of $(5x^3 - 4x^2 - 7)$ is $(-5x^3 + 4x^2 + 7)$.

$$(5x^3 + 9x^2 - 6) - (5x^3 - 4x^2 - 7) = (5x^3 + 9x^2 - 6) + (-5x^3 + 4x^2 + 7)$$
$$= (5 - 5)x^3 + (9 + 4)x^2 + (-6 + 7)$$
$$= 0x^3 + 13x^2 + 1$$
$$= 13x^2 + 1$$

CRITICAL THINKING

Is the sum of two quadratic polynomials always a quadratic polynomial? Explain.

The following summarizes addition and subtraction of polynomials.

ADDITION AND SUBTRACTION OF POLYNOMIALS

1. To *add* two polynomials, combine like terms.
2. To *subtract* two polynomials, add the first polynomial and the opposite of the second polynomial.

NOTE: The *opposite* of a polynomial is found by changing the sign of every term.

Polynomial Functions

The following expressions are examples of polynomials of one variable.

$$1 - 5x, \quad 3x^2 - 5x + 1, \quad \text{and} \quad x^3 + 5$$

As a result, we say that the following are *symbolic representations* of polynomial functions of one variable.

$$f(x) = 1 - 5x, \quad g(x) = 3x^2 - 5x + 1, \quad \text{and} \quad h(x) = x^3 + 5$$

Function f is a **linear function** because it has degree 1, function g is a **quadratic function** because it has degree 2, and function h is a **cubic function** because it has degree 3.

EXAMPLE 7 Identifying polynomial functions

Determine whether $f(x)$ represents a polynomial function. If possible, identify the type of polynomial function and its degree.

(a) $f(x) = 5x^3 - x + 10$ (b) $f(x) = x^{-2.5} + 1$

(c) $f(x) = 1 - 2x$ (d) $f(x) = \dfrac{3}{x - 1}$

Solution

(a) The expression $5x^3 - x + 10$ is a cubic polynomial, so $f(x)$ represents a cubic polynomial function. It has degree 3.
(b) $f(x)$ does not represent a polynomial function because the variables in a polynomial must have *nonnegative integer* exponents.
(c) $f(x) = 1 - 2x$ represents a polynomial function that is linear. It has degree 1.
(d) $f(x)$ does not represent a polynomial function because $\dfrac{3}{x-1}$ is not a polynomial.

Evaluating Polynomials

▶ **REAL-WORLD CONNECTION** Frequently, monomials and polynomials represent formulas that can be evaluated. This situation is illustrated in the next three examples.

EXAMPLE 8 Evaluating a polynomial function graphically and symbolically

A graph of $f(x) = 4x - x^3$ is shown in Figure 5.4. Evaluate $f(-1)$ graphically and check your result symbolically.

Solution

Graphical Evaluation To calculate $f(-1)$ graphically find -1 on the x-axis and move down until the graph of f is reached. Then move horizontally to the y-axis, as shown in Figure 5.5. Thus, when $x = -1$, $y = -3$ and $f(-1) = -3$.

Symbolic Evaluation Evaluation of $f(x) = 4x - x^3$ is performed as follows.

$$f(-1) = 4(-1) - (-1)^3 = -4 - (-1) = -3$$

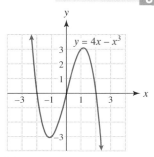

Figure 5.4

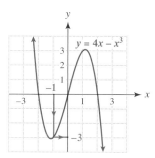

Figure 5.5 $f(-1) = -3$

EXAMPLE 9 Writing and evaluating a polynomial

Write a polynomial that represents the total volume of two identical boxes having square bases. Make a sketch to illustrate your formula. Find the total volume of the boxes if each base is 11 inches on a side and the height of each box is 5 inches.

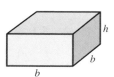

Figure 5.6 Volume: b^2h

Solution
Let b be the length (width) of the square base and h be the height of a side. The volume of one box is length times width times height, or b^2h, as illustrated in Figure 5.6, and the volume of two boxes is $2b^2h$. To calculate their total volumes let $b = 11$ and $h = 5$ in the expression $2b^2h$. Then

$$2 \cdot 11^2 \cdot 5 = 1210 \text{ cubic inches.}$$

EXAMPLE 10 Evaluating a polynomial function symbolically

Evaluate $f(x)$ at the given value of x.
(a) $f(x) = -3x^4 - 2, \quad x = 2$ **(b)** $f(x) = -2x^3 - 4x^2 + 5, \quad x = -3$

Solution
(a) Be sure to evaluate exponents before multiplying.
$$f(2) = -3(2)^4 - 2 = -3 \cdot 16 - 2 = -50$$
(b) $f(-3) = -2(-3)^3 - 4(-3)^2 + 5 = -2(-27) - 4(9) + 5 = 23$

Operations on Functions

▶ **REAL-WORLD CONNECTION** Frequently, we have found it necessary to add numbers or variables. Similarly, there may be a need to add two functions. For example, suppose that a student works road construction during summer break. On the taxable portion of this income, there might be a 6% state income tax and a 15% federal income tax. If x is the student's taxable income, then $S(x) = 0.06x$ calculates the state income tax and $F(x) = 0.15x$ calculates the federal income tax. Thus if $x = \$5000$, then the *total income tax T* is

$$T(5000) = S(5000) + F(5000)$$
$$= 0.06(5000) + 0.15(5000)$$
$$= \$300 + \$750$$
$$= \$1050.$$

That is, to find the total tax, we add state and federal taxes. This statement can be written in *function notation* for a taxable income x as

$$T(x) = S(x) + F(x).$$

Given two functions, f and g, we define the sum, $f + g$, and difference, $f - g$, as follows.

SUM AND DIFFERENCE OF TWO FUNCTIONS

The sum and difference of two functions f and g are defined as follows.
$$(f + g)(x) = f(x) + g(x)$$
$$(f - g)(x) = f(x) - g(x)$$
provided both $f(x)$ and $g(x)$ are defined.

EXAMPLE 11 Calculating sums and differences of functions

Let $f(x) = 2x^2 + 1$ and $g(x) = 5 - x^2$. Find each sum or difference.
(a) $(f + g)(1)$ **(b)** $(f - g)(-2)$
(c) $(f + g)(x)$ **(d)** $(f - g)(x)$

Solution
(a) Because $f(1) = 2 \cdot 1^2 + 1 = 3$ and $g(1) = 5 - 1^2 = 4$, it follows that
$$(f + g)(1) = f(1) + g(1) = 3 + 4 = 7.$$
(b) Because $f(-2) = 2 \cdot (-2)^2 + 1 = 9$ and $g(-2) = 5 - (-2)^2 = 1$, it follows that
$$(f - g)(-2) = f(-2) - g(-2) = 9 - 1 = 8.$$
(c) $(f + g)(x) = f(x) + g(x)$ Addition of functions
$\qquad\qquad\quad = (2x^2 + 1) + (5 - x^2)$ Substitute for $f(x)$ and $g(x)$.
$\qquad\qquad\quad = x^2 + 6$ Combine like terms.

(d) $(f - g)(x) = f(x) - g(x)$ Subtraction of functions
$\qquad\qquad\quad = (2x^2 + 1) - (5 - x^2)$ Substitute for $f(x)$ and $g(x)$.
$\qquad\qquad\quad = (2x^2 + 1) + (-5 + x^2)$ Add the opposite.
$\qquad\qquad\quad = 3x^2 - 4$ Combine like terms.

Applications and Models

▶ **REAL-WORLD CONNECTION** Polynomials are used to model a wide variety of data. A scatterplot of the cumulative number of reported AIDS cases in thousands from 1984 to 1994 is shown in Figure 5.7. In this graph $x = 4$ corresponds to 1984, $x = 5$ to 1985, and so on until $x = 14$ represents 1994. These data can be modeled with a quadratic function, as shown in Figure 5.8, where $f(x) = 4.1x^2 - 25x + 46$ is graphed with the data. Note that $f(x)$ was found by using a graphing calculator. (*Source:* U.S. Department of Health and Human Services.)

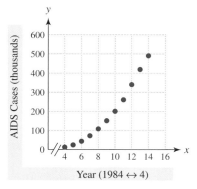

Figure 5.7 AIDS Cases (1984–1994)

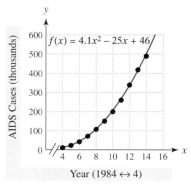

Figure 5.8 Modeling AIDS Cases

EXAMPLE 12 Modeling AIDS cases in the United States

Use $f(x) = 4.1x^2 - 25x + 46$ to model the number of AIDS cases.
(a) Estimate the number of AIDS cases reported by 1987. Compare it to the actual value of 71.4 thousand.
(b) By 1997, the total number of AIDS cases reported was 631 thousand. What estimate does $f(x)$ give? Discuss your result.

Solution
(a) The value $x = 7$ corresponds to 1987. Let $f(x) = 4.1x^2 - 25x + 46$ and evaluate $f(7)$, which gives
$$f(7) = 4.1(7)^2 - 25(7) + 46 = 71.9.$$
This model estimates that a cumulative total of 71.9 thousand AIDS cases were reported by 1987. This result compares favorably to the actual value of 71.4 thousand cases.

(b) To estimate the number in 1997, we evaluate $f(17)$ because $x = 17$ corresponds to 1997, obtaining
$$f(17) = 4.1(17)^2 - 25(17) + 46 = 805.9.$$
This result is considerably more than the actual value of 631 thousand. The reason is that f models data only from 1984 through 1994. After 1994, f gives estimates that are too large because the growth in AIDS cases has slowed more in recent years.

▶ **REAL-WORLD CONNECTION** A well-conditioned athlete's heart rate can reach 200 beats per minute during strenuous physical activity. Upon quitting, a typical heart rate decreases rapidly at first and then more gradually after a few minutes, as illustrated in the next example.

EXAMPLE 13 Modeling heart rate of an athlete

Let $P(t) = 1.875t^2 - 30t + 200$ model an athlete's heart rate (or pulse P) in beats per minute (bpm) t minutes after strenuous exercise has stopped, where $0 \leq t \leq 8$. (*Source: V. Thomas, Science and Sport.*)
(a) What is the initial heart rate when the athlete stops exercising?
(b) What is the heart rate after 8 minutes?
(c) A graph of P is shown in Figure 5.9. Interpret this graph.

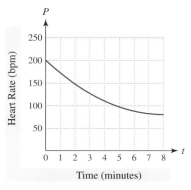

Figure 5.9

Solution
(a) To find the initial heart rate evaluate $P(t)$ at $t = 0$, or
$$P(0) = 1.875(0)^2 - 30(0) + 200 = 200.$$
When the athlete stops exercising, the heart rate is 200 beats per minute. (This result agrees with the graph.)
(b) $P(8) = 1.875(8)^2 - 30(8) + 200 = 80$ beats per minute.
(c) The heart rate does not drop at a constant rate; rather, it drops rapidly at first and then gradually begins to level off.

5.1 PUTTING IT ALL TOGETHER

In this section we introduced monomials, polynomials, and polynomial functions. A monomial is a number or a product of numbers and variables, where each variable has a *nonnegative integer* exponent. A polynomial is a monomial or a sum of monomials. Examples are given in the following table.

Concept	Examples
Addition of Polynomials	$(-3x^2 + 2x - 7) + (4x^2 - x + 1) = -3x^2 + 4x^2 + 2x - x - 7 + 1$ $= (-3 + 4)x^2 + (2 - 1)x + (-7 + 1)$ $= x^2 + x - 6$ $(x^4 + 8x^3 - 7x) + (5x^4 - 3x) = x^4 + 5x^4 + 8x^3 - 7x - 3x$ $= (1 + 5)x^4 + 8x^3 + (-7 - 3)x$ $= 6x^4 + 8x^3 - 10x$
Opposite of a Polynomial	The opposite of $-5x^2 + 3x - 5$ is $5x^2 - 3x + 5$. The opposite of $-6x^6 + 4x^4 - 8x^2 - 17$ is $6x^6 - 4x^4 + 8x^2 + 17$.
Subtraction of Polynomials (Add the Opposite.)	$(5x^2 - 6x + 1) - (-5x^2 + 3x - 5) = (5x^2 - 6x + 1) + (5x^2 - 3x + 5)$ $= (5 + 5)x^2 + (-6 - 3)x + (1 + 5)$ $= 10x^2 - 9x + 6$ $(x^4 - 6x^2 + 5x) - (x^4 - 5x + 7) = (x^4 - 6x^2 + 5x) + (-x^4 + 5x - 7)$ $= (1 - 1)x^4 - 6x^2 + (5 + 5)x - 7$ $= -6x^2 + 10x - 7$
Polynomial Functions	The following represent polynomial functions. $f(x) = 3$ Degree $= 0$ Constant $f(x) = 5x - 3$ Degree $= 1$ Linear $f(x) = x^2 - 2x - 1$ Degree $= 2$ Quadratic $f(x) = 3x^3 + 2x^2 - 6$ Degree $= 3$ Cubic
Evaluating a Polynomial Function	To evaluate $f(x) = -4x^2 + 3x - 1$ at $x = 2$, substitute 2 for x. $f(2) = -4(2)^2 + 3(2) - 1 = -16 + 6 - 1 = -11$
Sums and Differences of Functions	Let $f(x) = 2x + 3$ and $g(x) = x + 2$. $(f + g)(x) = (2x + 3) + (x + 2) = 3x + 5$ $(f - g)(x) = (2x + 3) - (x + 2) = x + 1$

5.2 MULTIPLICATION OF POLYNOMIALS

Review of Basic Properties ■ Multiplying Polynomials ■ Some Special Products ■ Multiplying Functions

A LOOK INTO MATH ▷

The study of polynomials dates back to Babylonian civilization in about 1800–1600 B.C. Much later, Gottfried Leibniz (1646–1716) was the first to generalize polynomial functions of degree n. Many eighteenth-century mathematicians devoted their entire careers to the study of polynomials. Their studies included multiplying and factoring polynomials. Both skills are used to solve equations. In this section we discuss the basics of polynomial multiplication. (*Sources: Historical Topics for the Mathematics Classroom, Thirty-first Yearbook*, NCTM; L. Motz and J. Weaver, *The Story of Mathematics*.)

Review of Basic Properties

Distributive properties are used frequently in the multiplication of polynomials. For all real numbers a, b, and c

$$a(b + c) = ab + ac \quad \text{and}$$
$$a(b - c) = ab - ac.$$

In the next example we use these distributive properties to multiply expressions.

EXAMPLE 1 Using distributive properties

Multiply.
(a) $4(5 + x)$ **(b)** $-3(x - 4y)$ **(c)** $(2x - 5)(6)$

Solution

(a) $4(5 + x) = 4 \cdot 5 + 4 \cdot x = 20 + 4x$

(b) $-3(x - 4y) = -3 \cdot x - (-3) \cdot (4y) = -3x + 12y$

(c) $(2x - 5)(6) = 2x \cdot 6 - 5 \cdot 6 = 12x - 30$

You can visualize the solution to part (a) of Example 1 by using areas of rectangles. If a rectangle has width 4 and length $5 + x$, its area is $20 + 4x$, as shown in Figure 5.10.

In Section 1.3 we discussed several properties of exponents. The following three properties of exponents are frequently used in the multiplication of polynomials.

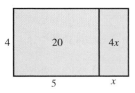

Figure 5.10 Area: $20 + 4x$

PROPERTIES OF EXPONENTS

For any numbers a and b and integers m and n,

$$a^m \cdot a^n = a^{m+n}, \quad (a^m)^n = a^{mn}, \quad \text{and} \quad (ab)^n = a^n b^n.$$

We use these three properties to simplify expressions in the next two examples.

EXAMPLE 2 Multiplying powers of variables

Multiply each expression.
(a) $-2x^3 \cdot 4x^5$ (b) $(3xy^3)(4x^2y^2)$ (c) $6y^3(2y - y^2)$ (d) $-mn(m^2 - n)$

Solution
(a) $-2x^3 \cdot 4x^5 = (-2)(4)x^3x^5 = -8x^{3+5} = -8x^8$
(b) $(3xy^3)(4x^2y^2) = (3)(4)xx^2y^3y^2 = 12x^{1+2}y^{3+2} = 12x^3y^5$

NOTE: We cannot simplify $12x^3y^5$ further because x^3 and y^5 have different bases.

(c) $6y^3(2y - y^2) = 6y^3 \cdot 2y - 6y^3 \cdot y^2 = 12y^4 - 6y^5$
(d) $-mn(m^2 - n) = -mn \cdot m^2 + mn \cdot n = -m^3n + mn^2$

EXAMPLE 3 Using properties of exponents

Simplify.
(a) $(x^2)^5$ (b) $(2x)^3$ (c) $(5x^3)^2$ (d) $(-mn^3)^2$

Solution
(a) $(x^2)^5 = x^{2 \cdot 5} = x^{10}$ (b) $(2x)^3 = 2^3x^3 = 8x^3$
(c) $(5x^3)^2 = 5^2(x^3)^2 = 25x^6$ (d) $(-mn^3)^2 = (-m)^2(n^3)^2 = m^2n^6$

Multiplying Polynomials

A polynomial with one term is a **monomial**, with two terms a **binomial**, and with three terms a **trinomial**. Examples are shown in Table 5.5.

TABLE 5.5

Monomials	$2x^2$	$-3x^4y$	9
Binomials	$3x - 1$	$2x^3 - x$	$x^2 + 5$
Trinomials	$x^2 - 3x + 5$	$5x^4 - 2x + 10$	$2x^3 - x^2 - 2$

In the next example we multiply binomials, using geometric and symbolic techniques.

EXAMPLE 4 Multiplying binomials

Multiply $(x + 1)(x + 3)$
(a) geometrically and (b) symbolically.

Solution
(a) To multiply $(x + 1)(x + 3)$ geometrically, draw a rectangle $x + 1$ wide and $x + 3$ long, as shown in Figure 5.11(a). The area of the rectangle equals the product $(x + 1)(x + 3)$. This large rectangle can be divided into four smaller rectangles as shown in Figure 5.11(b). The sum of the areas of these four smaller rectangles equals the area of the large rectangle. The smaller rectangles have areas of $x^2, x, 3x,$ and 3. Thus

$$(x + 1)(x + 3) = x^2 + x + 3x + 3$$
$$= x^2 + 4x + 3.$$

(b) To multiply $(x + 1)(x + 3)$ symbolically we apply the distributive property.

$$(x + 1)(x + 3) = (x + 1)(x) + (x + 1)(3)$$
$$= x \cdot x + 1 \cdot x + x \cdot 3 + 1 \cdot 3$$
$$= x^2 + x + 3x + 3$$
$$= x^2 + 4x + 3$$

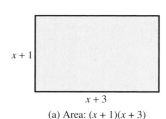
(a) Area: $(x + 1)(x + 3)$

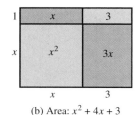
(b) Area: $x^2 + 4x + 3$

Figure 5.11

5.2 MULTIPLICATION OF POLYNOMIALS

GRAPHICAL AND NUMERICAL SUPPORT We can give graphical and numerical support to our result in Example 4 by letting $Y_1 = (X + 1)(X + 3)$ and by letting $Y_2 = X^2 + 4X + 3$. The graphs of y_1 and y_2 appear to be identical in Figures 5.12(a) and 5.12(b). Figure 5.12(c) shows that $y_1 = y_2$ for each value of x in the table.

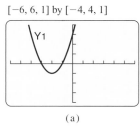
(a)

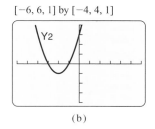
(b)

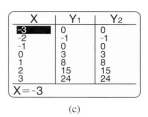
(c)

Figure 5.12

FOIL One way to multiply $(x + 1)$ by $(x + 3)$ is to multiply every term in $x + 1$ by every term in $x + 3$. That is,

$$(x + 1)(x + 3) = x^2 + 3x + x + 3$$
$$= x^2 + 4x + 3.$$

NOTE: This process of multiplying binomials is called *FOIL*. You may use it to remind yourself to multiply the first terms (F), outside terms (O), inside terms (I), and last terms (L).

Multiply the *First terms* to obtain x^2. $(x + 1)(x + 3)$

Multiply the *Outside terms* to obtain $3x$. $(x + 1)(x + 3)$

Multiply the *Inside terms* to obtain x. $(x + 1)(x + 3)$

Multiply the *Last terms* to obtain 3. $(x + 1)(x + 3)$

The following method summarizes how to multiply two polynomials in general.

MULTIPLICATION OF POLYNOMIALS

The product of two polynomials may be found by multiplying every term in the first polynomial by every term in the second polynomial.

EXAMPLE 5 Multiplying polynomials

Multiply each binomial.
(a) $(2x - 1)(x + 2)$ (b) $(1 - 3x)(2 - 4x)$ (c) $(x^2 + 1)(5x - 3)$

Solution

(a) $(2x - 1)(x + 2) = 2x \cdot x + 2x \cdot 2 - 1 \cdot x - 1 \cdot 2$
$= 2x^2 + 4x - x - 2$
$= 2x^2 + 3x - 2$

(b) $(1 - 3x)(2 - 4x) = 1 \cdot 2 - 1 \cdot 4x - 3x \cdot 2 + 3x \cdot 4x$
$= 2 - 4x - 6x + 12x^2$
$= 2 - 10x + 12x^2$

(c) $(x^2 + 1)(5x - 3) = x^2 \cdot 5x - x^2 \cdot 3 + 1 \cdot 5x - 1 \cdot 3$
$= 5x^3 - 3x^2 + 5x - 3$

CHAPTER 5 POLYNOMIAL EXPRESSIONS AND FUNCTIONS

EXAMPLE 6 Multiplying polynomials

Multiply each expression.
(a) $3x(x^2 + 5x - 4)$ (b) $-x^2(x^4 - 2x + 5)$ (c) $(x + 2)(x^2 + 4x - 3)$

Solution

(a) Multiply each term in the second polynomial by $3x$.

$$3x(x^2 + 5x - 4) = 3x \cdot x^2 + 3x \cdot 5x - 3x \cdot 4$$
$$= 3x^3 + 15x^2 - 12x$$

(b) $-x^2(x^4 - 2x + 5) = -x^2 \cdot x^4 + x^2 \cdot 2x - x^2 \cdot 5$
$$= -x^6 + 2x^3 - 5x^2$$

(c) $(x + 2)(x^2 + 4x - 3) = x \cdot x^2 + x \cdot 4x - x \cdot 3 + 2 \cdot x^2 + 2 \cdot 4x - 2 \cdot 3$
$$= x^3 + 4x^2 - 3x + 2x^2 + 8x - 6$$
$$= x^3 + 6x^2 + 5x - 6$$

EXAMPLE 7 Multiplying polynomials

Multiply each expression.
(a) $2ab^3(a^2 - 2ab + 3b^2)$ (b) $4m(mn^2 + 3m)(m^2n - 4n)$

Solution
(a) Multiply each term in the second polynomial by $2ab^3$.

$$2ab^3(a^2 - 2ab + 3b^2) = 2ab^3 \cdot a^2 - 2ab^3 \cdot 2ab + 2ab^3 \cdot 3b^2$$
$$= 2a^3b^3 - 4a^2b^4 + 6ab^5$$

(b) Start by multiplying every term in $(mn^2 + 3m)$ by $4m$.

$$4m(mn^2 + 3m)(m^2n - 4n) = (4m \cdot mn^2 + 4m \cdot 3m)(m^2n - 4n)$$
$$= (4m^2n^2 + 12m^2)(m^2n - 4n)$$
$$= 4m^2n^2 \cdot m^2n - 4m^2n^2 \cdot 4n + 12m^2 \cdot m^2n - 12m^2 \cdot 4n$$
$$= 4m^4n^3 - 16m^2n^3 + 12m^4n - 48m^2n$$

Sometimes it is convenient to multiply polynomials vertically. After multiplying, always place like terms in their respective columns. Leave blanks for missing terms.

EXAMPLE 8 Multiplying polynomials vertically

Multiply $(3x - 4y)(2x^2 + xy - 4y^2)$.

Solution
Start by stacking the polynomials as follows. Note that unlike terms can be multiplied but not added.

$$\begin{array}{r} 2x^2 + xy - 4y^2 \\ 3x - 4y \\ \hline -8x^2y - 4xy^2 + 16y^3 \\ 6x^3 + 3x^2y - 12xy^2 \\ \hline 6x^3 - 5x^2y - 16xy^2 + 16y^3 \end{array}$$

Multiply first row by $-4y$.
Multiply first row by $3x$.
Add columns.

▶ **REAL-WORLD CONNECTION** Polynomials frequently occur in business applications. For example, the **demand** D for buying a new video game might vary with its price. If the price is high, fewer games are sold and if the price is low, more games are sold. The revenue R from selling D games at price p is given by $R = pD$. This concept is used in the next example.

EXAMPLE 9 Using polynomials in a business application

Let the demand D, or number of games sold in thousands, be given by $D = 30 - \frac{1}{5}p$, where $p \geq 10$ is the price of the game in dollars.
(a) Find the demand when $p = \$30$ and when $p = \$60$.
(b) Write an expression for the revenue R. Multiply your expression.
(c) Find the revenue when $p = \$40$.

Solution
(a) When $p = 30$, $D = 30 - \frac{1}{5}(30) = 24$ thousand games.
   When $p = 60$, $D = 30 - \frac{1}{5}(60) = 18$ thousand games.
(b) $R = pD = p(30 - \frac{1}{5}p) = 30p - \frac{1}{5}p^2$
(c) When $p = 40$, $R = 30(40) - \frac{1}{5}(40)^2 = \$880$ thousand.

Some Special Products

The following special product often occurs in mathematics.

$$(a + b)(a - b) = a \cdot a - a \cdot b + b \cdot a - b \cdot b$$
$$= a^2 - ab + ba - b^2$$
$$= a^2 - b^2$$

That is, the product of a sum and difference equals the difference of their squares.

> **PRODUCT OF A SUM AND DIFFERENCE**
>
> For any real numbers a and b,
> $$(a + b)(a - b) = a^2 - b^2.$$

EXAMPLE 10 Finding the product of a sum and difference

Multiply.
(a) $(x + 3)(x - 3)$ (b) $(5 - 4x^2)(5 + 4x^2)$

Solution
(a) If we let $a = x$ and $b = 3$, we can apply the rule
$$(a + b)(a - b) = a^2 - b^2.$$
Thus
$$(x + 3)(x - 3) = (x)^2 - (3)^2$$
$$= x^2 - 9.$$

(b) Because $(a - b)(a + b) = a^2 - b^2$, we can multiply as follows.
$$(5 - 4x^2)(5 + 4x^2) = (5)^2 - (4x^2)^2$$
$$= 25 - 16x^4$$

EXAMPLE 11 Finding the product of sums and differences

Multiply each expression.
(a) $4rt(r - 4t)(r + 4t)$ (b) $(2z + 5k^4)(2z - 5k^4)$

Solution
(a) Start by finding the product of the difference and sum and then simplify.
$$4rt(r - 4t)(r + 4t) = 4rt(r^2 - 16t^2)$$
$$= 4rt \cdot r^2 - 4rt \cdot 16t^2$$
$$= 4r^3t - 64rt^3$$

(b) $(2z + 5k^4)(2z - 5k^4) = (2z)^2 - (5k^4)^2$
$$= 4z^2 - 25k^8$$

Two other special products involve *squaring a binomial:*

$(a + b)^2 = (a + b)(a + b)$ $\qquad$ $(a - b)^2 = (a - b)(a - b)$
$\quad = a^2 + ab + ba + b^2$ and $\quad = a^2 - ab - ba + b^2$
$\quad = a^2 + 2ab + b^2$ $\qquad$ $\quad = a^2 - 2ab + b^2.$

The first product is illustrated geometrically in Figure 5.13, where each side of a square has length $(a + b)$. The area of the square is
$$(a + b)(a + b) = (a + b)^2.$$

This area can also be computed by adding the areas of the four small rectangles.
$$a^2 + ab + ba + b^2 = a^2 + 2ab + b^2$$

Thus $(a + b)^2 = a^2 + 2ab + b^2$. *To obtain the middle term, multiply the two terms in the binomial and double the result.*

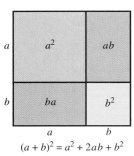

$(a + b)^2 = a^2 + 2ab + b^2$

Figure 5.13

SQUARE OF A BINOMIAL

For any real numbers a and b,
$$(a + b)^2 = a^2 + 2ab + b^2 \quad \text{and} \quad (a - b)^2 = a^2 - 2ab + b^2.$$

EXAMPLE 12 Squaring a binomial

Multiply.
(a) $(x + 5)^2$ (b) $(3 - 2x)^2$ (c) $(2m^3 + 4n)^2$ (d) $\bigl((2z - 3) + k\bigr)\bigl((2z - 3) - k\bigr)$

Solution
(a) If we let $a = x$ and $b = 5$, we can apply $(a + b)^2 = a^2 + 2ab + b^2$. Thus
$$(x + 5)^2 = (x)^2 + 2(x)(5) + (5)^2 \qquad \text{To find the middle term, multiply}$$
$$= x^2 + 10x + 25. \qquad\qquad x \text{ and 5 and double the result.}$$

(b) Applying the formula $(a - b)^2 = a^2 - 2ab + b^2$, we find
$$(3 - 2x)^2 = (3)^2 - 2(3)(2x) + (2x)^2$$
$$= 9 - 12x + 4x^2.$$

(c) $(2m^3 + 4n)^2 = (2m^3)^2 + 2(2m^3)(4n) + (4n)^2$
$$= 4m^6 + 16m^3n + 16n^2$$

(d) If $a = 2z - 3$ and $b = k$, we can use $(a + b)(a - b) = a^2 - b^2$.

$$\big((2z - 3) + k\big)\big((2z - 3) - k\big) = (2z - 3)^2 - k^2$$
$$= (2z)^2 - 2(2z)(3) + (3)^2 - k^2$$
$$= 4z^2 - 12z + 9 - k^2$$

CRITICAL THINKING

Suppose that a friend is convinced that the expressions

$(x + 3)^2$ and $x^2 + 9$

are equivalent. How could you convince your friend that $(x + 3)^2 \neq x^2 + 9$?

NOTE: If you forget these special products, you can still use techniques learned earlier to multiply the polynomials in Examples 10–12. For example,

$$(3 - 2x)^2 = (3 - 2x)(3 - 2x)$$
$$= 3 \cdot 3 - 3 \cdot 2x - 2x \cdot 3 + 2x \cdot 2x$$
$$= 9 - 6x - 6x + 4x^2$$
$$= 9 - 12x + 4x^2.$$

Multiplying Functions

In Section 5.1 we discussed how to add and subtract functions. These concepts can be extended to include multiplication of two functions, f and g, by using the following definition.

$$(fg)(x) = f(x)g(x)$$

NOTE: There is no operation symbol written between f and g in $(fg)(x)$, and so the operation is assumed to be multiplication, just as in the expression xy.

EXAMPLE 13 Multiplying polynomial functions

Let $f(x) = x^2 - x$ and $g(x) = 2x^2 + 4$. Find $(fg)(3)$ and $(fg)(x)$.

Solution
To find $(fg)(3)$, evaluate $f(3)$ and $g(3)$. Then multiply the results.

$$(fg)(3) = f(3)g(3) = (3^2 - 3)(2 \cdot 3^2 + 4) = (6)(22) = 132$$

To find $(fg)(x)$, multiply the formulas for $f(x)$ and $g(x)$.

$$(fg)(x) = f(x)g(x) = (x^2 - x)(2x^2 + 4) = 2x^4 - 2x^3 + 4x^2 - 4x$$

5.2 PUTTING IT ALL TOGETHER

The following table summarizes some important concepts in this section.

Concept	Explanation	Examples
Distributive Properties	For all real numbers a, b, and c, $a(b + c) = ab + ac$ and $a(b - c) = ab - ac$.	$4(3 + a) = 12 + 4a$, $5(x - 1) = 5x - 5$, and $-(b - 5) = -1(b - 5) = -b + 5$

continued on next page

Concept	Explanation	Examples
Multiplying Polynomials	The product of two polynomials may be found by multiplying every term in the first polynomial by every term in the second polynomial.	$(2x + 3)(x - 7) = 2x \cdot x - 2x \cdot 7 + 3 \cdot x - 3 \cdot 7$ $= 2x^2 - 14x + 3x - 21$ $= 2x^2 - 11x - 21$
Properties of Exponents	For numbers a and b and integers m and n, $a^m \cdot a^n = a^{m+n}$, $(a^m)^n = a^{mn}$, and $(ab)^n = a^n b^n$.	$5^2 \cdot 5^6 = 5^8$, $(2^3)^2 = 2^6$, $(5y^4)^2 = 25y^8$, $(a^2 b^3)^4 = a^8 b^{12}$, and $(-4rt^3)^2 = 16r^2 t^6$
Special Products of Binomials	Product of a sum and difference $(a + b)(a - b) = a^2 - b^2$ Squares of binomials $(a + b)^2 = a^2 + 2ab + b^2$ $(a - b)^2 = a^2 - 2ab + b^2$	$(x + 2)(x - 2) = x^2 - 4$, $(7y - 6z^2)(7y + 6z^2) = 49y^2 - 36z^4$, $(x + 4)^2 = x^2 + 8x + 16$, $(x - 4)^2 = x^2 - 8x + 16$, and $(2m - 3n^2)^2 = 4m^2 - 12mn^2 + 9n^4$
Multiplication of Functions	$(fg)(x) = f(x)g(x)$	If $f(x) = x^2$ and $g(x) = x + 2$, then $(fg)(x) = x^2(x + 2) = x^3 + 2x^2$.

5.3 FACTORING POLYNOMIALS

Common Factors ■ Factoring and Equations ■ Factoring by Grouping

A LOOK INTO MATH ▷ Video games involving sports frequently have to show a ball traveling through the air. To make the motion look realistic, the game must solve an equation in real time to determine when the ball hits the ground. For example, suppose that a golf ball is hit with an *upward* velocity of 88 feet per second, or 60 miles per hour. Then its height h in feet above the ground after t seconds is modeled by $h(t) = 88t - 16t^2$. To estimate when the ball strikes the ground, we can solve the *polynomial equation*

$$88t - 16t^2 = 0. \quad \text{At ground level, } h = 0.$$

One method for solving this equation is factoring (see Example 6). In this section we discuss factoring and how it is used to solve equations.

Common Factors

When factoring a polynomial, we first look for factors that are common to each term. By applying a distributive property, we can write a polynomial as two factors. For example, each term in $2x^2 + 4x$ contains a factor of $2x$.

$$2x^2 = 2x \cdot x$$
$$4x = 2x \cdot 2$$

Thus the polynomial $2x^2 + 4x$ can be factored as follows.

$$2x^2 + 4x = 2x(x + 2)$$

EXAMPLE 1 Finding common factors

Factor.
(a) $4x^2 + 5x$ **(b)** $12x^3 - 4x^2$ **(c)** $6z^3 - 2z^2 + 4z$ **(d)** $4x^3y^2 + x^2y^3$

Solution
(a) Both $4x^2$ and $5x$ contain a common factor of x. That is,
$$4x^2 = x \cdot 4x \quad \text{and} \quad 5x = x \cdot 5.$$
Thus $4x^2 + 5x = x(4x + 5)$.

(b) Both $12x^3$ and $4x^2$ contain a common factor of $4x^2$. That is,
$$12x^3 = 4x^2 \cdot 3x \quad \text{and} \quad 4x^2 = 4x^2 \cdot 1.$$
Thus $12x^3 - 4x^2 = 4x^2(3x - 1)$.

(c) Each of the terms $6z^3$, $2z^2$, and $4z$ contains a common factor of $2z$. That is,
$$6z^3 = 2z \cdot 3z^2, \quad 2z^2 = 2z \cdot z, \quad \text{and} \quad 4z = 2z \cdot 2.$$
Thus $6z^3 - 2z^2 + 4z = 2z(3z^2 - z + 2)$.

(d) Both $4x^3y^2$ and x^2y^3 contain a common factor of x^2y^2. That is,
$$4x^3y^2 = x^2y^2 \cdot 4x \quad \text{and} \quad x^2y^3 = x^2y^2 \cdot y.$$
Thus $4x^3y^2 + x^2y^3 = x^2y^2(4x + y)$.

Many times we factor out the *greatest common factor*. For example, the polynomial $15x^4 - 5x^2$ has a common factor of $5x$. We could write this polynomial as
$$15x^4 - 5x^2 = 5x(3x^3 - x).$$
However, we can also factor out $5x^2$ to obtain
$$15x^4 - 5x^2 = 5x^2(3x^2 - 1).$$
Because $5x^2$ is the common factor with the highest degree and largest (integer) coefficient, we say that $5x^2$ is the **greatest common factor** (GCF) of $15x^4 - 5x^2$. In Example 1, we factored out the GCF in each case.

EXAMPLE 2 Factoring greatest common factors

Factor.
(a) $24x^5 + 12x^3 - 6x^2$ **(b)** $6m^3n^2 - 3mn^2 + 9m$ **(c)** $-9x^3 + 6x^2 - 3x$

Solution
(a) The GCF of $24x^5$, $12x^3$, and $6x^2$ is $6x^2$.
$$24x^5 = 6x^2 \cdot 4x^3, \quad 12x^3 = 6x^2 \cdot 2x, \quad \text{and} \quad 6x^2 = 6x^2 \cdot 1$$
Thus $24x^5 + 12x^3 - 6x^2 = 6x^2(4x^3 + 2x - 1)$.

(b) The GCF of $6m^3n^2$, $3mn^2$, and $9m$ is $3m$.
$$6m^3n^2 = 3m \cdot 2m^2n^2, \quad 3mn^2 = 3m \cdot n^2, \quad \text{and} \quad 9m = 3m \cdot 3$$
Thus $6m^3n^2 - 3mn^2 + 9m = 3m(2m^2n^2 - n^2 + 3)$.

(c) Rather than factoring out $3x$, we can also factor out $-3x$ and make the leading coefficient of the remaining expression positive.
$$-9x^3 = -3x \cdot 3x^2, \quad 6x^2 = -3x \cdot -2x, \quad \text{and} \quad -3x = -3x \cdot 1$$
Thus $-9x^3 + 6x^2 - 3x = -3x(3x^2 - 2x + 1)$.

Factoring and Equations

To solve equations by using factoring, we use the **zero-product property**. It states that, if the product of two numbers is 0, then at least one of the numbers must equal 0.

> **ZERO-PRODUCT PROPERTY**
>
> For all real numbers a and b, if $ab = 0$, then $a = 0$ or $b = 0$ (or both).

NOTE: The zero-product property works only for 0. If $ab = 1$, then it does *not* follow that $a = 1$ or $b = 1$. For example, $a = \frac{1}{3}$ and $b = 3$ also satisfy the equation $ab = 1$.

Sometimes factoring needs to be performed on an equation before the zero-product property can be applied. For example, the left side of the equation

$$2x^2 + 4x = 0$$

may be factored to obtain

$$2x(x + 2) = 0.$$

Note that $2x$ times $(x + 2)$ equals 0. By the zero-product property, we must have either

$$2x = 0 \quad \text{or} \quad x + 2 = 0.$$

Solving each equation for x gives

$$x = 0 \quad \text{or} \quad x = -2.$$

The x-values of **0** and **−2** are called **zeros** of the polynomial $2x^2 + 4x$ because, when they are substituted in $2x^2 + 4x$, the result is **0**. That is, $2(\mathbf{0})^2 + 4(\mathbf{0}) = \mathbf{0}$ and $2(\mathbf{-2})^2 + 4(\mathbf{-2}) = \mathbf{0}$.

> **MAKING CONNECTIONS**
>
> **Expressions and Equations**
>
> There are important distinctions between expressions and equations. We often *factor expressions* and *solve equations*. For example, when we solved the *equation* $2x^2 + 4x = 0$, we started by factoring the *expression* $2x^2 + 4x$, as $2x(x + 2)$. The equivalent equation, $2x(x + 2) = 0$, resulted. Then the zero-product property was used to obtain the *solutions to the equation*: 0 and −2.
>
> An equation is a statement that two expressions are equal. Equations always have an equals sign. We often solve equations, but we do *not* solve expressions.

EXAMPLE 3 Applying the zero-product property

Solve.
(a) $x(x - 1) = 0$ (b) $2x(x + 3) = 0$ (c) $(2x - 1)(3x + 2) = 0$

Solution
(a) The expression $x(x - 1)$ is the product of x and $(x - 1)$. By the zero-product property, either $x = 0$ or $x - 1 = 0$. The solutions are 0 and 1.

(b)
$$2x(x + 3) = 0 \qquad \text{Given equation}$$
$$2x = 0 \quad \text{or} \quad x + 3 = 0 \qquad \text{Zero-product property}$$
$$x = 0 \quad \text{or} \quad x = -3 \qquad \text{Solve each equation.}$$

The solutions are 0 and −3.

(c)
$$(2x - 1)(3x + 2) = 0 \qquad \text{Given equation}$$
$$2x - 1 = 0 \quad \text{or} \quad 3x + 2 = 0 \qquad \text{Zero-product property}$$
$$2x = 1 \quad \text{or} \quad 3x = -2 \qquad \text{Add 1; Subtract 2.}$$
$$x = \frac{1}{2} \quad \text{or} \quad x = -\frac{2}{3} \qquad \text{Divide by 2; divide by 3.}$$

The solutions are $\frac{1}{2}$ and $-\frac{2}{3}$.

EXAMPLE 4   Solving polynomial equations with factoring

Solve each polynomial equation.
(a) $x^2 + 3x = 0$ (b) $4x^2 = 16x$ (c) $2x^3 + 2x = 0$

Solution

(a) We begin by factoring out the greatest common factor x.

$x^2 + 3x = 0$	Given equation
$x(x + 3) = 0$	Factor out x.
$x = 0$ or $x + 3 = 0$	Zero-product property
$x = 0$ or $x = -3$	Solve.

The solutions are 0 and -3.

(b) Write the equation so that there is a 0 on its right side before applying the zero-product property.

$4x^2 = 16x$	Given equation
$4x^2 - 16x = 0$	Subtract $16x$ from each side.
$4x(x - 4) = 0$	Factor out $4x$.
$4x = 0$ or $x - 4 = 0$	Zero-product property
$x = 0$ or $x = 4$	Solve.

The solutions are 0 and 4.

(c) Start by factoring out the GCF of $2x$.

$2x^3 + 2x = 0$	Given equation
$2x(x^2 + 1) = 0$	Factor out $2x$.
$2x = 0$ or $x^2 + 1 = 0$	Zero-product property
$x = 0$ or $x^2 = -1$	Solve.

Because $x^2 \geq 0$ for any x, it follows that no real number can satisfy $x^2 = -1$. The only solution is 0.

Polynomial equations can also be solved numerically and graphically. For example, to find a graphical solution to $x^2 - 2x = 0$, graph $y = x^2 - 2x$ by making a table of values, as shown in Table 5.6. When the points are plotted and connected a $\cup$-shaped curve results, with x-intercepts 0 and 2. This curve shown in Figure 5.14 is called a *parabola*.

NOTE: The zeros, 0 and 2, of $x^2 - 2x$ in Table 5.6, the x-intercepts, 0 and 2, in Figure 5.14, and the solutions, 0 and 2, to the equation $x^2 - 2x = 0$ are *identical*.

Numerical Solution

TABLE 5.6
$y = x^2 - 2x$

x	y
-1	3
0	0
1	-1
2	0
3	3

Graphical Solution

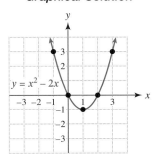

Figure 5.14 x-intercepts: 0 and 2

Symbolic Solution

$$x^2 - 2x = 0$$
$$x(x - 2) = 0$$
$$x = 0 \text{ or } x - 2 = 0$$
$$x = 0 \text{ or } x = 2$$

These concepts are generalized in the following.

MAKING CONNECTIONS

Zeros, x-Intercepts, and Solutions

The following statements are *equivalent*, where k is a real number and $P(x)$ is a polynomial.

1. A *zero* of a polynomial $P(x)$ is k.
2. An *x-intercept* on the graph of $y = P(x)$ is k.
3. A *solution* to the equation $P(x) = 0$ is k.

In the next example we solve an equation numerically, graphically, and symbolically.

EXAMPLE 5 Solving an equation

Solve the equation $4x - x^2 = 0$ numerically, graphically, and symbolically.

Solution

Numerical Solution Make a table of values for $y = 4x - x^2$, as shown in Table 5.7. Note that 0 and 4 are solutions to $4x - x^2 = 0$, because $4(0) - 0^2 = 0$ and $4(4) - 4^2 = 0$.

Graphical Solution Plot the points given in Table 5.7 and sketch a curve through them, as shown in Figure 5.15. The curve is a ∩-shaped graph, or parabola, with x-intercepts 0 and 4, which are the solutions to the given equation.

Numerical Solution
TABLE 5.7

x	y
-1	-5
0	**0**
1	3
2	4
3	3
4	**0**
5	-5

Graphical Solution

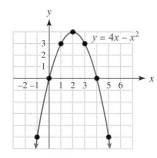

Figure 5.15

Symbolic Solution Start by factoring the left side of the equation.

$$4x - x^2 = 0 \qquad \text{Given equation}$$
$$x(4 - x) = 0 \qquad \text{Factor out } x.$$
$$x = 0 \quad \text{or} \quad 4 - x = 0 \qquad \text{Zero-product property}$$
$$x = 0 \quad \text{or} \quad x = 4 \qquad \text{Solve each equation.}$$

Note that the numerical and graphical solutions agree with the symbolic solutions.

GRAPH OF $y = ax^2 + bx + c$

The graph of $y = ax^2 + bx + c$ is a **parabola**. It is a U-shaped graph if $a > 0$ and a ∩-shaped graph if $a < 0$. This graph can have zero, one, or two x-intercepts.

TECHNOLOGY NOTE: Locating x-intercepts

CALCULATOR HELP

To find an x-intercept or zero, see the Appendix (page AP-10).

Many calculators have the capability to locate an x-intercept or zero. This method is illustrated in the accompanying figures, where the solution 2 to the equation $x^2 - 4 = 0$ is found.

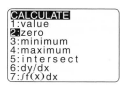
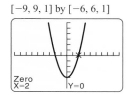

In the next example we use factoring to estimate when a golf ball strikes the ground.

EXAMPLE 6 Modeling the flight of a golf ball

If a golf ball is hit upward at 88 feet per second, or 60 miles per hour, its height h in feet after t seconds is modeled by $h(t) = 88t - 16t^2$.
(a) Use factoring to determine when the golf ball strikes the ground.
(b) Solve part (a) graphically and numerically.

Solution
(a) *Symbolic Solution* The golf ball strikes the ground when its height is 0.

$$88t - 16t^2 = 0 \qquad h(t) = 0$$
$$8t(11 - 2t) = 0 \qquad \text{Factor out } 8t.$$
$$8t = 0 \quad \text{or} \quad 11 - 2t = 0 \qquad \text{Zero-product property}$$
$$t = 0 \quad \text{or} \quad t = \frac{11}{2} \qquad \text{Solve for } t.$$

The ball strikes the ground after $\frac{11}{2}$, or 5.5 seconds. The solution of $t = 0$ is not used in this problem because it corresponds to when the ball is initially hit on the ground.

(b) *Graphical and Numerical Solutions* Graph $Y_1 = 88X - 16X^2$. Using a graphing calculator to find an x-intercept (or zero) yields the solution, 5.5. See Figure 5.16(a). Numerical support is shown in Figure 5.16(b), where $y_1 = 0$ when $x = 5.5$.

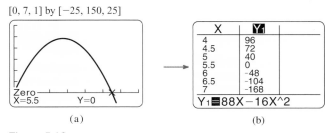

(a)   (b)

Figure 5.16

Factoring by Grouping

Factoring by grouping is a technique that makes use of the associative and distributive properties. The next example illustrates the first step in this factoring technique.

EXAMPLE 7 Factoring out binomials

Factor.
(a) $2x(x + 1) + 3(x + 1)$ (b) $2x^2(3x - 2) - x(3x - 2)$

Solution
(a) Both terms in the expression $2x(x + 1) + 3(x + 1)$ contain the binomial $(x + 1)$. Therefore the distributive property can be used to factor out this expression.

$$2x(x + 1) + 3(x + 1) = (2x + 3)(x + 1)$$

(b) Both terms in the expression $2x^2(3x - 2) - x(3x - 2)$ contain the binomial $(3x - 2)$. Therefore the distributive property can be used to factor this expression.

$$2x^2(3x - 2) - x(3x - 2) = (2x^2 - x)(3x - 2)$$

Now consider the polynomial

$$3t^3 + 6t^2 + 2t + 4.$$

We can factor this polynomial by first grouping it into two binomials.

$(3t^3 + 6t^2) + (2t + 4)$	Associative property
$3t^2(t + 2) + 2(t + 2)$	Factor out common factors.
$(3t^2 + 2)(t + 2)$	Factor out $(t + 2)$.

The following steps summarize factoring a polynomial with four terms by grouping.

FACTORING BY GROUPING

STEP 1: Use parentheses to group the terms into binomials with common factors. Begin by writing the expression with a plus sign between the binomials.

STEP 2: Factor out the common factor in each binomial.

STEP 3: Factor out the common binomial. If there is no common binomial, try a different grouping.

These steps are used in Example 8.

EXAMPLE 8 Factoring by grouping

Factor each polynomial.
(a) $x^3 - 2x^2 + 3x - 6$ (b) $12x^3 - 9x^2 - 8x + 6$ (c) $2x - 2y + ax - ay$

CRITICAL THINKING
Solve
$xy - 4x - 3y + 12 = 0.$

Solution

(a)
$x^3 - 2x^2 + 3x - 6 = (x^3 - 2x^2) + (3x - 6)$	Group terms.
$= x^2(x - 2) + 3(x - 2)$	Factor out x^2 and 3.
$= (x^2 + 3)(x - 2)$	Factor out $x - 2$.

(b)
$12x^3 - 9x^2 - 8x + 6 = (12x^3 - 9x^2) + (-8x + 6)$	Write with a plus sign between binomials.
$= 3x^2(4x - 3) - 2(4x - 3)$	Factor out $3x^2$ and -2.
$= (3x^2 - 2)(4x - 3)$	Factor out $4x - 3$.

(c)
$2x - 2y + ax - ay = (2x - 2y) + (ax - ay)$	Group terms.
$= 2(x - y) + a(x - y)$	Factor out 2 and a.
$= (2 + a)(x - y)$	Factor out $x - y$.

5.3 PUTTING IT ALL TOGETHER

The following table summarizes topics covered in this section.

Concept	Explanation	Examples
Greatest Common Factor (GCF)	We can sometimes factor the greatest common factor out of a polynomial. The GCF has the largest (integer) coefficient and greatest degree possible.	$3x^4 - 9x^3 + 12x^2 = 3x^2(x^2 - 3x + 4)$ The terms in $(x^2 - 3x + 4)$ have no obvious common factor, so $3x^2$ is called the *greatest common factor* of $3x^4 - 9x^3 + 12x^2$.
Zero-Product Property	If the product of two numbers is 0, then at least one of the numbers equals 0.	$xy = 0$ implies that $x = 0$ or $y = 0$. $x(2x + 1) = 0$ implies that $x = 0$ or $2x + 1 = 0$.
Factoring and Equations	Factoring may be used to solve equations.	$6x^2 - 9x = 0$ $3x(2x - 3) = 0$ Common factor, $3x$ $3x = 0$ or $2x - 3 = 0$ Zero-product property $x = 0$ or $x = \frac{3}{2}$ Solve for x.
Factoring by Grouping	Factoring by grouping is a method that can be used to factor four terms into a product of two binomials. It involves the associative and distributive properties.	$4x^3 + 6x^2 + 10x + 15 = (4x^3 + 6x^2) + (10x + 15)$ $= 2x^2(2x + 3) + 5(2x + 3)$ $= (2x^2 + 5)(2x + 3)$
Zeros, x-Intercepts, and Solutions	An x-intercept on the graph of $y = P(x)$ corresponds to a zero of $P(x)$ and to a solution to $P(x) = 0$. These three concepts are closely connected.	x-Intercepts: **0, 1** Zeros of $x^2 - x$: **0, 1** Solutions to $x^2 - x = 0$: **0, 1**

5.4 FACTORING TRINOMIALS

Factoring $x^2 + bx + c$ ■ Factoring Trinomials by Grouping ■
Factoring Trinomials with FOIL ■ Factoring with Graphs and Tables

A LOOK INTO MATH ▷

Items usually sell better at a lower price. At a higher price fewer items are sold, but more money is made on each item. For example, suppose that if concert tickets are priced at $100 no one will buy a ticket, but for each $1 reduction in price 100 additional tickets will be sold. If the promoters of the concert need to gross $240,000 from ticket sales, what ticket price accomplishes this goal? To solve this problem we need to set up and solve a polynomial equation. In this section we describe how to solve polynomial equations by factoring trinomials in the form $x^2 + bx + c$ and $ax^2 + bx + c$.

Factoring $x^2 + bx + c$

The product $(x + 3)(x + 4)$ can be found as follows.

$$(x + 3)(x + 4) = x^2 + 4x + 3x + 12$$
$$= x^2 + 7x + 12$$

The middle term $7x$ is found by calculating the sum $4x + 3x$, and the last term is found by calculating the product $3 \cdot 4 = 12$.

When we factor polynomials, we are *reversing* the process of multiplication. To factor $x^2 + 7x + 12$ we must find m and n that satisfy

$$x^2 + 7x + 12 = (x + m)(x + n).$$

Because

$$(x + m)(x + n) = x^2 + (m + n)x + mn,$$

it follows that $mn = 12$ and $m + n = 7$. To determine m and n we list factors of 12 and their sum, as shown in Table 5.8.

Because $3 \cdot 4 = 12$ and $3 + 4 = 7$, we can write the factored form as

$$x^2 + 7x + 12 = (x + 3)(x + 4).$$

This result can always be checked by multiplying the two binomials.

$$(x + 3)(x + 4) = x^2 + 7x + 12$$

The middle term checks.

TABLE 5.8
Factor Pairs for 12

Factors	Sum
1, 12	13
2, 6	8
3, 4	7

FACTORING $x^2 + bx + c$

To factor the trinomial $x^2 + bx + c$, find numbers m and n that satisfy

$$m \cdot n = c \quad \text{and} \quad m + n = b.$$

Then $x^2 + bx + c = (x + m)(x + n)$.

EXAMPLE 1 Factoring the form $x^2 + bx + c$

Factor each trinomial.
(a) $x^2 + 10x + 16$ (b) $x^2 - 5x - 24$ (c) $x^2 + 7x - 30$

TABLE 5.9
Factor Pairs for 16

Factors	Sum
1, 16	17
2, 8	10
4, 4	8

Solution
(a) We need to find a factor pair for 16 whose sum is 10. From Table 5.9 the required factor pair is $m = 2$ and $n = 8$. Thus
$$x^2 + 10x + 16 = (x + 2)(x + 8).$$
(b) Factors of -24 whose sum equals -5 are 3 and -8. Thus
$$x^2 - 5x - 24 = (x + 3)(x - 8).$$
(c) Factors of -30 whose sum equals 7 are -3 and 10. Thus
$$x^2 + 7x - 30 = (x - 3)(x + 10).$$

EXAMPLE 2 Removing common factors first

Factor completely.
(a) $3x^2 + 15x + 18$ (b) $5x^3 + 5x^2 - 60x$

Solution
(a) If we first factor out the common factor of 3, the resulting trinomial is easier to factor.
$$3x^2 + 15x + 18 = 3(x^2 + 5x + 6)$$
Now we find m and n such that $mn = 6$ and $m + n = 5$. These numbers are 2 and 3.
$$3x^2 + 15x + 18 = 3(x^2 + 5x + 6)$$
$$= 3(x + 2)(x + 3)$$
(b) First, we factor out the common factor of $5x$. Then we factor the resulting trinomial.
$$5x^3 + 5x^2 - 60x = 5x(x^2 + x - 12)$$
$$= 5x(x - 3)(x + 4)$$

▶ **REAL-WORLD CONNECTION** Factoring may be used to solve the problem presented in the introduction to this section. To do so we let x be the number of dollars that the price of a ticket is reduced below $100. Then $100 - x$ represents the price of a ticket. For each $1 reduction in price, 100 additional tickets will be sold, so the number of tickets sold is given by $100x$. Because the number of tickets sold times the price of each ticket equals the total sales, we need to solve
$$100x(100 - x) = 240{,}000$$
to determine when gross sales will reach $240,000. We cannot immediately apply the zero-product property because the product of $100x$ and $(100 - x)$ does not equal 0.

EXAMPLE 3 Determining ticket price

Solve the equation $100x(100 - x) = 240{,}000$ symbolically. What should be the price of the tickets and how many tickets will be sold? Support your answer graphically.

CRITICAL THINKING

In Example 3 what price will maximize the revenue from ticket sales? (*Hint:* See Figure 5.17.)

Solution

Symbolic Solution Begin by applying the distributive property.

$100x(100 - x) = 240{,}000$	Given equation
$10{,}000x - 100x^2 = 240{,}000$	Distributive property
$-100x^2 + 10{,}000x - 240{,}000 = 0$	Subtract 240,000.
$-100(x^2 - 100x + 2400) = 0$	Factor out -100.
$-100(x - 40)(x - 60) = 0$	Factor.
$x = 40 \quad \text{or} \quad x = 60$	Solve.

If $x = 40$, the price is $100 - 40 = \$60$ per ticket and $100 \cdot 40 = 4000$ tickets are sold at \$60 each. If $x = 60$, the price is $100 - 60 = \$40$ and $100 \cdot 60 = 6000$ tickets are sold at \$40 each. In either case, ticket sales are \$240,000.

Graphical Solution The graphs of $Y_1 = 100X(100 - X)$ and $Y_2 = 240000$ intersect in Figure 5.17 at $x = 40$ and $x = 60$, which agrees with the symbolic solution.

NOTE: These graphs also illustrate how revenue increases to a maximum value and then decreases as the price of the tickets decreases from \$100 to \$0.

CALCULATOR HELP

To find a point of intersection, see the Appendix (page AP-7).

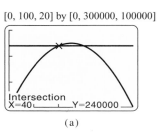

Figure 5.17

Factoring Trinomials by Grouping

In this subsection we use grouping to factor trinomials in the form $ax^2 + bx + c$ with $a \neq 1$. For example, one way to factor $3x^2 + 14x + 8$ is to find two numbers m and n such that $mn = 3 \cdot 8 = 24$ and $m + n = 14$. Because $2 \cdot 12 = 24$ and $2 + 12 = 14$, $m = 2$ and $n = 12$. Using grouping, we can now factor this trinomial.

$3x^2 + 14x + 8 = 3x^2 + 2x + 12x + 8$	Write $14x$ as $2x + 12x$.
$= (3x^2 + 2x) + (12x + 8)$	Associative property
$= x(3x + 2) + 4(3x + 2)$	Factor out x and 4.
$= (x + 4)(3x + 2)$	Factor out $3x + 2$.

> **FACTORING $ax^2 + bx + c$ BY GROUPING**
>
> To factor $ax^2 + bx + c$ perform the following steps. (Assume that a, b, and c have no factor in common.)
>
> 1. Find numbers m and n such that $mn = ac$ and $m + n = b$. (This step may require trial and error.)
> 2. Write the trinomial as $ax^2 + mx + nx + c$.
> 3. Use grouping to factor this expression as two binomials.

EXAMPLE 4   Factoring $ax^2 + bx + c$ by grouping

Factor each trinomial.
(a) $3x^2 + 17x + 10$ **(b)** $12y^2 + 5y - 3$ **(c)** $6r^2 - 19r + 10$

Solution
(a) In this trinomial $a = 3$, $b = 17$, and $c = 10$. Because $mn = ac$ and $m + n = b$, the numbers m and n satisfy $mn = 30$ and $m + n = 17$. Thus $m = \mathbf{2}$ and $n = \mathbf{15}$.

$$\begin{aligned}
3x^2 + \mathbf{17x} + 10 &= 3x^2 + \mathbf{2x + 15x} + 10 &&\text{Write } 17x \text{ as } 2x + 15x.\\
&= (3x^2 + 2x) + (15x + 10) &&\text{Associative property}\\
&= x(3x + 2) + 5(3x + 2) &&\text{Factor out } x \text{ and } 5.\\
&= (x + 5)(3x + 2) &&\text{Distributive property}
\end{aligned}$$

(b) In this trinomial $a = 12$, $b = 5$, and $c = -3$. Because $mn = ac$ and $m + n = b$, the numbers m and n satisfy $mn = -36$ and $m + n = 5$. Thus $m = \mathbf{9}$ and $n = \mathbf{-4}$.

$$\begin{aligned}
12y^2 + \mathbf{5y} - 3 &= 12y^2 + \mathbf{9y - 4y} - 3 &&\text{Write } 5y \text{ as } 9y - 4y.\\
&= (12y^2 + 9y) + (-4y - 3) &&\text{Associative property}\\
&= 3y(4y + 3) - 1(4y + 3) &&\text{Factor out } 3y \text{ and } -1.\\
&= (3y - 1)(4y + 3) &&\text{Distributive property}
\end{aligned}$$

(c) In this trinomial $a = 6$, $b = -19$, and $c = 10$. Because $mn = ac$ and $m + n = b$, the numbers m and n satisfy $mn = 60$ and $m + n = -19$. Thus $m = \mathbf{-4}$ and $n = \mathbf{-15}$.

$$\begin{aligned}
6r^2 - \mathbf{19r} + 10 &= 6r^2 - \mathbf{4r - 15r} + 10 &&\text{Write } -19r \text{ as } -4r - 15r.\\
&= (6r^2 - 4r) + (-15r + 10) &&\text{Associative property}\\
&= 2r(3r - 2) - 5(3r - 2) &&\text{Factor out } 2r \text{ and } -5.\\
&= (2r - 5)(3r - 2) &&\text{Distributive property}
\end{aligned}$$

Factoring Trinomials with FOIL

An alternative to factoring trinomials by grouping is to use FOIL in reverse. For example, the factors of $3x^2 + 7x + 2$ are two binomials.

$$3x^2 + 7x + 2 \stackrel{?}{=} (\underline{} + \underline{})(\underline{} + \underline{})$$

The expressions to be placed in the four blanks are yet to be found. By the FOIL method, we know that the product of the first terms is $3x^2$. Because $3x^2 = 3x \cdot x$, we can write

$$3x^2 + 7x + 2 \stackrel{?}{=} (\underline{\ 3x\ } + \underline{})(\underline{\ x\ } + \underline{}).$$

The product of the last terms in each binomial must equal 2. Because $2 = 1 \cdot 2$, we can put the 1 and 2 in the blanks, but we must be sure to place them correctly so that the product of the *outside terms* plus the product of the *inside terms* equals $7x$.

$$(3x + 1)(x + 2) = 3x^2 + 7x + 2$$

 $1x$
$+6x$
$\overline{7x}$ ⟵ Middle term checks.

If we had interchanged the 1 and 2, we would have obtained an incorrect result.

$$(3x + 2)(x + 1) = 3x^2 + 5x + 2$$

 $2x$
$+3x$
$\overline{5x}$ ⟵ Middle term is *not* $7x$.

In the next example we factor expressions of the form $ax^2 + bx + c$, where $a \neq 1$. In this situation, we may need to *guess and check* or use *trial and error* a few times before finding the correct factors.

EXAMPLE 5 Factoring the form $ax^2 + bx + c$

Factor each trinomial.
(a) $2x^2 + 9x + 4$ (b) $6x^2 - x - 2$ (c) $4x^3 - 14x^2 + 6x$

Solution
(a) The factors of $2x^2$ are $2x$ and x, so we begin by writing
$$2x^2 + 9x + 4 \stackrel{?}{=} (2x + \underline{})(x + \underline{}).$$
The factors of the last term, 4, are either 1 and 4 or 2 and 2. Selecting the factors 2 and 2 results in a middle term of $6x$ rather than $9x$.

$(2x + 2)(x + 2) = 2x^2 + 6x + 4$
  $\lfloor 2x \rfloor$
  $+4x$
  $\overline{6x}$ ← Middle term is *not* $9x$.

Next we try the factors 1 and 4.

$(2x + 4)(x + 1) = 2x^2 + 6x + 4$
  $\lfloor 4x \rfloor$
  $+2x$
  $\overline{6x}$ ← Middle term is *not* $9x$.

Again we obtain the wrong middle term. By interchanging the 1 and 4, we find the correct factorization.

$(2x + 1)(x + 4) = 2x^2 + 9x + 4$
  $\lfloor 1x \rfloor$
  $+8x$
  $\overline{9x}$ ← Middle term is $9x$.

NOTE: If a trinomial does not have any common factors, then its binomial factors will not have any common factors either. Thus $(2x + 2)$ and $(2x + 4)$ cannot be factors of $2x^2 + 9x + 4$. These factors can be eliminated without checking.

(b) The factors of $6x^2$ are either $2x$ and $3x$ or $6x$ and x. The factors of -2 are either -1 and 2 or 1 and -2. To obtain a middle term of $-x$ we use the following factors.

$(3x - 2)(2x + 1) = 6x^2 - x - 2$
  $\lfloor -4x \rfloor$
  $+3x$
  $\overline{-x}$ ← It checks.

To find the correct factorization we may need to guess and check a few times.

(c) Each term contains a common factor of $2x$, so we do the following step first.
$$4x^3 - 14x^2 + 6x = 2x(2x^2 - 7x + 3)$$
Next we factor $2x^2 - 7x + 3$. The factors of $2x^2$ are $2x$ and x. Because the middle term is negative, we use -1 and -3 for factors of 3.
$$4x^3 - 14x^2 + 6x = 2x(2x^2 - 7x + 3)$$
$$= 2x(2x - 1)(x - 3)$$

EXAMPLE 6　Estimating passing distance

A car traveling 48 miles per hour accelerates at a constant rate to pass a car in front of it. A no-passing zone begins 2000 feet away. The distance d traveled in feet by the car after t seconds is $d(t) = 3t^2 + 70t$. How long does it take the car to reach the no-passing zone?

Solution
We must determine the time when $d(t) = 2000$.

$$3t^2 + 70t = 2000 \quad \text{Equation to be solved}$$
$$3t^2 + 70t - 2000 = 0 \quad \text{Subtract 2000.}$$
$$(3t - 50)(t + 40) = 0 \quad \text{Factor.}$$
$$3t - 50 = 0 \quad \text{or} \quad t + 40 = 0 \quad \text{Zero-product property}$$
$$t = \frac{50}{3} \quad \text{or} \quad t = -40 \quad \text{Solve.}$$

The car reaches the no-passing zone after $\frac{50}{3} \approx 16.7$ seconds. (The solution $t = -40$ has no physical meaning in this problem.)

CRITICAL THINKING

Can every trinomial be factored with the methods discussed in this section? Try to factor the following trinomials and then make a conjecture.

$$x^2 + 2x + 2, \quad x^2 - x + 1, \quad \text{and} \quad 2x^2 + x + 2.$$

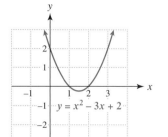

Figure 5.18

Factoring with Graphs and Tables

Polynomials can also be factored graphically. One way to do so is to use a graph of the polynomial to find its zeros. A number p is a *zero* if a value of 0 results when p is substituted in the polynomial. For example, both -2 and 2 are zeros of $x^2 - 4$ because

$$(-2)^2 - 4 = 0 \quad \text{and} \quad (2)^2 - 4 = 0.$$

Now consider the trinomial $x^2 - 3x + 2$, whose graph is shown in Figure 5.18. Its x-intercepts or zeros are 1 and 2, and this trinomial factors as

$$x^2 - 3x + 2 = (x - 1)(x - 2).$$

We generalize these concepts as follows.

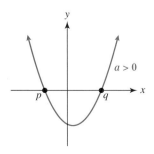

Figure 5.19

FACTORING WITH A GRAPH OR A TABLE OF VALUES

To factor the trinomial $x^2 + bx + c$, either graph or make a table of $y = x^2 + bx + c$. If the zeros of the trinomial are p and q, then the trinomial can be factored as

$$x^2 + bx + c = (x - p)(x - q).$$

If the trinomial $ax^2 + bx + c$ has zeros p and q, then it may be factored as

$$ax^2 + bx + c = a(x - p)(x - q).$$

See Figure 5.19.

EXAMPLE 7 Factoring with technology

Factor each trinomial graphically or numerically.
(a) $x^2 - 2x - 24$ (b) $2x^2 - 51x + 220$

Solution

(a) Graph $Y_1 = X{\wedge}2 - 2X - 24$. When the trace feature is used, the zeros of the trinomial are -4 and 6, as shown in Figure 5.20. Thus the trinomial factors as follows.

$$x^2 - 2x - 24 = (x - (-4))(x - 6)$$
$$= (x + 4)(x - 6)$$

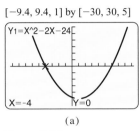
(a)

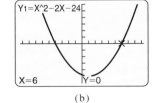
(b)

Figure 5.20

(b) Construct a table for $Y_1 = 2X{\wedge}2 - 51X + 220$. Figure 5.21 reveals that the zeros are **5.5** and **20**. The leading coefficient of $2x^2 - 51x + 220$ is 2, so we factor this expression as follows.

$$2x^2 - 51x + 220 = 2(x - 5.5)(x - 20)$$
$$= (2x - 11)(x - 20)$$

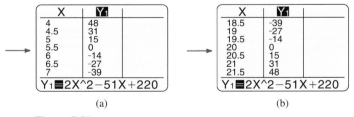

Figure 5.21

5.4 PUTTING IT ALL TOGETHER

Equations of the form $ax^2 + bx + c = 0$ occur in applications. One way to solve such equations is to factor the trinomial $ax^2 + bx + c$. This factorization may be done symbolically, graphically, or numerically. The following table includes an explanation and examples of each technique. To factor by grouping, see the box on page 224.

Technique	Explanation	Example
Symbolic Factoring	To factor $x^2 + bx + c$ find two factors of c whose sum is b.	To factor $x^2 - 3x - 4$ find two factors of -4 whose sum is -3. These factors are 1 and -4. $x^2 - 3x - 4 = (x + 1)(x - 4)$ It checks. $\longrightarrow -3x$
	To factor $ax^2 + bx + c$ find factors of ax^2 and c so that the middle term is bx. Grouping can also be used.	To factor $2x^2 + 3x - 2$ find factors of $2x^2$ and -2 so that the middle term is $3x$. $2x^2 + 3x - 2 = (2x - 1)(x + 2)$ It checks. $\longrightarrow 3x$
Graphical Factoring	If p and q are zeros of $x^2 + bx + c$, then the expression factors as $(x - p)(x - q)$. If p and q are zeros of $ax^2 + bx + c$, then the expression factors as $a(x - p)(x - q)$.	To factor $x^2 - 3x - 4$ ($a = 1$), graph $y = x^2 - 3x - 4$. The zeros (x-intercepts) are -1 and 4, and it follows that $x^2 - 3x - 4 = (x + 1)(x - 4)$.
Numerical Factoring	If p and q are zeros of $x^2 + bx + c$, then the expression factors as $(x - p)(x - q)$. If p and q are zeros of $ax^2 + bx + c$, then the expression factors as $a(x - p)(x - q)$.	To factor $2x^2 + 3x - 2$ make a table for $Y_1 = 2X^2 + 3X - 2$. The zeros are 0.5 and -2. $2x^2 + 3x - 2 = 2(x - 0.5)(x + 2)$ $= (2x - 1)(x + 2)$

5.5 SPECIAL TYPES OF FACTORING

Difference of Two Squares ▪ Perfect Square Trinomials ▪ Sum and Difference of Two Cubes

A LOOK INTO MATH ▷

Some polynomials can be factored with special methods. These factoring techniques are used in other mathematics courses to solve equations and simplify expressions. In this section we discuss these methods.

Difference of Two Squares

When we factor polynomials, we are *reversing* the process of multiplying polynomials. In Section 5.2 we discussed the equation

$$(a - b)(a + b) = a^2 - b^2.$$

We can use this equation to factor a difference of two squares. For example, if we want to factor $x^2 - 25$, we can substitute x for a and 5 for b in

$$a^2 - b^2 = (a - b)(a + b)$$

to get

$$x^2 - 5^2 = (x - 5)(x + 5).$$

> **DIFFERENCE OF TWO SQUARES**
>
> For any real numbers a and b,
>
> $$a^2 - b^2 = (a - b)(a + b).$$

NOTE: The sum of two squares *cannot* be factored (using real numbers). For example, $x^2 + y^2$ cannot be factored, whereas $x^2 - y^2$ can be factored. *It is important to remember that $x^2 + y^2 \neq (x + y)^2$.*

EXAMPLE 1 Factoring the difference of two squares

Factor each polynomial, if possible.
(a) $9x^2 - 64$ **(b)** $4x^2 + 9y^2$ **(c)** $9x^4 - y^6$ **(d)** $4a^3 - 4a$

Solution
(a) Note that $9x^2 = (3x)^2$ and $64 = 8^2$.
$$9x^2 - 64 = (3x)^2 - (8)^2$$
$$= (3x - 8)(3x + 8)$$

(b) Because $4x^2 + 9y^2$ is the *sum* of two squares, it *cannot* be factored.

(c) If we let $a^2 = 9x^4$ and $b^2 = y^6$, then $a = 3x^2$ and $b = y^3$. Thus
$$9x^4 - y^6 = (3x^2)^2 - (y^3)^2$$
$$= (3x^2 - y^3)(3x^2 + y^3).$$

(d) Start by factoring out the common factor of $4a$.
$$4a^3 - 4a = 4a(a^2 - 1)$$
$$= 4a(a - 1)(a + 1)$$

EXAMPLE 2   Applying the difference of two squares

Factor each expression.
(a) $(n + 1)^2 - 9$ (b) $x^4 - y^4$ (c) $6r^2 - 24t^4$ (d) $x^3 + 3x^2 - 4x - 12$

Solution
(a) Use $a^2 - b^2 = (a - b)(a + b)$, with $a = n + 1$ and $b = 3$.

$$\begin{aligned}
(n + 1)^2 - 9 &= (n + 1)^2 - 3^2 && 9 = 3^2 \\
&= \bigl((n + 1) - 3\bigr)\bigl((n + 1) + 3\bigr) && \text{Difference of squares} \\
&= (n - 2)(n + 4) && \text{Combine terms.}
\end{aligned}$$

(b) Use $a^2 - b^2 = (a - b)(a + b)$, with $a = x^2$ and $b = y^2$.

$$\begin{aligned}
x^4 - y^4 &= (x^2)^2 - (y^2)^2 && \text{Write as squares.} \\
&= (x^2 - y^2)(x^2 + y^2) && \text{Difference of squares} \\
&= (x - y)(x + y)(x^2 + y^2) && \text{Difference of squares again}
\end{aligned}$$

(c) Start by factoring out the common factor of 6.

$$\begin{aligned}
6r^2 - 24t^4 &= 6(r^2 - 4t^4) && \text{Factor out 6.} \\
&= 6\bigl(r^2 - (2t^2)^2\bigr) && \text{Write as squares.} \\
&= 6(r - 2t^2)(r + 2t^2) && \text{Difference of squares}
\end{aligned}$$

(d) Start factoring by using *grouping* and then factor the difference of squares.

$$\begin{aligned}
x^3 + 3x^2 - 4x - 12 &= (x^3 + 3x^2) + (-4x - 12) && \text{Associative property} \\
&= x^2(x + 3) - 4(x + 3) && \text{Factor out } x^2 \text{ and } -4. \\
&= (x^2 - 4)(x + 3) && \text{Distributive property} \\
&= (x - 2)(x + 2)(x + 3) && \text{Difference of squares}
\end{aligned}$$

Perfect Square Trinomials

In Section 5.2 we also showed how to multiply $(a + b)^2$ and $(a - b)^2$.

$$(a + b)^2 = a^2 + 2ab + b^2 \quad \text{and} \quad (a - b)^2 = a^2 - 2ab + b^2$$

The expressions $a^2 + 2ab + b^2$ and $a^2 - 2ab + b^2$ are **perfect square trinomials**. If we can recognize a perfect square trinomial, we can use these formulas to factor it.

PERFECT SQUARE TRINOMIALS

For any real numbers a and b,

$$a^2 + 2ab + b^2 = (a + b)^2 \quad \text{and}$$
$$a^2 - 2ab + b^2 = (a - b)^2.$$

EXAMPLE 3   Factoring perfect square trinomials

Factor.
(a) $x^2 + 6x + 9$ (b) $81x^2 - 72x + 16$

Solution
(a) Let $a^2 = x^2$ and $b^2 = 3^2$. For a *perfect square trinomial*, the middle term must be $2ab$.

$$2ab = 2(x)(3) = 6x,$$

which equals the given middle term. Thus $a^2 + 2ab + b^2 = (a + b)^2$ implies

$$x^2 + 6x + 9 = (x + 3)^2.$$

(b) Let $a^2 = (9x)^2$ and $b^2 = 4^2$. Again, the middle term must be $2ab$.

$$2ab = 2(9x)(4) = 72x,$$

which equals the given middle term. Thus $a^2 - 2ab + b^2 = (a-b)^2$ implies

$$81x^2 - 72x + 16 = (9x - 4)^2.$$

EXAMPLE 4 Factoring perfect square trinomials

Factor each expression.
(a) $4x^2 + 4xy + y^2$ **(b)** $9r^2 - 12rt + 4t^2$ **(c)** $25a^3 + 10a^2b + ab^2$

Solution
(a) Let $a^2 = (2x)^2$ and $b^2 = y^2$. The middle term must be $2ab$.

$$2ab = 2(2x)(y) = 4xy,$$

which equals the given middle term. Thus $a^2 + 2ab + b^2 = (a+b)^2$ implies

$$4x^2 + 4xy + y^2 = (2x + y)^2.$$

(b) Let $a^2 = (3r)^2$ and $b^2 = (2t)^2$. The middle term must be $2ab$.

$$2ab = 2(3r)(2t) = 12rt,$$

which equals the given middle term. Thus $a^2 - 2ab + b^2 = (a-b)^2$ implies

$$9r^2 - 12rt + 4t^2 = (3r - 2t)^2.$$

(c) Factor out the common factor of a. Then factor the resulting perfect square trinomial.

$$25a^3 + 10a^2b + ab^2 = a(25a^2 + 10ab + b^2)$$
$$= a(5a + b)^2$$

MAKING CONNECTIONS

Special Factoring and General Techniques

If you do *not* recognize a polynomial as being either the difference of two squares or a perfect square trinomial, then you can factor the polynomial by using the methods discussed in earlier sections.

Sum and Difference of Two Cubes

The sum or difference of two cubes may be factored. This fact is justified by the following two equations.

$$(a + b)(a^2 - ab + b^2) = a^3 + b^3 \quad \text{and}$$
$$(a - b)(a^2 + ab + b^2) = a^3 - b^3$$

These equations can be verified by multiplying the left side to obtain the right side. For example,

$$(a + b)(a^2 - ab + b^2) = a \cdot a^2 - a \cdot ab + a \cdot b^2 + b \cdot a^2 - b \cdot ab + b \cdot b^2$$
$$= a^3 - a^2b + ab^2 + a^2b - ab^2 + b^3$$
$$= a^3 + b^3.$$

SUM AND DIFFERENCE OF TWO CUBES

For any real numbers a and b,

$$a^3 + b^3 = (a + b)(a^2 - ab + b^2) \quad \text{and}$$
$$a^3 - b^3 = (a - b)(a^2 + ab + b^2).$$

EXAMPLE 5 Factoring the sum and difference of two cubes

Factor each polynomial.
(a) $x^3 + 8$ **(b)** $27x^3 - 64y^3$

Solution
(a) Because $x^3 = (x)^3$ and $8 = 2^3$, we let $a = x, b = 2$, and factor. Substituting in

$$a^3 + b^3 = (a + b)(a^2 - ab + b^2)$$

gives

$$x^3 + 2^3 = (x + 2)(x^2 - x \cdot 2 + 2^2)$$
$$= (x + 2)(x^2 - 2x + 4).$$

Note that the quadratic factor does not factor further.

(b) Here, $27x^3 = (3x)^3$ and $64y^3 = (4y)^3$, so

$$27x^3 - 64y^3 = (3x)^3 - (4y)^3.$$

Substituting $a = 3x$ and $b = 4y$ in

$$a^3 - b^3 = (a - b)(a^2 + ab + b^2)$$

gives

$$(3x)^3 - (4y)^3 = (3x - 4y)\big((3x)^2 + 3x \cdot 4y + (4y)^2\big)$$
$$= (3x - 4y)(9x^2 + 12xy + 16y^2).$$

EXAMPLE 6 Factoring the sum and difference of two cubes

Factor each expression.
(a) $x^6 + 8y^3$ **(b)** $27p^9 - 8q^6$

Solution
(a) Let $a^3 = (x^2)^3$ and $b^3 = (2y)^3$. Then $a^3 + b^3 = (a + b)(a^2 - ab + b^2)$ implies

$$x^6 + 8y^3 = (x^2 + 2y)(x^4 - 2x^2y + 4y^2).$$

(b) Let $a^3 = (3p^3)^3$ and $b^3 = (2q^2)^3$. Then $a^3 - b^3 = (a - b)(a^2 + ab + b^2)$ implies

$$27p^9 - 8q^6 = (3p^3 - 2q^2)(9p^6 + 6p^3q^2 + 4q^4).$$

5.5 PUTTING IT ALL TOGETHER

In this section we presented three special types of factoring, which are summarized in the following table.

Type of Factoring	Description	Example
The Difference of Two Squares	To factor the difference of two squares use $a^2 - b^2 = (a - b)(a + b).$ **Note:** $a^2 + b^2$ cannot be factored.	$25x^2 - 16 = (5x - 4)(5x + 4)$ $a = 5x, b = 4$ $4x^2 + y^2$ cannot be factored.

continued on next page

continued from previous page

Type of Factoring	Description	Example
A Perfect Square Trinomial	To factor a perfect square trinomial use $a^2 + 2ab + b^2 = (a + b)^2$ or $a^2 - 2ab + b^2 = (a - b)^2$.	$x^2 + 4x + 4 = (x + 2)^2 \quad a = x, b = 2$ $16x^2 - 24x + 9 = (4x - 3)^2 \quad a = 4x, b = 3$ Be sure to check the middle term.
The Sum and Difference of Two Cubes	The sum and difference of two cubes can be factored as $a^3 + b^3 = (a + b) \cdot (a^2 - ab + b^2)$ or $a^3 - b^3 = (a - b) \cdot (a^2 + ab + b^2)$.	$x^3 + 8y^3 = (x + 2y)(x^2 - x \cdot 2y + (2y)^2)$ $= (x + 2y)(x^2 - 2xy + 4y^2)$ $a = x, b = 2y$ $125x^3 - 64 = (5x - 4)((5x)^2 + 5x \cdot 4 + 4^2)$ $= (5x - 4)(25x^2 + 20x + 16)$ $a = 5x, b = 4$

5.6 SUMMARY OF FACTORING

Guidelines for Factoring Polynomials ▪ Factoring Polynomials

A LOOK INTO MATH ▷ Mathematics is being used more and more to model all types of phenomena. Modeling is not only used in science; it is also used to describe consumer behavior and to search the Internet. To model the world around us, mathematical equations need to be solved. One important technique for solving equations is to set an expression equal to 0 and then factor. Factoring is an important method for solving equations and for simplifying complicated expressions. In this section, we discuss general guidelines that can be used to factor polynomials.

Guidelines for Factoring Polynomials

The following guidelines can be used to factor polynomials in general.

FACTORING POLYNOMIALS

STEP 1: Factor out the greatest common factor, if possible.

STEP 2: **A.** If the polynomial has *four terms*, try factoring by grouping.
  B. If the polynomial is a *binomial*, try one of the following.
  1. $a^2 - b^2 = (a - b)(a + b)$ — Difference of two squares
  2. $a^3 - b^3 = (a - b)(a^2 + ab + b^2)$ — Difference of two cubes
  3. $a^3 + b^3 = (a + b)(a^2 - ab + b^2)$ — Sum of two cubes
  C. If the polynomial is a *trinomial*, check for a perfect square.
  1. $a^2 + 2ab + b^2 = (a + b)^2$ — Perfect square trinomial
  2. $a^2 - 2ab + b^2 = (a - b)^2$ — Perfect square trinomial

  Otherwise, apply FOIL or grouping, as described in Section 5.4.

STEP 3: Check to make sure that the polynomial is *completely* factored.

These guidelines give a general strategy for factoring a polynomial by hand. It is important to always perform Step 1 first. Factoring out the greatest common factor first usually makes it easier to factor the resulting polynomial. In the next subsection we apply these guidelines to several polynomials.

Factoring Polynomials

In the first example, we apply Step 1 to a polynomial with a common factor.

EXAMPLE 1 Factoring out a common factor

Factor $3x^3 - 9x^2$.

Solution

STEP 1: The greatest common factor is $3x^2$.
$$3x^3 - 9x^2 = 3x^2(x - 3)$$
STEP 2B: The binomial, $x - 3$, cannot be factored further.
STEP 3: The completely factored polynomial is $3x^2(x - 3)$.

In several of the next examples we apply more than one factoring technique.

EXAMPLE 2 Factoring a difference of squares

Factor $5x^3 - 20x$.

Solution

STEP 1: The greatest common factor is $5x$.
$$5x^3 - 20x = 5x(x^2 - 4)$$
STEP 2B: We can factor the binomial $x^2 - 4$ as a difference of squares.
$$5x(x^2 - 4) = 5x(x - 2)(x + 2)$$
STEP 3: The completely factored polynomial is $5x(x - 2)(x + 2)$.

EXAMPLE 3 Factoring a difference of squares

Factor $5x^4 - 80$.

Solution

STEP 1: The greatest common factor is 5.
$$5x^4 - 80 = 5(x^4 - 16)$$
STEP 2B: We can factor $x^4 - 16$ as a difference of squares *twice*.
$$5(x^4 - 16) = 5(x^2 - 4)(x^2 + 4) = 5(x - 2)(x + 2)(x^2 + 4)$$
The sum of squares, $x^2 + 4$, cannot be factored further.
STEP 3: The completely factored polynomial is $5(x - 2)(x + 2)(x^2 + 4)$.

EXAMPLE 4 Factoring a perfect square trinomial

Factor $12x^3 + 36x^2 + 27x$.

Solution

STEP 1: The greatest common factor is $3x$.
$$12x^3 + 36x^2 + 27x = 3x(4x^2 + 12x + 9)$$
STEP 2C: The trinomial $4x^2 + 12x + 9$ is a perfect square trinomial.
$$3x(4x^2 + 12x + 9) = 3x(2x + 3)(2x + 3)$$
STEP 3: The completely factored polynomial is $3x(2x + 3)^2$.

EXAMPLE 5 Factoring a sum of cubes

Factor $-8x^4 - 27x$.

Solution

STEP 1: The negative of the greatest common factor is $-x$.
$$-8x^4 - 27x = -x(8x^3 + 27)$$

STEP 2B: The binomial $8x^3 + 27$ can be factored as a sum of cubes.
$$-x(8x^3 + 27) = -x(2x + 3)(4x^2 - 6x + 9)$$

The trinomial, $4x^2 - 6x + 9$, cannot be factored further.

STEP 3: The completely factored polynomial is $-x(2x + 3)(4x^2 - 6x + 9)$.

EXAMPLE 6 Factoring by grouping

Factor $x^3 - 2x^2 + x - 2$.

Solution

STEP 1: There are no common factors.

STEP 2A: Because there are four terms, we apply grouping.

$$\begin{aligned} x^3 - 2x^2 + x - 2 &= (x^3 - 2x^2) + (x - 2) && \text{Associative property} \\ &= x^2(x - 2) + 1(x - 2) && \text{Distributive property} \\ &= (x^2 + 1)(x - 2) && \text{Distributive property} \end{aligned}$$

The sum of squares, $x^2 + 1$, cannot be factored further.

STEP 3: The completely factored polynomial is $(x^2 + 1)(x - 2)$.

EXAMPLE 7 Factoring a trinomial

Factor $2x^4 - 7x^3 + 6x^2$.

Solution

STEP 1: The greatest common factor is x^2.
$$2x^4 - 7x^3 + 6x^2 = x^2(2x^2 - 7x + 6)$$

STEP 2C: We can factor the resulting trinomial by using FOIL.
$$x^2(2x^2 - 7x + 6) = x^2(x - 2)(2x - 3)$$

STEP 3: The completely factored polynomial is $x^2(x - 2)(2x - 3)$.

EXAMPLE 8 Factoring with two variables

Factor $12x^2 - 27a^2$.

Solution

STEP 1: The greatest common factor is 3.
$$12x^2 - 27a^2 = 3(4x^2 - 9a^2)$$

STEP 2B: We can factor the binomial as the difference of squares.
$$3(4x^2 - 9a^2) = 3(2x - 3a)(2x + 3a)$$

STEP 3: The completely factored polynomial is $3(2x - 3a)(2x + 3a)$.

In the next example, we use common factors, grouping, the difference of two cubes, and the difference of two squares to factor a polynomial.

EXAMPLE 9 Applying several techniques

Factor $2x^5 - 8x^3 - 2x^2 + 8$.

Solution

STEP 1: The expression has a greatest common factor of 2.

$$2x^5 - 8x^3 - 2x^2 + 8 = 2(x^5 - 4x^3 - x^2 + 4)$$

STEP 2A: The resulting four terms can be factored by grouping.

$$
\begin{aligned}
2(x^5 - 4x^3 - x^2 + 4) &= 2((x^5 - 4x^3) + (-x^2 + 4)) &&\text{Associative property} \\
&= 2(x^3(x^2 - 4) - 1(x^2 - 4)) &&\text{Distributive property} \\
&= 2(x^3 - 1)(x^2 - 4) &&\text{Distributive property}
\end{aligned}
$$

STEP 2B: Both of the resulting binomials can be factored further as the difference of two cubes and the difference of two squares.

$$2(x^3 - 1)(x^2 - 4) = 2(x - 1)(x^2 + x + 1)(x - 2)(x + 2)$$

The trinomial, $x^2 + x + 1$, cannot be factored further.

STEP 3: The completely factored polynomial is $2(x - 1)(x^2 + x + 1)(x - 2)(x + 2)$.

5.6 PUTTING IT ALL TOGETHER

When using the guidelines presented in this section for general factoring strategy, the following rules can be helpful.

Concept	Explanation	Examples
Greatest Common Factor	Factor out the greatest common factor, or monomial, that occurs in each term.	$4x^2 - 8x = 4x(x - 2)$ $2x^2 - 4x + 8 = 2(x^2 - 2x + 4)$ $x^5 + x^3 = x^3(x^2 + 1)$
Factoring by Grouping	Use the associative and distributive properties to factor a polynomial with four terms.	$x^3 - 2x^2 + 3x - 6$ $= (x^3 - 2x^2) + (3x - 6)$ $= x^2(x - 2) + 3(x - 2)$ $= (x^2 + 3)(x - 2)$
Factoring Binomials	Use the difference of squares, the difference of cubes, or the sum of cubes.	$4x^2 - 9 = (2x - 3)(2x + 3)$ $x^3 - 8 = (x - 2)(x^2 + 2x + 4)$ $x^3 + 8 = (x + 2)(x^2 - 2x + 4)$
Factoring Trinomials	Use FOIL or grouping to factor a trinomial, after factoring out the greatest common factor.	$x^2 - 3x + 2 = (x - 1)(x - 2)$ $6x^2 + 13x + 5$ $= (6x^2 + 3x) + (10x + 5)$ $= 3x(2x + 1) + 5(2x + 1)$ $= (3x + 5)(2x + 1)$

5.7 POLYNOMIAL EQUATIONS

Quadratic Equations ▪ Higher Degree Equations ▪ Equations in Quadratic Form ▪ Applications

A LOOK INTO MATH ▷

Polynomials are used to solve important problems. One example is found in highway construction where the elevations of hills and valleys are modeled by quadratic polynomials. (*Source:* F. Mannering and W. Kilareski, *Principles of Highway Engineering and Traffic Analysis*.)

The quadratic polynomial

$$h(x) = -0.0001x^2 + 100$$

could model the height of a hill, where $x = 0$ corresponds to the peak or crest of the hill, as illustrated in Figure 5.22. To determine any locations where the height of the hill is 75 feet, we can solve the *quadratic equation*

$$-0.0001x^2 + 100 = 75.$$

Figure 5.22

Quadratic Equations

Any **quadratic equation** can be written as $ax^2 + bx + c = 0$, where a, b, and c are constants, with $a \neq 0$. Quadratic equations can sometimes be solved by factoring and then applying the zero-product property. They can also be solved graphically. These techniques are demonstrated in the next two examples.

EXAMPLE 1 Solving a quadratic equation

Solve $x^2 - 4 = 0$ symbolically and graphically.

Solution
Symbolic Solution Start by factoring the left side of the equation.

$x^2 - 4 = 0$	Given equation
$(x - 2)(x + 2) = 0$	Difference of squares
$x - 2 = 0$ or $x + 2 = 0$	Zero-product property
$x = 2$ or $x = -2$	Solve each equation.

The solutions are -2 and 2.

Graphical Solution Start by making Table 5.10. Plot these points and connect them with a smooth curve called a parabola, as shown in Figure 5.23. The *x*-intercepts, or zeros, are -2 and 2, which are also the solutions to the equation.

TABLE 5.10 $y = x^2 - 4$

x	-3	-2	-1	0	1	2	3
y	5	0	-3	-4	-3	0	5

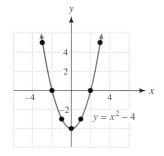

Figure 5.23

EXAMPLE 2   Solving quadratic equations

Solve each equation.
(a) $4x^2 - 20x + 25 = 0$ **(b)** $3x^2 + 11x = 20$ **(c)** $x^2 + 25 = 0$

Solution

(a) Start by factoring the left side, which is a perfect square trinomial.

$$4x^2 - 20x + 25 = 0 \quad \text{Given equation}$$
$$(2x - 5)(2x - 5) = 0 \quad \text{Factor.}$$
$$2x - 5 = 0 \quad \text{Zero-product property}$$
$$x = \tfrac{5}{2} \quad \text{Solve.}$$

The only solution is $\tfrac{5}{2}$.

(b) Start by subtracting 20 from each side to obtain a 0 on the right side.

$$3x^2 + 11x = 20 \quad \text{Given equation}$$
$$3x^2 + 11x - 20 = 0 \quad \text{Subtract 20.}$$
$$(x + 5)(3x - 4) = 0 \quad \text{Factor.}$$
$$x + 5 = 0 \quad \text{or} \quad 3x - 4 = 0 \quad \text{Zero-product property}$$
$$x = -5 \quad \text{or} \quad x = \tfrac{4}{3} \quad \text{Solve.}$$

The solutions are -5 and $\tfrac{4}{3}$.

(c) $x^2 + 25 = 0$ implies that $x^2 = -25$, which has no solutions because $x^2 \geq 0$ for all real numbers.

MAKING CONNECTIONS

Quadratic Equations and Solutions

Example 2 demonstrates that a quadratic equation can have no solutions, one solution, or two solutions. The following graphs illustrate this concept.

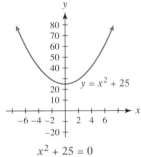

$x^2 + 25 = 0$
No solutions

$4x^2 - 20x + 25 = 0$
One solution

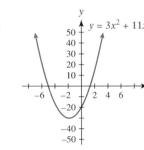

$3x^2 + 11x = 20$
Two solutions

EXAMPLE 3 Determining highway elevations

Solve the equation $-0.0001x^2 + 100 = 75$, which is discussed in the introduction to this section. See Figure 5.22 on page 238.

Solution
Start by subtracting 75 from each side. Then multiply each side by $-10,000$ to clear decimals.

$$-0.0001x^2 + 100 = 75 \quad \text{Given equation}$$
$$-0.0001x^2 + 25 = 0 \quad \text{Subtract 75.}$$
$$-10,000(-0.0001x^2 + 25) = -10,000(0) \quad \text{Multiply by } -10,000.$$
$$x^2 - 250,000 = 0 \quad \text{Distributive property}$$
$$(x - 500)(x + 500) = 0 \quad \sqrt{250,000} = 500$$
$$x - 500 = 0 \quad \text{or} \quad x + 500 = 0 \quad \text{Zero-product property}$$
$$x = 500 \quad \text{or} \quad x = -500 \quad \text{Solve each equation.}$$

Thus the highway elevation is 75 feet high a distance of 500 feet on either side of the crest of the hill.

Higher Degree Equations

Using the techniques of factoring that we discussed in earlier sections, we can solve equations that involve higher degree polynomials. They can have more than two solutions.

EXAMPLE 4 Solving higher degree equations

Solve each equation.
(a) $x^3 = 4x$ **(b)** $2x^3 + 2x^2 - 12x = 0$ **(c)** $x^3 - 5x^2 - x + 5 = 0$

Solution
(a) Start by subtracting $4x$ from each side of $x^3 = 4x$ to obtain a 0 on the right side. (Do *not* divide each side by x because the solution $x = 0$ will be lost.)

$$x^3 = 4x \quad \text{Given equation}$$
$$x^3 - 4x = 0 \quad \text{Subtract } 4x.$$
$$x(x^2 - 4) = 0 \quad \text{Factor out } x.$$
$$x(x - 2)(x + 2) = 0 \quad \text{Difference of squares}$$
$$x = 0 \quad \text{or} \quad x - 2 = 0 \quad \text{or} \quad x + 2 = 0 \quad \text{Zero-product property}$$
$$x = 0 \quad \text{or} \quad x = 2 \quad \text{or} \quad x = -2 \quad \text{Solve each equation.}$$

The solutions are $-2, 0,$ and 2.

(b) Start by factoring out the common factor $2x$.

$$2x^3 + 2x^2 - 12x = 0 \quad \text{Given equation}$$
$$2x(x^2 + x - 6) = 0 \quad \text{Factor out } 2x.$$
$$2x(x + 3)(x - 2) = 0 \quad \text{Factor the trinomial.}$$
$$2x = 0 \quad \text{or} \quad x + 3 = 0 \quad \text{or} \quad x - 2 = 0 \quad \text{Zero-product property}$$
$$x = 0 \quad \text{or} \quad x = -3 \quad \text{or} \quad x = 2 \quad \text{Solve each equation.}$$

The solutions are $-3, 0,$ and 2.

(c) To solve this equation use *grouping* to factor the left side.

$$x^3 - 5x^2 - x + 5 = 0 \quad \text{Given equation}$$
$$(x^3 - 5x^2) + (-x + 5) = 0 \quad \text{Associative property}$$
$$x^2(x - 5) - 1(x - 5) = 0 \quad \text{Factor out } x^2 \text{ and } -1.$$
$$(x^2 - 1)(x - 5) = 0 \quad \text{Distributive property}$$
$$(x - 1)(x + 1)(x - 5) = 0 \quad \text{Difference of squares}$$
$$x - 1 = 0 \quad \text{or} \quad x + 1 = 0 \quad \text{or} \quad x - 5 = 0 \quad \text{Zero-product property}$$
$$x = 1 \quad \text{or} \quad x = -1 \quad \text{or} \quad x = 5 \quad \text{Solve each equation.}$$

The solutions are -1, 1, and 5.

In the next example, we demonstrate how an equation can be solved symbolically, graphically, and numerically with the aid of a graphing calculator.

EXAMPLE 5 Solving a polynomial equation

Solve $16x^4 - 64x^3 + 64x^2 = 0$ symbolically, graphically, and numerically.

Solution

Symbolic Solution Begin by factoring out the common factor $16x^2$.

$$16x^4 - 64x^3 + 64x^2 = 0 \quad \text{Given equation}$$
$$16x^2(x^2 - 4x + 4) = 0 \quad \text{Factor out } 16x^2.$$
$$16x^2(x - 2)^2 = 0 \quad \text{Perfect square trinomial}$$

Solving results in

$$16x^2 = 0 \quad \text{or} \quad (x - 2)^2 = 0 \quad \text{Zero-product property}$$
$$x = 0 \quad \text{or} \quad x = 2. \quad \text{Solve.}$$

Graphical Solution Graph $Y_1 = 16X^4 - 64X^3 + 64X^2$, as shown in Figures 5.24(a) and 5.24(b). When the trace feature is used, solutions to $y_1 = 0$ occur when $x = 0$ and $x = 2$.

Numerical Solution Construct the table for $Y_1 = 16X^4 - 64X^3 + 64X^2$, as shown in Figure 5.24(c). Solutions occur at $x = 0$ and $x = 2$.

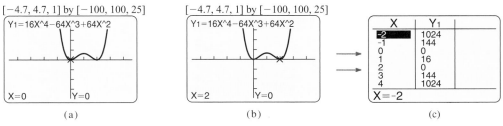

Figure 5.24

Equations in Quadratic Form

Sometimes equations do not appear to be quadratic, but they can be solved by using factoring techniques that we have already applied to quadratic equations. In these situations we often factor more than once, as demonstrated in the next example.

EXAMPLE 6 Solving equations in quadratic form

Solve each equation.
(a) $x^4 - 81 = 0$ (b) $x^4 - 5x^2 + 4 = 0$

Solution

(a) To solve this equation use the difference of squares twice.

$$(x^2)^2 - 9^2 = 0 \quad \text{Rewrite given equation.}$$
$$(x^2 - 9)(x^2 + 9) = 0 \quad \text{Difference of squares}$$
$$(x - 3)(x + 3)(x^2 + 9) = 0 \quad \text{Difference of squares again}$$
$$x - 3 = 0 \quad \text{or} \quad x + 3 = 0 \quad \text{or} \quad x^2 + 9 = 0 \quad \text{Zero-product property}$$
$$x = 3 \quad \text{or} \quad x = -3 \quad \text{or} \quad x^2 = -9 \quad \text{Solve each equation.}$$

The solutions are -3 and 3 because $x^2 = -9$ has no real number solutions.

(b) To solve this equation start by factoring the trinomial.

$$x^4 - 5x^2 + 4 = 0 \quad \text{Given equation}$$
$$(x^2 - 4)(x^2 - 1) = 0 \quad \text{Factor.}$$
$$(x - 2)(x + 2)(x - 1)(x + 1) = 0 \quad \text{Difference of squares twice}$$
$$x = 2 \quad \text{or} \quad x = -2 \quad \text{or} \quad x = 1 \quad \text{or} \quad x = -1 \quad \text{Solve.}$$

The solutions are $-2, -1, 1,$ and 2.

NOTE: To factor the equation in Example 6(a), it may be helpful to let $z = x^2$ and $z^2 = x^4$. Then the given equation becomes $z^2 - 81 = 0$, or $(z - 9)(z + 9) = 0$ after factoring. Substituting $z = x^2$ gives the factored equation $(x^2 - 9)(x^2 + 9) = 0$. Similar comments can be made for Example 6(b).

MAKING CONNECTIONS

Symbolic, Graphical, and Numerical Solutions

The following illustrates how to solve $x^3 - 4x = 0$ with each method. Note that the solutions to $x^3 - 4x = 0$, the x-intercepts of the graph of $y = x^3 - 4x$, and the zeros of $x^3 - 4x$ are *all identical*.

Symbolic Solution

$$x^3 - 4x = 0$$
$$x(x^2 - 4) = 0$$
$$x(x - 2)(x + 2) = 0$$
$$x = 0, x = 2, x = -2$$

Graphical Solution

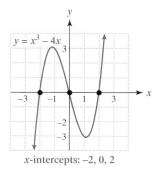

x-intercepts: $-2, 0, 2$

Numerical Solution

x	$x^3 - 4x$
-3	-15
-2	0
-1	3
0	0
1	-3
2	0
3	15

Applications

▶ **REAL-WORLD CONNECTION** Many times applications involving geometry make use of quadratic equations. In the next example we solve a quadratic equation to find the dimensions of a picture.

EXAMPLE 7 Finding the dimensions of a picture

A frame surrounding a picture is 3 inches wide. The picture inside the frame is 5 inches wider than it is high. If the overall area of the picture and frame is 336 square inches, find the dimensions of the picture inside the frame.

Solution

Let x be the height of the picture and $x + 5$ be its width. The dimensions of the picture and frame, in inches, are illustrated in Figure 5.25.

The overall area is given by $(x + 6)(x + 11)$, which equals 336 square inches.

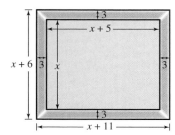

Figure 5.25

$$(x + 6)(x + 11) = 336 \quad \text{Equation to solve}$$
$$x^2 + 17x + 66 = 336 \quad \text{Multiply binomials.}$$
$$x^2 + 17x - 270 = 0 \quad \text{Subtract 336.}$$
$$(x - 10)(x + 27) = 0 \quad \text{Factor.}$$
$$x = 10 \quad \text{or} \quad x = -27 \quad \text{Zero-product property; solve.}$$

The only valid solution for x is 10 inches. Because the width is 5 inches more than the height, the dimensions are 10 inches by 15 inches.

▶ **REAL-WORLD CONNECTION** Highway engineers often use quadratic functions to calculate stopping distances for cars. The faster a car is traveling, the farther it takes for the car to stop. The stopping distance for a car can be used to estimate its speed, as demonstrated in the next example. (*Source:* F. Mannering.)

EXAMPLE 8 Finding the speed of a car

On wet, level pavement highway engineers sometimes estimate stopping distance d in feet for a car traveling at x miles per hour by

$$d(x) = \frac{1}{9}x^2 + \frac{11}{3}x.$$

Use this formula to approximate the speed of a car that takes 180 feet to stop.

Solution

We need to solve the equation $d(x) = 180$. Start by multiplying by 9 to clear fractions.

$$\frac{1}{9}x^2 + \frac{11}{3}x = 180 \quad \text{Equation to be solved}$$
$$9\left(\frac{1}{9}x^2 + \frac{11}{3}x\right) = 9 \cdot 180 \quad \text{Multiply by 9.}$$
$$x^2 + 33x = 1620 \quad \text{Distributive property}$$
$$x^2 + 33x - 1620 = 0 \quad \text{Subtract 1620.}$$
$$(x - 27)(x + 60) = 0 \quad \text{Factor.}$$
$$x = 27 \quad \text{or} \quad x = -60 \quad \text{Zero-product property; solve.}$$

The speed of the car is 27 miles per hour.

NOTE: To factor $x^2 + 33x - 1620$ we find values for m and n such that $mn = -1620$ and $m + n = 33$, which may require some trial and error.

EXAMPLE 9 Finding the dimensions of an iPod Mini

The width of an iPod Mini is 1.5 inches more than its thickness, and its length is 3 inches more than its thickness. If the volume of an iPod Mini is 3.5 cubic inches, find its dimensions graphically.

Solution

Let x be its thickness, $x + 1.5$ be its width, and $x + 3$ be its length. Because volume equals thickness times width times length, we have

$$x(x + 1.5)(x + 3) = 3.5.$$

The graphs of $Y_1 = X(X + 1.5)(X + 3)$ and $Y_2 = 3.5$ intersect at $(0.5, 3.5)$, as shown in Figure 5.26. Thus the iPod Mini is 0.5 inch thick, 2 inches wide, and 3.5 inches long.

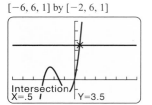

$[-6, 6, 1]$ by $[-2, 6, 1]$

Figure 5.26

5.7 PUTTING IT ALL TOGETHER

The following is a typical strategy for solving a polynomial equation. When you have finished, *be sure to check your answers*.

STEP 1: If necessary, rewrite the equation so that a 0 appears on one side of the equation.
STEP 2: Factor out common factors.
STEP 3: Factor the remaining polynomial, using the techniques presented in this chapter.
STEP 4: Apply the *zero-product* property.
STEP 5: Solve each resulting equation.

The following table summarizes how to solve three types of higher degree polynomial equations.

Concept	Explanation	Example
Higher Degree Equations	Factor out common factors; then use the techniques for factoring.	$2x^3 - 32x = 0$ $2x(x^2 - 16) = 0$ $2x(x - 4)(x + 4) = 0$ $x = 0$ or $x = 4$ or $x = -4$
Equations in Quadratic Form	Factor these forms as $ax^2 + bx + c$.	$x^4 - 13x^2 + 36 = 0$ $(x^2 - 4)(x^2 - 9) = 0$ $(x - 2)(x + 2)(x - 3)(x + 3) = 0$ $x = 2$ or $x = -2$ or $x = 3$ or $x = -3$
Solving by Grouping	Use grouping to help factor a cubic expression having four terms. Apply the distributive property.	$x^3 - 7x^2 - 16x + 112 = 0$ $(x^3 - 7x^2) + (-16x + 112) = 0$ $x^2(x - 7) - 16(x - 7) = 0$ $(x^2 - 16)(x - 7) = 0$ $(x - 4)(x + 4)(x - 7) = 0$ $x = 4$ or $x = -4$ or $x = 7$

CHAPTER 5 SUMMARY

SECTION 5.1 ■ POLYNOMIAL FUNCTIONS

Monomials and Polynomials A monomial is a term with only *nonnegative integer* exponents. A polynomial is either a monomial or a sum of monomials.

Examples: Monomial: $-4x^3y^2$ has degree 5 and coefficient -4

Polynomial: $3x^2 - 4x + 2$ has degree 2 and leading coefficient 3

Addition and Subtraction of Polynomials To add two polynomials, combine like terms. When subtracting two polynomials, add the first polynomial and the opposite of the second polynomial. Combine like terms.

Examples: $(3x^3 - 4x) + (-5x^3 + 7x) = (3 - 5)x^3 + (-4 + 7)x$
$= -2x^3 + 3x$

$(3xy^2 + 4x^2) - (4xy^2 - 6x) = (3xy^2 + 4x^2) + (-4xy^2 + 6x)$
$= 3xy^2 - 4xy^2 + 4x^2 + 6x$
$= -xy^2 + 4x^2 + 6x$

Polynomial Functions If the formula for a function f is a polynomial, then f is a polynomial function. The following are examples of polynomial functions.

$f(x) = 2x - 7$ Linear   Degree 1
$f(x) = 4x^2 - 5x + 1$ Quadratic   Degree 2
$f(x) = 9x^3 + 6x$ Cubic   Degree 3

Example: $f(x) = 4x^3 - 2x^2 + 4$ represents a polynomial function of one variable.
$f(2) = 4(2)^3 - 2(2)^2 + 4 = 32 - 8 + 4 = 28$

SECTION 5.2 ■ MULTIPLICATION OF POLYNOMIALS

Product of Two Polynomials The product of two polynomials may be found by multiplying every term in the first polynomial by every term in the second polynomial.

Example: $(2x + 3)(3x - 4) = 6x^2 - 8x + 9x - 12 = 6x^2 + x - 12$

Special Products

Product of a Sum and Difference $(a + b)(a - b) = a^2 - b^2$

Example: $(x + 3)(x - 3) = x^2 - 9$

Square of a Binomial $(a + b)^2 = a^2 + 2ab + b^2$
$(a - b)^2 = a^2 - 2ab + b^2$

Examples: $(x + 3)^2 = x^2 + 6x + 9$
$(x - 3)^2 = x^2 - 6x + 9$

SECTION 5.3 ■ FACTORING POLYNOMIALS

Common Factors   When factoring a polynomial, start by factoring out the greatest common factor (GCF).

Examples: $8z^3 - 12z^2 + 16z = 4z(2z^2 - 3z + 4)$
$18x^2y^3 - 12x^3y^2 = 6x^2y^2(3y - 2x)$

Zero-Product Property   For all real numbers a and b, if $ab = 0$, then $a = 0$ or $b = 0$ (or both).

Example: $(x - 3)(x + 2) = 0$ implies that either $x - 3 = 0$ or $x + 2 = 0$.

Solving Polynomial Equations by Factoring   Obtain a zero on one side of the equation. Factor the other side of the equation and apply the zero-product property.

Example: $x^2 - 7x = 0$
$x(x - 7) = 0$ Factor out common factor.
$x = 0$ or $x = 7$ Zero-product property; solve.

Factoring by Grouping   Grouping can be used to factor some polynomial expressions with four terms by applying the distributive and associative properties.

Example: $x^3 + x^2 + 5x + 5 = (x^3 + x^2) + (5x + 5)$ Associative property
$= x^2(x + 1) + 5(x + 1)$ Distributive property
$= (x^2 + 5)(x + 1)$ Distributive property

SECTION 5.4 ■ FACTORING TRINOMIALS

Factoring $x^2 + bx + c$ Find integers m and n that satisfy $mn = c$ and $m + n = b$. Then $x^2 + bx + c = (x + m)(x + n)$.

Example: $x^2 + 3x + 2 = (x + 1)(x + 2)$ $m = 1, n = 2$

Factoring $ax^2 + bx + c$ Either grouping or FOIL (in reverse) can be used to factor $ax^2 + bx + c$. Trinomials can also be factored graphically or numerically.

Examples: *Grouping*   Find m and n so that $mn = ac = 30$ and $m + n = b = 17$.

$6x^2 + 17x + 5 = (6x^2 + 2x) + (15x + 5)$ $m = 2, n = 15$
$= 2x(3x + 1) + 5(3x + 1)$
$= (2x + 5)(3x + 1)$

FOIL   Correctly factoring a trinomial requires checking the middle term.

$2x^2 + 3x - 14 = (2x + 7)(x - 2)$
It checks. $\longrightarrow \dfrac{\begin{array}{r} 7x \\ -4x \end{array}}{3x}$

Graphically or Numerically   The x-intercepts on the graph of $f(x) = 2x^2 + x - 1$ are $x = 0.5$ and $x = -1$. The zeros of f are 0.5 and -1. Thus $f(x)$ can be factored as

$2x^2 + x - 1 = 2(x - 0.5)(x + 1)$
$= (2x - 1)(x + 1)$.

SECTION 5.5 ■ SPECIAL TYPES OF FACTORING

Difference of Two Squares $\quad a^2 - b^2 = (a - b)(a + b)$

Example: $\quad 49x^2 - 36y^2 = (7x)^2 - (6y)^2 = (7x - 6y)(7x + 6y) \quad a = 7x, b = 6y$

Perfect Square Trinomials
$$a^2 + 2ab + b^2 = (a + b)^2 \quad \text{and} \quad a^2 - 2ab + b^2 = (a - b)^2$$

Examples: $\quad x^2 + 2xy + y^2 = (x + y)^2 \quad$ and $\quad 9r^2 - 12r + 4 = (3r - 2)^2$

Sum and Difference of Two Cubes
$$a^3 + b^3 = (a + b)(a^2 - ab + b^2) \quad \text{and}$$
$$a^3 - b^3 = (a - b)(a^2 + ab + b^2)$$

Examples: $\quad x^3 + y^3 = (x + y)(x^2 - xy + y^2) \qquad a = x, b = y$
$\qquad\qquad 8t^3 - 27r^6 = (2t - 3r^2)(4t^2 + 6tr^2 + 9r^4) \qquad a = 2t, b = 3r^2$

SECTION 5.6 ■ SUMMARY OF FACTORING

Factoring Polynomials The following steps can be used as general guidelines for factoring a polynomial.

STEP 1: Factor out the greatest common factor, if possible.

STEP 2: A. If the polynomial has four terms, try factoring by grouping.

B. If the polynomial is a binomial, consider the difference of two squares, difference of two cubes, or sum of two cubes.

C. If the polynomial is a trinomial, check for a perfect square. Otherwise, apply FOIL or grouping, as described in Section 5.4.

STEP 3: Check to make sure that the polynomial is *completely* factored.

SECTION 5.7 ■ POLYNOMIAL EQUATIONS

Solving Polynomial Equations Follow the strategy presented in Putting It All Together for Section 5.7.

Example:
$$x^4 = 2x^2 + 8 \qquad \text{Given equation}$$
$$x^4 - 2x^2 - 8 = 0 \qquad \text{Subtract } 2x^2 + 8 \text{ from each side.}$$
$$(x^2 - 4)(x^2 + 2) = 0 \qquad \text{Factor trinomial.}$$
$$(x - 2)(x + 2)(x^2 + 2) = 0 \qquad \text{Difference of squares}$$
$$x = 2 \quad \text{or} \quad x = -2 \qquad \text{Zero-product property; solve.}$$

Assignment Name _____ Name _____ Date _____
Show all work for these items: _____

# _____	# _____
# _____	# _____
# _____	# _____
# _____	# _____

CHAPTER 5   SHOW YOUR WORK   249

Assignment Name _____ Name _____ Date _____

Show all work for these items: _____

# _____	# _____
# _____	# _____
# _____	# _____
# _____	# _____

CHAPTER 6

Rational Expressions and Functions

On average, a person spends an estimated 45 to 60 minutes waiting each day. This waiting could be at the doctor's office, at the grocery store, in airports, or stuck in traffic. In a 70-year life span, a person might spend as much as 3 years waiting!

Motorists in the nation's top 68 urban areas spend about 50 percent more time stuck in traffic than they did 10 years ago. Traffic congestion delays travelers 3.7 billion hours and wastes 2.3 billion gallons of fuel at a cost of $63 billion each year. Many urban areas are not adding enough capacity to keep traffic moving.

The time spent waiting in lines is subject to a *nonlinear effect*. For example, you can put more cars on the road up to a point. Then, if the traffic intensity increases even slightly, congestions and long lines increase dramatically. This phenomenon cannot be modeled by linear or polynomial functions. Instead we need a new function, called a rational function, to describe this type of behavior. (See Example 5 in Section 6.4.) Rational functions can also be used to describe railroad track design, gravity, population growth, electricity, and aerial photography.

- **6.1** Introduction to Rational Functions and Equations
- **6.2** Multiplication and Division of Rational Expressions
- **6.3** Addition and Subtraction of Rational Expressions
- **6.4** Rational Equations
- **6.5** Complex Fractions
- **6.6** Modeling with Proportions and Variation
- **6.7** Division of Polynomials

The future belongs to those who believe in the beauty of their dreams.
—ELEANOR ROOSEVELT

Sources: Phillip J. Longman, "American Gridlock," *U.S. News & World Report*, May 28, 2001; answers.google.com; CBS/AP, 2007.

6.1 INTRODUCTION TO RATIONAL FUNCTIONS AND EQUATIONS

Recognizing and Using Rational Functions ▪ Solving Rational Equations ▪ Operations on Functions

A LOOK INTO MATH ▷ Suppose that a parking lot attendant can wait on 5 cars per minute and that cars are randomly leaving the parking lot at a rate of 4 cars per minute. Is it possible to predict how long the average driver will wait in line? Using rational functions, we can answer this question. In this section we introduce rational functions and equations.

Recognizing and Using Rational Functions

In Chapter 5 we discussed polynomials. Examples of polynomials include

$$4, \quad x, \quad x^2, \quad 2x - 5, \quad 3x^2 - 6x + 1, \quad \text{and} \quad 3x^4 - 7.$$

Rational expressions result when a polynomial is divided by a nonzero polynomial. Three examples of rational expressions are

$$\frac{4}{x}, \quad \frac{x^2}{2x - 5}, \quad \text{and} \quad \frac{3x^2 - 6x + 1}{3x^4 - 7}.$$

EXAMPLE 1 Recognizing a rational expression

Determine whether each expression is rational.

(a) $\dfrac{4 - x}{5x^2 + 3}$ (b) $\dfrac{x^2 - 1}{\sqrt{x} + 2x}$ (c) $\dfrac{6}{x - 1}$

Solution
(a) This expression is rational because both $4 - x$ and $5x^2 + 3$ are polynomials.
(b) This expression is not rational because $\sqrt{x} + 2x$ is not a polynomial.
(c) This expression is rational because both 6 and $x - 1$ are polynomials.

Rational expressions are used to define *rational functions*. For example, $\frac{4}{x+1}$ is a rational expression and so $f(x) = \frac{4}{x+1}$ represents a rational function.

> **RATIONAL FUNCTION**
>
> Let $p(x)$ and $q(x)$ be polynomials. Then a **rational function** is given by
>
> $$f(x) = \frac{p(x)}{q(x)}.$$
>
> The domain of f includes all x-values such that $q(x) \neq 0$.

EXAMPLE 2   Identifying the domain of rational functions

Identify the domain of each function.

(a) $f(x) = \dfrac{1}{x+2}$ (b) $g(x) = \dfrac{2x}{x^2 - 3x + 2}$ (c) $h(t) = \dfrac{4}{t^3 - t}$

Solution

(a) The domain of f includes all x-values for which the expression $f(x) = \dfrac{1}{x+2}$ is defined. Thus we must *exclude* values of x for which the denominator equals 0.

$\quad x + 2 = 0 \qquad$ Set denominator equal to 0.
$\quad x = -2 \qquad$ Subtract 2.

In set-builder notation, $D = \{x \mid x \neq -2\}$.

(b) The domain of g includes all values of x except where the denominator equals 0.

$\quad x^2 - 3x + 2 = 0 \qquad$ Set denominator equal to 0.
$\quad (x-1)(x-2) = 0 \qquad$ Factor.
$\quad x = 1 \quad \text{or} \quad x = 2 \qquad$ Zero-product property; solve.

Thus $D = \{x \mid x \neq 1, x \neq 2\}$.

(c) The domain of h includes all values of t except where the denominator equals 0.

$\quad t^3 - t = 0 \qquad$ Set denominator equal to 0.
$\quad t(t^2 - 1) = 0 \qquad$ Factor out t.
$\quad t(t-1)(t+1) = 0 \qquad$ Difference of squares
$\quad t = 0 \quad \text{or} \quad t = 1 \quad \text{or} \quad t = -1 \qquad$ Zero-product property; solve.

Thus $D = \{t \mid t \neq 0, t \neq 1, t \neq -1\}$.

Like other functions, rational functions have graphs. To graph a rational function by hand, we usually start by making a table of values, as demonstrated in the next example. Because the graphs of rational functions are typically nonlinear, it is a good idea to plot at least 3 points on each side of an x-value where the formula is undefined.

EXAMPLE 3   Graphing a rational function

Graph $f(x) = \dfrac{1}{x}$. State the domain of f.

Solution

Make a table of values for $f(x) = \dfrac{1}{x}$, as shown in Table 6.1. Notice that $x = 0$ is not in the domain of f, and a dash can be used to denote this undefined value. The domain of f is all real numbers such that $x \neq 0$. Start by picking three x-values on each side of 0.

TABLE 6.1

x	-2	-1	$-\tfrac{1}{2}$	0	$\tfrac{1}{2}$	1	2
$\tfrac{1}{x}$	$-\tfrac{1}{2}$	-1	-2	—	2	1	$\tfrac{1}{2}$

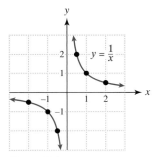

Figure 6.1

Plot the points shown in Table 6.1 and then connect the points with a smooth curve, as shown in Figure 6.1. Because $f(0)$ is undefined, the graph of $f(x) = \dfrac{1}{x}$ does not cross the line $x = 0$, the y-axis.

In the next example we evaluate a rational function in three ways.

EXAMPLE 4   Evaluating a rational function

Use Table 6.2, the formula for $f(x)$, and Figure 6.2 to evaluate $f(-1), f(1),$ and $f(2)$.

(a) TABLE 6.2

x	$f(x)$
-3	$\frac{3}{2}$
-2	$\frac{4}{3}$
-1	1
0	0
1	—
2	4
3	3

(b) $f(x) = \dfrac{2x}{x-1}$

(c)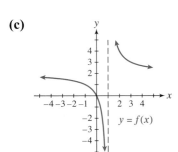

Figure 6.2

Solution

(a) *Numerical Evaluation*   Table 6.2 shows that

$$f(-1) = 1, \quad f(1) \text{ is undefined}, \quad \text{and} \quad f(2) = 4.$$

(b) *Symbolic Evaluation*

$$f(-1) = \frac{2(-1)}{-1-1} = 1$$

$$f(1) = \frac{2(1)}{1-1} = \frac{2}{0}, \text{ which is undefined. Input 1 is not in the domain of } f.$$

$$f(2) = \frac{2(2)}{2-1} = 4$$

(c) *Graphical Evaluation*   To evaluate $f(-1)$ graphically, find $x = -1$ on the x-axis and move upward to the graph of f. The y-value is 1 at the point of intersection, so $f(-1) = 1$, as shown in Figure 6.3(a). In Figure 6.3(b) the vertical line $x = 1$ is called a *vertical asymptote*. Because the graph of f does not intersect this line, $f(1)$ is undefined. Figure 6.3(c) reveals that $f(2) = 4$.

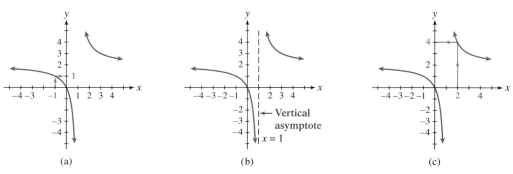

Figure 6.3

A **vertical asymptote** is a vertical line that typically occurs in (but is not part of) the graph of a rational function when the denominator of the rational expression is 0 but the numerator is *not* 0. The graph of a rational function *never* crosses a vertical asymptote. In Figure 6.1, the vertical asymptote is the y-axis or $x = 0$.

NOTE: In Figures 6.1 and 6.2 the y-values on the graph of f become very large (approach ∞) or become very small (approach −∞) for x-values near the vertical asymptote. This characteristic is true in general for vertical asymptotes.

TECHNOLOGY NOTE: Asymptotes, Dot Mode, and Decimal Windows

When rational functions are graphed on graphing calculators, pseudo-asymptotes often occur because the calculator is simply connecting dots to draw a graph. The accompanying figures show the graph of $y = \frac{2}{x-2}$ in connected mode, dot mode, and with a *decimal*, or *friendly*, *window*. In dot mode, pixels in the calculator screen are not connected. With dot mode (and sometimes with a decimal window) pseudo-asymptotes do not appear. To learn more about these features consult your owner's manual.

CALCULATOR HELP
To set a calculator in dot mode or to set a decimal window, see the Appendix (page AP-10).

[−6, 6, 1] by [−4, 4, 1]

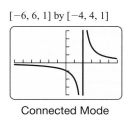

Connected Mode

[−6, 6, 1] by [−4, 4, 1]

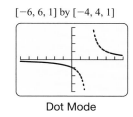

Dot Mode

[−4.7, 4.7, 1] by [−3.1, 3.1, 1]

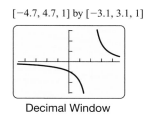

Decimal Window

▶ **REAL-WORLD CONNECTION** Applications involving rational functions are numerous. One instance is in the design of curves for train tracks, which we discuss in the next example.

EXAMPLE 5 Modeling a train track curve

When curves are designed for train tracks, sometimes the outer rail is elevated, or banked, so that a locomotive and its cars can safely negotiate the curve at a higher speed than if the tracks were level. Suppose that a circular curve with a radius of r feet is being designed for a train traveling 60 miles per hour. Then $f(r) = \frac{2540}{r}$ calculates the proper elevation y in inches for the outer rail, where $y = f(r)$. See Figure 6.4. (*Source:* L. Haefner, *Introduction to Transportation Systems*.)

(a) Evaluate $f(300)$ and interpret the result.
(b) A graph of f is shown in Figure 6.5. Discuss how the elevation of the outer rail changes as the radius r increases.

Figure 6.4

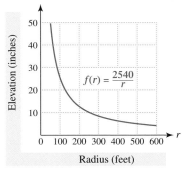

Figure 6.5

Solution
(a) $f(300) = \frac{2540}{300} \approx 8.5$. Thus the outer rail on a curve with a radius of 300 feet should be elevated about 8.5 inches for a train to safely travel through it at 60 miles per hour.
(b) As the radius increases (and the curve becomes less sharp), the outer rail needs less elevation.

256   CHAPTER 6   RATIONAL EXPRESSIONS AND FUNCTIONS

Solving Rational Equations

▶ **REAL-WORLD CONNECTION**   When working with rational functions, we commonly encounter rational equations. In Example 5, we used $f(r) = \frac{2540}{r}$ to calculate the elevation of the outer rail in a train track curve. Now suppose that the outer rail for a curve is elevated 6 inches. What should be the radius of the curve? To answer this question, we need to solve the **rational equation**

$$\frac{2540}{r} = 6.$$

We solve this equation in the next example.

EXAMPLE 6   Determining the proper radius for a train track curve

Solve the rational equation $\frac{2540}{r} = 6$ and interpret the result.

Solution
We begin by multiplying each side of the equation by r.

$$r \cdot \frac{2540}{r} = 6 \cdot r \qquad \text{Multiply by } r.$$
$$2540 = 6r \qquad \text{Simplify.}$$
$$\frac{2540}{6} = r \qquad \text{Divide by 6.}$$
$$r = 423.\overline{3} \qquad \text{Rewrite.}$$

A train track curve designed for 60 miles per hour and banked 6 inches should have a radius of about 423 feet.

Multiplication is often used as a first step when solving rational equations. To do so we apply the following property of rational expressions, with $D \neq 0$.

$$D \cdot \frac{C}{D} = C$$

Examples of this property include

$$5 \cdot \frac{7}{5} = 7 \quad \text{and} \quad (x - 1) \cdot \frac{4x}{x - 1} = 4x.$$

EXAMPLE 7   Solving rational equations

Solve each rational equation and check your answer.

(a) $\dfrac{3x}{2x - 1} = 3$ (b) $\dfrac{x + 1}{x - 2} = \dfrac{3}{x - 2}$ (c) $\dfrac{6}{x + 1} = x$

Solution
(a) First note that $\frac{1}{2}$ cannot be a solution to this equation because $x = \frac{1}{2}$ results in the left side of the equation being undefined; division by 0 is not possible. Multiply each side of the equation by the denominator, $2x - 1$.

$$(2x - 1) \cdot \frac{3x}{2x - 1} = 3 \cdot (2x - 1) \qquad \text{Multiply by } (2x - 1).$$
$$3x = 6x - 3 \qquad \text{Simplify.}$$
$$-3x = -3 \qquad \text{Subtract } 6x.$$
$$x = 1 \qquad \text{Divide by } -3.$$

To check your answer substitute **1** in the given equation for x.

$$\frac{3(1)}{2(1) - 1} = 3 \qquad \text{The answer checks.}$$

(b) Each side of the equation is undefined when $x = 2$. Thus 2 cannot be a solution. Begin by multiplying each side of the equation by the denominator, $x - 2$.

$$(x - 2) \cdot \frac{x + 1}{x - 2} = \frac{3}{x - 2} \cdot (x - 2) \qquad \text{Multiply by } x - 2.$$
$$x + 1 = 3 \qquad \text{Simplify.}$$
$$x = 2 \qquad \text{Subtract 1.}$$

In this case 2 is not a valid solution. There are no solutions. Instead, 2 is called an **extraneous solution**, which cannot be used. It is *important* to check your answers.

(c) The left side of the equation is undefined when $x = -1$. Thus -1 cannot be a solution. To solve this equation start by multiplying each side by the denominator, $x + 1$.

$$(x + 1) \cdot \frac{6}{x + 1} = x \cdot (x + 1) \qquad \text{Multiply by } x + 1.$$
$$6 = x^2 + x \qquad \text{Simplify.}$$
$$x^2 + x - 6 = 0 \qquad \text{Rewrite the equation.}$$
$$(x + 3)(x - 2) = 0 \qquad \text{Factor.}$$
$$x = -3 \quad \text{or} \quad x = 2 \qquad \text{Zero-product property}$$

Checking these results confirms that both -3 and 2 are solutions.

▶ **REAL-WORLD CONNECTION** The *grade* x of a hill is a measure of its steepness and corresponds to the slope of the road. For example, if a road rises 10 feet for every 100 feet of horizontal distance, it has an uphill grade of $x = \frac{10}{100}$, or 10%, as illustrated in Figure 6.6. The braking distance D in feet for a car traveling 60 miles per hour on a wet, uphill grade is given by

$$D(x) = \frac{3600}{30x + 9}.$$

In the next example we use this formula to determine the grade associated with a given braking distance. (*Source:* N. Garber and L. Hoel, *Traffic and Highway Engineering*.)

Figure 6.6

EXAMPLE 8 Solving a rational equation

The braking distance for a car traveling at 60 miles per hour on a wet, uphill grade is 250 feet. Use the formula $D(x) = \frac{3600}{30x + 9}$ to find the grade of the hill
(a) symbolically, **(b)** graphically, and **(c)** numerically.

Solution
(a) Symbolic Solution To solve the equation $\frac{3600}{30x + 9} = 250$ symbolically, we begin by multiplying each side by the denominator, $30x + 9$.

$$(30x + 9) \cdot \frac{3600}{30x + 9} = 250 \cdot (30x + 9) \qquad \text{Multiply by } (30x + 9).$$
$$3600 = 250(30x + 9) \qquad \text{Simplify.}$$
$$3600 = 7500x + 2250 \qquad \text{Distributive property}$$
$$1350 = 7500x \qquad \text{Subtract 2250.}$$
$$x = \frac{1350}{7500} \qquad \text{Divide by 7500; rewrite.}$$
$$x = 0.18 \qquad \text{Write } x \text{ in decimal form.}$$

Thus the grade is 0.18, or 18%.

(b) **Graphical Solution** To solve this equation graphically, we let $Y_1 = 3600/(30X + 9)$ and $Y_2 = 250$. Their graphs intersect at $(0.18, 250)$, as shown in Figure 6.7(a).

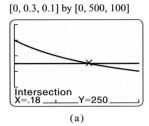
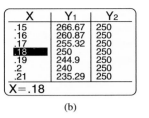

(a)   (b)

Figure 6.7

CALCULATOR HELP

To find a point of intersection, see the Appendix (page AP-7).

(c) **Numerical Solution** Figure 6.7(b) indicates that $y_1 = 250$ when $x = 0.18$.

MAKING CONNECTIONS

Symbolic, Numerical, and Graphical Solutions to a Rational Equation

Symbolic Solution

$$\frac{x+3}{2x} = 2$$

$$2x \cdot \frac{x+3}{2x} = 2 \cdot 2x$$

$$x + 3 = 4x$$

$$3 = 3x$$

$$x = 1$$

The solution is **1**.

Numerical Solution

x	$(x+3)/(2x)$
-2	-0.25
-1	-1
0	—
1	2
2	1.25

When $x = 1$, $\frac{x+3}{2x} = 2$.

Graphical Solution

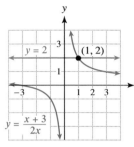

The graphs intersect at $(1, 2)$.

Graphing calculators are helpful for making tables or graphs of rational functions.

Operations on Functions

We have performed addition, subtraction, multiplication, and division on numbers and variables. In Chapter 5, we performed addition, subtraction, and multiplication on functions. In this subsection we will review these operations and also perform division on functions.

OPERATIONS ON FUNCTIONS

If $f(x)$ and $g(x)$ are both defined, then the sum, difference, product, and quotient of two functions f and g are defined by

$$(f + g)(x) = f(x) + g(x) \qquad \text{Sum}$$
$$(f - g)(x) = f(x) - g(x) \qquad \text{Difference}$$
$$(fg)(x) = f(x) \cdot g(x) \qquad \text{Product}$$
$$\left(\frac{f}{g}\right)(x) = \frac{f(x)}{g(x)}, \text{ where } g(x) \neq 0. \qquad \text{Quotient}$$

EXAMPLE 9 Performing arithmetic on functions

Use $f(x) = x^2$ and $g(x) = 2x - 4$ to evaluate each of the following.

(a) $(f + g)(3)$ (b) $(fg)(-1)$ (c) $\left(\frac{f}{g}\right)(0)$ (d) $(f/g)(2)$

Solution

(a) $(f + g)(3) = f(3) + g(3) = 3^2 + (2 \cdot 3 - 4) = 9 + 2 = 11$

(b) $(fg)(-1) = f(-1) \cdot g(-1) = (-1)^2 \cdot (2 \cdot (-1) - 4) = 1 \cdot (-6) = -6$

(c) $\left(\dfrac{f}{g}\right)(0) = \dfrac{f(0)}{g(0)} = \dfrac{0^2}{2 \cdot 0 - 4} = \dfrac{0}{-4} = 0$

(d) Note that $(f/g)(2)$ is equivalent to $\left(\dfrac{f}{g}\right)(2)$.

$$(f/g)(2) = \dfrac{f(2)}{g(2)} = \dfrac{2^2}{2 \cdot 2 - 4} = \dfrac{4}{0},$$

which is not possible. Thus $(f/g)(2)$ is undefined.

In the next example, we find the sum, difference, product, and quotient of two functions for a general x.

EXAMPLE 10 Performing arithmetic on functions

Use $f(x) = 4x - 5$ and $g(x) = 3x + 1$ to evaluate each of the following.

(a) $(f + g)(x)$ **(b)** $(f - g)(x)$ **(c)** $(fg)(x)$ **(d)** $\left(\dfrac{f}{g}\right)(x)$

Solution

(a) $(f + g)(x) = f(x) + g(x) = (4x - 5) + (3x + 1) = 7x - 4$

(b) $(f - g)(x) = f(x) - g(x) = (4x - 5) - (3x + 1) = x - 6$

(c) $(fg)(x) = f(x) \cdot g(x) = (4x - 5)(3x + 1) = 12x^2 - 11x - 5$

(d) $\left(\dfrac{f}{g}\right)(x) = \dfrac{f(x)}{g(x)} = \dfrac{4x - 5}{3x + 1}$

6.1 PUTTING IT ALL TOGETHER

The following table summarizes basic concepts about rational functions and equations.

Term	Explanation	Example
Rational Function	Let $p(x)$ and $q(x)$ be polynomials with $q(x) \neq 0$. Then a *rational function* is given by $$f(x) = \dfrac{p(x)}{q(x)}.$$	$f(x) = \dfrac{3x}{x + 2}$ The domain of f includes all real numbers except $x = -2$. A vertical asymptote occurs at $x = -2$ in the graph of f.
Rational Equation	An equation that contains rational expressions is a *rational equation*. When you are solving a rational equation, multiplication is often a good first step. Be sure to check your results.	To solve $\dfrac{3x}{x+1} = 6$ begin by multiplying each side by $x + 1$. $(x + 1) \cdot \dfrac{3x}{x + 1} = 6 \cdot (x + 1)$ $3x = 6x + 6$ $-3x = 6$ $x = -2$
Division of Functions	$(f/g)(x) = \dfrac{f(x)}{g(x)},\ g(x) \neq 0$	If $f(x) = x^2$ and $g(x) = x + 1$, then $(f/g)(x) = \dfrac{x^2}{x + 1}$.

6.2 MULTIPLICATION AND DIVISION OF RATIONAL EXPRESSIONS

Simplifying Rational Expressions ▪ Review of Multiplication and Division of Fractions ▪ Multiplication of Rational Expressions ▪ Division of Rational Expressions

A LOOK INTO MATH ▷ In previous chapters we demonstrated how to add, subtract, and multiply polynomials. In this section we show you how to multiply and divide rational expressions, and in the next section we discuss addition and subtraction of rational expressions. We start this section by discussing how to simplify rational expressions.

Simplifying Rational Expressions

When simplifying fractions, we often use the **basic principle of fractions**, which states that

$$\frac{a \cdot c}{b \cdot c} = \frac{a}{b}.$$

This principle holds because $\frac{c}{c} = 1$ and $\frac{a}{b} \cdot 1 = \frac{a}{b}$. That is,

$$\frac{a \cdot c}{b \cdot c} = \frac{a}{b} \cdot \frac{c}{c} = \frac{a}{b} \cdot 1 = \frac{a}{b}.$$

For example, this principle can be used to simplify a fraction.

$$\frac{6}{44} = \frac{3 \cdot 2}{22 \cdot 2} = \frac{3}{22}$$

This same principle can also be used to simplify rational expressions. For example,

$$\frac{(z + 1)(z + 3)}{z(z + 3)} = \frac{z + 1}{z}, \quad \text{provided } z \neq -3.$$

> **SIMPLIFYING RATIONAL EXPRESSIONS**
>
> The following principle can be used to simplify rational expressions, where A, B, and C are polynomials.
>
> $$\frac{A \cdot C}{B \cdot C} = \frac{A}{B} \qquad B \text{ and } C \text{ not zero}$$

To *simplify* a rational expression, we *completely factor* the numerator and the denominator and then apply the principle for simplifying rational expressions. These concepts are illustrated in the next example.

EXAMPLE 1 Simplifying rational expressions

Simplify each expression.

(a) $\dfrac{9x}{3x^2}$ (b) $\dfrac{5y - 10}{10y - 20}$ (c) $\dfrac{2z^2 - 3z - 9}{z^2 + 2z - 15}$ (d) $\dfrac{a^2 - b^2}{a + b}$

Solution

(a) First factor out the greatest common factor, $3x$, in the numerator and denominator.

$$\frac{9x}{3x^2} = \frac{3 \cdot 3x}{x \cdot 3x} = \frac{3}{x}$$

(b) Use the distributive property and then apply the basic principle of fractions.

$$\frac{5y-10}{10y-20} = \frac{5(y-2)}{10(y-2)} = \frac{5}{10} = \frac{1}{2}$$

(c) Start by factoring the numerator and denominator.

$$\frac{2z^2-3z-9}{z^2+2z-15} = \frac{(2z+3)(z-3)}{(z+5)(z-3)} = \frac{2z+3}{z+5}$$

(d) Start by factoring the numerator as the difference of squares.

$$\frac{a^2-b^2}{a+b} = \frac{(a-b)(a+b)}{a+b} = a-b$$

There are a number of ways that a negative sign can be placed in a fraction. For example,

$$-\frac{2}{3} = \frac{-2}{3} = \frac{2}{-3}$$

illustrates three fractions that are equal. This property can also be applied to rational expressions, as demonstrated in the next example.

EXAMPLE 2 Distributing a negative sign

Simplify each expression.

(a) $-\dfrac{1-z}{z-1}$ (b) $\dfrac{-y-2}{4y+8}$ (c) $\dfrac{5-x}{x-5}$

Solution

(a) Start by distributing the negative sign over the numerator.

$$-\frac{1-z}{z-1} = \frac{-(1-z)}{z-1} = \frac{-1+z}{z-1} = \frac{z-1}{z-1} = 1$$

Note that the negative sign could also be distributed over the denominator.

(b) Factor -1 out of the numerator and 4 out of the denominator.

$$\frac{-y-2}{4y+8} = \frac{-1(y+2)}{4(y+2)} = -\frac{1}{4}$$

(c) Start by factoring -1 out of the numerator.

$$\frac{5-x}{x-5} = \frac{-1(-5+x)}{x-5} = \frac{-1(x-5)}{x-5} = -1$$

MAKING CONNECTIONS

Negative Signs and Rational Expressions

In general, $(b-a)$ equals $-1(a-b)$. As a result, if $a \neq b$, then

$$\frac{b-a}{a-b} = -1.$$

See Example 2(c).

Review of Multiplication and Division of Fractions

Recall that to multiply two fractions we use the property

$$\frac{a}{b} \cdot \frac{c}{d} = \frac{ac}{bd}.$$

For example, $\frac{2}{5} \cdot \frac{3}{7} = \frac{2 \cdot 3}{5 \cdot 7} = \frac{6}{35}.$

EXAMPLE 3 Multiplying fractions

Multiply and simplify the product.

(a) $\frac{4}{9} \cdot \frac{3}{8}$ (b) $\frac{2}{3} \cdot \frac{3}{4} \cdot \frac{5}{6}$

Solution

(a) $\frac{4}{9} \cdot \frac{3}{8} = \frac{4 \cdot 3}{9 \cdot 8} = \frac{12}{72} = \frac{1 \cdot 12}{6 \cdot 12} = \frac{1}{6}$ (b) $\frac{2}{3} \cdot \frac{3}{4} \cdot \frac{5}{6} = \frac{6 \cdot 5}{12 \cdot 6} = \frac{5}{12}$

Recall that to divide two fractions we "invert and multiply." That is, we change a division problem to a multiplication problem, using the property

$$\frac{a}{b} \div \frac{c}{d} = \frac{a}{b} \cdot \frac{d}{c}.$$

For example, $\frac{3}{4} \div \frac{5}{4} = \frac{3}{4} \cdot \frac{4}{5} = \frac{3 \cdot 4}{5 \cdot 4} = \frac{3}{5}.$

Multiplication of Rational Expressions

Multiplying rational expressions is similar to multiplying fractions.

> **PRODUCTS OF RATIONAL EXPRESSIONS**
>
> To multiply two rational expressions, multiply numerators and multiply denominators.
>
> $$\frac{A}{B} \cdot \frac{C}{D} = \frac{AC}{BD} \qquad B \text{ and } D \text{ not zero}$$

EXAMPLE 4 Multiplying rational expressions

Multiply.

(a) $\frac{1}{x} \cdot \frac{x+1}{2x}$ (b) $\frac{x-1}{x} \cdot \frac{x-1}{x+2}$

Solution

(a) $\frac{1}{x} \cdot \frac{x+1}{2x} = \frac{1 \cdot (x+1)}{x \cdot 2x} = \frac{x+1}{2x^2}$

(b) $\frac{x-1}{x} \cdot \frac{x-1}{x+2} = \frac{(x-1)(x-1)}{x(x+2)}$

Many times the product of two rational expressions can be simplified. This process is demonstrated in the next two examples.

EXAMPLE 5 Multiplying rational expressions

Multiply and simplify.

(a) $\dfrac{5x}{8} \cdot \dfrac{4}{10x^2}$ (b) $\dfrac{x-2}{2x-1} \cdot \dfrac{x+1}{2x-4}$ (c) $\dfrac{x^2-1}{x^2-4} \cdot \dfrac{x+2}{x-1}$

Solution

(a) $\dfrac{5x}{8} \cdot \dfrac{4}{10x^2} = \dfrac{20x}{80x^2}$ Multiply rational expressions.

$\qquad\qquad\qquad = \dfrac{1}{4x}$ Simplify.

(b) $\dfrac{x-2}{2x-1} \cdot \dfrac{x+1}{2x-4} = \dfrac{(x-2)(x+1)}{(2x-1)(2x-4)}$ Multiply rational expressions.

$\qquad\qquad\qquad\qquad = \dfrac{(x-2)(x+1)}{(2x-1)2(x-2)}$ Factor $2x-4$.

$\qquad\qquad\qquad\qquad = \dfrac{(x+1)(x-2)}{2(2x-1)(x-2)}$ Commutative property

$\qquad\qquad\qquad\qquad = \dfrac{x+1}{2(2x-1)}$ Simplify.

(c) $\dfrac{x^2-1}{x^2-4} \cdot \dfrac{x+2}{x-1} = \dfrac{(x^2-1)(x+2)}{(x^2-4)(x-1)}$ Multiply rational expressions.

$\qquad\qquad\qquad\qquad = \dfrac{(x-1)(x+1)(x+2)}{(x-2)(x+2)(x-1)}$ Difference of squares

$\qquad\qquad\qquad\qquad = \dfrac{(x+1)(x-1)(x+2)}{(x-2)(x-1)(x+2)}$ Commutative property

$\qquad\qquad\qquad\qquad = \dfrac{x+1}{x-2}$ Simplify.

EXAMPLE 6 Multiplying rational expressions

Multiply and simplify.

(a) $\dfrac{2x^2 y^3}{3xy^2} \cdot \dfrac{(2x^3 y)^2}{2(xy)^3}$ (b) $\dfrac{x^3-x}{x-1} \cdot \dfrac{x+1}{x}$ (c) $\dfrac{a}{b} \cdot \dfrac{a^2-b^2}{2} \cdot \dfrac{b}{a+b}$

Solution

(a) $\dfrac{2x^2 y^3}{3xy^2} \cdot \dfrac{(2x^3 y)^2}{2(xy)^3} = \dfrac{2x^2 y^3 \cdot 4x^6 y^2}{3xy^2 \cdot 2x^3 y^3}$ Multiply; properties of exponents.

$\qquad\qquad\qquad\qquad = \dfrac{8x^8 y^5}{6x^4 y^5}$ Multiply; properties of exponents.

$\qquad\qquad\qquad\qquad = \dfrac{4}{3} x^4$ Simplify.

(b) $\dfrac{x^3 - x}{x - 1} \cdot \dfrac{x + 1}{x} = \dfrac{(x^3 - x)(x + 1)}{x(x - 1)}$ Multiply.

$= \dfrac{x(x^2 - 1)(x + 1)}{x(x - 1)}$ Factor out x.

$= \dfrac{x(x - 1)(x + 1)(x + 1)}{x(x - 1)}$ Difference of squares

$= (x + 1)(x + 1)$ Simplify.

$= (x + 1)^2$ Rewrite.

(c) $\dfrac{a}{b} \cdot \dfrac{a^2 - b^2}{2} \cdot \dfrac{b}{a + b} = \dfrac{a(a^2 - b^2)b}{b(2)(a + b)}$ Multiply fractions.

$= \dfrac{a(a - b)(a + b)b}{2(a + b)b}$ Difference of squares

$= \dfrac{a(a - b)}{2}$ Simplify.

Division of Rational Expressions

When dividing two rational expressions, *multiply* the first expression by the reciprocal of the second expression. This technique is similar to the division of fractions.

> **QUOTIENTS OF RATIONAL EXPRESSIONS**
>
> To divide two rational expressions, multiply by the reciprocal of the divisor.
>
> $$\dfrac{A}{B} \div \dfrac{C}{D} = \dfrac{A}{B} \cdot \dfrac{D}{C} \qquad B, C, \text{ and } D \text{ not zero}$$

NOTE: This technique of multiplying the first rational expression and the reciprocal of the second rational expression is sometimes summarized as "Invert and multiply."

The **reciprocal of a polynomial** $p(x)$ is $\dfrac{1}{p(x)}$, and the **reciprocal of a rational expression** $\dfrac{p(x)}{q(x)}$ is $\dfrac{q(x)}{p(x)}$. The next example demonstrates how to find reciprocals.

EXAMPLE 7 Finding reciprocals

Write the reciprocal of each expression.

(a) $3x + 4$ **(b)** $\dfrac{5}{x^2 + 1}$ **(c)** $\dfrac{x - 7}{x + 7}$

Solution

(a) The reciprocal of $3x + 4$ is $\dfrac{1}{3x + 4}$.

(b) The reciprocal of $\dfrac{5}{x^2 + 1}$ is $\dfrac{x^2 + 1}{5}$.

(c) The reciprocal of $\dfrac{x - 7}{x + 7}$ is $\dfrac{x + 7}{x - 7}$.

EXAMPLE 8 Dividing two rational expressions

Divide and simplify.

(a) $\dfrac{2}{x} \div \dfrac{2x - 1}{4x}$ **(b)** $\dfrac{x^2 - 1}{x^2 + x - 6} \div \dfrac{x - 1}{x + 3}$

Solution

(a)
$$\frac{2}{x} \div \frac{2x-1}{4x} = \frac{2}{x} \cdot \frac{4x}{2x-1}$$ "Invert and multiply."
$$= \frac{8x}{x(2x-1)}$$ Multiply.
$$= \frac{8}{2x-1}$$ Simplify.

(b) $\dfrac{x^2-1}{x^2+x-6} \div \dfrac{x-1}{x+3} = \dfrac{x^2-1}{x^2+x-6} \cdot \dfrac{x+3}{x-1}$ "Invert and multiply."
$$= \frac{(x+1)(x-1)}{(x-2)(x+3)} \cdot \frac{x+3}{x-1}$$ Factor completely.
$$= \frac{(x+1)(x-1)(x+3)}{(x-2)(x-1)(x+3)}$$ Commutative property
$$= \frac{x+1}{x-2}$$ Simplify.

EXAMPLE 9 Dividing rational expressions

Divide and simplify.

(a) $\dfrac{7a^2}{4b^3} \div \dfrac{21a}{8b^4}$ (b) $\dfrac{2x+2}{x-1} \div (x+1)$ (c) $\dfrac{x^2-25}{x^2+5x+4} \div \dfrac{x^2-10x+25}{2x^2+8x}$

Solution

(a)
$$\frac{7a^2}{4b^3} \div \frac{21a}{8b^4} = \frac{7a^2}{4b^3} \cdot \frac{8b^4}{21a}$$ "Invert and multiply."
$$= \frac{56a^2b^4}{84ab^3}$$ Multiply rational expressions.
$$= \frac{56}{84}a^{2-1}b^{4-3}$$ Properties of exponents
$$= \frac{2}{3}ab$$ Simplify.

(b) $\dfrac{2x+2}{x-1} \div (x+1) = \dfrac{2x+2}{x-1} \cdot \dfrac{1}{x+1}$ "Invert and multiply."
$$= \frac{2(x+1)}{x-1} \cdot \frac{1}{x+1}$$ Factor.
$$= \frac{2(x+1)}{(x-1)(x+1)}$$ Multiply rational expressions.
$$= \frac{2}{x-1}$$ Simplify.

(c) $\dfrac{x^2-25}{x^2+5x+4} \div \dfrac{x^2-10x+25}{2x^2+8x} = \dfrac{x^2-25}{x^2+5x+4} \cdot \dfrac{2x^2+8x}{x^2-10x+25}$
$$= \frac{(x-5)(x+5)}{(x+1)(x+4)} \cdot \frac{2x(x+4)}{(x-5)^2}$$
$$= \frac{2x(x+5)(x-5)(x+4)}{(x+1)(x-5)(x-5)(x+4)}$$
$$= \frac{2x(x+5)}{(x+1)(x-5)}$$

In the next example, we divide three rational expressions. It is important to realize that *the associative property does not apply to division*. For example,

$$(24 \div 6) \div 2 = 4 \div 2 = 2 \quad \text{but} \quad 24 \div (6 \div 2) = 24 \div 3 = 8.$$

The order in which we do division *does matter*. As a result, we perform division from *left to right* and

$$24 \div 6 \div 2 = 4 \div 2 = 2.$$

EXAMPLE 10 Dividing three fractions

Divide and simplify $\frac{4x}{y} \div \frac{x^2}{y} \div \frac{3y}{2x}$.

Solution

$$\frac{4x}{y} \div \frac{x^2}{y} \div \frac{3y}{2x} = \left(\frac{4x}{y} \div \frac{x^2}{y}\right) \div \frac{3y}{2x} \quad \text{Divide left to right.}$$

$$= \left(\frac{4x}{y} \cdot \frac{y}{x^2}\right) \div \frac{3y}{2x} \quad \text{"Invert and multiply."}$$

$$= \frac{4xy}{x^2 y} \div \frac{3y}{2x} \quad \text{Multiply fractions.}$$

$$= \frac{4xy}{x^2 y} \cdot \frac{2x}{3y} \quad \text{"Invert and multiply."}$$

$$= \frac{8x^2 y}{3x^2 y^2} \quad \text{Multiply fractions.}$$

$$= \frac{8}{3y} \quad \text{Simplify.}$$

In the next example, we find the length of a rectangle when its area and width are expressed as polynomials.

EXAMPLE 11 Finding the dimensions of a rectangle

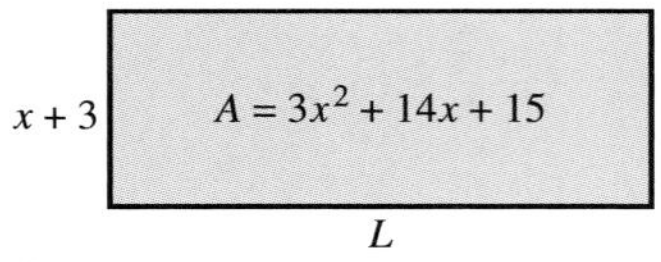

The area A of a rectangle is $3x^2 + 14x + 15$ and its width W is $x + 3$. See Figure 6.8.
(a) Find the length L of the rectangle.
(b) Find the length if the width is 12 inches.

Figure 6.8

Solution
(a) Because the area equals length times width, $A = LW$, length equals area divided by width, or $L = \frac{A}{W}$. To determine L, factor the expression for A and then simplify.

$$L = \frac{3x^2 + 14x + 15}{x + 3} = \frac{(3x + 5)(x + 3)}{x + 3} = 3x + 5.$$

The length of the rectangle is $3x + 5$.
(b) If the width is 12 inches, then $x + 3 = 12$ or $x = 9$. The length is

$$L = 3x + 5 = 3(9) + 5 = 32 \text{ inches.}$$

6.2 PUTTING IT ALL TOGETHER

The following table summarizes some important concepts found in this section.

Concept	Explanation	Examples
Simplifying Rational Expressions	Use the principle $$\frac{A \cdot C}{B \cdot C} = \frac{A}{B}.$$ B and C not zero	$\frac{5a(a+b)}{7b(a+b)} = \frac{5a}{7b}$ and $\frac{x^2-1}{x-1} = \frac{(x+1)(x-1)}{x-1} = x+1$
Multiplying Rational Expressions	To multiply two rational expressions, multiply numerators and multiply denominators. $$\frac{A}{B} \cdot \frac{C}{D} = \frac{AC}{BD}$$ B and D not zero	$\frac{5a}{4b^2} \cdot \frac{2b^3}{10a^3} = \frac{10ab^3}{40a^3b^2} = \frac{b}{4a^2}$
Dividing Rational Expressions	To divide two rational expressions, multiply by the reciprocal of the divisor. ("Invert and multiply.") $$\frac{A}{B} \div \frac{C}{D} = \frac{A}{B} \cdot \frac{D}{C}$$ B, C, and D not zero	$\frac{x}{x+1} \div \frac{x+2}{x+1} = \frac{x}{x+1} \cdot \frac{x+1}{x+2}$ $= \frac{x(x+1)}{(x+2)(x+1)}$ $= \frac{x}{x+2}$

6.3 ADDITION AND SUBTRACTION OF RATIONAL EXPRESSIONS

Least Common Multiples ■ Review of Addition and Subtraction of Fractions ■ Addition of Rational Expressions ■ Subtraction of Rational Expressions

A LOOK INTO MATH ▷ Rational expressions occur in many real-world applications. For example, addition of rational expressions is used in the design of electrical circuits. (See Example 8.) In this section we demonstrate how to add and subtract rational expressions. These techniques are similar to techniques used to add and subtract fractions. We begin by discussing least common multiples, which are used to find least common denominators.

Least Common Multiples

▶ **REAL-WORLD CONNECTION** Two friends work part-time at a store. The first person works every fourth day and the second person works every sixth day. If they are both working today, how many days pass before they both work on the same day again?

We can answer this question by listing the days that each person works.

First person: 4, 8, **12**, 16, 20, **24**, 28, 32, **36**, 40
Second person: 6, **12**, 18, **24**, 30, **36**, 42

After 12 days, the two friends work on the same day. The next time is after 24 days. The numbers 12 and 24 are *common multiples* of 4 and 6. (Find two more.) However, 12 is the **least common multiple** (LCM) of 4 and 6. Note that 12 is the smallest positive number that has both 4 and 6 as factors.

Another way to find the least common multiple for 4 and 6 is first to factor each number into prime numbers:

$$4 = 2 \cdot 2 \quad \text{and} \quad 6 = 2 \cdot 3.$$

To find the least common multiple, list each factor the *greatest* number of times that it occurs in either factorization. Then find the product of these numbers. For our example, the factor 2 occurs two times in the factorization of 4 and only once in the factorization of 6, so list 2 two times. The factor 3 appears only once in the factorization of 6 and not at all in the factorization of 4, so list it once:

$$2, 2, 3.$$

The least common multiple is their product: $2 \cdot 2 \cdot 3 = 12$. This same procedure can also be used to find the least common multiple for two (or more) polynomials.

> **FINDING THE LEAST COMMON MULTIPLE**
>
> The least common multiple (LCM) of two polynomials can be found as follows.
> **STEP 1:** Factor each polynomial completely.
> **STEP 2:** List each factor the greatest number of times that it occurs in either factorization.
> **STEP 3:** Find the product of this list of factors. The result is the LCM.

The next example illustrates how to use this procedure.

EXAMPLE 1 Finding least common multiples

Find the least common multiple for each pair of expressions.
(a) $4x, 5x^3$ (b) $x^2 - 2x, (x - 2)^2$
(c) $x + 2, x - 1$ (d) $x^2 + 4x + 4, x^2 + 3x + 2$

Solution
(a) **STEP 1:** Factor each polynomial completely.

$$4x = 2 \cdot 2 \cdot x \quad \text{and} \quad 5x^3 = 5 \cdot x \cdot x \cdot x$$

STEP 2: The factor 2 occurs twice, the factor 5 occurs once, and the factor x occurs at most three times. The list is $2, 2, 5, x, x,$ and x.
STEP 3: The LCM is the product $2 \cdot 2 \cdot 5 \cdot x \cdot x \cdot x$, or $20x^3$.

(b) **STEP 1:** Factor each polynomial completely.

$$x^2 - 2x = x(x - 2) \quad \text{and} \quad (x - 2)^2 = (x - 2)(x - 2)$$

STEP 2: The factor x occurs once, and the factor $(x - 2)$ occurs at most twice. The list of factors is $x, (x - 2),$ and $(x - 2)$.
STEP 3: The LCM is the product: $x(x - 2)^2$, which is left in factored form.

(c) **STEP 1:** Neither polynomial can be factored.
STEP 2: The list of factors is $(x + 2)$ and $(x - 1)$.
STEP 3: The LCM is the product $(x + 2)(x - 1)$, or $x^2 + x - 2$.

(d) **STEP 1:** Factor each polynomial as follows.

$$x^2 + 4x + 4 = (x + 2)(x + 2) \quad \text{and} \quad x^2 + 3x + 2 = (x + 1)(x + 2)$$

STEP 2: The factor $(x + 1)$ occurs once and $(x + 2)$ occurs at most twice.
STEP 3: The LCM is the product $(x + 1)(x + 2)^2$, which is left in factored form.

6.3 ADDITION AND SUBTRACTION OF RATIONAL EXPRESSIONS

Review of Addition and Subtraction of Fractions

Recall that to add two fractions we use the property

$$\frac{a}{c} + \frac{b}{c} = \frac{a+b}{c}.$$

This property requires that the fractions have like denominators. For example,

$$\frac{1}{5} + \frac{3}{5} = \frac{1+3}{5} = \frac{4}{5}.$$

When the denominators are not alike, we must find a common denominator. Before adding two fractions, such as $\frac{2}{3}$ and $\frac{1}{4}$, we write them with 12 as their common denominator. That is, we multiply each fraction by 1 written in an appropriate form. For example, to write $\frac{2}{3}$ with a denominator of 12 we multiply $\frac{2}{3}$ by 1, written as $\frac{4}{4}$.

$$\frac{2}{3} = \frac{2}{3} \cdot \frac{4}{4} = \frac{8}{12} \quad \text{and} \quad \frac{1}{4} = \frac{1}{4} \cdot \frac{3}{3} = \frac{3}{12}$$

Once the fractions have a common denominator, we can add them, as in

$$\frac{2}{3} + \frac{1}{4} = \frac{8}{12} + \frac{3}{12} = \frac{11}{12}.$$

Note that a **common denominator** for $\frac{2}{3}$ and $\frac{1}{4}$ is also a *common multiple* of 3 and 4. The **least common denominator** (LCD) for $\frac{2}{3}$ and $\frac{1}{4}$ is equal to the *least common multiple* (LCM) of 3 and 4. Thus the least common denominator is 12.

EXAMPLE 2 Adding fractions

Find the sum.

(a) $\dfrac{3}{4} + \dfrac{1}{8}$ (b) $\dfrac{3}{5} + \dfrac{2}{7}$

Solution
(a) The LCD is 8.

$$\frac{3}{4} + \frac{1}{8} = \frac{3}{4} \cdot \frac{2}{2} + \frac{1}{8} = \frac{6}{8} + \frac{1}{8} = \frac{7}{8}$$

(b) The LCD is 35.

$$\frac{3}{5} + \frac{2}{7} = \frac{3}{5} \cdot \frac{7}{7} + \frac{2}{7} \cdot \frac{5}{5} = \frac{21}{35} + \frac{10}{35} = \frac{31}{35}$$

Recall that subtraction is similar to addition. To subtract two fractions with like denominators we use the property

$$\frac{a}{c} - \frac{b}{c} = \frac{a-b}{c}.$$

For example, $\dfrac{3}{11} - \dfrac{7}{11} = \dfrac{3-7}{11} = -\dfrac{4}{11}.$

EXAMPLE 3 Subtracting fractions

Find the difference.

(a) $\dfrac{3}{10} - \dfrac{2}{15}$ (b) $\dfrac{3}{8} - \dfrac{5}{6}$

Solution

(a) The LCD is 30.

$$\frac{3}{10} - \frac{2}{15} = \frac{3}{10} \cdot \frac{3}{3} - \frac{2}{15} \cdot \frac{2}{2} = \frac{9}{30} - \frac{4}{30} = \frac{5}{30} = \frac{1}{6}$$

(b) The LCD is 24.

$$\frac{3}{8} - \frac{5}{6} = \frac{3}{8} \cdot \frac{3}{3} - \frac{5}{6} \cdot \frac{4}{4} = \frac{9}{24} - \frac{20}{24} = -\frac{11}{24}$$

Addition of Rational Expressions

Addition of rational expressions is similar to addition of fractions.

> **SUMS OF RATIONAL EXPRESSIONS**
>
> To add two rational expressions with like denominators, add their numerators. The denominator does not change.
>
> $$\frac{A}{C} + \frac{B}{C} = \frac{A + B}{C} \qquad C \text{ not zero}$$

EXAMPLE 4 Adding rational expressions with like denominators

Add and simplify.

(a) $\dfrac{x}{x+2} + \dfrac{3x-1}{x+2}$ (b) $\dfrac{x}{3x^2 + 4x - 4} + \dfrac{2}{3x^2 + 4x - 4}$

Solution

(a) The expressions have like denominators, so add the numerators.

$$\frac{x}{x+2} + \frac{3x-1}{x+2} = \frac{x + 3x - 1}{x + 2} \qquad \text{Add numerators.}$$

$$= \frac{4x - 1}{x + 2} \qquad \text{Combine like terms.}$$

(b) The expressions have like denominators, so add the numerators. However, the resulting sum can be simplified by factoring the denominator.

$$\frac{x}{3x^2 + 4x - 4} + \frac{2}{3x^2 + 4x - 4} = \frac{x + 2}{3x^2 + 4x - 4} \qquad \text{Add numerators.}$$

$$= \frac{x + 2}{(3x - 2)(x + 2)} \qquad \text{Factor denominator.}$$

$$= \frac{1}{3x - 2} \qquad \text{Simplify.}$$

To add rational expressions with unlike denominators, we must first write each expression so that it has the same common denominator. For example, the least common denominator for $\dfrac{1}{x+2}$ and $\dfrac{2}{x-1}$ equals the least common multiple of $x + 2$ and $x - 1$, which was shown to be their product $(x + 2)(x - 1)$ in Example 1(c). To rewrite $\dfrac{1}{x+2}$ with the new denominator we multiply it by 1, expressed as $\dfrac{x-1}{x-1}$.

$$\frac{1}{x+2} \cdot 1 = \frac{1}{x+2} \cdot \frac{x-1}{x-1} = \frac{x-1}{(x+2)(x-1)}$$

Similarly, we multiply $\frac{2}{x-1}$ by 1, expressed as $\frac{x+2}{x+2}$.

$$\frac{2}{x-1} \cdot 1 = \frac{2}{x-1} \cdot \frac{x+2}{x+2} = \frac{2x+4}{(x-1)(x+2)}$$

Now the two rational expressions have like denominators and can be added.

$$\frac{1}{x+2} + \frac{2}{x-1} = \frac{x-1}{(x+2)(x-1)} + \frac{2x+4}{(x-1)(x+2)} \qquad \text{Write with LCD.}$$

$$= \frac{x-1+2x+4}{(x+2)(x-1)} \qquad \text{Add numerators.}$$

$$= \frac{3x+3}{(x+2)(x-1)} \qquad \text{Combine like terms.}$$

EXAMPLE 5 Adding rational expressions with unlike denominators

Add and simplify.

(a) $\dfrac{1}{x} + \dfrac{2}{x^2}$ (b) $\dfrac{1}{x^2-9} + \dfrac{2}{x+3}$ (c) $\dfrac{a}{a-b} + \dfrac{b}{a+b}$ (d) $\dfrac{1}{x-1} + \dfrac{1}{1-x}$

Solution
(a) The LCD is x^2.

$$\frac{1}{x} \cdot \frac{x}{x} + \frac{2}{x^2} = \frac{x}{x^2} + \frac{2}{x^2} \qquad \text{Write with LCD.}$$

$$= \frac{x+2}{x^2} \qquad \text{Add numerators.}$$

(b) The LCD is $x^2 - 9 = (x-3)(x+3)$.

$$\frac{1}{x^2-9} + \frac{2}{x+3} \cdot \frac{x-3}{x-3} = \frac{1}{(x+3)(x-3)} + \frac{2x-6}{(x+3)(x-3)} \qquad \text{Write with LCD.}$$

$$= \frac{2x-5}{(x+3)(x-3)} \qquad \text{Add numerators.}$$

(c) The LCD is $(a-b)(a+b)$.

$$\frac{a}{a-b} \cdot \frac{a+b}{a+b} + \frac{b}{a+b} \cdot \frac{a-b}{a-b} = \frac{a^2+ab}{(a-b)(a+b)} + \frac{ab-b^2}{(a-b)(a+b)} \qquad \text{Write with LCD.}$$

$$= \frac{a^2+2ab-b^2}{(a-b)(a+b)} \qquad \text{Add numerators.}$$

(d) By multiplying the second expression by $\frac{-1}{-1}$, the LCD can be found.

$$\frac{1}{x-1} + \frac{1}{1-x} \cdot \frac{-1}{-1} = \frac{1}{x-1} + \frac{-1}{(1-x)(-1)} \qquad \text{Write with LCD.}$$

$$= \frac{1}{x-1} + \frac{-1}{x-1} \qquad \text{Distributive property}$$

$$= \frac{0}{x-1} \qquad \text{Add numerators.}$$

$$= 0 \qquad \text{Simplify.}$$

NOTE: When one denominator is the opposite of the other, try multiplying one expression by 1 in the form $\frac{-1}{-1}$.

Subtraction of Rational Expressions

Subtraction of rational expressions is similar to subtraction of fractions.

> **DIFFERENCES OF RATIONAL EXPRESSIONS**
>
> To subtract two rational expressions with like denominators, subtract their numerators. The denominator does not change.
>
> $$\frac{A}{C} - \frac{B}{C} = \frac{A - B}{C} \qquad C \text{ not zero}$$

When you are subtracting numerators of rational expressions, it is *essential* to apply the distributive property correctly: Subtract every term in the numerator of the second rational expression. For example,

$$\frac{1}{x+1} - \frac{2x-1}{x+1} = \frac{1-(2x-1)}{x+1} = \frac{1-2x+1}{x+1} = \frac{2-2x}{x+1}.$$

Be sure to place parentheses around the second numerator before subtracting.

EXAMPLE 6 Subtracting rational expressions with like denominators

Subtract and simplify.

(a) $\dfrac{3}{x^2} - \dfrac{x+3}{x^2}$ **(b)** $\dfrac{2x}{x^2-1} - \dfrac{x+1}{x^2-1}$

Solution

(a) The expressions have like denominators, so subtract the numerators.

$$\frac{3}{x^2} - \frac{x+3}{x^2} = \frac{3-(x+3)}{x^2} \qquad \text{Subtract numerators.}$$

$$= \frac{3-x-3}{x^2} \qquad \text{Distributive property}$$

$$= \frac{-x}{x^2} \qquad \text{Subtract.}$$

$$= -\frac{1}{x} \qquad \text{Simplify.}$$

(b) The expressions have like denominators, so subtract the numerators.

$$\frac{2x}{x^2-1} - \frac{x+1}{x^2-1} = \frac{2x-(x+1)}{x^2-1} \qquad \text{Subtract numerators.}$$

$$= \frac{2x-x-1}{x^2-1} \qquad \text{Distributive property}$$

$$= \frac{x-1}{(x+1)(x-1)} \qquad \text{Combine like terms; factor.}$$

$$= \frac{1}{x+1} \qquad \text{Simplify.}$$

6.3 ADDITION AND SUBTRACTION OF RATIONAL EXPRESSIONS

TECHNOLOGY NOTE: Graphical and Numerical Support

We can give support to our work in Example 6(a) by letting $Y_1 = 3/X^2 - (X + 3)/X^2$, the given expression, and $Y_2 = -1/X$, the simplified expression. In Figures 6.9(a) and 6.9(b) the graphs of y_1 and y_2 appear to be identical. In Figure 6.9(c) numerical support is given, where $y_1 = y_2$ for each value of x.

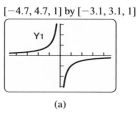
(a)

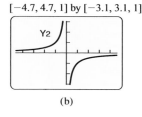
(b)

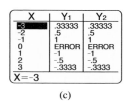
(c)

Figure 6.9

EXAMPLE 7 Subtracting rational expressions with unlike denominators

Subtract and simplify.

(a) $\dfrac{5a}{b^2} - \dfrac{4b}{a^2}$ 　　(b) $\dfrac{x-1}{x} - \dfrac{5}{x+5}$ 　　(c) $\dfrac{1}{x^2 - 3x + 2} - \dfrac{1}{x^2 - x - 2}$

Solution

(a) The LCD is a^2b^2.

$$\dfrac{5a}{b^2} \cdot \dfrac{a^2}{a^2} - \dfrac{4b}{a^2} \cdot \dfrac{b^2}{b^2} = \dfrac{5a^3}{b^2a^2} - \dfrac{4b^3}{a^2b^2} \qquad \text{Write with LCD.}$$

$$= \dfrac{5a^3 - 4b^3}{a^2b^2} \qquad \text{Subtract numerators.}$$

(b) The LCD is $x(x + 5)$.

$$\dfrac{x-1}{x} \cdot \dfrac{x+5}{x+5} - \dfrac{5}{x+5} \cdot \dfrac{x}{x} \qquad \text{Write with LCD.}$$

$$= \dfrac{(x-1)(x+5)}{x(x+5)} - \dfrac{5x}{x(x+5)} \qquad \text{Multiply.}$$

$$= \dfrac{(x-1)(x+5) - 5x}{x(x+5)} \qquad \text{Subtract numerators.}$$

$$= \dfrac{x^2 + 4x - 5 - 5x}{x(x+5)} \qquad \text{Multiply binomials.}$$

$$= \dfrac{x^2 - x - 5}{x(x+5)} \qquad \text{Combine like terms.}$$

(c) Because $x^2 - 3x + 2 = (x - 2)(x - 1)$ and $x^2 - x - 2 = (x - 2)(x + 1)$, the LCD is $(x - 1)(x + 1)(x - 2)$.

$$\dfrac{1}{(x-2)(x-1)} \cdot \dfrac{(x+1)}{(x+1)} - \dfrac{1}{(x-2)(x+1)} \cdot \dfrac{(x-1)}{(x-1)} \qquad \text{Write with LCD.}$$

$$= \dfrac{(x+1)}{(x-2)(x-1)(x+1)} - \dfrac{(x-1)}{(x-2)(x+1)(x-1)} \qquad \text{Multiply.}$$

$$= \dfrac{(x+1) - (x-1)}{(x-2)(x-1)(x+1)} \qquad \text{Subtract numerators.}$$

$$= \dfrac{x + 1 - x + 1}{(x-2)(x-1)(x+1)} \qquad \text{Distributive property}$$

$$= \dfrac{2}{(x-2)(x-1)(x+1)} \qquad \text{Simplify numerator.}$$

The following procedure summarizes how to add or subtract rational expressions.

> **STEPS FOR FINDING SUMS AND DIFFERENCES OF RATIONAL EXPRESSIONS**
>
> **STEP 1:** If the denominators are not common, multiply each expression by 1 written in the appropriate form to obtain the LCD.
>
> **STEP 2:** Add or subtract the numerators. Combine like terms.
>
> **STEP 3:** If possible, simplify the final expression.

▶ **REAL-WORLD CONNECTION** Sums of rational expressions occur in applications such as electrical circuits. The flow of electricity through a wire can be compared to the flow of water through a hose. Voltage is the force "pushing" the electricity and corresponds to water pressure in a hose. Resistance is the opposition to the flow of electricity, and more resistance results in less flow of electricity. Resistance corresponds to the diameter of a hose; if the diameter is smaller, less water flows. An ordinary light bulb is an example of a resistor in an electrical circuit.

Suppose that two light bulbs are wired in parallel so that electricity can flow through either light bulb, as depicted in Figure 6.10. If their resistances are R_1 and R_2, their combined resistance R can be computed by the equation

$$\frac{1}{R} = \frac{1}{R_1} + \frac{1}{R_2}.$$

Figure 6.10

Resistance is often measured in a unit called *ohms*. A standard 60-watt light bulb might have a resistance of about 200 ohms. (*Source:* R. Weidner and R. Sells, *Elementary Classical Physics, Vol. 2.*)

EXAMPLE 8 Modeling electrical resistance

A 100-watt light bulb with a resistance of $R_1 = 120$ ohms and a 75-watt light bulb with a resistance of $R_2 = 160$ ohms are placed in an electrical circuit, as shown in Figure 6.10. Find their combined resistance.

Solution
Let $R_1 = 120$ and $R_2 = 160$ in the given equation and solve for R.

$$\frac{1}{R} = \frac{1}{R_1} + \frac{1}{R_2} \quad \text{Given equation}$$

$$= \frac{1}{120} + \frac{1}{160} \quad \text{Substitute for } R_1 \text{ and } R_2.$$

$$= \frac{1}{120} \cdot \frac{4}{4} + \frac{1}{160} \cdot \frac{3}{3} \quad \text{LCD is 480.}$$

$$= \frac{4}{480} + \frac{3}{480} \quad \text{Multiply.}$$

$$= \frac{7}{480} \quad \text{Add.}$$

Because $\frac{1}{R} = \frac{7}{480}$, $R = \frac{480}{7} \approx 69$ ohms.

CRITICAL THINKING

If $\frac{1}{R} = \frac{1}{R_1} + \frac{1}{R_2}$, does it follow that $R = R_1 + R_2$? Explain your answer.

6.3 PUTTING IT ALL TOGETHER

The following table summarizes some important concepts from this section.

Concept	Explanation	Example
Least Common Multiple (LCM)	To find the LCM, follow the three-step procedure presented on page 268.	To find the LCM of $2x^2 - 2x$ and $8x^2$, factor each expression: $$2x^2 - 2x = 2 \cdot x \cdot (x-1) \text{ and}$$ $$8x^2 = 2 \cdot 2 \cdot 2 \cdot x \cdot x.$$ The LCM is $$2 \cdot 2 \cdot 2 \cdot x \cdot x \cdot (x-1) = 8x^2(x-1).$$
Least Common Denominator (LCD)	*The LCD equals the LCM of the denominators.*	The LCD for $\frac{3x}{2x^2 - 2x}$ and $\frac{5}{8x^2}$ is $8x^2(x-1)$.
Addition and Subtraction of Rational Expressions	To add (or subtract) two rational expressions with like denominators, add (or subtract) their numerators. The denominator does not change. $$\frac{A}{C} + \frac{B}{C} = \frac{A+B}{C}$$ $$\frac{A}{C} - \frac{B}{C} = \frac{A-B}{C} \quad C \text{ not zero}$$ If the denominators are *not alike*, write each term with the LCD first.	Like denominators $$\frac{x}{x+1} + \frac{3x}{x+1} = \frac{4x}{x+1}$$ Unlike denominators $$\frac{1}{x} - \frac{2x}{x+1} = \frac{1}{x} \cdot \frac{x+1}{x+1} - \frac{2x}{x+1} \cdot \frac{x}{x}$$ $$= \frac{x+1}{x(x+1)} - \frac{2x^2}{x(x+1)}$$ $$= \frac{-2x^2 + x + 1}{x(x+1)}$$

6.4 RATIONAL EQUATIONS

Solving Rational Equations ▪ Solving an Equation for a Variable

A LOOK INTO MATH ▷ In Section 6.1 we introduced rational equations. If an equation contains one or more rational expressions, it is often called a *rational equation*. Rational equations occur in applications involving distance and time. For example, if one athlete runs 2 miles per hour faster and finishes 3 minutes sooner than another athlete in a 3-mile race, then the speed of the slower athlete x can be found by solving the rational equation

$$\frac{3}{x+2} + \frac{1}{20} = \frac{3}{x}.$$

(To better understand why, refer to Example 7.) In this section we discuss how to solve different types of rational equations.

Solving Rational Equations

A rational expression is a ratio of two polynomials. In Sections 6.2 and 6.3, we *simplified* and *evaluated* rational expressions. Unlike a rational expression, a rational equation always contains an equals sign, and we *solve* this equation for any x-values that make the equation a true statement. One way to solve a rational equation is to *clear fractions* by multiplying each side of the equation by the least common denominator (LCD). For example, to solve

$$\frac{1}{2x} + \frac{2}{3} = \frac{5}{2x}$$

multiply each side by the LCD (LCM of 3 and $2x$), which is $6x$.

$$6x\left(\frac{1}{2x} + \frac{2}{3}\right) = 6x\left(\frac{5}{2x}\right) \quad \text{Multiply each side by } 6x.$$

$$\frac{6x}{2x} + \frac{12x}{3} = \frac{30x}{2x} \quad \text{Distributive property}$$

$$3 + 4x = 15 \quad \text{Simplify.}$$

$$4x = 12 \quad \text{Subtract 3 from each side.}$$

$$x = 3 \quad \text{Divide each side by 4.}$$

Check this answer to verify that 3 is indeed a solution. *When solving a rational equation, always check your answer.* If one of your *possible answers* makes the denominator of an expression in the given equation equal to 0, it cannot be a solution.

EXAMPLE 1 Solving rational equations

Solve each equation and check your answer.

(a) $\dfrac{x+1}{2x} - \dfrac{x-1}{4x} = \dfrac{1}{3}$ (b) $\dfrac{3}{x-2} + \dfrac{5}{x+2} = \dfrac{12}{x^2-4}$

Solution

(a) Note that the left side of the equation is undefined when $x = 0$. Thus 0 cannot be a solution. Start by determining the LCD, which is $12x$.

$$12x\left(\frac{x+1}{2x} - \frac{x-1}{4x}\right) = 12x\left(\frac{1}{3}\right) \quad \text{Multiply each side by } 12x.$$

$$\frac{12x(x+1)}{2x} - \frac{12x(x-1)}{4x} = \frac{12x}{3} \quad \text{Distributive property}$$

$$6(x+1) - 3(x-1) = 4x \quad \text{Simplify.}$$

$$6x + 6 - 3x + 3 = 4x \quad \text{Distributive property}$$

$$3x + 9 = 4x \quad \text{Combine like terms.}$$

$$x = 9 \quad \text{Solve for } x.$$

Check: $\dfrac{9+1}{2(9)} - \dfrac{9-1}{4(9)} \stackrel{?}{=} \dfrac{1}{3}$ Let $x = 9$ in given equation.

$\dfrac{10}{18} - \dfrac{8}{36} \stackrel{?}{=} \dfrac{1}{3}$ Simplify.

$\dfrac{1}{3} = \dfrac{1}{3}$ It checks.

(b) Note that both sides of the equation are undefined when $x = 2$ or $x = -2$. Thus neither 2 nor -2 can be a solution. The LCD is $x^2 - 4 = (x - 2)(x + 2)$.

$$\frac{3(x-2)(x+2)}{x-2} + \frac{5(x-2)(x+2)}{x+2} = \frac{12(x^2-4)}{x^2-4} \quad \text{Multiply each term by the LCD.}$$

$$3(x+2) + 5(x-2) = 12 \quad \text{Simplify.}$$
$$3x + 6 + 5x - 10 = 12 \quad \text{Distributive property}$$
$$8x - 4 = 12 \quad \text{Combine like terms.}$$
$$8x = 16 \quad \text{Add 4.}$$
$$x = 2 \quad \text{Divide by 8.}$$

Check: We already noted that 2 cannot be a solution. However, checking this answer gives the following result.

$$\frac{3}{2-2} + \frac{5}{2+2} \stackrel{?}{=} \frac{12}{2^2-4} \quad \text{Let } x = 2.$$

$$\frac{3}{0} + \frac{5}{4} \stackrel{?}{=} \frac{12}{0} \quad \text{Simplify fractions.}$$

Both sides are undefined because division by 0 is not possible. There are no solutions.

The next example illustrates how to solve a rational equation graphically. Because graphical solutions are only estimates, be sure to check them by substituting observed values in the given equation.

EXAMPLE 2 Solving a rational equation graphically

Solve the equation $\frac{4}{x-1} = 2x$ graphically. Check your results.

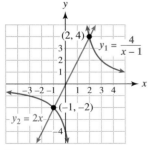

Figure 6.11

Solution
Start by graphing $y_1 = \frac{4}{x-1}$ and $y_2 = 2x$. The graph of y_2 is a line with slope 2 passing through the origin. To help graph y_1 make a table of values or use a graphing calculator. The two graphs intersect when $x = -1$ and $x = 2$, as shown in Figure 6.11.

Check: $\frac{4}{-1-1} \stackrel{?}{=} 2(-1)$ Let $x = -1$. $\frac{4}{2-1} \stackrel{?}{=} 2(2)$ Let $x = 2$.
 $-2 = -2$ It checks. $4 = 4$ It checks.

The next example illustrates how to solve two rational equations and support the results with a graphing calculator.

EXAMPLE 3　Solving rational equations

Solve each equation. Support your results either graphically or numerically.

(a) $\dfrac{1}{2} + \dfrac{x}{3} = \dfrac{x}{5}$　　(b) $\dfrac{3}{x+3} - 2 = \dfrac{x}{x+3}$　　(c) $\dfrac{3}{x-2} = \dfrac{5}{x+2}$

Solution
(a) The LCD is the product of 2, 3, and 5, which is 30.

$$30 \cdot \left(\dfrac{1}{2} + \dfrac{x}{3}\right) = \dfrac{x}{5} \cdot 30 \qquad \text{Multiply by the LCD.}$$

$$\dfrac{30}{2} + \dfrac{30x}{3} = \dfrac{30x}{5} \qquad \text{Distributive property}$$

$$15 + 10x = 6x \qquad \text{Simplify.}$$

$$4x = -15 \qquad \text{Subtract } 6x \text{ and } 15.$$

$$x = -\dfrac{15}{4} \qquad \text{Solve.}$$

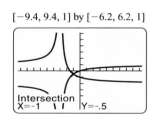

Figure 6.12

[−9, 9, 1] by [−6, 6, 1]

This result can be supported graphically by letting $Y_1 = 1/2 + X/3$ and $Y_2 = X/5$. Their graphs intersect at the point $(-3.75, -0.75)$, as shown in Figure 6.12. Therefore the solution to this equation is -3.75, or $-\dfrac{15}{4}$.

(b) Multiply each side by the LCD, which is $x + 3$.

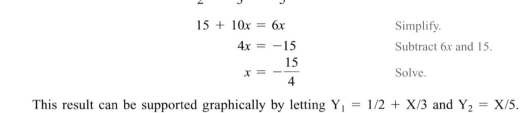

$$(x+3) \cdot \left(\dfrac{3}{x+3} - 2\right) = \dfrac{x}{x+3} \cdot (x+3) \qquad \text{Multiply by the LCD.}$$

$$3 - 2(x+3) = x \qquad \text{Distributive property}$$

$$3 - 2x - 6 = x \qquad \text{Distributive property}$$

$$-3 = 3x \qquad \text{Add } 2x \text{ to each side.}$$

$$x = -1 \qquad \text{Divide by 3; rewrite.}$$

[−9.4, 9.4, 1] by [−6.2, 6.2, 1]

Figure 6.13

Figure 6.13 supports that -1 is a solution by showing that the graphs of the equations $Y_1 = 3/(X+3) - 2$ and $Y_2 = X/(X+3)$ intersect at $(-1, -0.5)$.

NOTE: Because graphs of rational expressions are often nonlinear, it may be difficult to locate graphical solutions.

(c) Multiply each side by the LCD, which is $(x-2)(x+2)$.

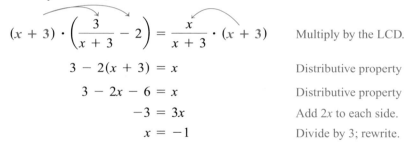

$$(x-2)(x+2) \cdot \dfrac{3}{x-2} = \dfrac{5}{x+2} \cdot (x-2)(x+2) \qquad \text{Multiply by the LCD.}$$

$$\dfrac{3(x-2)(x+2)}{x-2} = \dfrac{5(x-2)(x+2)}{x+2} \qquad \text{Multiply expressions.}$$

$$3(x+2) = 5(x-2) \qquad \text{Simplify.}$$

$$3x + 6 = 5x - 10 \qquad \text{Distributive property}$$

$$16 = 2x \qquad \text{Add 10 and subtract } 3x.$$

$$x = 8 \qquad \text{Divide by 2; rewrite.}$$

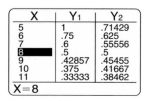

Figure 6.14

To support this result numerically, let $Y_1 = 3/(X-2)$ and $Y_2 = 5/(X+2)$. The table in Figure 6.14 shows that $y_1 = y_2 = 0.5$ when $x = 8$. The solution is 8.

TECHNOLOGY NOTE: Entering Rational Expressions

When entering a rational expression, use parentheses around the numerator and denominator, as in the following example. Enter

$$y = \frac{x+1}{2x-1} \quad \text{as} \quad Y_1 = (X + 1)/(2X - 1).$$

Sometimes solving a rational equation results in the need to solve a quadratic equation, which is demonstrated in the next example.

EXAMPLE 4 Solving rational equations

Solve the equation.

(a) $\dfrac{1}{x} + \dfrac{1}{x^2} = \dfrac{3}{4}$ (b) $\dfrac{2x}{x+2} + \dfrac{3x}{x-1} = 7$

Solution

(a) The LCD is $4x^2$. Note that 0 cannot be a solution.

$$4x^2 \cdot \left(\frac{1}{x} + \frac{1}{x^2}\right) = \frac{3}{4} \cdot 4x^2 \qquad \text{Multiply by } 4x^2.$$

$$\frac{4x^2}{x} + \frac{4x^2}{x^2} = \frac{12x^2}{4} \qquad \text{Distributive property}$$

$$4x + 4 = 3x^2 \qquad \text{Simplify.}$$

$$0 = 3x^2 - 4x - 4 \qquad \text{Subtract } 4x \text{ and } 4.$$

$$0 = (3x + 2)(x - 2) \qquad \text{Factor.}$$

$$x = -\frac{2}{3} \quad \text{or} \quad x = 2 \qquad \text{Solve.}$$

Both solutions check.

(b) The LCD is $(x + 2)(x - 1)$. Note that -2 and 1 cannot be solutions.

$$(x+2)(x-1)\left(\frac{2x}{x+2} + \frac{3x}{x-1}\right) = 7(x+2)(x-1) \qquad \text{Multiply by the LCD.}$$

$$\frac{2x(x+2)(x-1)}{x+2} + \frac{3x(x+2)(x-1)}{x-1} = 7(x+2)(x-1) \qquad \text{Distributive property.}$$

$$2x(x-1) + 3x(x+2) = 7(x+2)(x-1) \qquad \text{Simplify.}$$

$$2x^2 - 2x + 3x^2 + 6x = 7(x^2 + x - 2) \qquad \text{Multiply.}$$

$$5x^2 + 4x = 7x^2 + 7x - 14 \qquad \text{Simplify.}$$

$$0 = 2x^2 + 3x - 14 \qquad \text{Subtract } 5x^2 \text{ and } 4x.$$

$$0 = (2x + 7)(x - 2) \qquad \text{Factor.}$$

$$x = -\frac{7}{2} \quad \text{or} \quad x = 2 \qquad \text{Solve.}$$

Both solutions check.

▶ **REAL-WORLD CONNECTION** Rational equations are frequently used to estimate the time spent waiting in line whenever arrivals are random. This concept is illustrated in the next example.

EXAMPLE 5 Modeling waiting time

Suppose that cars are arriving randomly at a construction zone and that the flag person can instruct x drivers per minute. If cars arrive at an average rate of 7 cars per minute, the average time T in minutes for each driver to wait and talk to the flag person is

$$T(x) = \frac{1}{x - 7},$$

where $x > 7$. How many drivers per minute should the flag person be able to instruct to keep the average wait to 1 minute? (*Source:* N. Garber.)

Solution
We must determine x such that $T(x) = 1$.

$$\frac{1}{x-7} = 1 \qquad \text{Equation to be solved.}$$

$$(x - 7) \cdot \frac{1}{x - 7} = 1 \cdot (x - 7) \qquad \text{Multiply by the LCD.}$$

$$1 = x - 7 \qquad \text{Simplify.}$$

$$8 = x \qquad \text{Add 7 to each side.}$$

The flag person should be able to instruct 8 drivers per minute to limit the average wait to 1 minute.

▶ **REAL-WORLD CONNECTION** Rational equations occur in applications involving time and rate, as demonstrated in the next example.

EXAMPLE 6 Determining the time required to empty a pool

A pump can empty a swimming pool in 50 hours. To speed up the process a second pump is used that can empty the pool in 80 hours. How long will it take for both pumps working together to empty the pool?

Solution
Because the first pump can empty the entire pool in 50 hours, it can empty $\frac{1}{50}$ of the pool in 1 hour, $\frac{2}{50}$ of the pool in 2 hours, $\frac{3}{50}$ of the pool in 3 hours, and in general, it can empty $\frac{t}{50}$ of the pool in t hours. The second pump can empty the pool in 80 hours, so (using similar reasoning) it can empty $\frac{t}{80}$ of the pool in t hours.

Together the pumps can empty

$$\frac{t}{50} + \frac{t}{80}$$

of the pool in t hours. The job will be complete when the fraction of the pool that is empty equals 1. Thus we must solve the equation

$$\frac{t}{50} + \frac{t}{80} = 1.$$

To clear fractions we can multiply each side by $(50)(80)$. (We could also use the LCD of 400.)

$$(50)(80)\left(\frac{t}{50} + \frac{t}{80}\right) = 1(50)(80)$$

$$\frac{50 \cdot 80 \cdot t}{50} + \frac{50 \cdot 80 \cdot t}{80} = 4000 \qquad \text{Distributive property}$$

$$80t + 50t = 4000 \qquad \text{Simplify.}$$

$$130t = 4000 \qquad \text{Combine like terms.}$$

$$t = \frac{4000}{130} \approx 30.8 \text{ hours} \qquad \text{Solve.}$$

The two pumps can empty the pool in about 30.8 hours.

▶ **REAL-WORLD CONNECTION** If a person drives a car 60 miles per hour for 4 hours, the total distance d traveled is

$$d = 60 \cdot 4 = 240 \text{ miles.}$$

That is, distance equals the product of the rate (speed) and the elapsed time. This may be written as $d = rt$ and expressed verbally as "distance equals rate times time." This formula is used in Example 7.

CRITICAL THINKING

A bicyclist rides uphill at 6 miles per hour for 1 mile and then rides downhill at 12 miles per hour for 1 mile. What is the average speed of the bicyclist?

Figure 6.15

EXAMPLE 7 Solving an application

The winner of a 3-mile race finishes 3 minutes ahead of another runner. If the winner runs 2 miles per hour faster than the slower runner, find the average speed of each runner.

Solution
Let x be the speed of the slower runner. Then $x + 2$ represents the speed of the winner. See Figure 6.15. To determine the time for each runner to finish the race divide each side of $d = rt$ by r to obtain

$$t = \frac{d}{r}.$$

The slower runner runs 3 miles at x miles per hour, so the time is $\frac{3}{x}$; the winner runs 3 miles at $x + 2$ miles per hour, so the winning time is $\frac{3}{x+2}$. We summarize this in Table 6.3.

TABLE 6.3

	Distance (mi)	Rate (mi/hr)	Time (hr)
Winning Runner	3	$x + 2$	$\frac{3}{x+2}$
Other Runner	3	x	$\frac{3}{x}$

If you add 3 minutes (or equivalently $\frac{3}{60} = \frac{1}{20}$ hour) to the winner's time, it equals the slower runner's time, or

$$\frac{3}{x+2} + \frac{1}{20} = \frac{3}{x}.$$

NOTE: The runners' speeds are in miles per *hour*, so you need to keep time in hours.

Multiply each side by the LCD, which is $20x(x+2)$.

$$20x(x+2)\left(\frac{3}{x+2} + \frac{1}{20}\right) = 20x(x+2)\left(\frac{3}{x}\right) \quad \text{Multiply by the LCD.}$$
$$60x + x(x+2) = 60(x+2) \quad \text{Distributive property}$$
$$60x + x^2 + 2x = 60x + 120 \quad \text{Distributive property}$$
$$x^2 + 2x - 120 = 0 \quad \text{Rewrite the equation.}$$
$$(x+12)(x-10) = 0 \quad \text{Factor.}$$
$$x = -12 \quad \text{or} \quad x = 10 \quad \text{Solve.}$$

The only valid solution is 10 because a running speed cannot be negative. Thus the slower runner is running at 10 miles per hour, and the faster runner is running at 12 miles per hour.

Solving an Equation for a Variable

▶ **REAL-WORLD CONNECTION** Sometimes in science and other disciplines, it is necessary to solve a formula for a variable. These formulas are often rational equations. This technique is demonstrated in the next two examples.

EXAMPLE 8 Solving an equation for a variable

Solve the equation $P = \frac{nrT}{V}$ for V. (This formula is used to calculate the pressure of a gas.)

Solution
Start by multiplying each side of the equation by V.

$$V \cdot P = V\left(\frac{nrT}{V}\right) \quad \text{Multiply by } V.$$
$$PV = nrT \quad \text{Simplify.}$$
$$\frac{PV}{P} = \frac{nrT}{P} \quad \text{Divide each side by } P.$$
$$V = \frac{nrT}{P} \quad \text{Simplify.}$$

EXAMPLE 9 Solving an equation from electricity

Solve the equation $\frac{1}{R} = \frac{1}{R_1} + \frac{1}{R_2}$ for R_2.

Solution
Start by multiplying each side of the equation by the LCD: RR_1R_2.

$$(RR_1R_2)\frac{1}{R} = (RR_1R_2)\left(\frac{1}{R_1} + \frac{1}{R_2}\right) \quad \text{Multiply by the LCD.}$$
$$R_1R_2 = RR_2 + RR_1 \quad \text{Distributive property}$$
$$R_1R_2 - RR_2 = RR_1 \quad \text{Subtract } RR_2.$$
$$R_2(R_1 - R) = RR_1 \quad \text{Factor out } R_2.$$
$$R_2 = \frac{RR_1}{R_1 - R} \quad \text{Divide each side by } R_1 - R.$$

6.4 PUTTING IT ALL TOGETHER

The following table summarizes the basic steps for solving a rational equation.

Concept	Explanation	Example
Rational Equations	To solve a rational equation, first multiply each side of the equation by the LCD. Then solve the resulting polynomial equation. Be sure to check your answer(s). *The LCD is the LCM of the denominators.*	Multiply each term by $4x^2$. $\frac{1}{4x} + \frac{2}{x^2} = \frac{1}{x}$ $\frac{4x^2}{4x} + \frac{8x^2}{x^2} = \frac{4x^2}{x}$ $x + 8 = 4x$ $8 = 3x$ $x = \frac{8}{3}$

The following example illustrates how to solve the rational equation $\frac{2}{x} + 2 = 3$ symbolically, graphically, and numerically with the aid of a graphing calculator, where we let $Y_1 = 2/X + 2$ and $Y_2 = 3$.

Symbolic Solution

$\frac{2}{x} + 2 = 3$

$\frac{2}{x} = 1$

$2 = x$

The solution is **2**.

Numerical Solution

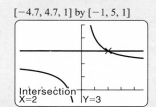

$y_1 = 3$ when $x = 2$.

Graphical Solution

$[-4.7, 4.7, 1]$ by $[-1, 5, 1]$

The graphs intersect at **(2, 3)**.

6.5 COMPLEX FRACTIONS

Basic Concepts ■ Simplifying Complex Fractions

A LOOK INTO MATH ▷ If two and a half pizzas are cut into fourths, then there are 10 pieces of pizza. This problem can be written as

$$\frac{2 + \frac{1}{2}}{\frac{1}{4}} = 10.$$

The expression on the left side of the equation is called a *complex fraction*. Typically, we want to simplify a complex fraction to a fraction with the standard form $\frac{a}{b}$. In this section we discuss how to simplify complex fractions.

Basic Concepts

A **complex fraction** is a rational expression that contains fractions in its numerator, denominator, or both. Examples of complex fractions include

$$\frac{1+\frac{1}{x}}{1-\frac{1}{x}}, \quad \frac{2x}{\frac{4}{x}+\frac{3}{y}}, \quad \text{and} \quad \frac{\frac{a}{3}+\frac{a}{4}}{a-\frac{1}{a-1}}.$$

Complex fractions involve division of fractions. When dividing two fractions, we can multiply the first fraction by the reciprocal of the second fraction.

SIMPLIFYING COMPLEX FRACTIONS

For any real numbers a, b, c, and d,

$$\frac{\frac{a}{b}}{\frac{c}{d}} = \frac{a}{b} \cdot \frac{d}{c},$$

where b, c, and d are not zero.

EXAMPLE 1 Simplifying basic types of complex fractions

Simplify.

(a) $\dfrac{\frac{7}{8}}{\frac{3}{4}}$ **(b)** $\dfrac{\frac{a}{4b}}{\frac{a}{6b^2}}$ **(c)** $\dfrac{2+\frac{1}{2}}{\frac{1}{4}}$

Solution

(a) Rather than dividing $\frac{7}{8}$ by $\frac{3}{4}$, multiply $\frac{7}{8}$ by $\frac{4}{3}$.

$$\frac{\frac{7}{8}}{\frac{3}{4}} = \frac{7}{8} \cdot \frac{4}{3} \qquad \text{"Invert and multiply."}$$

$$= \frac{7}{6} \qquad \text{Multiply and simplify.}$$

(b) Rather than dividing $\frac{a}{4b}$ by $\frac{a}{6b^2}$, multiply $\frac{a}{4b}$ by $\frac{6b^2}{a}$.

$$\frac{\frac{a}{4b}}{\frac{a}{6b^2}} = \frac{a}{4b} \cdot \frac{6b^2}{a} \qquad \text{"Invert and multiply."}$$

$$= \frac{6ab^2}{4ab} \qquad \text{Multiply fractions.}$$

$$= \frac{3b}{2} \qquad \text{Simplify.}$$

(c) Start by writing $2 + \frac{1}{2}$ as one fraction.

$$\frac{2 + \frac{1}{2}}{\frac{1}{4}} = \frac{\frac{5}{2}}{\frac{1}{4}} \qquad \text{Write as an improper fraction.}$$

$$= \frac{5}{2} \cdot \frac{4}{1} \qquad \text{"Invert and multiply."}$$

$$= 10 \qquad \text{Simplify.}$$

Simplifying Complex Fractions

There are two basic strategies for simplifying a complex fraction. The first is to simplify both the numerator and the denominator and then divide the resulting fractions as was done in Example 1(c). The second is to multiply the numerator and denominator by the least common denominator of the fractions in the numerator and denominator.

SIMPLIFYING THE NUMERATOR AND DENOMINATOR The next example illustrates the method whereby the numerator and denominator are simplified first.

EXAMPLE 2 Simplifying complex fractions

Simplify.

(a) $\dfrac{1 + \dfrac{1}{x}}{1 - \dfrac{1}{x}}$ **(b)** $\dfrac{2t - \dfrac{1}{2t}}{2t + \dfrac{1}{2t}}$ **(c)** $\dfrac{\dfrac{1}{z+1} - \dfrac{1}{z-1}}{\dfrac{2}{z+2} - \dfrac{2}{z-2}}$

Solution
(a) First, simplify the numerator by writing it as one term. The LCD is x.

$$1 + \frac{1}{x} = \frac{x}{x} + \frac{1}{x} = \frac{x+1}{x}$$

Second, simplify the denominator by writing it as one term.

$$1 - \frac{1}{x} = \frac{x}{x} - \frac{1}{x} = \frac{x-1}{x}$$

Finally, simplify the complex fraction.

$$\frac{1 + \dfrac{1}{x}}{1 - \dfrac{1}{x}} = \frac{\dfrac{x+1}{x}}{\dfrac{x-1}{x}} \qquad \text{Simplify.}$$

$$= \frac{x+1}{x} \cdot \frac{x}{x-1} \qquad \text{"Invert and multiply."}$$

$$= \frac{x+1}{x-1} \qquad \text{Multiply and simplify.}$$

(b) Combine the terms in the numerator and in the denominator. The LCD is $2t$.

$$\frac{2t - \dfrac{1}{2t}}{2t + \dfrac{1}{2t}} = \frac{\dfrac{4t^2}{2t} - \dfrac{1}{2t}}{\dfrac{4t^2}{2t} + \dfrac{1}{2t}} \qquad \text{Write with the LCD.}$$

$$= \frac{\dfrac{4t^2 - 1}{2t}}{\dfrac{4t^2 + 1}{2t}} \qquad \text{Combine terms.}$$

$$= \frac{4t^2 - 1}{2t} \cdot \frac{2t}{4t^2 + 1} \qquad \text{"Invert and multiply."}$$

$$= \frac{4t^2 - 1}{4t^2 + 1} \qquad \text{Multiply and simplify.}$$

(c) For the numerator the LCD is $(z + 1)(z - 1)$, and for the denominator the LCD is $(z + 2)(z - 2)$.

$$\frac{\dfrac{1}{z+1} - \dfrac{1}{z-1}}{\dfrac{2}{z+2} - \dfrac{2}{z-2}} = \frac{\dfrac{z-1}{(z+1)(z-1)} - \dfrac{z+1}{(z+1)(z-1)}}{\dfrac{2(z-2)}{(z+2)(z-2)} - \dfrac{2(z+2)}{(z+2)(z-2)}}$$

$$= \frac{\dfrac{-2}{(z+1)(z-1)}}{\dfrac{-8}{(z+2)(z-2)}}$$

$$= \frac{-2}{(z+1)(z-1)} \cdot \frac{(z+2)(z-2)}{-8}$$

$$= \frac{(z+2)(z-2)}{4(z+1)(z-1)}$$

MULTIPLYING BY THE LEAST COMMON DENOMINATOR A second strategy for simplifying a complex fraction is to multiply by 1 in the form $\frac{\text{LCD}}{\text{LCD}}$. To do this, we multiply the numerator and denominator by the LCD of the fractions in the numerator *and* denominator. For example, the LCD for the complex fraction

$$\frac{1 - \dfrac{1}{x}}{1 + \dfrac{1}{2x}}$$

is $2x$. To simplify, multiply the complex fraction by 1, expressed in the form $\frac{2x}{2x}$.

$$\frac{\left(1 - \dfrac{1}{x}\right) \cdot 2x}{\left(1 + \dfrac{1}{2x}\right) \cdot 2x} = \frac{2x - \dfrac{2x}{x}}{2x + \dfrac{2x}{2x}} \qquad \text{Distributive property}$$

$$= \frac{2x - 2}{2x + 1} \qquad \text{Simplify fractions.}$$

In the next example we use this method to simplify other complex fractions.

EXAMPLE 3 Simplifying complex fractions

Simplify.

(a) $\dfrac{1 + \dfrac{1}{x}}{1 + \dfrac{1}{y}}$ (b) $\dfrac{\dfrac{1}{x} - \dfrac{1}{y}}{x - y}$ (c) $\dfrac{\dfrac{3}{x-1} - \dfrac{2}{x}}{\dfrac{1}{x-1} + \dfrac{3}{x}}$ (d) $\dfrac{n^{-2} + m^{-2}}{1 + (nm)^{-2}}$

Solution

(a) The LCD for the numerator *and* the denominator is xy.

$$\dfrac{1 + \dfrac{1}{x}}{1 + \dfrac{1}{y}} = \dfrac{\left(1 + \dfrac{1}{x}\right) \cdot xy}{\left(1 + \dfrac{1}{y}\right) \cdot xy} \qquad \text{Multiply by the LCD.}$$

$$= \dfrac{xy + y}{xy + x} \qquad \text{Distributive property}$$

(b) The LCD is xy. Multiply the expression by $\dfrac{xy}{xy}$.

$$\dfrac{\left(\dfrac{1}{x} - \dfrac{1}{y}\right) \cdot xy}{(x-y) \cdot xy} = \dfrac{\dfrac{xy}{x} - \dfrac{xy}{y}}{xy(x-y)} \qquad \text{Distributive and commutative properties}$$

$$= \dfrac{y - x}{xy(x-y)} \qquad \text{Simplify fractions in numerator.}$$

$$= \dfrac{-1(x-y)}{xy(x-y)} \qquad \text{Factor out } -1;\text{ rewrite numerator.}$$

$$= -\dfrac{1}{xy} \qquad \text{Simplify.}$$

(c) The LCD is $x(x-1)$. Multiply the expression by $\dfrac{x(x-1)}{x(x-1)}$.

$$\dfrac{\left(\dfrac{3}{x-1} - \dfrac{2}{x}\right) \cdot x(x-1)}{\left(\dfrac{1}{x-1} + \dfrac{3}{x}\right) \cdot x(x-1)} = \dfrac{\dfrac{3x(x-1)}{x-1} - \dfrac{2x(x-1)}{x}}{\dfrac{x(x-1)}{x-1} + \dfrac{3x(x-1)}{x}} \qquad \text{Distributive property}$$

$$= \dfrac{3x - 2(x-1)}{x + 3(x-1)} \qquad \text{Simplify.}$$

$$= \dfrac{3x - 2x + 2}{x + 3x - 3} \qquad \text{Distributive property}$$

$$= \dfrac{x + 2}{4x - 3} \qquad \text{Combine like terms.}$$

(d) Start by rewriting the ratio with positive exponents.

$$\dfrac{n^{-2} + m^{-2}}{1 + (nm)^{-2}} = \dfrac{\dfrac{1}{n^2} + \dfrac{1}{m^2}}{1 + \dfrac{1}{n^2 m^2}}$$

CRITICAL THINKING

Does the expression
$\dfrac{\dfrac{x}{y}}{2 + \dfrac{x}{y}}$ equal $\dfrac{1}{2}$? Explain.

Does the expression
$\dfrac{2 + \dfrac{x}{y}}{\dfrac{x}{y}}$ equal 3? Explain.

The LCD for the numerator *and* the denominator is n^2m^2.

$$\frac{\frac{1}{n^2} + \frac{1}{m^2}}{1 + \frac{1}{n^2m^2}} = \frac{\left(\frac{1}{n^2} + \frac{1}{m^2}\right) \cdot n^2m^2}{\left(1 + \frac{1}{n^2m^2}\right) \cdot n^2m^2}$$ Multiply by LCD.

$$= \frac{m^2 + n^2}{n^2m^2 + 1}$$ Distributive property

6.5 PUTTING IT ALL TOGETHER

Important concepts from this section are summarized in the following table.

Concept	Explanation	Examples
Complex Fraction	A rational expression that contains fractions in its numerator, denominator, or both	$\dfrac{\frac{1}{x} + \frac{1}{2x+1}}{1 - \frac{1}{2x+1}}$ and $\dfrac{\frac{a}{2b} - \frac{b}{2a}}{\frac{a}{2b} + \frac{b}{2a}}$
Simplifying Basic Complex Fractions	$\dfrac{\frac{a}{b}}{\frac{c}{d}} = \dfrac{a}{b} \cdot \dfrac{d}{c}$	$\dfrac{\frac{2y}{z}}{\frac{y}{z-1}} = \dfrac{2y}{z} \cdot \dfrac{z-1}{y} = \dfrac{2(z-1)}{z}$
Method I: Simplifying the Numerator and Denominator First and Then Dividing	Combine the terms in the numerator, combine the terms in the denominator, and then invert and multiply.	$\dfrac{\frac{1}{2b} + \frac{1}{2a}}{\frac{1}{2b} - \frac{1}{2a}} = \dfrac{\frac{a+b}{2ab}}{\frac{a-b}{2ab}}$ $= \dfrac{a+b}{2ab} \cdot \dfrac{2ab}{a-b}$ $= \dfrac{a+b}{a-b}$
Method II: Multiplying the Numerator and Denominator by the LCD	Start by multiplying the numerator and denominator by the LCD of the numerator *and* the denominator.	$\dfrac{\frac{1}{2b} + \frac{1}{2a}}{\frac{1}{2b} - \frac{1}{2a}} = \dfrac{\left(\frac{1}{2b} + \frac{1}{2a}\right) \cdot 2ab}{\left(\frac{1}{2b} - \frac{1}{2a}\right) \cdot 2ab}$ $= \dfrac{a+b}{a-b}$ Note that the LCD is $2ab$.

6.6 MODELING WITH PROPORTIONS AND VARIATION

Proportions ▪ Direct Variation ▪ Inverse Variation ▪ Joint Variation

A LOOK INTO MATH ▷ Proportions are used frequently to solve problems. The following are examples.

- If someone earns $100 per day, that person can earn $500 in 5 days.
- If a car goes 210 miles on 10 gallons of gasoline, the car can go 420 miles on 20 gallons of gasoline.
- If a person walks a mile in 16 minutes, that person can walk a half mile in 8 minutes.

Many applications involve proportions or variation. In this section we discuss some of them.

Proportions

▶ **REAL-WORLD CONNECTION** A 650-megabyte compact disc (CD) can store about 74 minutes of music. (*Source:* Maxell Corporation.) Suppose that we have already selected some music to record on the CD and 256 megabytes are still available. Using proportions we can determine how many more minutes of music could be recorded. A **proportion** is a statement (equation) that two ratios (fractions) are equal.

Let x be the number of minutes available on the CD. Then **74** minutes *are to* **650** megabytes *as* x minutes *are to* **256** megabytes, which can be written as the proportion

$$\frac{74}{650} = \frac{x}{256}. \qquad \frac{\text{Minutes}}{\text{Megabytes}} = \frac{\text{Minutes}}{\text{Megabytes}}$$

Solving this equation for x gives

$$x = \frac{74 \cdot 256}{650} \approx 29.1 \text{ minutes.}$$

About 29 minutes are still available on the CD.

MAKING CONNECTIONS

Proportions and Fractional Parts

We could have solved the preceding problem by noting that the fraction of the CD still available for recording music is $\frac{256}{650}$. So $\frac{256}{650}$ of 74 minutes equals

$$\frac{256}{650} \cdot 74 \approx 29.1 \text{ minutes.}$$

The following property is a convenient way to solve proportions:

$$\frac{a}{b} = \frac{c}{d} \quad \text{is equivalent to} \quad ad = bc,$$

provided $b \neq 0$ and $d \neq 0$. This is a result of multiplying each side of the equation by a common denominator bd, and is sometimes referred to as *clearing fractions*.

$$bd \cdot \frac{a}{b} = \frac{c}{d} \cdot bd \qquad \text{Multiply by } bd.$$

$$\frac{bda}{b} = \frac{cbd}{d} \qquad \text{Property of multiplying fractions}$$

$$ad = bc \qquad \text{Simplify.}$$

For example, the proportion

$$\frac{6}{5} = \frac{8}{x}$$

is equivalent to

$$6x = 5 \cdot 8 \quad \text{or} \quad x = \frac{40}{6} = \frac{20}{3}.$$

EXAMPLE 1 Calculating the water content in snow

Six inches of light, fluffy snow is equivalent to about half an inch of rain in terms of water content. If 21 inches of snow fall, use proportions to estimate the water content.

Solution
Let x be the equivalent amount of rain. Then 6 inches of snow is to $\frac{1}{2}$ inch of rain as 21 inches of snow is to x inches of rain, which can be written as the proportion

$$\frac{6}{\frac{1}{2}} = \frac{21}{x}. \qquad \frac{\text{Snow}}{\text{Rain}} = \frac{\text{Snow}}{\text{Rain}}$$

Solving this equation gives

$$6x = \frac{21}{2} \quad \text{or} \quad x = \frac{21}{12} = 1.75.$$

Thus 21 inches of light, fluffy snow is equivalent to about 1.75 inches of rain.

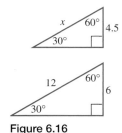

Figure 6.16

▶ **REAL-WORLD CONNECTION** Proportions frequently occur in geometry when we work with similar figures. Two triangles are similar if the measures of their corresponding angles are equal. Corresponding sides of similar triangles are proportional. Figure 6.16 shows two right triangles that are similar because each has angles of 30°, 60°, and 90°.

We can find the length of side x by using proportions. Side x is to 12 as 4.5 is to 6, which can be written as the proportion

$$\frac{x}{12} = \frac{4.5}{6}. \qquad \frac{\text{Hypotenuse}}{\text{Hypotenuse}} = \frac{\text{Shorter leg}}{\text{Shorter leg}}$$

Solving yields the equation

$$6x = 4.5(12) \qquad \text{Clear fractions.}$$
$$x = 9. \qquad \text{Divide by 6.}$$

EXAMPLE 2 Calculating the height of a tree

A 6–foot tall person casts a 4–foot long shadow. If a nearby tree casts a 44–foot long shadow, estimate the height of the tree. See Figure 6.17.

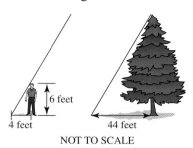

NOT TO SCALE

Figure 6.17

Solution

The triangles in Figure 6.18 are similar because the measures of corresponding angles are equal. Therefore corresponding sides are proportional. Let h be the height of the tree.

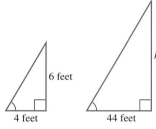

Figure 6.18

$$\frac{h}{6} = \frac{44}{4} \qquad \frac{\text{Height}}{\text{Height}} = \frac{\text{Shadow length}}{\text{Shadow length}}$$

$$4h = 6(44) \qquad \text{Clear fractions.}$$

$$h = \frac{6(44)}{4} \qquad \text{Divide by 4.}$$

$$h = 66 \qquad \text{Simplify.}$$

The tree is 66 feet tall.

▶ **REAL-WORLD CONNECTION** Biologists sometimes use proportions to determine fish populations. They tag a small number of a particular species of fish and then release them. It is assumed that over time these tagged fish will distribute themselves evenly throughout the lake. Later, a sample of fish is taken and the number of tagged fish is compared to the total number of fish in the sample. These numbers can be used to estimate the population of a particular species of fish in the lake. (*Source:* Minnesota Department of Natural Resources.)

EXAMPLE 3 Estimating fish population

In May, 500 tagged largemouth bass are released into a lake. Later in the summer, a sample of 194 largemouth bass from the lake contains 28 tagged fish. Estimate the population of largemouth bass in the lake to the nearest hundred.

Solution

The sample contains 28 tagged bass out of a total of 194 bass. The ratio of tagged bass to the total number of bass in this sample is $\frac{28}{194}$. Let B be the total number of largemouth bass in the lake. Then the ratio of tagged bass to the total number of bass in the lake is $\frac{500}{B}$. If the bass are evenly distributed throughout the lake, then these two ratios are equal and form the following proportion.

Tagged bass in sample ⟶ $\dfrac{28}{194} = \dfrac{500}{B}$ ⟵ Total number of tagged bass

Total number of bass in sample ⟶   ⟵ Total number of bass in the lake

To solve this proportion we clear fractions to obtain the following equation.

$$28B = 194 \cdot 500 \qquad \text{Clear fractions.}$$

$$B = \frac{194 \cdot 500}{28} \approx 3464 \qquad \text{Divide by 28 and approximate.}$$

To the nearest hundred there are approximately 3500 largemouth bass in the lake.

Direct Variation

▶ **REAL-WORLD CONNECTION** If your wage is $9 per hour, the amount you earn is proportional to the number of hours that you work. If you worked H hours, your total pay P satisfies the equation

$$\frac{P}{H} = \frac{9}{1}, \qquad \frac{\text{Pay}}{\text{Hours}}$$

or, equivalently,

$$P = 9H.$$

We say that your pay P is *directly proportional* to the number of hours H worked, and the *constant of proportionality* is 9.

DIRECT VARIATION

Let x and y denote two quantities. Then y is **directly proportional** to x, or y **varies directly** with x, if there is a nonzero number k such that

$$y = kx.$$

The number k is called the **constant of proportionality**, or the **constant of variation**.

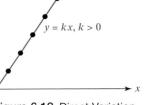

Figure 6.19 Direct Variation

The graph of $y = kx$ is a line passing through the origin, as illustrated in Figure 6.19. Sometimes data in a scatterplot indicate that two quantities are directly proportional. The constant of proportionality k corresponds to the slope of the graph.

EXAMPLE 4 Modeling college tuition

Table 6.4 lists the tuition for taking various numbers of credits.

TABLE 6.4

Credits	3	5	8	11	17
Tuition	$189	$315	$504	$693	$1071

(a) A scatterplot of the data is shown in Figure 6.20. Could the data be modeled using a line?

TABLE 6.4 (Repeated)

Credits	Tuition
3	$189
5	$315
8	$504
11	$693
17	$1071

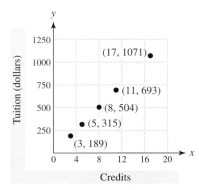

Figure 6.20

(b) Explain why tuition is directly proportional to the number of credits taken.
(c) Find the constant of proportionality. Interpret your result.
(d) Predict the cost of taking 15 credits.

Solution

(a) The data are linear and suggest a line passing through the origin.
(b) Because the data can be modeled by a line passing through the origin, tuition is directly proportional to the number of credits taken. Hence doubling the credits will double the tuition and tripling the credits will triple the tuition.
(c) The slope of the line equals the constant of proportionality k. If we use the first and last data points $(3, 189)$ and $(17, 1071)$, the slope is

$$k = \frac{1071 - 189}{17 - 3} = 63.$$

That is, tuition is **$63** per credit. If we graph the line $y = 63x$, it models the data as shown in Figure 6.21. This graph can also be created with a graphing calculator.

(d) If y represents tuition and x represents the credits taken, 15 credits would cost

$$y = 63(15) = \$945.$$

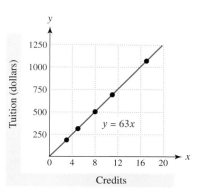

Figure 6.21

MAKING CONNECTIONS

Ratios and the Constant of Proportionality

The constant of proportionality in Example 4 can also be found by calculating the ratios $\frac{y}{x}$, where y is the tuition and x is the credits taken. Note that each ratio in the table equals 63 because the equation $y = 63x$ is equivalent to the equation $\frac{y}{x} = 63$.

x	3	5	8	11	17
y	189	315	504	693	1071
$\frac{y}{x}$	63	63	63	63	63

Inverse Variation

▶ **REAL-WORLD CONNECTION** When two quantities vary inversely, an increase in one quantity results in a decrease in the second quantity. For example, at 25 miles per hour a car travels 100 miles in 4 hours, whereas at 50 miles per hour the car travels 100 miles in 2 hours. Doubling the speed (or rate) decreases the travel time by half. Distance equals rate times time, so $d = rt$. Thus

$$100 = rt, \quad \text{or equivalently,} \quad t = \frac{100}{r}.$$

We say that the time t to travel 100 miles is *inversely proportional* to the speed or rate r. The constant of proportionality or constant of variation is 100.

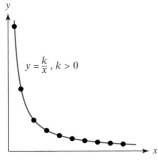

Figure 6.22 Inverse Variation

> ### INVERSE VARIATION
>
> Let x and y denote two quantities. Then y is **inversely proportional** to x, or y **varies inversely** with x, if there is a nonzero number k such that
>
> $$y = \frac{k}{x}.$$

NOTE: We assume that the constant k is positive.

The data shown in Figure 6.22 represent inverse variation and are modeled by $y = \frac{k}{x}$. Note that as x increases, y decreases.

▶ **REAL-WORLD CONNECTION** A wrench is commonly used to loosen a nut on a bolt. See Figure 6.23. If the nut is difficult to loosen, a wrench with a longer handle is often helpful because a longer wrench requires less force. The length of the wrench and the force needed to loosen a nut on a bolt are inversely proportional, as illustrated in the next example.

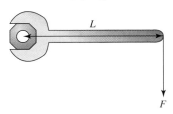

Figure 6.23

EXAMPLE 5 Loosening a nut on a bolt

Table 6.5 lists the force F necessary to loosen a particular nut using wrenches of different lengths L.
(a) Make a scatterplot of the data and discuss the graph. Are the data linear?
(b) Explain why the force F is inversely proportional to the handle length L. Find k so that $F = \frac{k}{L}$ models the data.
(c) Predict the force needed to loosen the nut using an 8-inch wrench.

TABLE 6.5

L (inches)	6	10	12	15	20
F (pounds)	10	6	5	4	3

TECHNOLOGY NOTE:
Scatterplots and Graphs
A scatterplot of the data in Table 6.5 is shown in the first figure. In the second figure the data and the equation $y = \frac{60}{x}$ are graphed.

[0, 24, 4] by [0, 12, 4]

[0, 24, 4] by [0, 12, 4]

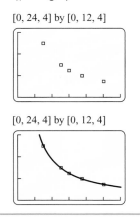

Solution
(a) The scatterplot shown in Figure 6.24 reveals that the data are nonlinear. As the length L of the wrench increases, the force F necessary to loosen the nut decreases.

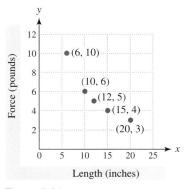

Figure 6.24

(b) If F is inversely proportional to L, then $F = \frac{k}{L}$, or $FL = k$. That is, the product of F and L equals the constant of proportionality k. In Table 6.5, the product of F and L always equals 60 for each data point. Thus F is inversely proportional to L with constant of proportionality $k = 60$.

(c) If $L = 8$, then $F = \frac{60}{8} = 7.5$. A wrench with an 8-inch handle requires a force of 7.5 pounds to loosen the nut.

CALCULATOR HELP
To make a scatterplot, see the Appendix (page AP-4).

EXAMPLE 6 Analyzing data

Determine whether the data in each table represent direct variation, inverse variation, or neither. For direct and inverse variation, find an equation. Graph the data and equation.

(a)

x	4	5	10	20
y	50	40	20	10

(b)

x	2	5	9	11
y	14	35	63	77

(c)

x	2	4	6	8
y	10	16	24	48

Solution

(a) As x increases, y decreases. Because $xy = 200$ for each data point, the equation $y = \frac{200}{x}$ models the data. The data represent inverse variation. The data and equation are graphed in Figure 6.25(a).

(b) As $\frac{y}{x} = 7$ for each data point in the table, the equation $y = 7x$ models the data. These data represent direct variation. The data and equation are graphed in Figure 6.25(b).

(c) Neither the product xy nor the ratio $\frac{y}{x}$ are constant for the data in the table. Therefore these data represent neither direct nor inverse variation. The data are plotted in Figure 6.25(c). Note that the data values increase and are nonlinear.

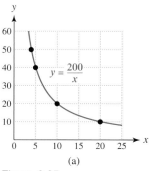
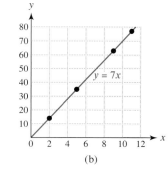
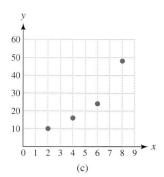

Figure 6.25

Joint Variation

▶ **REAL-WORLD CONNECTION** In many applications a quantity depends on more than one variable. In *joint variation* a quantity varies with the product of more than one variable. For example, the formula for the area A of a rectangle is given by

$$A = WL,$$

where W and L are the width and length, respectively. Thus the area of a rectangle varies jointly with the width and length.

JOINT VARIATION

Let x, y, and z denote three quantities. Then z **varies jointly** with x and y if there is a nonzero number k such that

$$z = kxy.$$

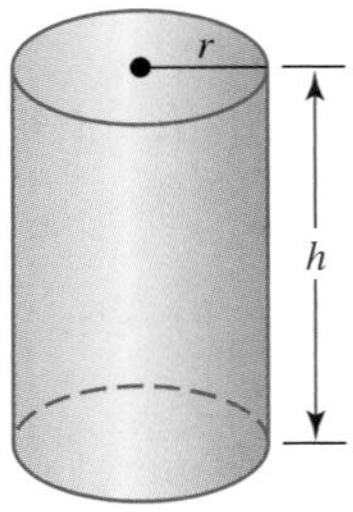

Sometimes joint variation can involve a power of a variable. For example, the volume V of a cylinder is given by $V = \pi r^2 h$, where r is its radius and h is its height, as illustrated in Figure 6.26. In this case we say that the volume varies jointly with the height and the *square* of the radius. The constant of variation is $k = \pi$.

Figure 6.26

EXAMPLE 7 Finding the strength of a rectangular beam

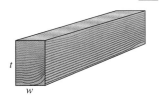

Figure 6.27

The strength S of a rectangular beam varies jointly as its width w and the square of its thickness t. See Figure 6.27. If a beam 3 inches wide and 5 inches thick supports 750 pounds, how much can a similar beam 2 inches wide and 6 inches thick support?

Solution
The strength of a beam is modeled by $S = kwt^2$, where k is a constant of variation. We can find k by substituting $S = 750$, $w = 3$, and $t = 5$ in the formula.

$$750 = k \cdot 3 \cdot 5^2 \quad \text{Substitute in } S = kwt^2.$$
$$k = \frac{750}{3 \cdot 5^2} \quad \text{Solve for } k; \text{ rewrite.}$$
$$= 10 \quad \text{Simplify.}$$

Thus $S = 10wt^2$ models the strength of this type of beam. When $w = 2$ and $t = 6$, the beam can support

$$S = 10 \cdot 2 \cdot 6^2 = 720 \text{ pounds.}$$

CRITICAL THINKING

Compare the increased strength of a beam if the width doubles and if the thickness doubles. What happens to the strength of a beam if both the width and thickness triple?

6.6 PUTTING IT ALL TOGETHER

In this section we introduced some basic concepts of proportion and variation. They are summarized in the following table.

Concept	Explanation	Examples
Proportion	A statement (equation) that two ratios (fractions) are equal The proportion $\frac{a}{b} = \frac{c}{d}$ can be expressed verbally as "a is to b as c is to d."	10 is to 13 as 32 is to x. $\frac{10}{13} = \frac{32}{x}$ x is to 7 as 2 is to 5. $\frac{x}{7} = \frac{2}{5}$

continued on next page

continued from previous page

Concept	Explanation	Examples
Direct Variation	Two quantities x and y vary according to the equation $y = kx$, where k is a nonzero constant. The constant of proportionality (or variation) is k.	$y = 3x$ or $\frac{y}{x} = 3$ | x | 1 | 2 | 4 | | y | 3 | 6 | 12 | If x doubles, then y also doubles.
Inverse Variation	Two quantities x and y vary according to the equation $y = \frac{k}{x}$, where k is a nonzero constant. The constant of proportionality (or variation) is k.	$y = \frac{2}{x}$ or $xy = 2$ | x | 1 | 2 | 4 | | y | 2 | 1 | $\frac{1}{2}$ | If x doubles, then y decreases by half.
Joint Variation	Three quantities x, y, and z vary according to the equation $z = kxy$, where k is a constant.	The area A of a triangle varies jointly as b and h according to the equation $A = \frac{1}{2}bh$, where b is its base and h is its height. The constant of variation is $k = \frac{1}{2}$.

6.7 DIVISION OF POLYNOMIALS

Division by a Monomial ▪ Division by a Polynomial ▪ Synthetic Division

Girolamo Cardano (1501–1576)

A LOOK INTO MATH ▷ The study of polynomials has occupied the minds of mathematicians for centuries. During the sixteenth century, Italian mathematicians, such as Cardano, discovered how to solve higher degree polynomial equations. In this section we demonstrate symbolic methods for dividing polynomials. Division is often needed to factor higher degree polynomials and to solve polynomial equations. (*Source:* H. Eves, *An Introduction to the History of Mathematics.*)

Division by a Monomial

To divide a polynomial by a monomial we use the two properties

$$\frac{a+b}{c} = \frac{a}{c} + \frac{b}{c} \quad \text{and} \quad \frac{a-b}{c} = \frac{a}{c} - \frac{b}{c}.$$

For example,

$$\frac{3x^2 + x}{x} = \frac{3x^2}{x} + \frac{x}{x} = 3x + 1. \qquad \text{Divide each term by } x.$$

When dividing natural numbers, we can check our work by using multiplication. Because

$$\frac{6}{2} = 3,$$

$2 \cdot 3 = 6$. Similarly, to check whether

$$\frac{3x^2 + x}{x} = 3x + 1,$$

we multiply x and $3x + 1$.

$$x(3x + 1) = x \cdot 3x + x \cdot 1 \quad \text{Distributive property}$$
$$= 3x^2 + x \quad \text{It checks.}$$

Graphical support is shown in Figures 6.28(a) and 6.28(b), where the graphs of $Y_1 = (3X^2 + X)/X$ and $Y_2 = 3X + 1$ appear to be identical. In Figure 6.28(c) numerical support is given, where $y_1 = y_2$ for all x-values except 0. Note that we assumed that $x \neq 0$ in $\frac{3x^2 + x}{x}$ because 0 would result in this expression being undefined.

[−4.7, 4.7, 1] by [−3.1, 3.1, 1]   [−4.7, 4.7, 1] by [−3.1, 3.1, 1]

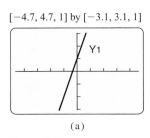
(a)

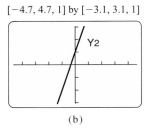
(b)

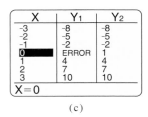
(c)

Figure 6.28

EXAMPLE 1 Dividing by a monomial

Divide and check.

(a) $\dfrac{5x^3 - 15x^2}{15x}$ (b) $\dfrac{4x^2 + 8x - 12}{4x^2}$ (c) $\dfrac{5x^2y + 10xy^2}{5xy}$

Solution

(a) $\dfrac{5x^3 - 15x^2}{15x} = \dfrac{5x^3}{15x} - \dfrac{15x^2}{15x} = \dfrac{x^2}{3} - x$

Check: $15x\left(\dfrac{x^2}{3} - x\right) = \dfrac{15x \cdot x^2}{3} - 15x \cdot x$
$= 5x^3 - 15x^2$

(b) $\dfrac{4x^2 + 8x - 12}{4x^2} = \dfrac{4x^2}{4x^2} + \dfrac{8x}{4x^2} - \dfrac{12}{4x^2} = 1 + \dfrac{2}{x} - \dfrac{3}{x^2}$

Check: $4x^2\left(1 + \dfrac{2}{x} - \dfrac{3}{x^2}\right) = 4x^2 \cdot 1 + \dfrac{4x^2 \cdot 2}{x} - \dfrac{4x^2 \cdot 3}{x^2}$
$= 4x^2 + 8x - 12$

(c) $\dfrac{5x^2y + 10xy^2}{5xy} = \dfrac{5x^2y}{5xy} + \dfrac{10xy^2}{5xy} = x + 2y$

Check: $5xy(x + 2y) = 5xy \cdot x + 5xy \cdot 2y$
$= 5x^2y + 10xy^2$

MAKING CONNECTIONS

Division and "Canceling" Incorrectly

When dividing expressions, students commonly "cancel" incorrectly. For example,

$$\dfrac{x - 3x^3}{x} \neq 1 - 3x^3 \quad \text{and} \quad \dfrac{x - 3x^3}{x} \neq -3x^2.$$

Rather, $\dfrac{x - 3x^3}{x} = \dfrac{x}{x} - \dfrac{3x^3}{x} = 1 - 3x^2.$

Divide the monomial into *every* term in the numerator.

Division by a Polynomial

First, we briefly review division of natural numbers.

$$
\begin{array}{r}
\text{Quotient} \longrightarrow 58 \text{ R } 1 \longleftarrow \text{Remainder} \\
\text{Divisor} \longrightarrow 3\overline{)175} \longleftarrow \text{Dividend} \\
\underline{15} \\
25 \\
\underline{24} \\
1
\end{array}
$$

Because $3 \cdot 58 + 1 = 175$, the answer checks. The quotient and remainder also can be expressed as $58\frac{1}{3}$. Division of polynomials is similar to long division of natural numbers.

EXAMPLE 2   **Dividing polynomials**

Divide $4x^2 - 2x + 1$ by $2x + 1$ and check.

Solution

Begin by dividing $2x$ into $4x^2$.

$$
\begin{array}{r}
2x \\
2x + 1 \overline{)4x^2 - 2x + 1} \\
\underline{4x^2 + 2x} \\
-4x + 1
\end{array}
\qquad
\begin{array}{l}
\frac{4x^2}{2x} = 2x \\
2x(2x + 1) = 4x^2 + 2x \\
\text{Subtract } -2x - 2x = -4x. \\
\text{Bring down the 1.}
\end{array}
$$

Next, divide $2x$ into $-4x$.

$$
\begin{array}{r}
2x - 2 \\
2x + 1 \overline{)4x^2 - 2x + 1} \\
\underline{4x^2 + 2x} \\
-4x + 1 \\
\underline{-4x - 2} \\
3
\end{array}
\qquad
\begin{array}{l}
\frac{-4x}{2x} = -2 \\
\\
\\
\\
-2(2x + 1) = -4x - 2 \\
\text{Subtract } 1 - (-2) = 3. \\
\text{Remainder is 3.}
\end{array}
$$

The quotient is $2x - 2$ with remainder 3, which can also be written as

$$2x - 2 + \frac{3}{2x + 1}, \qquad \text{Quotient} + \frac{\text{Remainder}}{\text{Divisor}}$$

in the same manner as 58 R 1 was expressed as $58\frac{1}{3}$. Polynomial division is checked by multiplying the divisor and the quotient and adding the remainder.

$$
\underbrace{(2x + 1)}_{\text{Divisor}} \underbrace{(2x - 2)}_{\text{Quotient}} + \underbrace{3}_{\text{Remainder}} = 4x^2 - 4x + 2x - 2 + 3
$$
$$
= 4x^2 - 2x + 1 \quad \leftarrow \text{Dividend}
$$

EXAMPLE 3 Dividing polynomials

Divide $2x^3 - 5x^2 + 4x - 1$ by $x - 1$.

Solution
Begin by dividing x into $2x^3$.

$$\begin{array}{r}
2x^2 \\
x - 1 \overline{\smash{)}2x^3 - 5x^2 + 4x - 1} \\
\underline{2x^3 - 2x^2 } \\
-3x^2 + 4x
\end{array}$$

$\dfrac{2x^3}{x} = 2x^2$

$2x^2(x - 1) = 2x^3 - 2x^2$
Subtract $-5x^2 - (-2x^2) = -3x^2$.
Bring down $4x$.

Next, divide x into $-3x^2$.

$$\begin{array}{r}
2x^2 - 3x \\
x - 1 \overline{\smash{)}2x^3 - 5x^2 + 4x - 1} \\
\underline{2x^3 - 2x^2 } \\
-3x^2 + 4x \\
\underline{-3x^2 + 3x } \\
x - 1
\end{array}$$

$\dfrac{-3x^2}{x} = -3x$

$-3x(x - 1) = -3x^2 + 3x$
Subtract $4x - 3x = x$.
Bring down -1.

Finally, divide x into x.

$$\begin{array}{r}
2x^2 - 3x + 1 \\
x - 1 \overline{\smash{)}2x^3 - 5x^2 + 4x - 1} \\
\underline{2x^3 - 2x^2 } \\
-3x^2 + 4x \\
\underline{-3x^2 + 3x } \\
x - 1 \\
\underline{x - 1} \\
0
\end{array}$$

$\dfrac{x}{x} = 1$

$1(x - 1) = x - 1$
Subtract. Remainder is 0.

The quotient is $2x^2 - 3x + 1$, and the remainder is 0.

EXAMPLE 4 Dividing by a quadratic divisor

Divide $3x^3 - 2x^2 - 4x + 4$ by $x^2 - 1$.

Solution
Begin by writing $x^2 - 1$ as $x^2 + 0x - 1$.

$$\begin{array}{r}
3x - 2 \\
x^2 + 0x - 1 \overline{\smash{)}3x^3 - 2x^2 - 4x + 4} \\
\underline{3x^3 + 0x^2 - 3x } \\
-2x^2 - x + 4 \\
\underline{-2x^2 - 0x + 2} \\
-x + 2
\end{array}$$

The quotient is $3x - 2$ with remainder $-x + 2$, or $2 - x$, which can be written as

$$3x - 2 + \frac{2 - x}{x^2 - 1}.$$

EXAMPLE 5 Dividing into a polynomial with missing terms

Divide $2x^3 - 3$ by $x + 1$.

Solution
Begin by writing $2x^3 - 3$ as $2x^3 + 0x^2 + 0x - 3$.

$$
\begin{array}{r}
2x^2 - 2x + 2 \\
x + 1 \overline{) 2x^3 + 0x^2 + 0x - 3} \\
\underline{2x^3 + 2x^2} \\
-2x^2 + 0x \\
\underline{-2x^2 - 2x} \\
2x - 3 \\
\underline{2x + 2} \\
-5
\end{array}
$$

The quotient is $2x^2 - 2x + 2$ with remainder -5, which can be written as

$$2x^2 - 2x + 2 - \frac{5}{x+1}.$$

Synthetic Division

A shortcut called **synthetic division** can be used to divide $x - k$, where k is a constant, into a polynomial. For example, to divide $x - 2$ into $3x^3 - 8x^2 + 7x - 6$, we do the following (with the equivalent long division shown at the right).

Synthetic Division

$$
\begin{array}{r|rrrr}
2 & 3 & -8 & 7 & -6 \\
 & & 6 & -4 & 6 \\
\hline
 & 3 & -2 & 3 & 0
\end{array}
$$

Long Division of Polynomials

$$
\begin{array}{r}
3x^2 - 2x + 3 \\
x - 2 \overline{) 3x^3 - 8x^2 + 7x - 6} \\
\underline{3x^3 - 6x^2} \\
-2x^2 + 7x \\
\underline{-2x^2 + 4x} \\
3x - 6 \\
\underline{3x - 6} \\
0
\end{array}
$$

Note that the blue numbers in the expression for long division correspond to the third row in synthetic division. The remainder is 0, which is the last number in the third row. The quotient is $3x^2 - 2x + 3$. Its coefficients are 3, -2, and 3 and are located in the third row. To divide $x - 2$ into $3x^3 - 8x^2 + 7x - 6$ with synthetic division use the following.

STEP 1: In the top row write 2 (the value of k) on the left and then write the coefficients of the dividend $3x^3 - 8x^2 + 7x - 6$.

STEP 2: (a) Copy the leading coefficient 3 of $3x^3 - 8x^2 + 7x - 6$ in the third row and multiply it by 2 (the value of k). Write the result 6 in the second row below -8. Add -8 and 6 in the second column to obtain the -2 in the third row.
(b) Repeat the process by multiplying -2 by 2 and place the result -4 below 7. Then add 7 and -4 to obtain 3.
(c) Multiply 3 by 2 and place the result 6 below the -6. Adding 6 and -6 gives 0.

STEP 3: The last number in the third row is 0, which is the remainder. The other numbers in the third row are the coefficients of the quotient, which is $3x^2 - 2x + 3$.

EXAMPLE 6 Performing synthetic division

Use synthetic division to divide $x^4 - 5x^3 + 9x^2 - 10x + 3$ by $x - 3$.

Solution
Because the divisor is $x - 3$, the value of k is 3.

$$\begin{array}{r|rrrrr} 3 & 1 & -5 & 9 & -10 & 3 \\ & & 3 & -6 & 9 & -3 \\ \hline & 1 & -2 & 3 & -1 & 0 \end{array}$$

The quotient is $x^3 - 2x^2 + 3x - 1$ and the remainder is 0. This result is expressed by

$$\frac{x^4 - 5x^3 + 9x^2 - 10x + 3}{x - 3} = x^3 - 2x^2 + 3x - 1 + \frac{0}{x - 3}, \quad \text{or}$$

$$\frac{x^4 - 5x^3 + 9x^2 - 10x + 3}{x - 3} = x^3 - 2x^2 + 3x - 1.$$

MAKING CONNECTIONS

Factors and Remainders

Multiplying the last equation in the solution to Example 6 by $x - 3$ gives

$$x^4 - 5x^3 + 9x^2 - 10x + 3 = (x - 3)(x^3 - 2x^2 + 3x - 1).$$

That is, $x - 3$ is a *factor* of $x^4 - 5x^3 + 9x^2 - 10x + 3$ because the remainder is 0.

This concept is true in general: If a polynomial $p(x)$ is divided by $x - k$ and the remainder is 0, then $x - k$ is a factor of $p(x)$.

CRITICAL THINKING

Suppose that $p(x)$ is a polynomial and that $(x - k)$ divides into $p(x)$ with remainder 0.

1. Evaluate $p(k)$. 0
2. Give one x-intercept on the graph of $p(x)$. k

EXAMPLE 7 Performing synthetic division

Use synthetic division to divide $2x^3 - x + 5$ by $x + 1$.

Solution
Write $2x^3 - x + 5$ as $2x^3 + 0x^2 - x + 5$. The divisor $x + 1$ can be written as

$$x + 1 = x - (-1),$$

so we let $k = -1$.

$$\begin{array}{r|rrrr} -1 & 2 & 0 & -1 & 5 \\ & & -2 & 2 & -1 \\ \hline & 2 & -2 & 1 & 4 \end{array}$$

The remainder is 4, and the quotient is $2x^2 - 2x + 1$. This result can also be expressed as

$$\frac{2x^3 - x + 5}{x + 1} = 2x^2 - 2x + 1 + \frac{4}{x + 1}.$$

6.7 PUTTING IT ALL TOGETHER

The following table summarizes division of polynomials.

Type of Division	Explanation	Example
By a Monomial	Use the property $\frac{a \pm b}{c} = \frac{a}{c} \pm \frac{b}{c}$ to divide a polynomial by a monomial. Be sure to divide the denominator into every term of the numerator.	$\frac{8a^3 - 4a^2}{2a} = \frac{8a^3}{2a} - \frac{4a^2}{2a} = 4a^2 - 2a$
By a Polynomial	Division by a polynomial may be done in a manner similar to long division of natural numbers. See Examples 2–5.	When $2x^2 - 7x + 4$ is divided by $2x - 1$, the quotient is $x - 3$ and the remainder is 1. This result may be expressed as $$\frac{2x^2 - 7x + 4}{2x - 1} = x - 3 + \frac{1}{2x - 1}.$$
Synthetic	Synthetic division is a fast way to divide a polynomial by a divisor in the form $x - k$.	Divide $4x^2 + 7x - 14$ by $x + 3$. $\begin{array}{r|rrr} -3 & 4 & 7 & -14 \\ & & -12 & 15 \\ \hline & 4 & -5 & 1 \end{array}$ The quotient is $4x - 5$ with remainder 1.

CHAPTER 6 SUMMARY

SECTION 6.1 ■ INTRODUCTION TO RATIONAL FUNCTIONS AND EQUATIONS

Rational Functions A rational function is given by $f(x) = \frac{p(x)}{q(x)}$, where $p(x)$ and $q(x)$ are polynomials. The domain of f includes all x-values such that $q(x) \neq 0$.

Example: $f(x) = \frac{4}{x - 2}$ defines a rational function with domain $\{x \mid x \neq 2\}$.

Graphs of Rational Functions Graphs of rational functions are usually curves. A vertical asymptote typically occurs at x-values where the denominator equals zero, but the numerator does not. The graph of a rational function never crosses a vertical asymptote. Graphing a rational function by hand may require plotting several points on each side of a vertical asymptote.

Example: $f(x) = \dfrac{2x}{x - 2}$

A vertical asymptote occurs at $x = 2$. The graph does not cross this vertical asymptote.

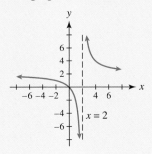

Rational Equations A rational equation contains one or more rational expressions. To solve a rational equation in the form $\frac{A}{B} = C$, start by multiplying each side of the equation by B to obtain $A = BC$. This step clears fractions from the equation.

Example: To solve $\frac{2x}{x-3} = 8$, start by multiplying each side by $x - 3$.

$$(x - 3)\frac{2x}{x - 3} = (x - 3)8$$
$$2x = 8x - 24$$
$$x = 4 \qquad \text{Check your answer.}$$

Operations on Functions If $f(x)$ and $g(x)$ are both defined, then the sum, difference, product, and quotient of two functions f and g are defined by

$$(f + g)(x) = f(x) + g(x) \qquad \text{Sum}$$
$$(f - g)(x) = f(x) - g(x) \qquad \text{Difference}$$
$$(fg)(x) = f(x) \cdot g(x) \qquad \text{Product}$$
$$\left(\frac{f}{g}\right)(x) = \frac{f(x)}{g(x)}, \text{ where } g(x) \neq 0. \qquad \text{Quotient}$$

Example: Let $f(x) = x^2 - 1$ and $g(x) = x^2 + 1$.

$$(f + g)(x) = f(x) + g(x) = (x^2 - 1) + (x^2 + 1) = 2x^2$$
$$(f - g)(x) = f(x) - g(x) = (x^2 - 1) - (x^2 + 1) = -2$$
$$(fg)(x) = f(x) \cdot g(x) = (x^2 - 1)(x^2 + 1) = x^4 - 1$$
$$\left(\frac{f}{g}\right)(x) = \frac{f(x)}{g(x)} = \frac{x^2 - 1}{x^2 + 1}$$

SECTION 6.2 ■ MULTIPLICATION AND DIVISION OF RATIONAL EXPRESSIONS

Simplifying a Rational Expression To simplify a rational expression, factor the numerator and the denominator. Then apply the following principle.

$$\frac{PR}{QR} = \frac{P}{Q}$$

Example: $\dfrac{x^2 - 1}{x^2 + 2x + 1} = \dfrac{(x - 1)(x + 1)}{(x + 1)(x + 1)} = \dfrac{x - 1}{x + 1}$

Multiplying Rational Expressions To multiply two rational expressions, multiply numerators and multiply denominators. Simplify the result, if possible.

$$\frac{A}{B} \cdot \frac{C}{D} = \frac{AC}{BD} \qquad B \text{ and } D \text{ not zero}$$

Example: $\dfrac{3}{x - 1} \cdot \dfrac{x - 1}{x + 1} = \dfrac{3(x - 1)}{(x - 1)(x + 1)} = \dfrac{3}{x + 1}$

Dividing Rational Expressions To divide two rational expressions, multiply by the reciprocal of the divisor. Simplify the result, if possible.

$$\frac{A}{B} \div \frac{C}{D} = \frac{A}{B} \cdot \frac{D}{C} = \frac{AD}{BC} \qquad B, C, \text{ and } D \text{ not zero}$$

Example: $\dfrac{x+1}{x^2-4} \div \dfrac{x+1}{x-2} = \dfrac{x+1}{(x-2)(x+2)} \cdot \dfrac{x-2}{x+1}$

$= \dfrac{(x+1)(x-2)}{(x-2)(x+2)(x+1)} = \dfrac{1}{x+2}$

SECTION 6.3 ■ ADDITION AND SUBTRACTION OF RATIONAL EXPRESSIONS

Addition and Subtraction of Rational Expressions with Like Denominators To add (subtract) two rational expressions with like denominators, add (subtract) their numerators. The denominator does not change.

$$\dfrac{A}{C} + \dfrac{B}{C} = \dfrac{A+B}{C} \quad \text{and} \quad \dfrac{A}{C} - \dfrac{B}{C} = \dfrac{A-B}{C}$$

Examples: $\dfrac{2x}{2x+1} + \dfrac{3x}{2x+1} = \dfrac{5x}{2x+1}$ and

$\dfrac{x}{x^2-1} - \dfrac{1}{x^2-1} = \dfrac{x-1}{x^2-1} = \dfrac{x-1}{(x-1)(x+1)} = \dfrac{1}{x+1}$

Finding the Least Common Denominator The least common denominator (LCD) is the least common multiple (LCM) of the denominators.

Example: The LCD for $\dfrac{1}{x(x+1)}$ and $\dfrac{1}{x^2}$ is $x^2(x+1)$.

Addition and Subtraction of Rational Expressions with Unlike Denominators First write each rational expression with the LCD. Then add or subtract the numerators.

Examples: $\dfrac{2}{x+1} - \dfrac{1}{x} = \dfrac{2x}{x(x+1)} - \dfrac{x+1}{x(x+1)} = \dfrac{2x-(x+1)}{x(x+1)} = \dfrac{x-1}{x(x+1)}$

$\dfrac{1}{x+1} + \dfrac{2}{x-1} = \dfrac{1(x-1)}{(x+1)(x-1)} + \dfrac{2(x+1)}{(x-1)(x+1)} = \dfrac{3x+1}{(x-1)(x+1)}$

SECTION 6.4 ■ RATIONAL EQUATIONS

Solving Rational Equations A first step in solving a rational equation is to multiply each side by the LCD of the rational expressions to clear fractions from the equation. *Be sure to check all answers.*

Example: The LCD for the equation $\dfrac{2}{x} + 1 = \dfrac{4-x}{x}$ is x.

$\dfrac{2}{x} + 1 = \dfrac{4-x}{x}$ Given equation

$x\left(\dfrac{2}{x} + 1\right) = \left(\dfrac{4-x}{x}\right)x$ Multiply by x.

$2 + x = 4 - x$ Clear fractions.

$2x = 2$ Add x and subtract 2.

$x = 1$ Divide by 2. Check your answer.

SECTION 6.5 ■ COMPLEX FRACTIONS

Complex Fractions A complex fraction is a rational expression that contains fractions in its numerator, denominator, or both. The following equation can be used to help simplify basic complex fractions.

$$\frac{\frac{a}{b}}{\frac{c}{d}} = \frac{a}{b} \cdot \frac{d}{c} \quad \text{Invert and multiply.}$$

Example: $\dfrac{\frac{1-x}{2}}{\frac{4}{1+x}} = \dfrac{1-x}{2} \cdot \dfrac{1+x}{4} = \dfrac{(1-x)(1+x)}{(2)(4)} = \dfrac{1-x^2}{8}$

Simplifying Complex Fractions

Method I: Combine terms in the numerator, combine terms in the denominator, and simplify the resulting expression.

Method II: Multiply the numerator and the denominator by the LCD and simplify the resulting expression.

Example: *Method I:* $\dfrac{2 - \frac{1}{b}}{2 + \frac{1}{b}} = \dfrac{\frac{2b-1}{b}}{\frac{2b+1}{b}} = \dfrac{2b-1}{b} \cdot \dfrac{b}{2b+1} = \dfrac{2b-1}{2b+1}$

Method II: The LCD is b.

$$\dfrac{\left(2 - \frac{1}{b}\right)b}{\left(2 + \frac{1}{b}\right)b} = \dfrac{2b - \frac{b}{b}}{2b + \frac{b}{b}} = \dfrac{2b-1}{2b+1}$$

SECTION 6.6 ■ MODELING WITH PROPORTIONS AND VARIATION

Proportions A proportion is a statement (equation) that two ratios (fractions) are equal.

Example: $\frac{5}{x} = \frac{4}{7}$ (in words, 5 *is to* x *as* 4 *is to* 7.)

Similar Triangles Two triangles are similar if the measures of their corresponding angles are equal. Corresponding sides of similar triangles are proportional.

Example: A right triangle has legs with lengths 3 and 4. A similar right triangle has a shorter leg with length 6. Its longer leg can be found by solving the proportion $\frac{3}{6} = \frac{4}{x}$, or $x = 8$.

Direct Variation A quantity y is *directly proportional* to a quantity x, or y *varies directly* with x, if there is a nonzero constant k such that $y = kx$. The number k is called the *constant of proportionality* or the *constant of variation*.

Example: If y varies directly with x, then the ratio $\frac{y}{x}$ always equals k. The following data satisfy $\frac{y}{x} = 4$, so the constant of variation is 4.

x	1	2	3	4
y	4	8	12	16

Inverse Variation A quantity y is *inversely proportional* to a quantity x, or y *varies inversely* with x, if there is a nonzero constant k such that $y = \frac{k}{x}$. (We assume that $k > 0$.)

Example: If y varies inversely with x, then $xy = k$. The following data satisfy $xy = 12$, so the constant of variation is 12.

x	1	2	4	6
y	12	6	3	2

Joint Variation The quantity z *varies jointly* with x and y if $z = kxy$, $k \neq 0$.

Example: The area A of a rectangle varies jointly, with the width W and length L because $A = LW$. Note that $k = 1$ in this example.

SECTION 6.7 ■ DIVISION OF POLYNOMIALS

Division of a Polynomial by a Monomial Divide the monomial in the denominator into *every* term of the numerator.

Example:
$$\frac{4x^4 - 8x^3 + 16x^2}{8x^2} = \frac{4x^4}{8x^2} - \frac{8x^3}{8x^2} + \frac{16x^2}{8x^2} = \frac{x^2}{2} - x + 2$$

Division of a Polynomial by a Polynomial Division of polynomials is similar to long division of natural numbers.

Example: Divide $2x^3 - 3x + 3$ by $x - 1$. (Be sure to include $0x^2$.)

$$\begin{array}{r}
2x^2 + 2x - 1 \\
x - 1 \overline{)2x^3 + 0x^2 - 3x + 3} \\
\underline{2x^3 - 2x^2} \\
2x^2 - 3x \\
\underline{2x^2 - 2x} \\
-x + 3 \\
\underline{-x + 1} \\
2
\end{array}$$

The quotient is $2x^2 + 2x - 1$ with remainder 2, which can be written as

$$2x^2 + 2x - 1 + \frac{2}{x - 1}.$$

Synthetic Division Synthetic division is a fast way to divide a polynomial by an expression in the form $x - k$, where k is a constant.

Example: Divide $3x^3 + 4x^2 - 7x - 1$ by $x + 2$.

$$\begin{array}{r|rrrr}
-2 & 3 & 4 & -7 & -1 \\
   &   & -6 & 4 & 6 \\ \hline
   & 3 & -2 & -3 & 5
\end{array}$$

The quotient is $3x^2 - 2x - 3$ with remainder 5.

Assignment Name _____ Name _____ Date _____
Show all work for these items: _____

_____

_____

_____

_____

_____

_____

_____

_____

6.1 INTRODUCTION TO RATIONAL FUNCTIONS AND EQUATIONS

Assignment Name _____ Name _____ Date _____

Show all work for these items: _____

_____

_____

_____

_____

_____

_____

_____

_____

CHAPTER 7
Radical Expressions and Functions

Throughout history, people have created new numbers. Often these new numbers were met with resistance and regarded as being imaginary or unreal. The number 0 was not invented at the same time as the natural numbers. There was no Roman numeral for 0, which is one reason why our calendar started with A.D. 1 and, as a result, the twenty-first century began in 2001. No doubt there were skeptics during the time of the Roman Empire who questioned why anyone needed a number to represent nothing. Negative numbers also met strong resistance. After all, how could anyone possibly have −6 apples?

In this chapter we describe a new number system called *complex numbers*, which involve square roots of negative numbers. The Italian mathematician Cardano (1501–1576) was one of the first mathematicians to work with complex numbers and called them useless. René Descartes (1596–1650) originated the term *imaginary number*, which is associated with complex numbers. However, today complex numbers are used in many applications, such as electricity, fiber optics, and the design of airplanes. We are privileged to study in a period of days what took people centuries to discover.

7.1 Radical Expressions and Rational Exponents
7.2 Simplifying Radical Expressions
7.3 Operations on Radical Expressions
7.4 Radical Functions
7.5 Equations Involving Radical Expressions
7.6 Complex Numbers

Bear in mind that the wonderful things you learn in schools are the work of many generations, produced by enthusiastic effort and infinite labor in every country.

—ALBERT EINSTEIN

Source: Historical Topics for the Mathematics Classroom, Thirty-first Yearbook, NCTM.

7.1 RADICAL EXPRESSIONS AND RATIONAL EXPONENTS

Radical Notation ■ Rational Exponents ■ Properties of Rational Exponents

A LOOK INTO MATH ▷

Cellular phone technology has become a part of everyday life. In order to have cellular phone coverage, transmission towers, or cellular sites, are spread throughout a region. To estimate the minimum broadcasting distance for each cellular site, *radical expressions* are needed. (See Example 4.) In this section we discuss radical expressions and rational exponents and show how to manipulate them symbolically. (**Source:** C. Smith, *Practical Cellular & PCS Design*.)

Radical Notation

Recall the definition of the square root of a number a.

> **SQUARE ROOT**
>
> The number b is a *square root* of a if $b^2 = a$.

EXAMPLE 1 Finding square roots

Find the square roots of 100.

Solution
The square roots of 100 are **10** *and* **−10** because $10^2 = 100$ and $(-10)^2 = 100$.

Every positive number a has two square roots, one positive and one negative. Recall that the *positive* square root is called the *principal square root* and is denoted $\sqrt{a}$. The *negative* square root is denoted $-\sqrt{a}$. To identify both square roots we write $\pm\sqrt{a}$. The symbol $\pm$ is read "plus or minus." The symbol $\sqrt{}$ is called the **radical sign**. The expression under the radical sign is called the **radicand**, and an expression containing a radical sign is called a **radical expression**. Examples of radical expressions include

$$\sqrt{6}, \quad 5 + \sqrt{x+1}, \quad \text{and} \quad \sqrt{\frac{3x}{2x-1}}.$$

> **MAKING CONNECTIONS**
>
> **Expressions and Equations**
>
> Expressions and equations are *different* mathematical concepts. An expression does not contain an equals sign, whereas an equation *always* contains an equals sign. *An equation is a statement that two expressions are equal.* For example,
>
> $$\sqrt{x+1} \quad \text{and} \quad \sqrt{5-x}$$
>
> are two different expressions, and
>
> $$\sqrt{x+1} = \sqrt{5-x}$$
>
> is an equation. We often *solve an equation*, but we *do not solve an expression*. Instead, we simplify and evaluate expressions.

In the next example we show how to find the principal square root of an expression.

EXAMPLE 2 Finding principal square roots

Evaluate each square root.
(a) $\sqrt{25}$ (b) $\sqrt{0.49}$ (c) $\sqrt{\frac{4}{9}}$ (d) $\sqrt{c^2}, c > 0$

Solution
(a) Because $5 \cdot 5 = 25$, the principal, or *positive*, square root of 25 is $\sqrt{25} = 5$.
(b) Because $(0.7)(0.7) = 0.49$, the principal square root of 0.49 is $\sqrt{0.49} = 0.7$.
(c) Because $\frac{2}{3} \cdot \frac{2}{3} = \frac{4}{9}$, the principal square root of $\frac{4}{9}$ is $\sqrt{\frac{4}{9}} = \frac{2}{3}$.
(d) The principal square root of c^2 is $\sqrt{c^2} = c$, as c is positive.

EXAMPLE 3 Approximating a square root

Approximate $\sqrt{17}$ to the nearest thousandth.

Solution
Figure 7.1 shows that $\sqrt{17} \approx 4.123$, rounded to the nearest thousandth. This result means that $4.123 \times 4.123 \approx 17$.

```
√(17)
       4.123105626
```

Figure 7.1

NOTE: Calculators often give decimal approximations rather than exact answers when evaluating radical expressions.

▶ **REAL-WORLD CONNECTION** In the next example we use the principal square root to estimate the minimum transmission distance for a cellular site.

EXAMPLE 4 Estimating cellular phone transmission distance

If the ground is level, a cellular transmission tower will broadcast its signal in roughly a circular pattern, whose radius can be altered by changing the strength of its signal. See Figure 7.2. Suppose that a city has an area of 50 square miles and that there are 10 identical transmission towers spread evenly throughout the city. Estimate a *minimum* transmission radius R for each tower. (Note that a larger radius would probably be necessary to adequately cover the city.) (*Source:* C. Smith.)

Figure 7.2

Solution
The circular area A covered by one transmission tower is $A = \pi R^2$. The total area covered by 10 towers is $10\pi R^2$, which must equal *at least* 50 square miles.

$$10\pi R^2 = 50$$
$$R^2 = \frac{50}{10\pi} \quad \text{Divide by } 10\pi.$$
$$R^2 = \frac{5}{\pi} \quad \text{Simplify.}$$
$$R = \sqrt{\frac{5}{\pi}} \approx 1.26 \quad \text{Take principal square root.}$$

Each transmission tower must broadcast with a minimum radius of about 1.26 miles.

Another common radical expression is the cube root of a number a, denoted $\sqrt[3]{a}$.

CUBE ROOT

The number b is a *cube root* of a if $b^3 = a$.

Although the square root of a negative number is *not* a real number, the cube root of a negative number is a negative real number. *Every real number has one real cube root.*

EXAMPLE 5 Finding cube roots

Evaluate the cube root. Approximate your answer to the nearest hundredth when appropriate.

(a) $\sqrt[3]{8}$ (b) $\sqrt[3]{-27}$ (c) $\sqrt[3]{\frac{1}{64}}$ (d) $\sqrt[3]{d^6}$ (e) $\sqrt[3]{16}$

Solution

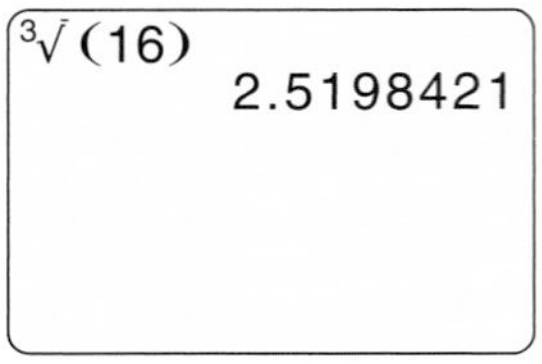

(a) $\sqrt[3]{8} = 2$ because $2^3 = 2 \cdot 2 \cdot 2 = 8$.
(b) $\sqrt[3]{-27} = -3$ because $(-3)^3 = (-3)(-3)(-3) = -27$.
(c) $\sqrt[3]{\frac{1}{64}} = \frac{1}{4}$ because $\left(\frac{1}{4}\right)^3 = \frac{1}{4} \cdot \frac{1}{4} \cdot \frac{1}{4} = \frac{1}{64}$.
(d) $\sqrt[3]{d^6} = d^2$ because $(d^2)^3 = d^2 \cdot d^2 \cdot d^2 = d^{2+2+2} = d^6$.
(e) $\sqrt[3]{16}$ is not an integer. Figure 7.3 shows that $\sqrt[3]{16} \approx 2.52$.

```
³√(16)
         2.5198421
```

Figure 7.3

CALCULATOR HELP
To calculate a cube root, see the Appendix (page AP-1).

NOTE: $\sqrt[3]{-b} = -\sqrt[3]{b}$ for any real number b. That is, the cube root of a negative is the negative of the cube root. For example, $\sqrt[3]{-8} = -\sqrt[3]{8} = -2$.

We can generalize square roots and cube roots to include *n*th roots of a number a. The number b is an **nth root** of a if $b^n = a$, where n is a positive integer. For example, $2^5 = 32$ and so the 5th root of 32 is 2 and can be written as $\sqrt[5]{32} = 2$.

THE NOTATION $\sqrt[n]{a}$

The equation $\sqrt[n]{a} = b$ means that $b^n = a$, where n is a natural number called the **index**. If n is odd, we are finding an **odd root** and if n is even, we are finding an **even root**.

1. If $a > 0$, then $\sqrt[n]{a}$ is a positive number.
2. If $a < 0$ and
   (a) n is odd, then $\sqrt[n]{a}$ is a negative number.
   (b) n is even, then $\sqrt[n]{a}$ is *not* a real number.

If $a > 0$ and n is even, then a has two real *n*th roots: one positive and one negative. In this case, the positive root is denoted $\sqrt[n]{a}$ and called the **principal nth root of a**. For example, $(-3)^4 = 81$ *and* $3^4 = 81$, but $\sqrt[4]{81} = 3$ in the same way *principal square roots* are calculated.

EXAMPLE 6 Finding nth roots

Find each root, if possible.

(a) $\sqrt[4]{16}$ (b) $\sqrt[5]{-32}$ (c) $\sqrt[4]{-81}$

Solution

(a) $\sqrt[4]{16} = 2$ because $2^4 = 2 \cdot 2 \cdot 2 \cdot 2 = 16$.
(b) $\sqrt[5]{-32} = -2$ because $(-2)^5 = (-2)(-2)(-2)(-2)(-2) = -32$.
(c) An *even* root of a *negative* number is *not* a real number.

Consider the calculations

$$\sqrt{3^2} = \sqrt{9} = 3, \quad \sqrt{(-4)^2} = \sqrt{16} = 4, \quad \text{and} \quad \sqrt{(-6)^2} = \sqrt{36} = 6.$$

In general, the expression $\sqrt{x^2}$ equals $|x|$. Graphical support is shown in Figure 7.4, where the graphs of $Y_1 = \sqrt{(X^2)}$ and $Y_2 = \text{abs}(X)$ appear to be identical.

CRITICAL THINKING

Evaluate $\sqrt[6]{(-2)^6}$ and $\sqrt[3]{(-2)^3}$. Now simplify $\sqrt[n]{x^n}$ when n is even and when n is odd.

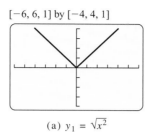

(a) $y_1 = \sqrt{x^2}$ (b) $y_2 = |x|$

Figure 7.4

THE EXPRESSION $\sqrt{x^2}$

For every real number x, $\sqrt{x^2} = |x|$.

EXAMPLE 7 Simplifying expressions

Write each expression in terms of an absolute value.
(a) $\sqrt{(-3)^2}$ (b) $\sqrt{(x+1)^2}$ (c) $\sqrt{z^2 - 4z + 4}$

Solution
(a) $\sqrt{x^2} = |x|$, so $\sqrt{(-3)^2} = |-3| = 3$
(b) $\sqrt{(x+1)^2} = |x+1|$
(c) $\sqrt{z^2 - 4z + 4} = \sqrt{(z-2)^2} = |z-2|$

Rational Exponents

When m and n are integers, the product rule states that $a^m a^n = a^{m+n}$. This rule can be extended to include exponents that are fractions. For example,

$$4^{1/2} \cdot 4^{1/2} = 4^{1/2 + 1/2} = 4^1 = 4.$$

That is, if we multiply the expression $4^{1/2}$ by itself, the result is 4. Because we also know that $\sqrt{4} \cdot \sqrt{4} = 4$, this discussion suggests that $4^{1/2} = \sqrt{4}$ and motivates the following definition.

THE EXPRESSION $a^{1/n}$

If n is an integer greater than 1 and a is a real number, then

$$a^{1/n} = \sqrt[n]{a}.$$

NOTE: If $a < 0$ and n is an even positive integer, then $a^{1/n}$ is not a real number.

In the next two examples, we show how to interpret rational exponents.

EXAMPLE 8 Interpreting rational exponents

Write each expression in radical notation. Then evaluate the expression and round to the nearest hundredth when appropriate.

(a) $36^{1/2}$ (b) $23^{1/5}$ (c) $x^{1/3}$ (d) $(5x)^{1/2}$

Solution

(a) The exponent $\frac{1}{2}$ indicates a square root. Thus $36^{1/2} = \sqrt{36}$, which evaluates to 6.
(b) The exponent $\frac{1}{5}$ indicates a fifth root. Thus $23^{1/5} = \sqrt[5]{23}$, which is not an integer. Figure 7.5 shows this expression approximated in both exponential and radical notation. In either case $23^{1/5} \approx 1.87$.
(c) The exponent $\frac{1}{3}$ indicates a cube root, so $x^{1/3} = \sqrt[3]{x}$.
(d) The exponent $\frac{1}{2}$ indicates a square root, so $(5x)^{1/2} = \sqrt{5x}$.

```
23^(1/5)
        1.872171231
5ˣ√(23)
        1.872171231
```

Figure 7.5

CALCULATOR HELP

To calculate other roots, see the Appendix (page AP-1).

Suppose that we want to define the expression $8^{2/3}$. On the one hand, using properties of exponents we have

$$8^{1/3} \cdot 8^{1/3} = 8^{1/3+1/3} = 8^{2/3}.$$

On the other hand, we have

$$8^{1/3} \cdot 8^{1/3} = \sqrt[3]{8} \cdot \sqrt[3]{8} = 2 \cdot 2 = 4.$$

Thus $8^{2/3} = 4$, and that value is obtained whether we interpret $8^{2/3}$ as either

$$8^{2/3} = (8^{1/3})^2 = (\sqrt[3]{8})^2 = 2^2 = 4$$

or

$$8^{2/3} = (8^2)^{1/3} = \sqrt[3]{8^2} = \sqrt[3]{64} = 4.$$

This result suggests the following definition.

> **THE EXPRESSION $a^{m/n}$**
>
> If m and n are positive integers with $\frac{m}{n}$ in lowest terms, then
>
> $$a^{m/n} = \sqrt[n]{a^m} = (\sqrt[n]{a})^m.$$
>
> **NOTE:** If $a < 0$ and n is an even integer, then $a^{m/n}$ is *not* a real number.

The exponent $\frac{m}{n}$ indicates that we either take the nth root and then calculate the mth power of the result or calculate the mth power and then take the nth root. For example, $4^{3/2}$ means that we can either take the square root of 4 and then cube the result or we cube 4 and then take the square root of the result. In either case the result is the same: $4^{3/2} = 8$. This concept is illustrated in the next example.

EXAMPLE 9 Interpreting rational exponents

Write each expression in radical notation. Evaluate the expression by hand when possible.

(a) $(-27)^{2/3}$ (b) $12^{3/5}$

Solution

(a) The exponent $\frac{2}{3}$ indicates that we either take the cube root of -27 and then square it or that we square -27 and then take the cube root. Thus

$$(-27)^{2/3} = (\sqrt[3]{-27})^2 = (-3)^2 = 9$$

or

$$(-27)^{2/3} = \sqrt[3]{(-27)^2} = \sqrt[3]{729} = 9.$$

Same result

(b) The exponent $\frac{3}{5}$ indicates that we either take the fifth root of 12 and then cube it or that we cube 12 and then take the fifth root. Thus

$$12^{3/5} = (\sqrt[5]{12})^3 \quad \text{or} \quad 12^{3/5} = \sqrt[5]{12^3}.$$

This result cannot be evaluated by hand.

TECHNOLOGY NOTE: Rational Exponents

When evaluating expressions with rational (fractional) exponents, be sure to put parentheses around the fraction. For example, most calculators will evaluate 8^(2/3) and 8^2/3 differently. The accompanying figure shows evaluation of $8^{2/3}$ input correctly, 8^(2/3), as 4 but shows evaluation of $8^{2/3}$ input incorrectly, 8^2/3, as $\frac{8^2}{3} = 21.\overline{3}$.

```
Correct  →  8^(2/3)
                         4
Incorrect →  8^2/3
                21.33333333
```

From properties of exponents we know that $a^{-n} = \frac{1}{a^n}$, where n is a positive integer. We now define this property for negative rational exponents.

THE EXPRESSION $a^{-m/n}$

If m and n are positive integers with $\frac{m}{n}$ in lowest terms, then

$$a^{-m/n} = \frac{1}{a^{m/n}}, \qquad a \neq 0.$$

EXAMPLE 10 Interpreting negative rational exponents

Write each expression in radical notation and then evaluate.
(a) $64^{-1/3}$ **(b)** $81^{-3/4}$

Solution

(a) $64^{-1/3} = \dfrac{1}{64^{1/3}} = \dfrac{1}{\sqrt[3]{64}} = \dfrac{1}{4}.$

(b) $81^{-3/4} = \dfrac{1}{81^{3/4}} = \dfrac{1}{(\sqrt[4]{81})^3} = \dfrac{1}{3^3} = \dfrac{1}{27}.$

▶ **REAL-WORLD CONNECTION** In the next example, we use a formula from biology that involves a rational exponent.

EXAMPLE 11 Analyzing stepping frequency

When smaller (four-legged) animals walk, they tend to take faster, shorter steps, whereas larger animals tend to take slower, longer steps. If an animal is h feet high at the shoulder, then the number N of steps per second that the animal takes *while walking* can be estimated by $N(h) = 1.6h^{-1/2}$. Use a calculator to estimate N for an elephant that is 10 feet high at the shoulder. (*Source:* C. Pennycuick, *Newton Rules Biology.*)

Solution

$$N(10) = 1.6(10)^{-1/2} = \frac{1.6}{\sqrt{10}} \approx 0.51$$

The elephant takes about $\frac{1}{2}$ step per second while walking or about 1 step every 2 seconds.

Properties of Rational Exponents

Any rational number can be written as a ratio of two integers. That is, if p is a rational number, then $p = \frac{m}{n}$, where m and n are integers. Properties for integer exponents also apply to rational exponents, with one exception. If n is even in the expression $a^{m/n}$ and $\frac{m}{n}$ is written in lowest terms, then a must be nonnegative (not negative) for the result to be a real number.

> **PROPERTIES OF EXPONENTS**
>
> Let p and q be rational numbers written in lowest terms. For all real numbers a and b for which the expressions are real numbers the following properties hold.
>
> 1. $a^p \cdot a^q = a^{p+q}$ Product rule for exponents
> 2. $a^{-p} = \dfrac{1}{a^p}, \quad \dfrac{1}{a^{-p}} = a^p$ Negative exponents
> 3. $\left(\dfrac{a}{b}\right)^{-p} = \left(\dfrac{b}{a}\right)^{p}$ Negative exponents for quotients
> 4. $\dfrac{a^p}{a^q} = a^{p-q}$ Quotient rule for exponents
> 5. $(a^p)^q = a^{pq}$ Power rule for exponents
> 6. $(ab)^p = a^p b^p$ Power rule for products
> 7. $\left(\dfrac{a}{b}\right)^p = \dfrac{a^p}{b^p}$ Power rule for quotients

In the next two examples, we apply these properties.

EXAMPLE 12 Applying properties of exponents

Write each expression using rational exponents and simplify. Write the answer with a positive exponent. Assume that all variables are positive numbers.

(a) $\sqrt{x} \cdot \sqrt[3]{x}$ **(b)** $\sqrt[3]{27x^2}$ **(c)** $\dfrac{\sqrt[4]{16x}}{\sqrt[3]{x}}$ **(d)** $\left(\dfrac{x^2}{81}\right)^{-1/2}$

Solution

(a) $\sqrt{x} \cdot \sqrt[3]{x} = x^{1/2} \cdot x^{1/3}$ Use rational exponents.

$\qquad\qquad\quad = x^{1/2 + 1/3}$ Product rule for exponents

$\qquad\qquad\quad = x^{5/6}$ Simplify: $\frac{1}{2} + \frac{1}{3} = \frac{3}{6} + \frac{2}{6} = \frac{5}{6}$.

(b) $\sqrt[3]{27x^2} = (27x^2)^{1/3}$ Use rational exponents.

$\qquad\qquad = 27^{1/3}(x^2)^{1/3}$ Power rule for products

$\qquad\qquad = 3x^{2/3}$ Simplify; power rule for exponents

(c) $\dfrac{\sqrt[4]{16x}}{\sqrt[3]{x}} = \dfrac{(16x)^{1/4}}{x^{1/3}}$ Use rational exponents.

$\qquad\quad = \dfrac{16^{1/4} x^{1/4}}{x^{1/3}}$ Power rule for products

$\qquad\quad = 16^{1/4} x^{1/4 - 1/3}$ Quotient rule for exponents

$\qquad\quad = 2x^{-1/12}$ Simplify: $\frac{1}{4} - \frac{1}{3} = \frac{3}{12} - \frac{4}{12} = -\frac{1}{12}$.

$\qquad\quad = \dfrac{2}{x^{1/12}}$ Negative exponents

(d) $\left(\dfrac{x^2}{81}\right)^{-1/2} = \left(\dfrac{81}{x^2}\right)^{1/2}$ Negative exponents for quotients

$= \dfrac{(81)^{1/2}}{(x^2)^{1/2}}$ Power rule for quotients

$= \dfrac{9}{x}$ Simplify; power rule for exponents

EXAMPLE 13   Applying properties of exponents

Write each expression with positive rational exponents and simplify, if possible.

(a) $\sqrt[3]{\sqrt{x+1}}$ (b) $\sqrt[5]{c^{15}}$ (c) $\dfrac{y^{-1/2}}{x^{-1/3}}$ (d) $\sqrt{x}(\sqrt{x} - 1)$

Solution

(a) $\sqrt[3]{\sqrt{x+1}} = ((x+1)^{1/2})^{1/3} = (x+1)^{1/6}$

(b) $\sqrt[5]{c^{15}} = c^{15/5} = c^3$

(c) $\dfrac{y^{-1/2}}{x^{-1/3}} = \dfrac{x^{1/3}}{y^{1/2}}$

(d) $\sqrt{x}(\sqrt{x} - 1) = x^{1/2}(x^{1/2} - 1) = x^{1/2}x^{1/2} - x^{1/2} = x - x^{1/2}$

7.1 PUTTING IT ALL TOGETHER

Properties of radicals and rational exponents are summarized in the following table.

Concept	Explanation	Examples
nth Root of a Real Number	An nth root of a real number a is b if $b^n = a$ and the (principal) nth root is denoted $\sqrt[n]{a}$. If $a < 0$ and n is even, $\sqrt[n]{a}$ is not a real number.	The square roots of 25 are 5 and -5. The principal square root is $\sqrt{25} = 5$. $\sqrt[3]{-125} = -5$ because $(-5)^3 = -125$.
Rational Exponents	If m and n are positive integers with $\dfrac{m}{n}$ in lowest terms, $a^{m/n} = \sqrt[n]{a^m} = \left(\sqrt[n]{a}\right)^m$. If $a < 0$ and n is even, $a^{m/n}$ is not a real number.	$8^{4/3} = \left(\sqrt[3]{8}\right)^4 = 2^4 = 16$ $(-27)^{3/4} = \left(\sqrt[4]{-27}\right)^3$ is *not* a real number.
Properties of Exponents	Let p and q be rational numbers. 1. $a^p \cdot a^q = a^{p+q}$ 2. $a^{-p} = \dfrac{1}{a^p}$, $\dfrac{1}{a^{-p}} = a^p$ 3. $\left(\dfrac{a}{b}\right)^{-p} = \left(\dfrac{b}{a}\right)^p$ 4. $\dfrac{a^p}{a^q} = a^{p-q}$ 5. $(a^p)^q = a^{pq}$ 6. $(ab)^p = a^p b^p$ 7. $\left(\dfrac{a}{b}\right)^p = \dfrac{a^p}{b^p}$	1. $2^{1/3} \cdot 2^{2/3} = 2^{1/3+2/3} = 2^1 = 2$ 2. $2^{-1/2} = \dfrac{1}{2^{1/2}}$, $\dfrac{1}{3^{-1/4}} = 3^{1/4}$ 3. $\left(\dfrac{3}{4}\right)^{-4/5} = \left(\dfrac{4}{3}\right)^{4/5}$ 4. $\dfrac{7^{2/3}}{7^{1/3}} = 7^{2/3-1/3} = 7^{1/3}$ 5. $(8^{2/3})^{1/2} = 8^{(2/3)\cdot(1/2)} = 8^{1/3} = 2$ 6. $(2x)^{1/3} = 2^{1/3} x^{1/3}$ 7. $\left(\dfrac{x}{y}\right)^{1/6} = \dfrac{x^{1/6}}{y^{1/6}}$

7.2 SIMPLIFYING RADICAL EXPRESSIONS

Product Rule for Radical Expressions ▪ Quotient Rule for Radical Expressions

A LOOK INTO MATH ▷ Radical expressions often occur in biology. For example, heavier birds tend to have larger wings. The relationship between the weight of a bird and the surface area of its wings can be modeled by a radical expression. In this section we discuss performing arithmetic operations with radical expressions.

Product Rule for Radical Expressions

Consider the following examples of multiplying radical expressions. The equations

$$\sqrt{4} \cdot \sqrt{25} = 2 \cdot 5 = 10 \quad \text{and} \quad \sqrt{4 \cdot 25} = \sqrt{100} = 10$$

imply that

$$\sqrt{4} \cdot \sqrt{25} = \sqrt{4 \cdot 25} \quad \text{(see Figure 7.6(a))}.$$

Similarly, the equations

$$\sqrt[3]{8} \cdot \sqrt[3]{27} = 2 \cdot 3 = 6 \quad \text{and} \quad \sqrt[3]{8 \cdot 27} = \sqrt[3]{216} = 6$$

imply that

$$\sqrt[3]{8} \cdot \sqrt[3]{27} = \sqrt[3]{8 \cdot 27} \quad \text{(see Figure 7.6(b))}.$$

These examples suggest that *the product of the roots is equal to the root of the product.*

(a)

(b)

Figure 7.6

> **PRODUCT RULE FOR RADICAL EXPRESSIONS**
>
> Let a and b be real numbers, where $\sqrt[n]{a}$ and $\sqrt[n]{b}$ are both defined. Then
> $$\sqrt[n]{a} \cdot \sqrt[n]{b} = \sqrt[n]{a \cdot b}.$$

NOTE: The product rule only works when the radicals have the *same index*. For example, the product $\sqrt{2} \cdot \sqrt[3]{4}$ cannot be simplified because the indexes are 2 and 3. (However, by using rational exponents, we can simplify this product. See Example 5(b).)

We apply the product rule in the next two examples.

EXAMPLE 1 Multiplying radical expressions

Multiply each radical expression.
(a) $\sqrt{5} \cdot \sqrt{20}$ (b) $\sqrt[3]{-3} \cdot \sqrt[3]{9}$ (c) $\sqrt[4]{\frac{1}{3}} \cdot \sqrt[4]{\frac{1}{9}} \cdot \sqrt[4]{\frac{1}{3}}$

Solution
(a) $\sqrt{5} \cdot \sqrt{20} = \sqrt{5 \cdot 20} = \sqrt{100} = 10$
(b) $\sqrt[3]{-3} \cdot \sqrt[3]{9} = \sqrt[3]{-3 \cdot 9} = \sqrt[3]{-27} = -3$
(c) The product rule can also be applied to three or more factors. Thus

$$\sqrt[4]{\frac{1}{3}} \cdot \sqrt[4]{\frac{1}{9}} \cdot \sqrt[4]{\frac{1}{3}} = \sqrt[4]{\frac{1}{3} \cdot \frac{1}{9} \cdot \frac{1}{3}} = \sqrt[4]{\frac{1}{81}} = \frac{1}{3}$$

because $\frac{1}{3} \cdot \frac{1}{3} \cdot \frac{1}{3} \cdot \frac{1}{3} = \frac{1}{81}$.

EXAMPLE 2 Multiplying radical expressions containing variables

Multiply each radical expression. Assume that all variables are positive.

(a) $\sqrt{x} \cdot \sqrt{x^3}$ (b) $\sqrt[3]{2a} \cdot \sqrt[3]{5a}$ (c) $\sqrt{11} \cdot \sqrt{xy}$ (d) $\sqrt[5]{\dfrac{2x}{y}} \cdot \sqrt[5]{\dfrac{16y}{x}}$

Solution
(a) $\sqrt{x} \cdot \sqrt{x^3} = \sqrt{x \cdot x^3} = \sqrt{x^4} = x^2$
(b) $\sqrt[3]{2a} \cdot \sqrt[3]{5a} = \sqrt[3]{2a \cdot 5a} = \sqrt[3]{10a^2}$
(c) $\sqrt{11} \cdot \sqrt{xy} = \sqrt{11xy}$
(d) $\sqrt[5]{\dfrac{2x}{y}} \cdot \sqrt[5]{\dfrac{16y}{x}} = \sqrt[5]{\dfrac{2x}{y} \cdot \dfrac{16y}{x}}$ Product rule

$= \sqrt[5]{\dfrac{32xy}{xy}}$ Multiply fractions.

$= \sqrt[5]{32}$ Simplify.
$= 2$ $2^5 = 32$

An integer a is a **perfect nth power** if there exists an integer b such that $b^n = a$. Thus 36 is a **perfect square** because $6^2 = 36$. Similarly, 8 is a **perfect cube** because $2^3 = 8$, and 81 is a *perfect fourth power* because $3^4 = 81$.

The product rule for radicals can be used to simplify radical expressions. For example, because the largest perfect square factor of 50 is 25, the expression $\sqrt{50}$ can be simplified as

$$\sqrt{50} = \sqrt{25 \cdot 2} = \sqrt{25} \cdot \sqrt{2} = 5\sqrt{2}.$$

This procedure is generalized as follows.

SIMPLIFYING RADICALS (nth ROOTS)

STEP 1: Determine the largest perfect nth power factor of the radicand.
STEP 2: Use the product rule to factor out and simplify this perfect nth power.

EXAMPLE 3 Simplifying radical expressions

Simplify each expression.

(a) $\sqrt{300}$ (b) $\sqrt[3]{16}$ (c) $\sqrt{54}$ (d) $\sqrt[4]{512}$

Solution
(a) First note that $300 = 100 \cdot 3$ and that 100 is the largest perfect square factor of 300.
$$\sqrt{300} = \sqrt{100} \cdot \sqrt{3} = 10\sqrt{3}$$
(b) The largest perfect cube factor of 16 is 8. Thus $\sqrt[3]{16} = \sqrt[3]{8} \cdot \sqrt[3]{2} = 2\sqrt[3]{2}$.
(c) $\sqrt{54} = \sqrt{9} \cdot \sqrt{6} = 3\sqrt{6}$
(d) $\sqrt[4]{512} = \sqrt[4]{256} \cdot \sqrt[4]{2} = 4\sqrt[4]{2}$ because $4^4 = 256$.

NOTE: To simplify a cube root of a negative number we factor out the negative of the largest perfect cube factor. For example, $-16 = -8 \cdot 2$, so $\sqrt[3]{-16} = \sqrt[3]{-8} \cdot \sqrt[3]{2} = -2\sqrt[3]{2}$. This procedure can be used with any odd root of a negative number.

EXAMPLE 4 Simplifying radical expressions

Simplify each expression. Assume that all variables are positive.

(a) $\sqrt{25x^4}$ (b) $\sqrt{32n^3}$ (c) $\sqrt[3]{-16x^3y^5}$ (d) $\sqrt[3]{2a} \cdot \sqrt[3]{4a^2b}$

Solution

(a) $\sqrt{25x^4} = 5x^2$ Perfect square: $(5x^2)^2 = 25x^4$

(b) $\sqrt{32n^3} = \sqrt{(16n^2)2n}$ $16n^2$ is the largest perfect square factor.

$= \sqrt{16n^2} \cdot \sqrt{2n}$ Product rule

$= 4n\sqrt{2n}$ $(4n)^2 = 16n^2$

(c) $\sqrt[3]{-16x^3y^5} = \sqrt[3]{(-8x^3y^3)2y^2}$ $8x^3y^3$ is the largest perfect cube factor.

$= \sqrt[3]{-8x^3y^3} \cdot \sqrt[3]{2y^2}$ Product rule

$= -2xy\sqrt[3]{2y^2}$ $(-2xy)^3 = -8x^3y^3$

(d) $\sqrt[3]{2a} \cdot \sqrt[3]{4a^2b} = \sqrt[3]{(2a)(4a^2b)}$ Product rule

$= \sqrt[3]{(8a^3)b}$ $8a^3$ is the largest perfect cube factor.

$= \sqrt[3]{8a^3} \cdot \sqrt[3]{b}$ Product rule

$= 2a\sqrt[3]{b}$ $(2a)^3 = 8a^3$

The product rule for radical expressions cannot be used if the radicals do not have the same indexes. In this case we use rational exponents, as illustrated in the next example.

EXAMPLE 5 Multiplying radicals with different indexes

Simplify each expression. Write your answer in radical notation.

(a) $\sqrt{5} \cdot \sqrt[4]{5}$ (b) $\sqrt{2} \cdot \sqrt[3]{4}$ (c) $\sqrt[3]{x} \cdot \sqrt[4]{x}$

Solution

(a) The given expression is $\sqrt{5} \cdot \sqrt[4]{5}$. Because $\sqrt{5} = 5^{1/2}$ and $\sqrt[4]{5} = 5^{1/4}$,

$$\sqrt{5} \cdot \sqrt[4]{5} = 5^{1/2} \cdot 5^{1/4} = 5^{1/2+1/4} = 5^{3/4}.$$

In radical notation, $5^{3/4} = \sqrt[4]{5^3} = \sqrt[4]{125}$.

(b) The given expression is $\sqrt{2} \cdot \sqrt[3]{4}$. Because $\sqrt[3]{4} = \sqrt[3]{2^2} = 2^{2/3}$,

$$\sqrt{2} \cdot \sqrt[3]{4} = 2^{1/2} \cdot 2^{2/3} = 2^{1/2+2/3} = 2^{7/6}.$$

In radical notation, $2^{7/6} = \sqrt[6]{2^7} = \sqrt[6]{2^6 \cdot 2^1} = \sqrt[6]{2^6} \cdot \sqrt[6]{2} = 2\sqrt[6]{2}$.

(c) $\sqrt[3]{x} \cdot \sqrt[4]{x} = x^{1/3} \cdot x^{1/4} = x^{7/12} = \sqrt[12]{x^7}$

Quotient Rule for Radical Expressions

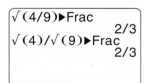

Figure 7.7

CALCULATOR HELP

To use the Frac feature, see the Appendix (page AP-2).

Consider the following examples of dividing radical expressions. The equations

$$\sqrt{\frac{4}{9}} = \sqrt{\frac{2}{3} \cdot \frac{2}{3}} = \frac{2}{3} \quad \text{and} \quad \frac{\sqrt{4}}{\sqrt{9}} = \frac{2}{3}$$

imply that

$$\sqrt{\frac{4}{9}} = \frac{\sqrt{4}}{\sqrt{9}} \quad \text{(see Figure 7.7)}.$$

These examples suggest that *the root of a quotient is equal to the quotient of the roots.*

7.2 SIMPLIFYING RADICAL EXPRESSIONS

> **QUOTIENT RULE FOR RADICAL EXPRESSIONS**
>
> Let a and b be real numbers, where $\sqrt[n]{a}$ and $\sqrt[n]{b}$ are both defined and $b \neq 0$. Then
> $$\sqrt[n]{\frac{a}{b}} = \frac{\sqrt[n]{a}}{\sqrt[n]{b}}.$$

EXAMPLE 6 Simplifying quotients

Simplify each radical expression. Assume that all variables are positive.

(a) $\sqrt[3]{\frac{5}{8}}$ (b) $\sqrt[4]{\frac{x}{16}}$ (c) $\sqrt{\frac{16}{y^2}}$

Solution

(a) $\sqrt[3]{\frac{5}{8}} = \frac{\sqrt[3]{5}}{\sqrt[3]{8}} = \frac{\sqrt[3]{5}}{2}$ (b) $\sqrt[4]{\frac{x}{16}} = \frac{\sqrt[4]{x}}{\sqrt[4]{16}} = \frac{\sqrt[4]{x}}{2}$

(c) $\sqrt{\frac{16}{y^2}} = \frac{\sqrt{16}}{\sqrt{y^2}} = \frac{4}{y}$ because $y > 0$.

EXAMPLE 7 Simplifying quotients

Simplify each radical expression. Assume that all variables are positive.

(a) $\frac{\sqrt{40}}{\sqrt{10}}$ (b) $\frac{\sqrt[3]{2}}{\sqrt[3]{16}}$ (c) $\frac{\sqrt{x^2 y}}{\sqrt{y}}$

Solution

(a) $\frac{\sqrt{40}}{\sqrt{10}} = \sqrt{\frac{40}{10}} = \sqrt{4} = 2$

(b) $\frac{\sqrt[3]{2}}{\sqrt[3]{16}} = \sqrt[3]{\frac{2}{16}} = \sqrt[3]{\frac{1}{8}} = \frac{1}{2}$ because $\frac{1}{2} \cdot \frac{1}{2} \cdot \frac{1}{2} = \frac{1}{8}$.

(c) $\frac{\sqrt{x^2 y}}{\sqrt{y}} = \sqrt{\frac{x^2 y}{y}} = \sqrt{x^2} = x$ because $x > 0$.

NOTE: When we simplify a radical *expression* we do *not* set the expression equal to 0 and solve the equation. Instead we write a sequence of equivalent expressions.

> **MAKING CONNECTIONS**
>
> **Rules for Radical Expressions and Rational Exponents**
>
> The rules for radical expressions are a result of the properties of rational exponents.
>
> $\sqrt[n]{a \cdot b} = \sqrt[n]{a} \cdot \sqrt[n]{b}$ is equivalent to $(a \cdot b)^{1/n} = a^{1/n} \cdot b^{1/n}$.
>
> $\sqrt[n]{\frac{a}{b}} = \frac{\sqrt[n]{a}}{\sqrt[n]{b}}$ is equivalent to $\left(\frac{a}{b}\right)^{1/n} = \frac{a^{1/n}}{b^{1/n}}$.

EXAMPLE 8 Simplifying radical expressions

Simplify each radical expression. Assume that all variables are positive.

(a) $\sqrt[4]{\dfrac{16x^3}{y^4}}$ (b) $\sqrt{\dfrac{5a^2}{8}} \cdot \sqrt{\dfrac{5a^3}{2}}$

Solution

(a) To simplify this expression we first use the quotient rule for radical expressions and then apply the product rule for radical expressions.

$$\sqrt[4]{\dfrac{16x^3}{y^4}} = \dfrac{\sqrt[4]{16x^3}}{\sqrt[4]{y^4}} \qquad \text{Quotient rule}$$

$$= \dfrac{\sqrt[4]{16}\,\sqrt[4]{x^3}}{\sqrt[4]{y^4}} \qquad \text{Product rule}$$

$$= \dfrac{2\sqrt[4]{x^3}}{y} \qquad \text{Evaluate 4th roots.}$$

(b) To simplify this expression, we use both the product and quotient rules.

$$\sqrt{\dfrac{5a^2}{8}} \cdot \sqrt{\dfrac{5a^3}{2}} = \sqrt{\dfrac{25a^5}{16}} \qquad \text{Product rule}$$

$$= \dfrac{\sqrt{25a^5}}{\sqrt{16}} \qquad \text{Quotient rule}$$

$$= \dfrac{\sqrt{25a^4 \cdot a}}{\sqrt{16}} \qquad \text{Factor out largest perfect square.}$$

$$= \dfrac{\sqrt{25a^4} \cdot \sqrt{a}}{\sqrt{16}} \qquad \text{Product rule}$$

$$= \dfrac{5a^2\sqrt{a}}{4} \qquad (5a^2)^2 = 25a^4$$

EXAMPLE 9 Simplifying products and quotients of roots

Simplify each expression. Assume all radicands are positive.

(a) $\sqrt{x-3} \cdot \sqrt{x+3}$ (b) $\dfrac{\sqrt[3]{x^2 + 3x + 2}}{\sqrt[3]{x+1}}$

Solution

(a) Start by applying the product rule for radical expressions.

$$\sqrt{x-3} \cdot \sqrt{x+3} = \sqrt{(x-3)(x+3)} \qquad \text{Product rule}$$
$$= \sqrt{x^2 - 9} \qquad \text{Multiply binomials.}$$

NOTE: The expression $\sqrt{x^2 - 9}$ is *not* simplified further. It is important to realize that

$$\sqrt{x^2 - 9} \neq \sqrt{x^2} - \sqrt{9} = x - 3.$$

For example, $\sqrt{5^2 - 3^2} = \sqrt{16} = 4$, but $\sqrt{5^2} - \sqrt{3^2} = 5 - 3 = 2$.

(b) Start by applying the quotient rule for radical expressions.

$$\frac{\sqrt[3]{x^2 + 3x + 2}}{\sqrt[3]{x+1}} = \sqrt[3]{\frac{x^2 + 3x + 2}{x+1}} \quad \text{Quotient rule}$$

$$= \sqrt[3]{\frac{(x+1)(x+2)}{x+1}} \quad \text{Factor trinomial.}$$

$$= \sqrt[3]{x+2} \quad \text{Simplify quotient.}$$

7.2 PUTTING IT ALL TOGETHER

In this section we discussed how to simplify radical expressions. Results are summarized in the following table.

Procedure	Explanation	Examples
Product Rule for Radical Expressions	Let a and b be real numbers, where $\sqrt[n]{a}$ and $\sqrt[n]{b}$ are both defined. Then $\sqrt[n]{a} \cdot \sqrt[n]{b} = \sqrt[n]{a \cdot b}.$	$\sqrt{2} \cdot \sqrt{32} = \sqrt{64} = 8$ $\sqrt{500} = \sqrt{100} \cdot \sqrt{5} = 10\sqrt{5}$
Quotient Rule for Radical Expressions	Let a and b be real numbers, where $\sqrt[n]{a}$ and $\sqrt[n]{b}$ are both defined and $b \neq 0$. Then $\sqrt[n]{\frac{a}{b}} = \frac{\sqrt[n]{a}}{\sqrt[n]{b}}.$	$\frac{\sqrt{60}}{\sqrt{15}} = \sqrt{\frac{60}{15}} = \sqrt{4} = 2 \quad \text{and}$ $\sqrt[3]{\frac{x^2}{-27}} = \frac{\sqrt[3]{x^2}}{\sqrt[3]{-27}} = \frac{\sqrt[3]{x^2}}{-3} = -\frac{\sqrt[3]{x^2}}{3}$
Simplifying Radicals (nth roots)	**STEP 1:** Find the largest nth power factor of the radicand. **STEP 2:** Factor out and simplify this perfect nth power.	$\sqrt{12} = \sqrt{4 \cdot 3} = \sqrt{4} \cdot \sqrt{3} = 2\sqrt{3}$ $\sqrt[3]{81} = \sqrt[3]{27 \cdot 3} = \sqrt[3]{27} \cdot \sqrt[3]{3} = 3\sqrt[3]{3}$ $\sqrt[4]{x^5} = \sqrt[4]{x^4 \cdot x} = \sqrt[4]{x^4} \cdot \sqrt[4]{x} = x\sqrt[4]{x},$ provided $x \geq 0$.

7.3 OPERATIONS ON RADICAL EXPRESSIONS

Addition and Subtraction ■ Multiplication ■ Rationalizing the Denominator

A LOOK INTO MATH ▷ So far we have discussed how to add, subtract, multiply, and divide numbers and variables. In this section we extend these operations to radical expressions. In doing so, we use many of the techniques discussed in Sections 7.1 and 7.2.

Addition and Subtraction

We can add $2x^2$ and $5x^2$ to obtain $7x^2$ because they are *like* terms. That is,

$$2x^2 + 5x^2 = (2 + 5)x^2 = 7x^2.$$

We can also add and subtract *like radicals*. **Like radicals** have the same index and the same radicand. For example, we can add $3\sqrt{2}$ and $5\sqrt{2}$ because they are like radicals.

$$3\sqrt{2} + 5\sqrt{2} = (3 + 5)\sqrt{2} = 8\sqrt{2}$$

EXAMPLE 1 Adding like radicals

If possible, add the expressions and simplify.
(a) $10\sqrt{11} + 4\sqrt{11}$ (b) $5\sqrt[3]{6} + \sqrt[3]{6}$
(c) $4 + 5\sqrt{3}$ (d) $\sqrt{7} + \sqrt{11}$

Solution
(a) These terms are like radicals because they have the same index, 2, and the same radicand, 11.
$$10\sqrt{11} + 4\sqrt{11} = (10 + 4)\sqrt{11} = 14\sqrt{11}$$

(b) These terms are like radicals because they have the same index, 3, and the same radicand, 6. Note that the coefficient on the second term is understood to be 1.
$$5\sqrt[3]{6} + 1\sqrt[3]{6} = (5 + 1)\sqrt[3]{6} = 6\sqrt[3]{6}$$

(c) The expression $4 + 5\sqrt{3}$ can be written as $4\sqrt{1} + 5\sqrt{3}$. These terms cannot be added because they are not like radicals.
NOTE: $4 + 5\sqrt{3} \ne 9\sqrt{3}$

(d) The expression $\sqrt{7} + \sqrt{11}$ contains unlike radicals that *cannot* be added.

Sometimes two radical expressions that are not alike can be added by changing them to like radicals. For example, $\sqrt{20}$ and $\sqrt{5}$ are unlike radicals. However,
$$\sqrt{20} = \sqrt{4 \cdot 5} = \sqrt{4} \cdot \sqrt{5} = 2\sqrt{5},$$
so
$$\sqrt{20} + \sqrt{5} = 2\sqrt{5} + 1\sqrt{5} = 3\sqrt{5}.$$

We cannot combine $x + x^2$ because they are unlike terms. Similarly, we cannot combine $\sqrt{2} + \sqrt{5}$ because they are unlike radicals. When combining radicals, the first step is to see if we can write pairs of terms as like radicals, as demonstrated in the next example.

EXAMPLE 2 Finding like radicals

Write each pair of terms as like radicals, if possible.
(a) $\sqrt{45}, \sqrt{20}$ (b) $\sqrt{27}, \sqrt{5}$ (c) $5\sqrt[3]{16}, 4\sqrt[3]{54}$

Solution
(a) The expressions $\sqrt{45}$ and $\sqrt{20}$ are unlike radicals. However, they can be changed to like radicals as follows.
$$\sqrt{45} = \sqrt{9 \cdot 5} = \sqrt{9} \cdot \sqrt{5} = 3\sqrt{5} \quad \text{and}$$
$$\sqrt{20} = \sqrt{4 \cdot 5} = \sqrt{4} \cdot \sqrt{5} = 2\sqrt{5}$$

The expressions $3\sqrt{5}$ and $2\sqrt{5}$ are like radicals.

(b) Because $\sqrt{27} = \sqrt{9 \cdot 3} = \sqrt{9} \cdot \sqrt{3} = 3\sqrt{3}$, the given expressions $\sqrt{27}$ and $\sqrt{5}$ are unlike radicals and cannot be written as like radicals.

(c) $5\sqrt[3]{16} = 5\sqrt[3]{8 \cdot 2} = 5\sqrt[3]{8} \cdot \sqrt[3]{2} = 5 \cdot 2 \cdot \sqrt[3]{2} = 10\sqrt[3]{2}$ and
$4\sqrt[3]{54} = 4\sqrt[3]{27 \cdot 2} = 4\sqrt[3]{27} \cdot \sqrt[3]{2} = 4 \cdot 3 \cdot \sqrt[3]{2} = 12\sqrt[3]{2}$

The expressions $10\sqrt[3]{2}$ and $12\sqrt[3]{2}$ are like radicals.

7.3 OPERATIONS ON RADICAL EXPRESSIONS

We use these techniques to add radical expressions in the next two examples.

EXAMPLE 3 Adding radical expressions

Add the expressions and simplify.
(a) $\sqrt{12} + 7\sqrt{3}$ (b) $\sqrt[3]{16} + \sqrt[3]{2}$ (c) $3\sqrt{2} + \sqrt{8} + \sqrt{18}$

Solution
(a) $\sqrt{12} + 7\sqrt{3} = \sqrt{4 \cdot 3} + 7\sqrt{3}$
$= \sqrt{4} \cdot \sqrt{3} + 7\sqrt{3}$
$= 2\sqrt{3} + 7\sqrt{3}$
$= 9\sqrt{3}$

(b) $\sqrt[3]{16} + \sqrt[3]{2} = \sqrt[3]{8 \cdot 2} + \sqrt[3]{2}$
$= \sqrt[3]{8} \cdot \sqrt[3]{2} + \sqrt[3]{2}$
$= 2\sqrt[3]{2} + 1\sqrt[3]{2}$
$= (2 + 1)\sqrt[3]{2}$
$= 3\sqrt[3]{2}$

(c) $3\sqrt{2} + \sqrt{8} + \sqrt{18} = 3\sqrt{2} + \sqrt{4 \cdot 2} + \sqrt{9 \cdot 2}$
$= 3\sqrt{2} + \sqrt{4} \cdot \sqrt{2} + \sqrt{9} \cdot \sqrt{2}$
$= 3\sqrt{2} + 2\sqrt{2} + 3\sqrt{2}$
$= 8\sqrt{2}$

EXAMPLE 4 Adding radical expressions

Add the expressions and simplify. Assume that all variables are positive.
(a) $\sqrt[4]{32} + 3\sqrt[4]{2}$ (b) $-2\sqrt{4x} + \sqrt{x}$ (c) $3\sqrt{3k} + 5\sqrt{12k} + 9\sqrt{48k}$

Solution
(a) Because $\sqrt[4]{32} = \sqrt[4]{16 \cdot 2} = \sqrt[4]{16} \cdot \sqrt[4]{2} = 2\sqrt[4]{2}$, we can add and simplify as follows.
$$\sqrt[4]{32} + 3\sqrt[4]{2} = 2\sqrt[4]{2} + 3\sqrt[4]{2} = 5\sqrt[4]{2}$$

(b) Note that $\sqrt{4x} = \sqrt{4} \cdot \sqrt{x} = 2\sqrt{x}$.
$-2\sqrt{4x} + \sqrt{x} = -2(2\sqrt{x}) + \sqrt{x} = -4\sqrt{x} + 1\sqrt{x} = -3\sqrt{x}$

(c) Note that $\sqrt{12k} = \sqrt{4} \cdot \sqrt{3k} = 2\sqrt{3k}$ and that $\sqrt{48k} = \sqrt{16} \cdot \sqrt{3k} = 4\sqrt{3k}$.
$3\sqrt{3k} + 5\sqrt{12k} + 9\sqrt{48k} = 3\sqrt{3k} + 5(2\sqrt{3k}) + 9(4\sqrt{3k})$
$= (3 + 10 + 36)\sqrt{3k}$
$= 49\sqrt{3k}$

Subtraction of radical expressions is similar to addition, as illustrated in the next three examples.

EXAMPLE 5 Subtracting like radicals

Simplify the expressions.

(a) $5\sqrt{7} - 3\sqrt{7}$ (b) $8\sqrt[3]{5} - 3\sqrt[3]{5} + \sqrt[3]{11}$ (c) $5\sqrt{z} + \sqrt[3]{z} - 2\sqrt{z}$

Solution

(a) $\quad 5\sqrt{7} - 3\sqrt{7} = (5-3)\sqrt{7} = 2\sqrt{7}$

(b) $8\sqrt[3]{5} - 3\sqrt[3]{5} + \sqrt[3]{11} = (8-3)\sqrt[3]{5} + \sqrt[3]{11} = 5\sqrt[3]{5} + \sqrt[3]{11}$

(c) $\quad 5\sqrt{z} + \sqrt[3]{z} - 2\sqrt{z} = 5\sqrt{z} - 2\sqrt{z} + \sqrt[3]{z}$ Commutative property

$\qquad\qquad\qquad\qquad\quad = (5-2)\sqrt{z} + \sqrt[3]{z}$ Distributive property

$\qquad\qquad\qquad\qquad\quad = 3\sqrt{z} + \sqrt[3]{z}$ Subtract.

NOTE: We cannot combine $3\sqrt{z} + \sqrt[3]{z}$ because their indexes are different. That is, one term is a square root and the other is a cube root.

EXAMPLE 6 Subtracting radical expressions

Subtract and simplify. Assume that all variables are positive.

(a) $3\sqrt[3]{xy^2} - 2\sqrt[3]{xy^2}$ (b) $\sqrt{16x^3} - \sqrt{x^3}$ (c) $\sqrt[3]{\dfrac{5x}{27}} - \dfrac{\sqrt[3]{5x}}{6}$

Solution

(a) $\quad 3\sqrt[3]{xy^2} - 2\sqrt[3]{xy^2} = (3-2)\sqrt[3]{xy^2} = \sqrt[3]{xy^2}$

(b) $\quad \sqrt{16x^3} - \sqrt{x^3} = \sqrt{16x^2} \cdot \sqrt{x} - \sqrt{x^2} \cdot \sqrt{x}$ Factor out perfect squares.

$\qquad\qquad\qquad\quad = 4x\sqrt{x} - x\sqrt{x}$ Simplify.

$\qquad\qquad\qquad\quad = (4x - x)\sqrt{x}$ Distributive property

$\qquad\qquad\qquad\quad = 3x\sqrt{x}$ Subtract.

(c) $\quad \sqrt[3]{\dfrac{5x}{27}} - \dfrac{\sqrt[3]{5x}}{6} = \dfrac{\sqrt[3]{5x}}{\sqrt[3]{27}} - \dfrac{\sqrt[3]{5x}}{6}$ Quotient rule for radical expressions

$\qquad\qquad\qquad\quad = \dfrac{\sqrt[3]{5x}}{3} - \dfrac{\sqrt[3]{5x}}{6}$ Evaluate $\sqrt[3]{27} = 3$.

$\qquad\qquad\qquad\quad = \dfrac{2\sqrt[3]{5x}}{6} - \dfrac{\sqrt[3]{5x}}{6}$ Find a common denominator.

$\qquad\qquad\qquad\quad = \dfrac{2\sqrt[3]{5x} - \sqrt[3]{5x}}{6}$ Subtract numerators.

$\qquad\qquad\qquad\quad = \dfrac{\sqrt[3]{5x}}{6}$ Simplify.

7.3 OPERATIONS ON RADICAL EXPRESSIONS

EXAMPLE 7 Subtracting radical expressions

Subtract and simplify. Assume that all variables are positive.

(a) $\dfrac{5\sqrt{2}}{3} - \dfrac{2\sqrt{2}}{4}$ (b) $\sqrt[4]{81a^5b^6} - \sqrt[4]{16ab^2}$ (c) $3\sqrt[3]{\dfrac{n^5}{27}} - 2\sqrt[3]{n^2}$

Solution

(a)
$$\dfrac{5\sqrt{2}}{3} - \dfrac{2\sqrt{2}}{4} = \dfrac{5\sqrt{2}}{3} \cdot \dfrac{4}{4} - \dfrac{2\sqrt{2}}{4} \cdot \dfrac{3}{3}$$ LCD is 12.

$$= \dfrac{20\sqrt{2}}{12} - \dfrac{6\sqrt{2}}{12}$$ Multiply fractions.

$$= \dfrac{14\sqrt{2}}{12}$$ Subtract numerators.

$$= \dfrac{7\sqrt{2}}{6}$$ Simplify.

(b) $\sqrt[4]{81a^5b^6} - \sqrt[4]{16ab^2} = \sqrt[4]{81a^4b^4} \cdot \sqrt[4]{ab^2} - \sqrt[4]{16} \cdot \sqrt[4]{ab^2}$ Factor out perfect powers.

$\qquad = 3ab\sqrt[4]{ab^2} - 2\sqrt[4]{ab^2}$ Simplify.

$\qquad = (3ab - 2)\sqrt[4]{ab^2}$ Distributive property

(c) $3\sqrt[3]{\dfrac{n^5}{27}} - 2\sqrt[3]{n^2} = 3\sqrt[3]{\dfrac{n^3}{27}} \cdot \sqrt[3]{n^2} - 2\sqrt[3]{n^2}$ Factor out perfect cube.

$\qquad = \dfrac{3\sqrt[3]{n^3}}{\sqrt[3]{27}} \cdot \sqrt[3]{n^2} - 2\sqrt[3]{n^2}$ Quotient rule

$\qquad = n\sqrt[3]{n^2} - 2\sqrt[3]{n^2}$ Simplify.

$\qquad = (n - 2)\sqrt[3]{n^2}$ Distributive property

Radicals often occur in geometry. In the next example, we find the perimeter of a triangle by adding radical expressions.

EXAMPLE 8 Finding the perimeter of a triangle

Find the *exact* perimeter of the right triangle shown in Figure 7.8. Then approximate your answer to the nearest hundredth of a foot.

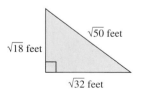

Figure 7.8

Solution
The sum of the lengths of the sides of the triangle is
$$\sqrt{18} + \sqrt{32} + \sqrt{50} = 3\sqrt{2} + 4\sqrt{2} + 5\sqrt{2} = 12\sqrt{2}.$$
The perimeter is $12\sqrt{2} \approx 16.97$ feet.

Multiplication

Some types of radical expressions can be multiplied like binomials. For example, because
$$(x + 1)(x + 2) = x^2 + 3x + 2,$$
we have $(\sqrt{x} + 1)(\sqrt{x} + 2) = (\sqrt{x})^2 + 2\sqrt{x} + 1\sqrt{x} + 2 = x + 3\sqrt{x} + 2,$

provided that $x \geq 0$. The next example demonstrates this technique.

EXAMPLE 9 Multiplying radical expressions

Multiply and simplify.
(a) $(\sqrt{b} - 4)(\sqrt{b} + 5)$
(b) $(4 + \sqrt{3})(4 - \sqrt{3})$

Solution
(a) This expression can be multiplied and then simplified.

$$(\sqrt{b} - 4)(\sqrt{b} + 5) = \sqrt{b} \cdot \sqrt{b} + 5\sqrt{b} - 4\sqrt{b} - 4 \cdot 5$$
$$= b + \sqrt{b} - 20$$

Compare this product with $(b - 4)(b + 5) = b^2 + b - 20$.

(b) This expression is in the form $(a + b)(a - b)$, which equals $a^2 - b^2$.

$$(4 + \sqrt{3})(4 - \sqrt{3}) = (4)^2 - (\sqrt{3})^2$$
$$= 16 - 3$$
$$= 13$$

NOTE: Example 9(b) illustrates a special case for multiplying radicals. In general,

$$(\sqrt{a} + \sqrt{b})(\sqrt{a} - \sqrt{b}) = (\sqrt{a})^2 - (\sqrt{b})^2 = a - b,$$

provided a and b are nonnegative.

Rationalizing the Denominator

In mathematics it is common to write expressions without radicals in the denominator. Quotients containing radical expressions in the numerator or denominator can appear to be different but actually be equal. For example, $\frac{1}{\sqrt{3}}$ and $\frac{\sqrt{3}}{3}$ represent the same real number even though they look like they are unequal. To show that they are equal, we multiply the first quotient by 1 in the form $\frac{\sqrt{3}}{\sqrt{3}}$:

$$\frac{1}{\sqrt{3}} \cdot \frac{\sqrt{3}}{\sqrt{3}} = \frac{1 \cdot \sqrt{3}}{\sqrt{3} \cdot \sqrt{3}} = \frac{\sqrt{3}}{3}.$$

If the denominator of a quotient contains only one term with one square root, then we can rationalize the denominator by multiplying the numerator and denominator by this square root. For example, the denominator of $\frac{1}{\sqrt{3}}$ contains one term, which is $\sqrt{3}$. Therefore, we multiplied the $\frac{1}{\sqrt{3}}$ by $\frac{\sqrt{3}}{\sqrt{3}}$ to rationalize the denominator.

NOTE: $\sqrt{b} \cdot \sqrt{b} = \sqrt{b^2} = b$ for any *positive* number b.

One way to *standardize* radical expressions is to remove any radical expressions from the denominator. This process is called **rationalizing the denominator**. The next example demonstrates how to rationalize the denominator of several quotients.

7.3 OPERATIONS ON RADICAL EXPRESSIONS

EXAMPLE 10 Rationalizing the denominator

Rationalize each denominator. Assume that all variables are positive.

(a) $\dfrac{1}{\sqrt{2}}$ (b) $\dfrac{3}{5\sqrt{3}}$ (c) $\sqrt{\dfrac{x}{24}}$ (d) $\dfrac{xy}{\sqrt{y^3}}$

Solution

(a) We start by multiplying this expression by 1 in the form $\dfrac{\sqrt{2}}{\sqrt{2}}$:

$$\dfrac{1}{\sqrt{2}} \cdot \dfrac{\sqrt{2}}{\sqrt{2}} = \dfrac{\sqrt{2}}{\sqrt{4}} = \dfrac{\sqrt{2}}{2}.$$

Note that the expression $\dfrac{\sqrt{2}}{2}$ does not have a radical in the denominator.

(b) We multiply this expression by 1 in the form $\dfrac{\sqrt{3}}{\sqrt{3}}$:

$$\dfrac{3}{5\sqrt{3}} \cdot \dfrac{\sqrt{3}}{\sqrt{3}} = \dfrac{3\sqrt{3}}{5\sqrt{9}} = \dfrac{3\sqrt{3}}{5 \cdot 3} = \dfrac{\sqrt{3}}{5}.$$

(c) Because $\sqrt{24} = \sqrt{4} \cdot \sqrt{6} = 2\sqrt{6}$, we start by simplifying the expression.

$$\sqrt{\dfrac{x}{24}} = \dfrac{\sqrt{x}}{\sqrt{24}} = \dfrac{\sqrt{x}}{2\sqrt{6}}$$

To rationalize the denominator we multiply this expression by 1 in the form $\dfrac{\sqrt{6}}{\sqrt{6}}$:

$$\dfrac{\sqrt{x}}{2\sqrt{6}} = \dfrac{\sqrt{x}}{2\sqrt{6}} \cdot \dfrac{\sqrt{6}}{\sqrt{6}} = \dfrac{\sqrt{6x}}{12}.$$

(d) Because $\sqrt{y^3} = \sqrt{y^2} \cdot \sqrt{y} = y\sqrt{y}$, we start by simplifying the expression.

$$\dfrac{xy}{\sqrt{y^3}} = \dfrac{xy}{y\sqrt{y}} = \dfrac{x}{\sqrt{y}}$$

To rationalize the denominator we multiply by 1 in the form $\dfrac{\sqrt{y}}{\sqrt{y}}$:

$$\dfrac{x}{\sqrt{y}} \cdot \dfrac{\sqrt{y}}{\sqrt{y}} = \dfrac{x\sqrt{y}}{y}.$$

When the denominator is either a sum or difference containing a square root, we multiply the numerator and denominator by the *conjugate* of the denominator. In this case, the **conjugate** of the denominator is found by changing a + sign to a − sign or vice versa. Table 7.1 lists examples of conjugates.

TABLE 7.1

Expression	$1 + \sqrt{2}$	$\sqrt{3} - 2$	$\sqrt{x} + 7$	$\sqrt{a} - \sqrt{b}$
Conjugate	$1 - \sqrt{2}$	$\sqrt{3} + 2$	$\sqrt{x} - 7$	$\sqrt{a} + \sqrt{b}$

In the next two examples we illustrate this method to rationalize a denominator.

CHAPTER 7 RADICAL EXPRESSIONS AND FUNCTIONS

EXAMPLE 11 Using a conjugate to rationalize the denominator

Rationalize the denominator of $\frac{1}{1+\sqrt{2}}$.

Solution
From Table 7.1, the conjugate of $1 + \sqrt{2}$ is $1 - \sqrt{2}$.

$$\frac{1}{1+\sqrt{2}} = \frac{1}{1+\sqrt{2}} \cdot \frac{(1-\sqrt{2})}{(1-\sqrt{2})} \quad \text{Multiply numerator and denominator by the conjugate.}$$

$$= \frac{1-\sqrt{2}}{(1)^2 - (\sqrt{2})^2} \quad (a+b)(a-b) = a^2 - b^2$$

$$= \frac{1-\sqrt{2}}{1-2} \quad \text{Simplify.}$$

$$= \frac{1-\sqrt{2}}{-1} \quad \text{Subtract.}$$

$$= \frac{1}{-1} - \frac{\sqrt{2}}{-1} \quad \frac{a-b}{c} = \frac{a}{c} - \frac{b}{c}$$

$$= -1 + \sqrt{2} \quad \text{Simplify.}$$

EXAMPLE 12 Rationalizing the denominator

Rationalize the denominator.

(a) $\dfrac{3+\sqrt{5}}{2-\sqrt{5}}$ (b) $\dfrac{\sqrt{x}}{\sqrt{x}-2}$

Solution
(a) The conjugate of the denominator is $2 + \sqrt{5}$.

$$\frac{3+\sqrt{5}}{2-\sqrt{5}} = \frac{(3+\sqrt{5})}{(2-\sqrt{5})} \cdot \frac{(2+\sqrt{5})}{(2+\sqrt{5})} \quad \text{Multiply by 1.}$$

$$= \frac{6 + 3\sqrt{5} + 2\sqrt{5} + (\sqrt{5})^2}{(2)^2 - (\sqrt{5})^2} \quad \text{Multiply.}$$

$$= \frac{11 + 5\sqrt{5}}{4 - 5} \quad \text{Combine terms.}$$

$$= -11 - 5\sqrt{5} \quad \text{Simplify.}$$

(b) The conjugate of the denominator is $\sqrt{x} + 2$.

$$\frac{\sqrt{x}}{\sqrt{x}-2} = \frac{\sqrt{x}}{(\sqrt{x}-2)} \cdot \frac{(\sqrt{x}+2)}{(\sqrt{x}+2)} \quad \text{Multiply by 1.}$$

$$= \frac{x + 2\sqrt{x}}{x - 4} \quad \text{Multiply.}$$

EXAMPLE 13 Rationalizing a denominator having a cube root

Rationalize the denominator of $\dfrac{5}{\sqrt[3]{x}}$.

Solution

The expression $\dfrac{5}{\sqrt[3]{x}}$ is equal to $\dfrac{5}{x^{1/3}}$. To rationalize the denominator, $x^{1/3}$, we can multiply it by $x^{2/3}$ because $x^{1/3} \cdot x^{2/3} = x^{1/3+2/3} = x^1$.

$$\dfrac{5}{x^{1/3}} = \dfrac{5}{x^{1/3}} \cdot \dfrac{x^{2/3}}{x^{2/3}}$$ Multiply by 1.

$$= \dfrac{5x^{2/3}}{x^{1/3+2/3}}$$ Product rule

$$= \dfrac{5\sqrt[3]{x^2}}{x}$$ Add; write in radical notation.

7.3 PUTTING IT ALL TOGETHER

In this section we discussed how to add, subtract, and multiply radical expressions. Rationalization of the denominator is a technique that can sometimes be helpful when dividing rational expressions. The following table summarizes important topics in this section.

Concept	Explanation	Examples
Like Radicals	Like radicals have the same index and the same radicand.	$7\sqrt{5}$ and $3\sqrt{5}$ are like radicals. $5\sqrt[3]{ab}$ and $\sqrt[3]{ab}$ are like radicals. $\sqrt[3]{5}$ and $\sqrt[3]{4}$ are unlike radicals. $\sqrt[3]{7}$ and $\sqrt{7}$ are unlike radicals.
Adding and Subtracting Radical Expressions	Combine like radicals when adding or subtracting.	$6\sqrt{13} + \sqrt{13} = (6 + 1)\sqrt{13} = 7\sqrt{13}$
	We cannot combine unlike radicals such as $\sqrt{2}$ and $\sqrt{5}$. But sometimes we can rewrite radicals and then combine.	$\sqrt{40} - \sqrt{10} = \sqrt{4} \cdot \sqrt{10} - \sqrt{10}$ $= 2\sqrt{10} - \sqrt{10}$ $= \sqrt{10}$
Multiplying Radical Expressions	Radical expressions can sometimes be multiplied like binomials.	$(\sqrt{a} - 5)(\sqrt{a} + 5) = a - 25$ and $(\sqrt{x} - 3)(\sqrt{x} + 1) = x - 2\sqrt{x} - 3$
Conjugate	The conjugate is found by changing a $+$ sign to a $-$ sign or vice versa.	*Expression* *Conjugate* $\sqrt{x} + 7$ $\sqrt{x} - 7$ $\sqrt{a} - 2\sqrt{b}$ $\sqrt{a} + 2\sqrt{b}$
Rationalizing a Denominator Having One Term	Write the quotient without a radical expression in the denominator.	To rationalize $\dfrac{5}{\sqrt{7}}$, multiply the expression by 1 in the form $\dfrac{\sqrt{7}}{\sqrt{7}}$: $\dfrac{5}{\sqrt{7}} \cdot \dfrac{\sqrt{7}}{\sqrt{7}} = \dfrac{5\sqrt{7}}{\sqrt{49}} = \dfrac{5\sqrt{7}}{7}.$

continued on next page

Concept	Explanation	Examples
Rationalizing a Denominator Having Two Terms Containing Square Roots	Multiply the numerator and denominator by the conjugate of the denominator.	$\dfrac{1}{2 - \sqrt{3}} = \dfrac{1}{2 - \sqrt{3}} \cdot \dfrac{(2 + \sqrt{3})}{(2 + \sqrt{3})}$ $= \dfrac{2 + \sqrt{3}}{4 - 3} = 2 + \sqrt{3}$

7.4 RADICAL FUNCTIONS

The Square Root Function ■ The Cube Root Function ■ Power Functions ■ Modeling with Power Functions (Optional)

A LOOK INTO MATH ▷ A good punter can kick a football so that it has a long *hang time*. Hang time is the length of time that the football is in the air. Long hang time gives the kicking team time to run down the field and stop the punt return. Using square roots, we can derive a function that calculates hang time. (See the discussion before Example 3.) First, we discuss the square root function.

The Square Root Function

The square root function is given by $f(x) = \sqrt{x}$. The domain of the square root function is all nonnegative real numbers because we have not defined the square root of a negative number. Table 7.2 lists three points that lie on the graph of $f(x) = \sqrt{x}$. In Figure 7.9 these points are plotted and the graph of $y = \sqrt{x}$ has been sketched. The graph does not appear to the left of the origin because $f(x) = \sqrt{x}$ is undefined for negative inputs.

TABLE 7.2

x	$\sqrt{x}$
0	0
1	1
4	2

Figure 7.9 Square Root Function

TECHNOLOGY NOTE: Square Roots of Negative Numbers
If a table of values for $y_1 = \sqrt{x}$ includes both negative and positive values for x, then many calculators give error messages when x is negative, as shown in the accompanying figure.

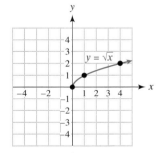

EXAMPLE 1 Evaluating functions involving square roots

If possible, evaluate $f(1)$ and $f(-2)$ for each $f(x)$.
(a) $f(x) = \sqrt{2x - 1}$ (b) $f(x) = \sqrt{4 - x^2}$

Solution
(a) $f(1) = \sqrt{2(1) - 1} = \sqrt{1} = 1$
$f(-2) = \sqrt{2(-2) - 1} = \sqrt{-5}$, which does not equal a real number.
(b) $f(1) = \sqrt{4 - (1)^2} = \sqrt{3}$; $f(-2) = \sqrt{4 - (-2)^2} = \sqrt{0} = 0$

In the next example we find a formula, a table of values, and a graph for a function.

EXAMPLE 2 Finding symbolic, numerical, and graphical representations

A function f takes the square root of x and then multiplies the result by 2. Give symbolic, numerical, and graphical representations of f.

Solution
First find a formula for f. The square root of x multiplied by 2 equals $2\sqrt{x}$, so $f(x) = 2\sqrt{x}$. A table of values is shown in Table 7.3. A curve is sketched through these points to obtain the graph of f in Figure 7.10. The domain of f includes all nonnegative numbers.

Symbolic Representation

$f(x) = 2\sqrt{x}$

Numerical Representation

TABLE 7.3

x	$2\sqrt{x}$
0	0
1	2
4	4
9	6

Graphical Representation

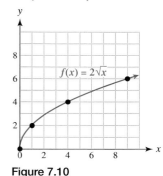

Figure 7.10

▶ **REAL-WORLD CONNECTION** To derive a function that calculates the hang time of a football we need two facts from physics. First, when a ball is kicked into the air, half the hang time of the ball is spent going up and the other half is spent coming down. Second, the time t in seconds required for a ball to fall from a height of h feet is modeled by the equation $16t^2 = h$. Solving this equation for t gives half the hang time.

$16t^2 = h$ Given equation

$t^2 = \dfrac{h}{16}$ Divide by 16.

$t = \sqrt{\dfrac{h}{16}}$ Take the principal square root.

$t = \dfrac{\sqrt{h}}{4}$ Simplify.

Half the hang time is $\frac{\sqrt{h}}{4}$, so the total hang time T in seconds is given by

$$T(h) = 2 \cdot \frac{\sqrt{h}}{4} = \frac{\sqrt{h}}{2},$$

where h is the maximum height of the ball.

In the next example, we use this formula to calculate hang time.

EXAMPLE 3 Calculating hang time

If a football is kicked 50 feet into the air, estimate the hang time. Does the hang time double for a football kicked 100 feet into the air?

Solution
A football kicked 50 feet into the air has a hang time of $T(50) = \frac{\sqrt{50}}{2} \approx 3.5$ seconds. For 100 feet, the hang time is $T(100) = \frac{\sqrt{100}}{2} = 5$ seconds. The hang time does not double.

CRITICAL THINKING

How high would a football have to be kicked to have twice the hang time of a football kicked 50 feet into the air?

EXAMPLE 4 Finding the domain of a function

Let $f(x) = \sqrt{x - 1}$.
(a) Find the domain of f. Write your answer in interval notation.
(b) Graph $y = f(x)$ and compare it to the graph of $y = \sqrt{x}$.

Solution
(a) For $f(x)$ to be defined, $x - 1$ cannot be negative. Thus valid inputs for x must satisfy

$$x - 1 \geq 0 \quad \text{or} \quad x \geq 1.$$

The domain is $[1, \infty)$.
(b) Table 7.4 lists points that lie on the graph of $y = \sqrt{x - 1}$. Note in Figure 7.11 that the graph appears only when $x \geq 1$. This graph is similar to $y = \sqrt{x}$ (see Figure 7.9) except that it is shifted one unit to the right.

TABLE 7.4

x	$\sqrt{x - 1}$
1	0
2	1
5	2

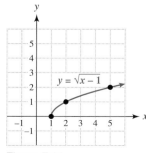

Figure 7.11

EXAMPLE 5 Finding the domains of functions

Find the domain of each function. Write your answer in interval notation.
(a) $f(x) = \sqrt{4 - 2x}$ (b) $g(x) = \sqrt{x^2 + 1}$

Solution
(a) To determine when $f(x)$ is defined, we must solve the inequality $4 - 2x \geq 0$.

$4 - 2x \geq 0$	Inequality to be solved
$4 \geq 2x$	Add $2x$ to each side.
$2 \geq x$	Divide each side by 2.

The domain is $(-\infty, 2]$.

(b) Regardless of the value of x, the expression $x^2 + 1$ is always positive because $x^2 \geq 0$. Thus $g(x)$ is defined for all real numbers, and its domain is $(-\infty, \infty)$.

MAKING CONNECTIONS

Domains of Functions and Their Graphs

In Example 5, the domains of f and g were found *symbolically*. Notice that the graph of f does not appear to the right of $x = 2$ because the domain of f is $(-\infty, 2]$, whereas the graph of g appears for all values of x because the domain of g is $(-\infty, \infty)$.

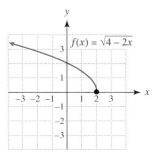
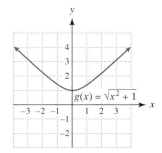

The Cube Root Function

In Section 7.1, we discussed the cube root of a number. We can define the **cube root function** by $f(x) = \sqrt[3]{x}$. Cube roots are defined for both positive and negative numbers, so *the domain of the cube root function includes all real numbers*. Table 7.5 lists points that lie on the graph of the cube root function. Figure 7.12 shows a graph of $y = \sqrt[3]{x}$.

TABLE 7.5 Cube Root Function

x	-27	-8	-1	0	1	8	27
$\sqrt[3]{x}$	-3	-2	-1	0	1	2	3

Figure 7.12 Cube Root Function

MAKING CONNECTIONS

Domains of the Square Root and Cube Root Functions

$f(x) = \sqrt{x}$ equals a real number for any nonnegative x. Thus $D = [0, \infty)$.

$f(x) = \sqrt[3]{x}$ equals a real number for any x. Thus $D = (-\infty, \infty)$.

Examples: $\sqrt{4} = 2$ but $\sqrt{-4}$ is *not* a real number. $\sqrt[3]{8} = 2$ and $\sqrt[3]{-8} = -2$.

EXAMPLE 6 Evaluating functions involving cube roots

Evaluate $f(1)$ and $f(-3)$ for each $f(x)$.
(a) $f(x) = \sqrt[3]{x^2 - 1}$ (b) $f(x) = \sqrt[3]{2 - x^2}$

Solution
(a) $f(1) = \sqrt[3]{1^2 - 1} = \sqrt[3]{0} = 0$; $f(-3) = \sqrt[3]{(-3)^2 - 1} = \sqrt[3]{8} = 2$
(b) $f(1) = \sqrt[3]{2 - 1^2} = \sqrt[3]{1} = 1$; $f(-3) = \sqrt[3]{2 - (-3)^2} = \sqrt[3]{-7}$ or $-\sqrt[3]{7}$

EXAMPLE 7 Finding symbolic, numerical, and graphical representations

A function f takes the cube root of twice x. Give symbolic, numerical, and graphical representations of f.

Solution
First find a formula for f. Twice x is $2x$, so $f(x) = \sqrt[3]{2x}$. A table of values is shown in Table 7.6. A curve is sketched through these points to obtain the graph of f in Figure 7.13.

Symbolic Representation

$f(x) = \sqrt[3]{2x}$

Numerical Representation

TABLE 7.6

x	$\sqrt[3]{2x}$
-4	-2
$-\frac{1}{2}$	-1
0	0
$\frac{1}{2}$	1
4	2

Graphical Representation

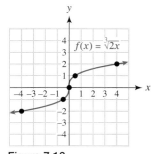

Figure 7.13

Power Functions

Power functions are a generalization of root functions. Examples of power functions include

$$f(x) = x^{1/2}, \quad g(x) = x^{2/3}, \quad \text{and} \quad h(x) = x^{-3/5}.$$

The exponents for power functions can be rational numbers. Any rational number can be written in lowest terms as $\frac{m}{n}$, where m and n are integers. When n is an odd integer, then $f(x) = x^{m/n}$ is defined for all real numbers. When n is even, $f(x) = x^{m/n}$ is defined only for nonnegative real numbers.

POWER FUNCTION

If a function f can be represented by

$$f(x) = x^p,$$

where p is a rational number, then f is a **power function**. If $p = \frac{1}{n}$, where $n \geq 2$ is an integer, then f is also a **root function**, which is given by

$$f(x) = \sqrt[n]{x}.$$

EXAMPLE 8   Evaluating power functions

If possible, evaluate $f(x)$ at the given value of x.
(a) $f(x) = x^{0.75}$ at $x = 16$ (b) $f(x) = x^{1/4}$ at $x = -81$

Solution
(a) $0.75 = \frac{3}{4}$, so $f(x) = x^{3/4}$. Thus $f(16) = 16^{3/4} = (16^{1/4})^3 = 2^3 = 8$.

NOTE: $16^{1/4} = \sqrt[4]{16} = 2$.

(b) $f(-81) = (-81)^{1/4} = \sqrt[4]{-81}$, which is undefined. There is no real number a such that $a^4 = -81$ because a^4 is never negative.

▶ **REAL-WORLD CONNECTION**   The surface area of the skin covering the human body is influenced by both the height and weight of a person. A taller person tends to have a larger surface area, as does a heavier person. In the next example, we use a power function from biology to model this situation.

EXAMPLE 9   Modeling surface area of the human body

The surface area of a person who is 66 inches tall and weighs w pounds can be estimated by $S(w) = 327w^{0.425}$, where S is in square inches.   (**Source:** H. Lancaster, Quantitative *Methods in Biological and Medical Sciences*.)
(a) Find S if this person weighs 130 pounds.
(b) If the person gains 20 pounds, by how much does the person's surface area increase?

Solution
(a) $S(130) = 327(130)^{0.425} \approx 2588$ square inches
(b) Because $S(150) = 327(150)^{0.425} \approx 2750$ square inches, the surface area of the person increases by about $2750 - 2588 = 162$ square inches.

In the next example, we investigate the graph of $y = x^p$ for different values of p.

EXAMPLE 10   Graphing power functions

The graphs of three power functions,
$$f(x) = x^{1/3}, \quad g(x) = x^{0.75}, \quad \text{and} \quad h(x) = x^{1.4},$$
are shown in Figure 7.14. Discuss how the value of p affects the graph of $y = x^p$ when $x > 1$ and when $0 < x < 1$.

Solution
First note that $g(x)$ and $h(x)$ can be written as $g(x) = x^{3/4}$ and $h(x) = x^{7/5}$. When $x > 1$, $h(x) > g(x) > f(x)$ and the graphs increase (rise) faster for larger values of p. When $0 < x < 1$, $h(x) < g(x) < f(x)$. Thus smaller values of p result in larger y-values when $0 < x < 1$. All three graphs appear to intersect at the points $(0, 0)$ and $(1, 1)$.

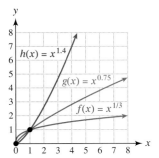
Figure 7.14 Power Functions

Modeling with Power Functions (Optional)

▶ **REAL-WORLD CONNECTION**   Allometry is the study of the relative sizes of different characteristics of an organism. For example, the weight of a bird is related to the surface area of its wings: Heavier birds tend to have larger wings. Allometric relations are often modeled with $f(x) = kx^p$, where k and p are constants.   (**Source:** C. Pennycuick, *Newton Rules Biology*.)

EXAMPLE 11 Modeling surface area of wings

The surface area A of a bird's wings with weight w is shown in Table 7.7.

TABLE 7.7 Weight and Wing Size

w (kilograms)	0.5	2.0	3.5	5.0
A (square meters)	0.069	0.175	0.254	0.325

CALCULATOR HELP

To make a scatterplot, see the Appendix (page AP-4).

[0, 6, 1] by [0, 0.4, 0.1]

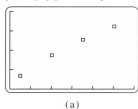

(a)

[0, 6, 1] by [0, 0.4, 0.1]

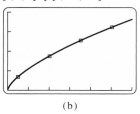

(b)

Figure 7.15

(a) Make a scatterplot of the data. Discuss any trends in the data.
(b) Biologists modeled the data with $A(w) = kw^{2/3}$, where k is a constant. Find k.
(c) Graph A and the data in the same viewing rectangle.
(d) Estimate the area of the wings of a 3-kilogram bird.

Solution
(a) A scatterplot of the data is shown in Figure 7.15(a). As the weight of a bird increases so does the surface area of its wings.
(b) To determine k, substitute one of the data points into $A(w)$. We use $(2.0, 0.175)$.

$$A(w) = kw^{2/3} \qquad \text{Given formula}$$
$$0.175 = k(2)^{2/3} \qquad \text{Let } w = 2 \text{ and } A(w) = 0.175$$
$$\qquad\qquad\qquad\quad \text{(any data point could be used).}$$
$$k = \frac{0.175}{2^{2/3}} \qquad \text{Solve for } k; \text{ rewrite equation.}$$
$$k \approx 0.11 \qquad \text{Approximate } k.$$

Thus $A(w) = 0.11w^{2/3}$.

(c) The data and graph of $Y_1 = 0.11X^\wedge(2/3)$ are shown in Figure 7.15(b). Note that the graph appears to pass through each data point.
(d) $A(3) = 0.11(3)^{2/3} \approx 0.23$ square meter

7.4 PUTTING IT ALL TOGETHER

In this section we discussed some new functions involving radicals and rational exponents. Properties of these functions are summarized in the following table.

Function	Explanation
Square Root and Cube Root	$f(x) = \sqrt{x}$ and $g(x) = \sqrt[3]{x}$ The cube root function g is defined for all inputs, whereas the square root function f is defined only for nonnegative inputs. Square Root Function Cube Root Function

continued on next page

continued from previous page

Function	Explanation
Power and Root	$f(x) = x^p$, where p is a rational number, is a power function. If $p = \frac{1}{n}$, where $n \geq 2$ is an integer, f is also a root function, given by $f(x) = \sqrt[n]{x}$. **Examples:** $f(x) = x^{5/3}$ Power function $g(x) = x^{1/4} = \sqrt[4]{x}$ Root and power function

7.5 EQUATIONS INVOLVING RADICAL EXPRESSIONS

Solving Radical Equations ▪ The Distance Formula ▪ Solving the Equation $x^n = k$

A LOOK INTO MATH ▷ In Section 7.4 we showed that for some types of birds there is a relationship between weight and the size of the wings—heavier birds tend to have larger wings. This relationship can sometimes be modeled by $A = 100\sqrt[3]{W^2}$, where W is weight in pounds and A is area in square inches. Suppose that we want to estimate the weight of a bird whose wings have an area of 600 square inches. To do so, we would need to solve the equation

$$600 = 100\sqrt[3]{W^2}$$

for W. This equation contains a radical expression. In this section we explain how to solve this type of equation. (*Source:* C. Pennycuick, *Newton Rules Biology.*)

Solving Radical Equations

Many times, equations contain either radical expressions or rational exponents. Examples include

$$\sqrt{x} = 6, \quad 5x^{1/2} = 1, \quad \text{and} \quad \sqrt[3]{x} - 1 = 3.$$

One strategy for solving an equation containing a square root is to isolate the square root (if necessary) and then square each side of the equation. This technique is an example of the *power rule for solving equations.*

> **POWER RULE FOR SOLVING EQUATIONS**
>
> If each side of an equation is raised to the same positive integer power, then any solutions to the given equation are among the solutions to the new equation. That is, the solutions to the equation $a = b$ are among the solutions to $a^n = b^n$.

We must check our solutions when applying the power rule. For example, consider the equation $2x = 6$. If we square each side, we obtain $4x^2 = 36$. Solving this second equation gives $x^2 = 9$, or $x = \pm 3$. Here, 3 is a solution to both equations, but -3 is an **extraneous solution** that satisfies the second equation but not the given equation. That is, $4(-3)^2 = 36$ is a *true* statement, but $2(-3) = 6$ is a *false* statement. Thus -3 is *not* a solution to the given equation.

We illustrate the power rule in the next example.

EXAMPLE 1 Solving a radical equation symbolically

Solve $\sqrt{2x-1} = 3$. Check your solution.

Solution
Begin by squaring each side of the equation.

$$\sqrt{2x-1} = 3 \qquad \text{Given equation}$$
$$(\sqrt{2x-1})^2 = 3^2 \qquad \text{Square each side.}$$
$$2x - 1 = 9 \qquad \text{Simplify.}$$
$$2x = 10 \qquad \text{Add 1.}$$
$$x = 5 \qquad \text{Divide by 2.}$$

To check this answer we substitute $x = 5$ in the given equation.

$$\sqrt{2(5) - 1} \stackrel{?}{=} 3$$
$$3 = 3 \qquad \text{It checks.}$$

NOTE: To simplify $(\sqrt{2x-1})^2$ in Example 1, we used the fact that

$$(\sqrt{a})^2 = \sqrt{a} \cdot \sqrt{a} = a.$$

The following steps can be used to solve a radical equation.

SOLVING A RADICAL EQUATION

STEP 1: Isolate a radical term on one side of the equation.

STEP 2: Apply the power rule by raising each side of the equation to the power equal to the index of the isolated radical term.

STEP 3: Solve the equation. If it still contains a radical, repeat Steps 1 and 2.

STEP 4: Check your answers by substituting each result in the *given* equation.

NOTE: In Sections 7.1–7.3, we *simplified expressions*. In this section we *solve equations*. Equations contain equals signs and when we solve an equation, we try to find values of the variable that make the equation a true statement.

In the next example, we apply these steps to a radical equation.

EXAMPLE 2 Isolating the radical term

Solve $\sqrt{4-x} + 5 = 8$.

Solution

STEP 1: To isolate the radical term, we subtract 5 from each side of the equation.

$$\sqrt{4-x} + 5 = 8 \qquad \text{Given equation}$$
$$\sqrt{4-x} = 3 \qquad \text{Subtract 5.}$$

STEP 2: The isolated term involves a square root, so we must square each side.

$$(\sqrt{4-x})^2 = (3)^2 \qquad \text{Square each side.}$$

STEP 3: Next we solve the resulting equation. (It is not necessary to repeat Steps 1 and 2 because the resulting equation does not contain any radical expressions.)

$$4 - x = 9 \qquad \text{Simplify.}$$
$$-x = 5 \qquad \text{Subtract 4.}$$
$$x = -5 \qquad \text{Multiply by } -1.$$

STEP 4: To check this answer we substitute $x = -5$ in the given equation.

$$\sqrt{4 - (-5)} + 5 \stackrel{?}{=} 8$$
$$\sqrt{9} + 5 \stackrel{?}{=} 8$$
$$8 = 8 \qquad \text{It checks.}$$

Example 3 shows that we must check our answers when squaring each side of an equation to identify extraneous solutions.

EXAMPLE 3 Solving a radical equation

Solve $\sqrt{3x + 3} = 2x - 1$. Check your results and then solve the equation graphically.

Solution

Symbolic Solution Begin by squaring each side of the equation.

$$\sqrt{3x + 3} = 2x - 1 \qquad \text{Given equation}$$
$$(\sqrt{3x + 3})^2 = (2x - 1)^2 \qquad \text{Square each side.}$$
$$3x + 3 = 4x^2 - 4x + 1 \qquad \text{Multiply.}$$
$$0 = 4x^2 - 7x - 2 \qquad \text{Subtract } 3x + 3.$$
$$0 = (4x + 1)(x - 2) \qquad \text{Factor.}$$
$$x = -\frac{1}{4} \quad \text{or} \quad x = 2 \qquad \text{Solve for } x.$$

To check these values substitute $x = -\frac{1}{4}$ and $x = 2$ in the given equation.

$$\sqrt{3\left(-\frac{1}{4}\right) + 3} \stackrel{?}{=} 2\left(-\frac{1}{4}\right) - 1$$
$$\sqrt{2.25} \stackrel{?}{=} -1.5$$
$$1.5 \neq -1.5 \qquad \text{It does not check.}$$

Thus $-\frac{1}{4}$ is an *extraneous solution*. Next substitute $x = 2$ in the given equation.

$$\sqrt{3 \cdot 2 + 3} \stackrel{?}{=} 2 \cdot 2 - 1$$
$$\sqrt{9} \stackrel{?}{=} 3$$
$$3 = 3 \qquad \text{It checks.}$$

The only solution is 2.

Graphical Solution The solution 2 is supported graphically in Figure 7.16, where the graphs of $Y_1 = \sqrt{(3X + 3)}$ and $Y_2 = 2X - 1$ intersect at the point $(2, 3)$. *Note that the graphical solution does not give an extraneous solution.*

Example 3 demonstrates that *checking a solution is essential when you are squaring each side of an equation.* Squaring may introduce extraneous solutions, which are solutions to the resulting equation but are not solutions to the given equation.

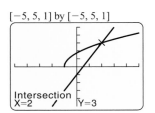

Figure 7.16

CALCULATOR HELP

To find a point of intersection, see the Appendix (page AP-7).

CRITICAL THINKING

Will a numerical solution give extraneous solutions?

When an equation contains two or more terms with square roots, it may be necessary to square each side of the equation more than once. In these situations, isolate one of the square roots and then square each side of the equation. If a radical term remains after simplifying, repeat these steps. We apply this technique in the next example.

EXAMPLE 4 Squaring twice

Solve $\sqrt{2x} - 1 = \sqrt{x + 1}$.

Solution
Begin by squaring each side of the equation.

$$\sqrt{2x} - 1 = \sqrt{x + 1} \quad \text{Given equation}$$
$$(\sqrt{2x} - 1)^2 = (\sqrt{x + 1})^2 \quad \text{Square each side.}$$
$$(\sqrt{2x})^2 - 2(\sqrt{2x})(1) + 1^2 = x + 1 \quad (a - b)^2 = a^2 - 2ab + b^2$$
$$2x - 2\sqrt{2x} + 1 = x + 1 \quad \text{Simplify.}$$
$$2x - 2\sqrt{2x} = x \quad \text{Subtract 1.}$$
$$x = 2\sqrt{2x} \quad \text{Subtract } x \text{ and add } 2\sqrt{2x}.$$
$$x^2 = 4(2x) \quad \text{Square each side again.}$$
$$x^2 - 8x = 0 \quad \text{Subtract } 8x.$$
$$x(x - 8) = 0 \quad \text{Factor.}$$
$$x = 0 \quad \text{or} \quad x = 8 \quad \text{Solve.}$$

To check these answers substitute $x = 0$ and $x = 8$ in the given equation.

$$\sqrt{2 \cdot 0} - 1 \stackrel{?}{=} \sqrt{0 + 1} \qquad \sqrt{2 \cdot 8} - 1 \stackrel{?}{=} \sqrt{8 + 1}$$
$$-1 \neq 1 \quad \text{It does not check.} \qquad 3 = 3 \quad \text{It checks.}$$

The only solution is 8.

In the next example we apply the power rule to an equation that contains a cube root.

EXAMPLE 5 Solving an equation containing a cube root

Solve $\sqrt[3]{4x - 7} = 4$.

Solution
STEP 1: The cube root term is already isolated, so we proceed to Step 2.
STEP 2: Because the index is 3, we cube each side of the equation.

$$\sqrt[3]{4x - 7} = 4 \quad \text{Given equation}$$
$$(\sqrt[3]{4x - 7})^3 = (4)^3 \quad \text{Cube each side.}$$

STEP 3: We solve the resulting equation.

$$4x - 7 = 64 \quad \text{Simplify.}$$
$$4x = 71 \quad \text{Add 7 to each side.}$$
$$x = \frac{71}{4} \quad \text{Divide each side by 4.}$$

STEP 4: To check this answer we substitute $x = \frac{71}{4}$ in the given equation.

$$\sqrt[3]{4\left(\frac{71}{4}\right) - 7} \stackrel{?}{=} 4$$
$$\sqrt[3]{64} \stackrel{?}{=} 4$$
$$4 = 4 \quad \text{It checks.}$$

▶ **REAL-WORLD CONNECTION** In the next example we solve the equation presented in the introduction to this section that relates a bird's weight and wing surface area.

EXAMPLE 6 Finding the weight of a bird

Solve the equation $600 = 100\sqrt[3]{W^2}$ to determine the weight in pounds of a bird having wings with an area of 600 square inches.

Solution
Begin by dividing each side of the equation by 100 to isolate the radical term.

$$\frac{600}{100} = \sqrt[3]{W^2} \qquad \text{Divide each side by 100.}$$
$$(6)^3 = (\sqrt[3]{W^2})^3 \qquad \text{Cube each side.}$$
$$216 = W^2 \qquad \text{Simplify.}$$
$$\sqrt{216} = W \qquad \text{Take principal square root, } W > 0.$$
$$W \approx 14.7 \qquad \text{Approximate.}$$

The weight of the bird is approximately 14.7 pounds.

TECHNOLOGY NOTE: Graphing Radical Expressions

The equation in Example 6 can be solved graphically. Sometimes it is more convenient to use rational exponents than radical notation. Thus $y = 100\sqrt[3]{W^2}$ can be entered as $Y_1 = 100X\wedge(2/3)$. (Be sure to include parentheses around the 2/3.) The accompanying figure shows y_1 intersecting the line $y_2 = 600$ near the point (14.7, 600), which supports our symbolic result.

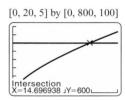

[0, 20, 5] by [0, 800, 100]

In the next example we solve an equation that would be difficult to solve symbolically, but an *approximate* solution can be found graphically.

EXAMPLE 7 Solving an equation with rational exponents

Solve $x^{2/3} = 3 - x^2$ graphically.

Solution
Graph $Y_1 = X\wedge(2/3)$ and $Y_2 = 3 - X\wedge2$. Their graphs intersect near $(-1.34, 1.21)$ and $(1.34, 1.21)$, as shown in Figure 7.17. Thus the solutions are given by $x \approx \pm 1.34$.

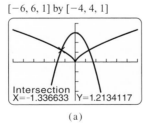
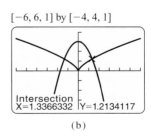

(a)   (b)

Figure 7.17

The Distance Formula

▶ **REAL-WORLD CONNECTION** One of the most famous theorems in mathematics is the **Pythagorean theorem**. It states that if a right triangle has legs a and b with hypotenuse c (see Figure 7.18), then

$$a^2 + b^2 = c^2.$$

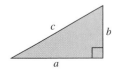

Figure 7.18
$a^2 + b^2 = c^2$

For example, if the legs of a right triangle are $a = 3$ and $b = 4$, the hypotenuse is $c = 5$ because $3^2 + 4^2 = 5^2$. Also, if the sides of a triangle satisfy $a^2 + b^2 = c^2$, it is a right triangle.

EXAMPLE 8 Applying the Pythagorean theorem

A rectangular television screen has a width of 20 inches and a height of 15 inches. Find the diagonal of the television. Why is it called a 25-inch television?

Solution
In Figure 7.19, let $a = 20$ and $b = 15$. Then the diagonal of the television corresponds to the hypotenuse of a right triangle with legs of 20 inches and 15 inches.

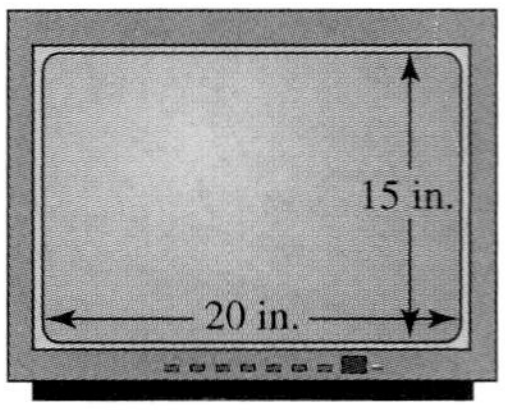

$$c^2 = a^2 + b^2 \quad \text{Pythagorean theorem}$$
$$c = \sqrt{a^2 + b^2} \quad \text{Take the principal square root, } c > 0.$$
$$c = \sqrt{20^2 + 15^2} \quad \text{Substitute } a = 20 \text{ and } b = 15.$$
$$c = 25 \quad \text{Simplify.}$$

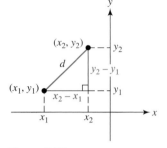

Figure 7.19

A 25-inch television has a diagonal of 25 inches.

The Pythagorean theorem can be used to determine the distance between two points. Suppose that a line segment has endpoints (x_1, y_1) and (x_2, y_2), as illustrated in Figure 7.20. The lengths of the legs of the right triangle are $x_2 - x_1$ and $y_2 - y_1$. The distance d is the hypotenuse of a right triangle. Applying the Pythagorean theorem, we have

$$d^2 = (x_2 - x_1)^2 + (y_2 - y_1)^2.$$

Distance is nonnegative, so we let d be the principal square root and obtain

$$d = \sqrt{(x_2 - x_1)^2 + (y_2 - y_1)^2}.$$

Figure 7.20

DISTANCE FORMULA

The **distance** d between the points (x_1, y_1) and (x_2, y_2) in the xy-plane is
$$d = \sqrt{(x_2 - x_1)^2 + (y_2 - y_1)^2}.$$

EXAMPLE 9 Finding distance between points

Find the distance between the points $(-2, 3)$ and $(1, -4)$.

Solution
Start by letting $(x_1, y_1) = (-2, 3)$ and $(x_2, y_2) = (1, -4)$. Then substitute these values into the distance formula.

$$d = \sqrt{(x_2 - x_1)^2 + (y_2 - y_1)^2} \quad \text{Distance formula}$$
$$= \sqrt{(1 - (-2))^2 + (-4 - 3)^2} \quad \text{Substitute.}$$
$$= \sqrt{9 + 49} \quad \text{Simplify.}$$
$$= \sqrt{58} \quad \text{Add.}$$
$$\approx 7.62 \quad \text{Approximate.}$$

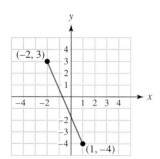

Figure 7.21

The distance between the points, as shown in Figure 7.21, is exactly $\sqrt{58}$ units, or about 7.62 units. Note that we would obtain the same result if we let $(x_1, y_1) = (1, -4)$ and $(x_2, y_2) = (-2, 3)$.

NOTE: In Example 9, $\sqrt{9 + 49} \neq \sqrt{9} + \sqrt{49} = 3 + 7 = 10$. In general, for any a and b, $\sqrt{a^2 + b^2} \neq a + b$.

▶ **REAL-WORLD CONNECTION** In the next example the distance formula is applied to road construction.

EXAMPLE 10 Designing a highway curve

Figure 7.22 shows a circular highway curve joining a straight section of road. A surveyor is trying to locate the x-coordinate of the *point of curvature PC* where the two sections of the highway meet. The distance between the surveyor and *PC* should be 400 feet. Estimate the x-coordinate of *PC* if x is positive.

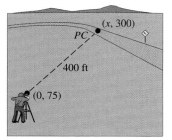

Figure 7.22

Solution

In Figure 7.22, the distance between the points $(0, 75)$ and $(x, 300)$ is 400 feet. We can apply the distance formula and solve for x.

$$d = \sqrt{(x_2 - x_1)^2 + (y_2 - y_1)^2} \qquad \text{Distance formula}$$
$$400 = \sqrt{(x - 0)^2 + (300 - 75)^2} \qquad \text{Substitute.}$$
$$400^2 = (x - 0)^2 + (300 - 75)^2 \qquad \text{Square each side.}$$
$$160{,}000 = x^2 + 50{,}625 \qquad \text{Simplify.}$$
$$x^2 = 109{,}375 \qquad \text{Solve for } x^2.$$
$$x = \sqrt{109{,}375} \qquad \text{Take the principal square root, } x > 0.$$
$$x \approx 330.7 \qquad \text{Approximate.}$$

The x-coordinate is about 330.7.

Solving the Equation $x^n = k$

The equation $x^n = k$, where n is a *positive integer*, can be solved by taking the nth root of each side of the equation. The following technique allows us to find all *real* solutions to this equation.

> **SOLVING THE EQUATION $x^n = k$**
>
> Take the nth root of each side of $x^n = k$ to obtain $\sqrt[n]{x^n} = \sqrt[n]{k}$.
>
> 1. If n is odd, then $\sqrt[n]{x^n} = x$ and the equation becomes $x = \sqrt[n]{k}$.
> 2. If n is even and $k > 0$, then $\sqrt[n]{x^n} = |x|$ and the equation becomes $|x| = \sqrt[n]{k}$.
>
> (If $k < 0$, there are no real solutions.)

To understand this technique better, consider the following examples. First let $x^3 = 8$, so that n is odd. Taking the cube root of each side gives

$$\sqrt[3]{x^3} = \sqrt[3]{8}, \quad \text{which is equivalent to} \quad x = \sqrt[3]{8}, \quad \text{or} \quad x = 2.$$

Next let $x^2 = 4$, so that n is even. Taking the square root of each side gives

$$\sqrt{x^2} = \sqrt{4}, \quad \text{which is equivalent to} \quad |x| = \sqrt{4}, \quad \text{or} \quad |x| = 2.$$

The solutions to $|x| = 2$ are -2 or 2, which can be written ± 2.

EXAMPLE 11 Solving the equation $x^n = k$

Solve each equation.
(a) $x^3 = -64$ (b) $x^2 = 12$ (c) $2(x - 1)^4 = 32$

Solution
(a) Taking the cube root of each side of $x^3 = -64$ gives
$$\sqrt[3]{x^3} = \sqrt[3]{-64} \quad \text{or} \quad x = -4.$$
(b) Taking the square root of each side of $x^2 = 12$ gives
$$\sqrt{x^2} = \sqrt{12} \quad \text{or} \quad |x| = \sqrt{12}.$$
The equation $|x| = \sqrt{12}$ is equivalent to $x = \pm\sqrt{12}$.
(c) First divide each side of $2(x - 1)^4 = 32$ by 2 to isolate the power of $x - 1$.

$(x - 1)^4 = 16$	Divide each side by 2.
$\sqrt[4]{(x - 1)^4} = \sqrt[4]{16}$	Take the 4th root of each side.
$\|x - 1\| = 2$	Simplify: $\sqrt[4]{y^4} = \|y\|$.
$x - 1 = -2$ or $x - 1 = 2$	Solve the absolute value equation.
$x = -1$ or $x = 3$	Add 1 to each side.

▶ **REAL-WORLD CONNECTION**   In some parts of the United States, wind power is used to generate electricity. Suppose that the diameter of the circular path created by the blades of a wind-powered generator is 8 feet. Then the wattage W generated by a wind velocity of v miles per hour is modeled by
$$W(v) = 2.4v^3.$$
If the wind blows at 10 miles per hour, the generator can produce about
$$W(10) = 2.4 \cdot 10^3 = 2400 \text{ watts}.$$

(*Source: Conquering the Sciences*, Sharp Electronics.)

EXAMPLE 12 Modeling a wind generator

The formula $W(v) = 2.4v^3$ is used to calculate the watts generated when there is a wind velocity of v miles per hour.
(a) Find a function f that calculates the wind velocity when W watts are being produced.
(b) If the wattage doubles, has the wind velocity also doubled? Explain.

Solution
(a) Given W we need a formula to find v, so solve $W = 2.4v^3$ for v.

$W = 2.4v^3$	Given formula
$\dfrac{W}{2.4} = v^3$	Divide by 2.4.
$\sqrt[3]{\dfrac{W}{2.4}} = \sqrt[3]{v^3}$	Take the cube root of each side.
$v = \sqrt[3]{\dfrac{W}{2.4}}$	Simplify and rewrite equation.

Thus $f(W) = \sqrt[3]{\dfrac{W}{2.4}}$.

(b) Suppose that the power generated is 1000 watts. Then the wind speed is

$$f(1000) = \sqrt[3]{\frac{1000}{2.4}} \approx 7.5 \text{ miles per hour.}$$

If the power doubles to 2000 watts, then the wind speed is

$$f(2000) = \sqrt[3]{\frac{2000}{2.4}} \approx 9.4 \text{ miles per hour.}$$

Thus for the wattage to double, the wind speed does not need to double.

7.5 PUTTING IT ALL TOGETHER

In this section we focused on equations that contain either radical expressions or rational exponents. A good strategy for solving an equation containing radical expressions *symbolically* is to raise each side of the equation to the same positive integer power. However, checking answers is important to eliminate extraneous solutions. Note that these extraneous solutions do not occur when an equation is being solved graphically or numerically.

Concept	Description	Example
Power Rule for Solving Equations	If each side of an equation is raised to the same positive integer power, any solutions to the given equation are among the solutions to the new equation.	$\sqrt{2x} = x$ $2x = x^2$ Square each side. $x^2 - 2x = 0$ Rewrite equation. $x = 0$ or $x = 2$ Factor and solve. Be sure to check any solutions.
Pythagorean Theorem	If c is the hypotenuse of a right triangle and a and b are its legs, then $a^2 + b^2 = c^2$.	If the sides of the right triangle are $a = 5$, $b = 12$, and $c = 13$, then they satisfy $a^2 + b^2 = c^2$ or $5^2 + 12^2 = 13^2$.
Distance Formula	The distance d between the points (x_1, y_1) and (x_2, y_2) is $$d = \sqrt{(x_2 - x_1)^2 + (y_2 - y_1)^2}.$$	The distance between the points $(2, 3)$ and $(-3, 4)$ is $d = \sqrt{(-3 - 2)^2 + (4 - 3)^2}$ $= \sqrt{(-5)^2 + (1)^2} = \sqrt{26}.$
Solving the Equation $x^n = k$, Where n Is a Positive Integer	Take the nth root of each side to obtain $\sqrt[n]{x^n} = \sqrt[n]{k}$. Then, **1.** $x = \sqrt[n]{k}$, if n is odd. **2.** $x = \pm\sqrt[n]{k}$, if n is even and $k \geq 0$.	**1.** n odd: If $x^5 = 32$, then $x = \sqrt[5]{32} = 2$. **2.** n even: If $x^4 = 81$, then $x = \pm\sqrt[4]{81} = \pm 3$.

continued on next page

Equations involving radical expressions can be solved symbolically, numerically, and graphically. All symbolic solutions must be checked; *numerical and graphical methods do not give extraneous solutions.* These concepts are illustrated for the equation $\sqrt{x+2} = x$, where $Y_1 = \sqrt{(X+2)}$ and $Y_2 = X$.

Symbolic Solution

$$\sqrt{x+2} = x$$
$$x + 2 = x^2$$
$$x^2 - x - 2 = 0$$
$$(x-2)(x+1) = 0$$
$$x = 2 \text{ or } x = -1$$

Check: $\sqrt{2+2} = 2$
$\sqrt{-1+2} \neq -1$

The only solution is 2.

Numerical Solution

X	Y₁	Y₂
-2	0	-2
-1	1	-1
0	1.4142	0
1	1.7321	1
2	2	2
3	2.2361	3
4	2.4495	4

X=2

$y_1 = y_2$ when $x = 2$, so 2 is a solution. Note that -1 is *not* a solution.

Graphical Solution

$[-4.7, 4.7, 1]$ by $[-3.1, 3.1, 1]$

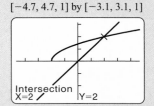

Intersection X=2, Y=2

The graphs intersect at $(2, 2)$. Note that there is no point of intersection when $x = -1$.

7.6 COMPLEX NUMBERS

Basic Concepts ■ Addition, Subtraction, and Multiplication ■ Powers of *i* ■ Complex Conjugates and Division

A LOOK INTO MATH ▷

Mathematics is both applied and theoretical. A common misconception is that abstract or theoretical mathematics is unimportant in today's world. Many new ideas with great practical importance were first developed as abstract concepts with no particular application in mind. For example, complex numbers, which are related to square roots of negative numbers, started as an abstract concept to solve equations. Today complex numbers are used in many sophisticated applications, such as the design of electrical circuits, ships, and airplanes. Even the *fractal image* shown to the left would not have been discovered without complex numbers.

Basic Concepts

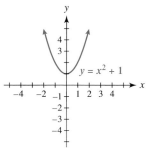

Figure 7.23

A graph of $y = x^2 + 1$ is shown in Figure 7.23. There are no *x*-intercepts, so the equation $x^2 + 1 = 0$ has no real number solutions.

If we try to solve $x^2 + 1 = 0$ by subtracting 1 from each side, the result is $x^2 = -1$. Because $x^2 \geq 0$ for any real number x, there are no real solutions. However, mathematicians have invented solutions.

$$x^2 = -1$$
$$x = \pm\sqrt{-1} \quad \text{Solve for } x.$$

We now define a number called the **imaginary unit**, denoted *i*.

PROPERTIES OF THE IMAGINARY UNIT *i*

$$i = \sqrt{-1} \quad \text{and} \quad i^2 = -1$$

By creating the number i, the solutions to the equation $x^2 + 1 = 0$ are i and $-i$. Using the real numbers and the imaginary unit i, we can define a new set of numbers called the *complex numbers*. A **complex number** can be written in **standard form**, as $a + bi$, where a and b are real numbers. The **real part** is a and the **imaginary part** is b. Every real number a is also a complex number because it can be written $a + 0i$. A complex number $a + bi$ with $b \neq 0$ is an **imaginary number**. A complex number $a + bi$ with $a = 0$ and $b \neq 0$ is sometimes called a **pure imaginary number**. Examples include $4i$ and $-2i$. Table 7.8 lists several complex numbers with their real and imaginary parts.

TABLE 7.8

Complex Number: $a + bi$	$-3 + 2i$	5	$-3i$	$-1 + 7i$	$-5 - 2i$	$4 + 6i$
Real Part: a	-3	5	0	-1	-5	4
Imaginary Part: b	2	0	-3	7	-2	6

Figure 7.24 shows how different sets of numbers are related. Note that *the set of complex numbers contains the set of real numbers.*

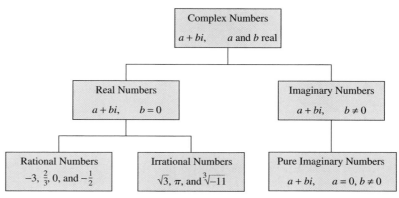

Figure 7.24

Using the imaginary unit i, we may write the square root of a negative number as a complex number. For example, $\sqrt{-2} = i\sqrt{2}$, and $\sqrt{-4} = i\sqrt{4} = 2i$. This method is summarized as follows.

CALCULATOR HELP

To set your calculator in $a + bi$ mode, see the Appendix (page AP-11).

> **THE EXPRESSION $\sqrt{-a}$**
>
> If $a > 0$, then $\sqrt{-a} = i\sqrt{a}$.

NOTE: Although it is standard for a complex number to be expressed as $a + bi$, we often write $\sqrt{2}i$ as $i\sqrt{2}$ so that it is clear that i is *not* under the square root. Similarly $\frac{1}{2}\sqrt{2}i$ is sometimes written $\frac{1}{2}i\sqrt{2}$ or $\frac{i\sqrt{2}}{2}$.

EXAMPLE 1 Writing the square root of a negative number

Write each square root using the imaginary unit i.

(a) $\sqrt{-25}$ (b) $\sqrt{-7}$ (c) $\sqrt{-20}$

Solution
(a) $\sqrt{-25} = i\sqrt{25} = 5i$ (b) $\sqrt{-7} = i\sqrt{7}$
(c) $\sqrt{-20} = i\sqrt{20} = i\sqrt{4}\sqrt{5} = 2i\sqrt{5}$

Addition, Subtraction, and Multiplication

Addition can be defined for complex numbers in a similar manner to how we add binomials. For example,
$$(-3 + 2x) + (2 - x) = (-3 + 2) + (2x - x) = -1 + x.$$

ADDITION AND SUBTRACTION To add the complex numbers $(-3 + 2i)$ and $(2 - i)$ add the real parts and then add the imaginary parts.
$$\begin{aligned}(-3 + 2i) + (2 - i) &= (-3 + 2) + (2i - i) \\ &= (-3 + 2) + (2 - 1)i \\ &= -1 + i\end{aligned}$$

This same process works for subtraction.
$$\begin{aligned}(6 - 3i) - (2 + 5i) &= (6 - 2) + (-3i - 5i) \\ &= (6 - 2) + (-3 - 5)i \\ &= 4 - 8i\end{aligned}$$

This method is summarized as follows.

> **SUM OR DIFFERENCE OF COMPLEX NUMBERS**
>
> Let $a + bi$ and $c + di$ be two complex numbers. Then
>
> $(a + bi) + (c + di) = (a + c) + (b + d)i$ Sum
>
> and $(a + bi) - (c + di) = (a - c) + (b - d)i.$ Difference

EXAMPLE 2 Adding and subtracting complex numbers

Write each sum or difference in standard form.
(a) $(-7 + 2i) + (3 - 4i)$ (b) $3i - (5 - i)$

Solution
(a) $(-7 + 2i) + (3 - 4i) = (-7 + 3) + (2 - 4)i = -4 - 2i$
(b) $3i - (5 - i) = 3i - 5 + i = -5 + (3 + 1)i = -5 + 4i$

TECHNOLOGY NOTE: Complex Numbers

Many calculators can perform arithmetic with complex numbers. The figure shows a calculator display for the results in Example 2.

```
(-7+2i)+(3-4i)
              -4-2i
3i-(5-i)
              -5+4i
```

CALCULATOR HELP
To access the imaginary unit i, see the Appendix (page AP-11).

MULTIPLICATION We multiply two complex numbers in the same way that we multiply binomials and then we apply the property $i^2 = -1$. For example,
$$(2 - 3x)(1 + 4x) = 2 + 8x - 3x - 12x^2 = 2 + 5x - 12x^2.$$

In the next example we find the product of $2 - 3i$ and $1 + 4i$ in a similar manner.

EXAMPLE 3 Multiplying complex numbers

Write each product in standard form.
(a) $(2 - 3i)(1 + 4i)$ (b) $(5 - 2i)(5 + 2i)$

Solution

(a) Multiply the complex numbers like binomials.

$$(2 - 3i)(1 + 4i) = (2)(1) + (2)(4i) - (3i)(1) - (3i)(4i)$$
$$= 2 + 8i - 3i - 12i^2$$
$$= 2 + 5i - 12(-1)$$
$$= 14 + 5i$$

(b)
$$(5 - 2i)(5 + 2i) = (5)(5) + (5)(2i) - (2i)(5) - (2i)(2i)$$
$$= 25 + 10i - 10i - 4i^2$$
$$= 25 - 4(-1)$$
$$= 29$$

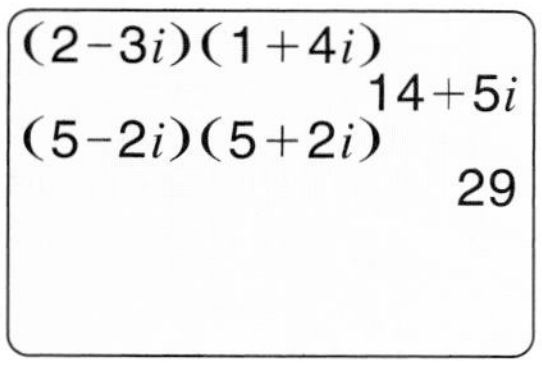

```
(2-3i)(1+4i)
              14+5i
(5-2i)(5+2i)
                 29
```

Figure 7.25

These results are supported in Figure 7.25.

Powers of i

An interesting pattern appears when powers of i are calculated.

$$i^1 = i$$
$$i^2 = -1$$
$$i^3 = i^2 \cdot i = -1 \cdot i = -i$$
$$i^4 = i^2 \cdot i^2 = (-1)(-1) = \mathbf{1}$$
$$i^5 = i^4 \cdot i = (1)i = i$$
$$i^6 = i^4 \cdot i^2 = (1)(-1) = -1$$
$$i^7 = i^4 \cdot i^3 = (1)(-i) = -i$$
$$i^8 = i^4 \cdot i^4 = (1)(1) = \mathbf{1}$$

The powers of i cycle with the pattern i, -1, $-i$, and $\mathbf{1}$. These examples suggest the following method for calculating powers of i.

> **POWERS OF i**
>
> The value of i^n can be found by dividing n (a positive integer) by 4. If the remainder is r, then
> $$i^n = i^r.$$
> Note that $i^0 = 1$, $i^1 = i$, $i^2 = -1$, and $i^3 = -i$.

EXAMPLE 4 Calculating powers of i

Evaluate each expression.
(a) i^9 **(b)** i^{19} **(c)** i^{40}

Solution
(a) When 9 is divided by 4, the result is 2 with remainder **1**. Thus $i^9 = i^1 = i$.
(b) When 19 is divided by 4, the result is 4 with remainder **3**. Thus $i^{19} = i^3 = -i$.
(c) When 40 is divided by 4, the result is 10 with remainder **0**. Thus $i^{40} = i^0 = 1$.

Complex Conjugates and Division

The **complex conjugate** of $a + bi$ is $a - bi$. To find the conjugate, we change the sign of the imaginary part b. Table 7.9 contains some complex numbers and their conjugates.

TABLE 7.9 Complex Conjugates

Number	$2 + 5i$	$6 - 3i$	$-2 + 7i$	$-1 - i$	5	$-4i$
Conjugate	$2 - 5i$	$6 + 3i$	$-2 - 7i$	$-1 + i$	5	$4i$

NOTE: The product of a complex number and its conjugate is a real number. That is,

$$(a + bi)(a - bi) = a^2 + b^2.$$

For example, $(3 + 4i)(3 - 4i) = 3^2 + 4^2 = 25$, which is a real number.

This property of complex conjugates is used to divide two complex numbers. To convert the quotient $\frac{2 + 3i}{3 - i}$ into standard form $a + bi$, we multiply the numerator and the denominator by the complex conjugate of the *denominator*, which is $3 + i$. The next example illustrates this method.

EXAMPLE 5   Dividing complex numbers

Write each quotient in standard form.

(a) $\dfrac{2 + 3i}{3 - i}$ (b) $\dfrac{4}{2i}$

Solution

(a) Multiply the numerator and denominator by $3 + i$.

$$\frac{2 + 3i}{3 - i} = \frac{(2 + 3i)(3 + i)}{(3 - i)(3 + i)} \qquad \text{Multiply by 1.}$$

$$= \frac{2(3) + (2)(i) + (3i)(3) + (3i)(i)}{(3)(3) + (3)(i) - (i)(3) - (i)(i)} \qquad \text{Multiply.}$$

$$= \frac{6 + 2i + 9i + 3i^2}{9 + 3i - 3i - i^2} \qquad \text{Simplify.}$$

$$= \frac{6 + 11i + 3(-1)}{9 - (-1)} \qquad i^2 = -1$$

$$= \frac{3 + 11i}{10} \qquad \text{Simplify.}$$

$$= \frac{3}{10} + \frac{11}{10}i \qquad \frac{a + bi}{c} = \frac{a}{c} + \frac{b}{c}i$$

(b) Multiply the numerator and denominator by $-2i$.

$$\frac{4}{2i} = \frac{(4)(-2i)}{(2i)(-2i)} \qquad \text{Multiply by 1.}$$

$$= \frac{-8i}{-4i^2} \qquad \text{Simplify.}$$

$$= \frac{-8i}{-4(-1)} \qquad i^2 = -1$$

$$= \frac{-8i}{4} \qquad \text{Simplify.}$$

$$= -2i \qquad \text{Divide.}$$

These results are supported in Figure 7.26.

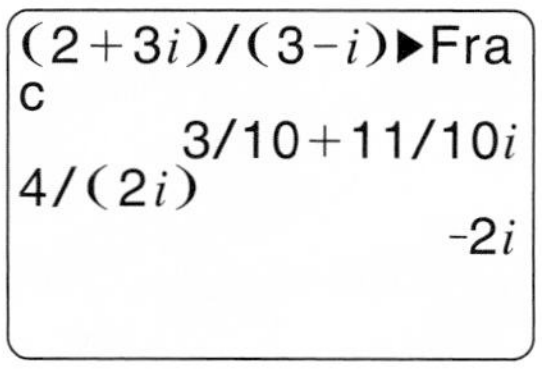

```
(2+3i)/(3-i)▶Fra
c
           3/10+11/10i
4/(2i)
                  -2i
```

Figure 7.26

7.6 PUTTING IT ALL TOGETHER

In this section we discussed complex numbers and how to perform arithmetic operations with them. Complex numbers allow us to solve equations that could not be solved only with real numbers. The following table summarizes the important concepts in the section.

Concept	Explanation	Examples
Complex Numbers	A complex number can be expressed as $a + bi$, where a and b are real numbers. The imaginary unit i satisfies $i = \sqrt{-1}$ and $i^2 = -1$. As a result, we can write $\sqrt{-a} = i\sqrt{a}$ if $a > 0$.	The real part of $5 - 3i$ is 5 and the imaginary part is -3. $$\sqrt{-13} = i\sqrt{13} \quad \text{and} \quad \sqrt{-9} = 3i$$
Addition, Subtraction, and Multiplication	To add (subtract) complex numbers, add (subtract) the real parts and then add (subtract) the imaginary parts. Multiply complex numbers in a similar manner to how *FOIL* is used to multiply binomials. Then apply the property $i^2 = -1$.	$(3 + 6i) + (-1 + 2i)$ Sum $= (3 + -1) + (6 + 2)i$ $= 2 + 8i$ $(2 - 5i) - (1 + 4i)$ Difference $= (2 - 1) + (-5 - 4)i$ $= 1 - 9i$ $(-1 + 2i)(3 + i)$ Product $= (-1)(3) + (-1)(i) + (2i)(3) + (2i)(i)$ $= -3 - i + 6i + 2i^2$ $= -3 + 5i + 2(-1)$ $= -5 + 5i$
Complex Conjugates	The conjugate of $a + bi$ is $a - bi$.	The conjugate of $3 - 5i$ is $3 + 5i$. The conjugate of $2i$ is $-2i$.
Division	To simplify a quotient, multiply the numerator and denominator by the complex conjugate of the *denominator*. Then simplify the expression and write it in standard form as $a + bi$.	$\dfrac{10}{1 + 2i} = \dfrac{10(1 - 2i)}{(1 + 2i)(1 - 2i)}$ Quotient $= \dfrac{10 - 20i}{5}$ $= 2 - 4i$

CHAPTER 7 SUMMARY

SECTION 7.1 ■ RADICAL EXPRESSIONS AND RATIONAL EXPONENTS

Radicals and Radical Notation

Square Root b is a square root of a if $b^2 = a$.

Principal Square Root $\sqrt{a} = b$ if $b^2 = a$ and $b \geq 0$.

Examples: $\sqrt{16} = 4$, $-\sqrt{9} = -3$, and $\pm\sqrt{36} = \pm 6$

Cube Root	b is a cube root of a if $b^3 = a$.
	Examples: $\sqrt[3]{27} = 3$, $\sqrt[3]{-8} = -2$
nth Root	b is an nth root of a if $b^n = a$.
	Example: $\sqrt[4]{16} = 2$ because $2^4 = 16$.
	NOTE: An *even* root of a *negative* number is not a real number. Also, $\sqrt[n]{a}$ denotes the *principal* nth root.
Absolute Value	The expressions $|x|$ and $\sqrt{x^2}$ are equivalent.
	Example: $\sqrt{(x+y)^2} = |x+y|$
The Expression $a^{1/n}$	$a^{1/n} = \sqrt[n]{a}$ if n is an integer greater than 1.
	Examples: $5^{1/2} = \sqrt{5}$ and $64^{1/3} = \sqrt[3]{64} = 4$
The Expression $a^{m/n}$	$a^{m/n} = \sqrt[n]{a^m}$ or $a^{m/n} = (\sqrt[n]{a})^m$
	Examples: $8^{2/3} = \sqrt[3]{8^2} = \sqrt[3]{64} = 4$ and
	$8^{2/3} = (\sqrt[3]{8})^2 = (2)^2 = 4$

Properties of Exponents

Product Rule	$a^p a^q = a^{p+q}$
Negative Exponents	$a^{-p} = \dfrac{1}{a^p}, \dfrac{1}{a^{-p}} = a^p$
Negative Exponents for Quotients	$\left(\dfrac{a}{b}\right)^{-p} = \left(\dfrac{b}{a}\right)^p$
Quotient Rule for Exponents	$\dfrac{a^p}{a^q} = a^{p-q}$
Power Rule for Exponents	$(a^p)^q = a^{pq}$
Power Rule for Products	$(ab)^p = a^p b^p$
Power Rule for Quotients	$\left(\dfrac{a}{b}\right)^p = \dfrac{a^p}{b^p}$

SECTION 7.2 ■ SIMPLIFYING RADICAL EXPRESSIONS

Product Rule for Radical Expressions Provided each expression is defined,
$$\sqrt[n]{a} \cdot \sqrt[n]{b} = \sqrt[n]{a \cdot b}.$$

Example: $\sqrt[3]{3} \cdot \sqrt[3]{9} = \sqrt[3]{27} = 3$

Perfect nth Power An integer a is a perfect nth power if $b^n = a$ for some integer b.

Examples: 25 is a perfect square, 8 is a perfect cube, and 16 is a perfect fourth power.

Quotient Rule for Radical Expressions Provided each expression is defined,
$$\sqrt[n]{\dfrac{a}{b}} = \dfrac{\sqrt[n]{a}}{\sqrt[n]{b}}.$$

Example: $\dfrac{\sqrt[3]{24}}{\sqrt[3]{3}} = \sqrt[3]{\dfrac{24}{3}} = \sqrt[3]{8} = 2$

SECTION 7.3 ■ OPERATIONS ON RADICAL EXPRESSIONS

Addition and Subtraction   Combine like radicals.

Examples: $2\sqrt[3]{4} + 3\sqrt[3]{4} = 5\sqrt[3]{4}$ and $\sqrt{5} - 2\sqrt{5} = -\sqrt{5}$

Multiplication   Sometimes radical expressions can be multiplied like binomials.

Examples: $(4 - \sqrt{2})(2 + \sqrt{2}) = 8 + 4\sqrt{2} - 2\sqrt{2} - 2 = 6 + 2\sqrt{2}$
$(5 - \sqrt{3})(5 + \sqrt{3}) = (5)^2 - (\sqrt{3})^2 = 25 - 3 = 22$ because
$(a - b)(a + b) = a^2 - b^2.$

Rationalizing the Denominator   One technique is to multiply the numerator and denominator by the conjugate of the denominator if the denominator is a binomial containing square roots.

Examples: $\dfrac{1}{4 + \sqrt{2}} = \dfrac{1}{(4 + \sqrt{2})} \cdot \dfrac{(4 - \sqrt{2})}{(4 - \sqrt{2})} = \dfrac{4 - \sqrt{2}}{(4)^2 - (\sqrt{2})^2} = \dfrac{4 - \sqrt{2}}{14}$

$\dfrac{4}{\sqrt{7}} = \dfrac{4}{\sqrt{7}} \cdot \dfrac{\sqrt{7}}{\sqrt{7}} = \dfrac{4\sqrt{7}}{7}$

SECTION 7.4 ■ RADICAL FUNCTIONS

The Square Root Function   The square root function is denoted $f(x) = \sqrt{x}$. Its domain is $\{x \mid x \geq 0\}$ and its graph is shown in the figure.

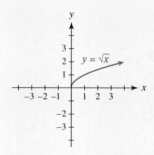

The Cube Root Function

The cube root function is denoted $f(x) = \sqrt[3]{x}$. Its domain is all real numbers and its graph is shown in the figure.

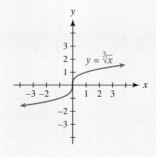

Power Functions If a function can be defined by $f(x) = x^p$, where p is a rational number, then it is a power function.

Examples: $f(x) = x^{4/5}$ and $g(x) = x^{2.3}$

SECTION 7.5 ■ EQUATIONS INVOLVING RADICAL EXPRESSIONS

Power Rule for Solving Radical Equations The solutions to $a = b$ are among the solutions to $a^n = b^n$, where n is a positive integer.

Example: The solutions to the equation $\sqrt{3x + 3} = 2x - 1$ are among the solutions to $3x + 3 = (2x - 1)^2$.

Solving Radical Equations

STEP 1: Isolate a radical term on one side of the equation.

STEP 2: Apply the power rule by raising each side of the equation to the power equal to the index of the isolated radical term.

STEP 3: Solve the equation. If it still contains a radical, repeat Steps 1 and 2.

STEP 4: Check your answers by substituting each result in the *given* equation.

Example: To isolate the radical in $\sqrt{x + 1} + 4 = 6$ subtract 4 from each side to obtain $\sqrt{x + 1} = 2$. Next square each side, which gives $x + 1 = 4$ or $x = 3$. Checking verifies that 3 is a solution.

Pythagorean Theorem If a right triangle has legs a and b with hypotenuse c, then
$$a^2 + b^2 = c^2.$$

Example: If a right triangle has legs 8 and 15, then the hypotenuse equals
$$c = \sqrt{8^2 + 15^2} = \sqrt{289} = 17.$$

The Distance Formula The distance d between (x_1, y_1) and (x_2, y_2) is
$$d = \sqrt{(x_2 - x_1)^2 + (y_2 - y_1)^2}.$$

Example: The distance between $(-1, 3)$ and $(4, 5)$ is
$$d = \sqrt{(4 - (-1))^2 + (5 - 3)^2} = \sqrt{25 + 4} = \sqrt{29}.$$

Solving the Equation $x^n = k$ Let n be a positive integer. Take the nth root of each side of $x^n = k$ to obtain $\sqrt[n]{x^n} = \sqrt[n]{k}$.

1. If n is *odd*, then $\sqrt[n]{x^n} = x$ and the equation becomes $x = \sqrt[n]{k}$.
2. If n is *even* and $k \geq 0$, then $\sqrt[n]{x^n} = |x|$ and the equation becomes $|x| = \sqrt[n]{k}$.

Examples: $x^3 = -27$ implies that $x = \sqrt[3]{-27} = -3$.
$x^4 = 81$ implies that $|x| = \sqrt[4]{81} = 3$ or $x = \pm 3$.

SECTION 7.6 ■ COMPLEX NUMBERS

Complex Numbers

Imaginary Unit $\qquad\qquad\qquad i = \sqrt{-1}$ and $i^2 = -1$

Standard Form $\qquad\qquad\qquad a + bi$, where a and b are real numbers

$\qquad\qquad\qquad\qquad\qquad$ **Examples:** $4 + 3i$, $5 - 6i$, 8, and $-2i$

Real Part	The real part of $a + bi$ is a.
	Example: The real part of $3 - 2i$ is 3.
Imaginary Part	The imaginary part of $a + bi$ is b.
	Example: The imaginary part of $2 - i$ is -1.
Arithmetic Operations	Arithmetic operations are similar to arithmetic operations on binomials.
	Examples: $(2 + 2i) + (3 - i) = 5 + i$,
	$(1 - i) - (1 - 2i) = i$,
	$(1 - i)(1 + i) = 1^2 - i^2 = 1 - (-1) = 2$, and
	$\dfrac{2}{1-i} = \dfrac{2}{1-i} \cdot \dfrac{1+i}{1+i} = \dfrac{2+2i}{2} = 1 + i$
Powers of i	The value of i^n equals i^r, where r is the remainder when n is divided by 4. Note that $i^0 = 1$, $i^1 = i$, $i^2 = -1$, and $i^3 = -i$.
	Example: $i^{21} = i^1 = i$ because when 21 is divided by 4 the remainder is 1.

Assignment Name _____ Name _____ Date _____

Show all work for these items: _____

# _____	# _____
# _____	# _____
# _____	# _____
# _____	# _____

CHAPTER 7 SHOW YOUR WORK   361

Assignment Name _____ Name _____ Date _____
Show all work for these items: _____

# _____	# _____
# _____	# _____
# _____	# _____
# _____	# _____

CHAPTER 8
Quadratic Functions and Equations

- 8.1 Quadratic Functions and Their Graphs
- 8.2 Parabolas and Modeling
- 8.3 Quadratic Equations
- 8.4 The Quadratic Formula
- 8.5 Quadratic Inequalities
- 8.6 Equations in Quadratic Form

What size television should you buy? Should you buy a 32-inch screen or a 50-inch screen? According to a home entertainment article in *Money* magazine, the answer depends on how far you sit from your television. The farther you sit from your television, the larger the television that is recommended. For example, if you sit only 6 feet from the screen, then a 32-inch television would be adequate; if you sit 10 feet from the screen, then a 50-inch television is more appropriate.

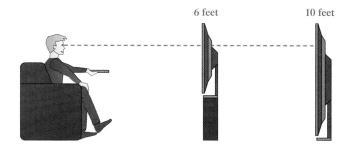

We can use a *quadratic function* to calculate the size S of the television screen needed for a person who sits x feet from the screen, where $6 \leq x \leq 15$. This function is given by

$$S(x) = -0.227x^2 + 8.155x - 8.8.$$

You might want to use S to determine the size of screen recommended when you sit 8 feet from a television. Quadratic functions are a special type of polynomial function that occur frequently in applications involving economics, road construction, falling objects, geometry, and modeling real-world data. In this chapter we discuss quadratic functions and equations.

> *There is no branch of mathematics, however abstract, which may not some day be applied to the real world.*
>
> —NIKOLAI LOBACHEVSKY

Source: Money, January 2007, p. 107; hdguru.com.

8.1 QUADRATIC FUNCTIONS AND THEIR GRAPHS

Graphs of Quadratic Functions ▪ Basic Transformations of Graphs ▪ More About Graphing Quadratic Functions (Optional) ▪ Min–Max Applications

A LOOK INTO MATH ▷

Suppose that a hotel is considering giving a group discount on room rates. The regular price is $80, but for each room rented the price decreases by $2. On the one hand, if the hotel rents one room, it makes only $78. On the other hand, if the hotel rents 40 rooms, the rooms are all free and the hotel makes nothing. Is there an optimal number of rooms between 1 and 40 that should be rented to maximize the revenue? In this section we use quadratic functions and their graphs to answer this question.

Graphs of Quadratic Functions

In Chapter 5 we discussed how a quadratic function could be represented by a polynomial of degree 2. We now give an alternative definition of a quadratic function.

> **QUADRATIC FUNCTION**
>
> A **quadratic function** can be written in the form
> $$f(x) = ax^2 + bx + c,$$
> where a, b, and c are constants with $a \neq 0$.

The graph of *any* quadratic function is a *parabola*. Recall that a parabola is a ∪-shaped graph that either opens upward or downward. The graph of the simple quadratic function $y = x^2$ is a parabola that opens upward, with its *vertex* located at the origin, as shown in Figure 8.1(a). The **vertex** is the *lowest* point on the graph of a parabola that opens upward and the *highest* point on the graph of a parabola that opens downward. A parabola opening downward is shown in Figure 8.1(b). Its vertex is the point (0, 2) and is the highest point on the graph. If we were to fold the xy-plane along the y-axis, the left and right sides of the graph would match. That is, the graph is symmetric with respect to the y-axis. In this case the y-axis is the **axis of symmetry** for the graph. Figure 8.1(c) shows a parabola that opens upward with vertex $(2, -1)$ and axis of symmetry $x = 2$.

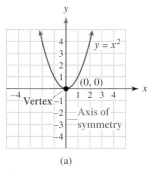
(a)

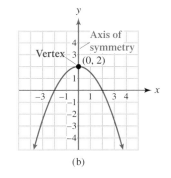
(b)

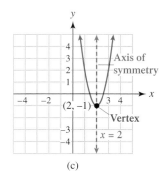
(c)

Figure 8.1

INCREASING AND DECREASING Suppose that the graph of the equation $y = x^2$ shown in Figure 8.2 represents a valley. If we walk from *left to right*, the valley "goes down" and then "goes up." Mathematically, we say that the graph of $y = x^2$ is *decreasing* when

$x \leq 0$ and *increasing* when $x \geq 0$. The vertex represents the point at which the graph switches from decreasing to increasing. In Figure 8.1(b), the graph increases when $x \leq 0$ and decreases when $x \geq 0$, and in Figure 8.1(c) the graph decreases when $x \leq 2$ and increases when $x \geq 2$.

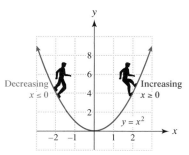

Figure 8.2

NOTE: When determining where a graph is increasing and where it is decreasing, we must "walk" along the graph *from left to right*. (We read English from left to right, which might help you remember.)

EXAMPLE 1 Graphing quadratic functions

Graph each quadratic function. Identify the vertex and the axis of symmetry. Then state where the graph is increasing and where it is decreasing.
(a) $f(x) = x^2 - 1$ **(b)** $f(x) = -(x + 1)^2$ **(c)** $f(x) = x^2 + 4x + 3$

Solution

(a) Begin by making a convenient table of values (see Table 8.1). Then plot the points and sketch a smooth ∪-shaped curve that opens upward, as shown in Figure 8.3. The lowest point on this graph is $(0, -1)$, which is the vertex. The axis of symmetry is the vertical line $x = 0$, which passes through the vertex and coincides with the y-axis. Note also the symmetry of the y-values in Table 8.1 about the point $(0, -1)$. This graph is decreasing when $x \leq 0$ and increasing when $x \geq 0$.

TABLE 8.1

	x	$y = x^2 - 1$	
	-2	3	
	-1	0	
Vertex →	0	-1	Equal
	1	0	
	2	3	

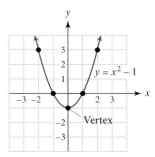

Figure 8.3

(b) Make a table of values (see Table 8.2). Plot the points and sketch a smooth ∩-shaped curve opening downward, as shown in Figure 8.4. The highest point on this graph is $(-1, 0)$, which is the vertex. The axis of symmetry is the vertical line $x = -1$, which passes through the vertex. This graph is increasing when $x \leq -1$ and decreasing when $x \geq -1$.

TABLE 8.2

x	$y = -(x+1)^2$
−3	−4
−2	−1
Vertex → −1	0
0	−1
1	−4

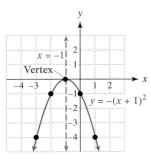

Figure 8.4

(c) Table 8.3 lists points on the graph of f, which is shown in Figure 8.5. The vertex is $(-2, -1)$ and the axis of symmetry is $x = -2$. The graph is decreasing when $x \leq -2$ and increasing when $x \geq -2$.

TABLE 8.3

x	$y = x^2 + 4x + 3$
−5	8
−4	3
−3	0
Vertex → −2	−1
−1	0
0	3
1	8

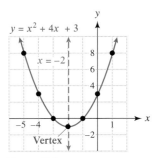

Figure 8.5

THE VERTEX FORMULA Rather than finding a vertex by graphing, we can use the following formula. This formula can be derived by completing the square, a technique discussed in the next section.

VERTEX FORMULA

The x-coordinate of the vertex of the graph of $y = ax^2 + bx + c$, $a \neq 0$, is given by

$$x = -\frac{b}{2a}.$$

To find the y-coordinate of the vertex, substitute this x-value in the equation.

NOTE: The equation of the axis of symmetry for $f(x) = ax^2 + bx + c$ is $x = -\frac{b}{2a}$, and the vertex is the point $\left(-\frac{b}{2a}, f\left(-\frac{b}{2a}\right)\right)$.

We apply the vertex formula in the next example.

EXAMPLE 2 Finding the vertex of a parabola

Find the vertex for the graph of $f(x) = 2x^2 - 4x + 1$. Support your answer graphically.

Solution
For $f(x) = 2x^2 - 4x + 1$, $a = 2$ and $b = -4$. The x-coordinate of the vertex is

$$x = -\frac{b}{2a} = -\frac{(-4)}{2(2)} = 1.$$

To find the y-coordinate of the vertex, substitute $x = 1$ in the given formula.

$$f(1) = 2(1)^2 - 4(1) + 1 = -1.$$

Thus the vertex is located at $(1, -1)$, which is supported by Figure 8.6.

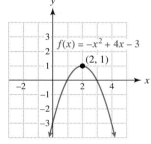
Figure 8.6

The vertex formula can help us find maximum y-values on a graph.

EXAMPLE 3 Finding a maximum y-value

Find the maximum y-value on the graph of $f(x) = -x^2 + 4x - 3$. Then state where the graph of f is increasing and where it is decreasing.

Solution
For $f(x) = -x^2 + 4x - 3$, $a = -1$ and $b = 4$. The graph of $f(x) = -x^2 + 4x - 3$ is a parabola that *opens downward* because $a < 0$. The highest point on the graph of f is the vertex and the y-coordinate of the vertex corresponds to the maximum y-coordinate. (A graph of f is shown in Figure 8.7, but it is not necessary to graph f.) The x-coordinate of the vertex is

$$x = -\frac{b}{2a} = -\frac{4}{2(-1)} = 2.$$

The corresponding y-coordinate of the vertex is $f(2) = -(2)^2 + 4(2) - 3 = 1$. The maximum y-value is 1. Because the graph opens downward, f is increasing when $x \leq 2$ and decreasing when $x \geq 2$.

Figure 8.7 Maximum y-value: 1

Basic Transformations of Graphs

THE GRAPH OF $y = ax^2$, $a > 0$ First, graph $y_1 = \frac{1}{2}x^2$, $y_2 = x^2$, and $y_3 = 2x^2$, as shown in Figure 8.8(a). Note that $a = \frac{1}{2}$, $a = 1$, and $a = 2$, respectively, and that as a increases, the resulting parabola becomes narrower. The graph of $y_1 = \frac{1}{2}x^2$ is wider than the graph of $y_2 = x^2$, and the graph of $y_3 = 2x^2$ is narrower than the graph of $y_2 = x^2$. In general, the graph of $y = ax^2$ is wider than the graph of $y = x^2$ when $0 < a < 1$ and narrower than the graph of $y = x^2$ when $a > 1$. When $a > 0$, the graph of $y = ax^2$ never lies *below* the x-axis.

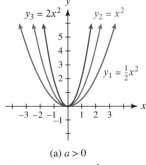
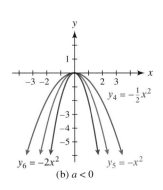

(a) $a > 0$ (b) $a < 0$

Figure 8.8 $y = ax^2$

THE GRAPH OF $y = ax^2$, $a < 0$ When $a < 0$, the graph of $y = ax^2$ never lies *above* the x-axis because, for any input x, the product $ax^2 \leq 0$. The graphs of $y_4 = -\frac{1}{2}x^2$, $y_5 = -x^2$, and $y_6 = -2x^2$ are shown in Figure 8.8(b). The graph of $y_4 = -\frac{1}{2}x^2$ is wider than the graph of $y_5 = -x^2$ and the graph of $y_6 = -2x^2$ is narrower than the graph of $y_5 = -x^2$.

> **THE GRAPH OF $y = ax^2$**
>
> The graph of $y = ax^2$ is a parabola with the following characteristics.
>
> 1. The vertex is (0, 0), and the axis of symmetry is given by $x = 0$.
> 2. It opens upward if $a > 0$ and opens downward if $a < 0$.
> 3. It is wider than the graph of $y = x^2$, if $0 < |a| < 1$. It is narrower than the graph of $y = x^2$, if $|a| > 1$.

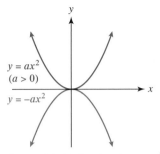

Figure 8.9 Reflection Across the x-Axis

REFLECTIONS OF $y = ax^2$ In Figure 8.8 the graph of $y_1 = \frac{1}{2}x^2$ can be *transformed* into the graph of $y_4 = -\frac{1}{2}x^2$ by *reflecting* it across the x-axis. The graph of y_4 is a **reflection** of the graph of y_1 across the x-axis. In general, *the graph of $y = -ax^2$ is a reflection of the graph of $y = ax^2$ across the x-axis*, as shown in Figure 8.9. That is, if we folded the xy-plane along the x-axis the two graphs would match.

EXAMPLE 4   Graphing $y = ax^2$

Compare the graph of $g(x) = -3x^2$ to the graph of $f(x) = x^2$. Then graph both functions on the same coordinate axes.

Solution
Both graphs are parabolas. However, the graph of g opens downward and is narrower than the graph of f. Their graphs are shown in Figure 8.10.

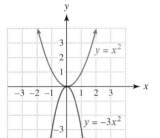

Figure 8.10

More About Graphing Quadratic Functions (Optional)

The graph of a quadratic function is a parabola, and any quadratic function f can be written as $f(x) = ax^2 + bx + c$, where a, b, and c are constants. As a result, the values of a, b, and c determine both the shape and position of the parabola in the xy-plane. In this subsection, we summarize some of the effects that these constants have on the graph of f.

THE EFFECTS OF a   The effects of a on the graph of $y = ax^2$ were discussed extensively in the previous subsection, and these effects can be generalized to include the graph of $y = ax^2 + bx + c$.

1. **Width:** The graph of $f(x) = ax^2 + bx + c$ is wider than the graph of $y = x^2$ if $0 < |a| < 1$ and narrower if $|a| > 1$. See Figures 8.11(a) and (b).
2. **Opening:** The graph of $f(x) = ax^2 + bx + c$ opens upward if $a > 0$ and downward if $a < 0$. See Figure 8.11(c).

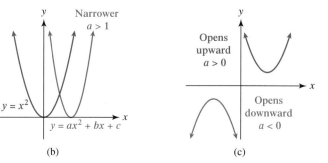

Figure 8.11 The Effects of a

8.1 QUADRATIC FUNCTIONS AND THEIR GRAPHS

THE EFFECTS OF c The value of c affects the vertical placement of the parabola in the xy-plane.

1. **y-intercept:** Because $f(0) = a(0)^2 + b(0) + c = c$, it follows that the y-intercept for the graph of f is c. See Figure 8.12(a).
2. **Vertical Shifts:** The graph of $f(x) = ax^2 + bx + c$ is shifted vertically c units compared to the graph of $y = ax^2 + bx$. If $c < 0$, the shift is downward; if $c > 0$, the shift is upward. See Figures 8.12(b) and (c). The parabolas in both figures have identical shapes.

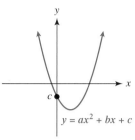
(a) y-intercept: c

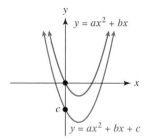
(b) Shifted downward: $c < 0$

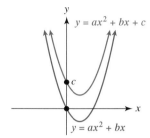
(c) Shifted upward: $c > 0$

Figure 8.12 The Effects of c

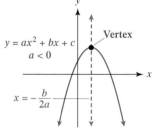

Figure 8.13 The Combined Effects of a and b

THE COMBINED EFFECTS OF a AND b The combined values of a and b determine the x-coordinate of the vertex and the equation of the axis of symmetry.

1. **Vertex:** The x-coordinate of the vertex is $-\frac{b}{2a}$.
2. **Axis of Symmetry:** The axis of symmetry is given by $x = -\frac{b}{2a}$.

Figure 8.13 illustrates these concepts.

EXAMPLE 5 Analyzing the graph of $y = ax^2 + bx + c$

Let $f(x) = -\frac{1}{2}x^2 + x + \frac{3}{2}$.

(a) Does the graph of f open upward or downward? Is this graph wider or narrower than the graph of $y = x^2$?
(b) Find the axis of symmetry and the vertex.
(c) Find the y-intercept and any x-intercepts.
(d) Sketch a graph of f.

Solution

(a) If $f(x) = -\frac{1}{2}x^2 + x + \frac{3}{2}$, then $a = -\frac{1}{2}$, $b = 1$, and $c = \frac{3}{2}$. Because $a = -\frac{1}{2} < 0$, the parabola opens downward. Also, because $0 < |a| < 1$, the graph is wider than the graph of $y = x^2$.

(b) The axis of symmetry is $x = -\frac{b}{2a} = -\frac{1}{2\left(-\frac{1}{2}\right)} = 1$, or $x = 1$. Because

$$f(1) = -\frac{1}{2}(1)^2 + (1) + \frac{3}{2} = -\frac{1}{2} + 1 + \frac{3}{2} = 2,$$

the vertex is $(1, 2)$.

(c) The y-intercept equals c, or $\frac{3}{2}$. To find x-intercepts we let $y = 0$ and solve for x.

$-\frac{1}{2}x^2 + x + \frac{3}{2} = 0$	Equation to be solved
$x^2 - 2x - 3 = 0$	Multiply by -2; clear fractions.
$(x + 1)(x - 3) = 0$	Factor.
$x + 1 = 0$ or $x - 3 = 0$	Zero-product property
$x = -1$ or $x = 3$	Solve.

The x-intercepts are -1 and 3.

(d) Start by plotting the vertex and intercepts, as shown in Figure 8.14. Then sketch a smooth, ∩-shaped graph that connects these points.

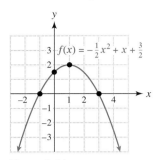

Figure 8.14

Min–Max Applications

Sometimes when a quadratic function f is used to model real-world data, the vertex provides important information. The reason is that the y-coordinate of the vertex gives either the maximum value of $f(x)$ or the minimum value of $f(x)$. For example, Figure 8.15(a) shows a parabola that opens upward. The minimum y-value on this graph is 1 and occurs at the vertex $(2, 1)$. Similarly, Figure 8.15(b) shows a parabola that opens downward. The maximum y-value on this graph is 3 and occurs at the vertex $(-1, 3)$.

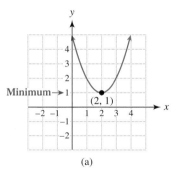

Figure 8.15

▶ **REAL-WORLD CONNECTION** In the next example, we demonstrate finding a maximum height reached by a baseball.

EXAMPLE 6 Finding maximum height

A baseball is hit into the air and its height h in feet after t seconds can be calculated by $h(t) = -16t^2 + 96t + 3$.
(a) What is the height of the baseball when it is hit?
(b) Determine the maximum height of the baseball.

Solution
(a) The baseball is hit when $t = 0$, so $h(0) = -16(0)^2 + 96(0) + 3 = 3$ feet.
(b) The graph of h opens downward because $a = -16 < 0$. Thus the maximum height of the baseball occurs at the vertex. To find the vertex, we apply the vertex formula with $a = -16$ and $b = 96$ because $h(t) = -16t^2 + 96t + 3$.

$$t = -\frac{b}{2a} = -\frac{96}{2(-16)} = 3 \text{ seconds}$$

The maximum height of the baseball occurs at $t = 3$ seconds and is

$$h(3) = -16(3)^2 + 96(3) + 3 = 147 \text{ feet.}$$

▶ **REAL-WORLD CONNECTION** In the next example, we answer the question presented in the introduction to this section.

EXAMPLE 7   Maximizing revenue

A hotel is considering giving the following group discount on room rates. The regular price for a room is $80, but for each room rented the price decreases by $2. A graph of the revenue received from renting x rooms is shown in Figure 8.16.

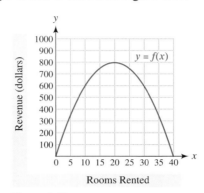

Figure 8.16

(a) Interpret the graph.
(b) What is the maximum revenue? How many rooms should be rented to receive the maximum revenue?
(c) Write a formula for $y = f(x)$ whose graph is shown in Figure 8.16.
(d) Use $f(x)$ to determine symbolically the maximum revenue and the number of rooms that should be rented.

Solution
(a) The revenue increases at first, reaches a maximum, which corresponds to the vertex, and then decreases.
(b) In Figure 8.16 the vertex is (20, 800). Thus the maximum revenue of $800 occurs when 20 rooms are rented.
(c) If x rooms are rented, the price for each room is $80 - 2x$. The revenue equals the number of rooms rented times the price of each room. Thus $f(x) = x(80 - 2x)$.
(d) First, multiply $x(80 - 2x)$ to obtain $80x - 2x^2$ and then let $f(x) = -2x^2 + 80x$. The x-coordinate of the vertex is

$$x = -\frac{b}{2a} = -\frac{80}{2(-2)} = 20.$$

The y-coordinate is $f(20) = -2(20)^2 + 80(20) = 800$. These calculations verify our results in part (b).

TECHNOLOGY NOTE: Locating a Vertex

Some graphing calculators can locate a vertex on a parabola with either the MAXIMUM or MINIMUM utility. The maximum for the graph in Example 7 is found in the accompanying figure. This utility is typically more accurate than the TRACE utility.

[0, 50, 10] by [0, 1000, 100]

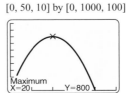

CALCULATOR HELP
To find a minimum or maximum, see the Appendix (page AP-12).

▶ **REAL-WORLD CONNECTION** In the next example, we minimize a quadratic function that models percentages of births by cesarean section (C-section).

EXAMPLE 8 Analyzing percentage of births by cesarean section

The percentage P of births performed by cesarean section between 1991 and 2001 is modeled by the formula $P(t) = 0.105t^2 - 1.08t + 23.6$, where $t = 1$ corresponds to 1991, $t = 2$ to 1992, and so on. (*Source:* The National Center for Health Statistics.)
(a) Estimate the year when the percentage of births by cesarean section was minimum.
(b) What is this minimum percentage?

Solution
(a) The graph of P is a parabola that opens upward because $a = 0.105 > 0$. Therefore the t-coordinate of the vertex represents the year when the percentage of births done by cesarean section was minimum, or

$$t = -\frac{b}{2a} = -\frac{-1.08}{2(0.105)} \approx 5.1.$$

Because $t = 5$ corresponds to 1995, the minimum percentage occurred during 1995.
(b) In 1995, this percentage was about $P(5) = 0.105(5)^2 - 1.08(5) + 23.6 \approx 20.8\%$.

8.1 PUTTING IT ALL TOGETHER

The following table summarizes some of the important topics in this section.

Concept	Explanation	Examples
Quadratic Function	Can be written as $f(x) = ax^2 + bx + c, a \neq 0$	$f(x) = x^2 + x - 2$ and $g(x) = -2x^2 + 4$ ($b = 0$)
Graph of a Quadratic Function	Its graph is a parabola that opens upward if $a > 0$ and downward if $a < 0$. The value of $|a|$ affects the width of the parabola. The vertex can be used to determine the maximum or minimum output of a quadratic function.	The graph of $y = -\frac{1}{4}x^2$ opens downward, and is wider than the graph of $y = x^2$ as shown in the figure. Each graph has its vertex at $(0, 0)$.

continued on next page

continued from previous page

Concept	Explanation	Examples
Vertex of a Parabola	The x-coordinate of the vertex for the function $f(x) = ax^2 + bx + c$ with $a \neq 0$ is given by $$x = -\frac{b}{2a}.$$ The y-coordinate of the vertex is found by substituting this x-value in the equation. Hence the vertex is $\left(-\frac{b}{2a}, f\left(-\frac{b}{2a}\right)\right)$.	If $f(x) = -2x^2 + 8x - 7$, then $$x = -\frac{8}{2(-2)} = 2$$ and $$f(2) = -2(2)^2 + 8(2) - 7 = 1.$$ The vertex is $(2, 1)$. The graph of f opens downward because $a < 0$.

Like other functions that we have studied, quadratic functions also have symbolic, numerical, and graphical representations. The following gives these representations for the quadratic function f that *squares the input x and then subtracts 1.*

Symbolic Representation

$f(x) = x^2 - 1$

Numerical Representation

x	$f(x) = x^2 - 1$
-2	3
-1	0
0	-1
1	0
2	3

Equal

Graphical Representation

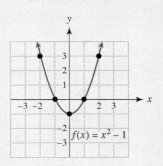

8.2 PARABOLAS AND MODELING

Vertical and Horizontal Translations ▪ Vertex Form ▪ Modeling with Quadratic Functions (Optional)

A LOOK INTO MATH ▷

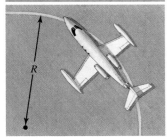

A taxiway used by an airplane to exit a runway often contains curves. A curve that is too sharp for the speed of the plane is a safety hazard. The scatterplot shown in Figure 8.17 gives an appropriate radius R of a curve designed for an airplane taxiing x miles per hour. The data are nonlinear because they do not lie on a line. In this section we explain how a quadratic function may be used to model such data. First, we discuss translations of parabolas.

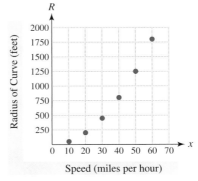

Figure 8.17

Vertical and Horizontal Translations

The graph of $y = x^2$ is a parabola opening upward with vertex $(0, 0)$. Suppose that we graph $y_1 = x^2$, $y_2 = x^2 + 1$, and $y_3 = x^2 - 2$ in the same xy-plane, as calculated for Table 8.4 and shown in Figure 8.18. All three graphs have the same shape. However, compared to the graph of $y_1 = x^2$, the graph of $y_2 = x^2 + 1$ is shifted *upward* 1 unit and the graph of $y_3 = x^2 - 2$ is shifted *downward* 2 units. Such shifts are called **translations** because they do not change the shape of a graph—only its position.

TABLE 8.4

x	$y_1 = x^2$	$y_2 = x^2 + 1$	$y_3 = x^2 - 2$
-2	4	5	2
-1	1	2	-1
0	0	1	-2
1	1	2	-1
2	4	5	2

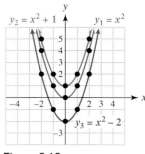
Figure 8.18

Next, suppose that we graph $y_1 = x^2$ and $y_2 = (x - 1)^2$ in the same xy-plane. Compare Tables 8.5 and 8.6. Note that the y-values are equal when the x-value for y_2 is 1 unit *larger* than the x-value for y_1. For example, $y_1 = 4$ when $x = -2$ and $y_2 = 4$ when $x = -1$. Thus the graph of $y_2 = (x - 1)^2$ has the same shape as the graph of $y_1 = x^2$, except that it is translated *horizontally to the right* 1 unit, as illustrated in Figure 8.19.

TABLE 8.5

x	$y_1 = x^2$
-2	4
-1	1
0	0
1	1
2	4

TABLE 8.6

x	$y_2 = (x - 1)^2$
-1	4
0	1
1	0
2	1
3	4

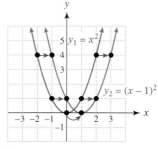
Figure 8.19

The graphs $y_1 = x^2$ and $y_2 = (x + 2)^2$ are shown in Figure 8.20. Note that Tables 8.7 and 8.8 show their y-values to be equal when the x-value for y_2 is 2 units *smaller* than the x-value for y_1. As a result, the graph of $y_2 = (x + 2)^2$ has the same shape as the graph of $y_1 = x^2$ except that it is translated *horizontally to the left* 2 units.

TABLE 8.7

x	$y_1 = x^2$
-2	4
-1	1
0	0
1	1
2	4

TABLE 8.8

x	$y_2 = (x + 2)^2$
-4	4
-3	1
-2	0
-1	1
0	4

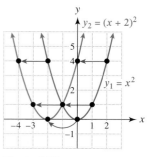
Figure 8.20

These results are summarized as follows.

VERTICAL AND HORIZONTAL TRANSLATIONS OF PARABOLAS

Let h and k be positive numbers.

To graph	shift the graph of $y = x^2$ by k units
$y = x^2 + k$	upward.
$y = x^2 - k$	downward.

To graph	shift the graph of $y = x^2$ by h units
$y = (x - h)^2$	right.
$y = (x + h)^2$	left.

The next example demonstrates this method.

EXAMPLE 1 Translating the graph $y = x^2$

Sketch the graph of the equation and identify the vertex.
(a) $y = x^2 + 2$ (b) $y = (x + 3)^2$ (c) $y = (x - 2)^2 - 3$

Solution
(a) The graph of $y = x^2 + 2$ is similar to the graph of $y = x^2$ except that it has been translated upward 2 units, as shown in Figure 8.21(a). The vertex is $(0, 2)$.
(b) The graph of $y = (x + 3)^2$ is similar to the graph of $y = x^2$ except that it has been translated *left* 3 units, as shown in Figure 8.21(b). The vertex is $(-3, 0)$.

NOTE: If you are thinking that the graph should be shifted right (instead of left) 3 units, try graphing $y = (x + 3)^2$ on a graphing calculator.

(c) The graph of $y = (x - 2)^2 - 3$ is similar to the graph of $y = x^2$ except that it has been translated downward 3 units *and* right 2 units, as shown in Figure 8.21(c). The vertex is $(2, -3)$.

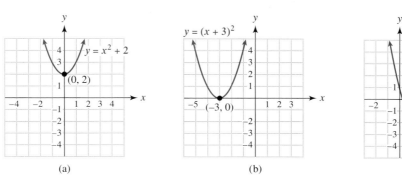

Figure 8.21

Vertex Form

Suppose that a parabola has the equation $y = ax^2 + bx + c$ with vertex (h, k). We can write this equation in a different form called the *vertex form* by transforming the graph of $y = x^2$. The vertex for $y = x^2$ is $(0, 0)$ so we need to translate it h units horizontally and k units vertically. Thus $y = (x - h)^2 + k$ has vertex (h, k). For the graph of our new equation to open correctly and have the same shape as $y = ax^2 + bx + c$, we must be sure that their leading coefficients are identical. That is, the graph of $y = a(x - h)^2 + k$ is identical to $y = ax^2 + bx + c$, provided that the vertex for the second equation is (h, k). This discussion is summarized as follows.

VERTEX FORM

The **vertex form** of the equation of a parabola with vertex (h, k) is

$$y = a(x - h)^2 + k,$$

where $a \neq 0$ is a constant. If $a > 0$, the parabola opens upward; if $a < 0$, the parabola opens downward.

NOTE: Vertex form is sometimes called **standard form for a parabola with a vertical axis**.

In the next three examples, we demonstrate the graphing of parabolas in vertex form, finding their equations, and writing vertex forms of equations.

EXAMPLE 2 Graphing parabolas in vertex form

Compare the graph of $y = f(x)$ to the graph of $y = x^2$. Then sketch a graph of $y = f(x)$ and $y = x^2$ in the same xy-plane.
(a) $f(x) = \frac{1}{2}(x - 5)^2 + 2$ (b) $f(x) = -3(x + 5)^2 - 3$

Solution
(a) Compared to the graph of $y = x^2$, the graph of $y = f(x)$ is translated 5 units right and 2 units upward. The vertex for $f(x)$ is $(5, 2)$, whereas the vertex of $y = x^2$ is $(0, 0)$. Because $a = \frac{1}{2}$, the graph of $y = f(x)$ opens upward and is wider than the graph of $y = x^2$. These graphs are shown in Figure 8.22(a).
(b) Compared to the graph of $y = x^2$, the graph of $y = f(x)$ is translated 5 units left and 3 units downward. The vertex for $f(x)$ is $(-5, -3)$. Because $a = -3$, the graph of $y = f(x)$ opens downward and is narrower than the graph of $y = x^2$. These graphs are shown in Figure 8.22(b).

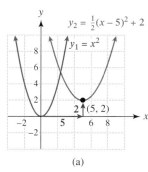
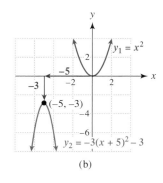

Figure 8.22

EXAMPLE 3 Finding equations of parabolas

Write the vertex form of a parabola with $a = 2$ and vertex $(-2, 1)$. Then express this equation in the form $y = ax^2 + bx + c$.

Solution
The vertex form of a parabola is $y = a(x - h)^2 + k$, where the vertex is (h, k). For $a = 2, h = -2$, and $k = 1$, the equation becomes

$$y = 2(x - (-2))^2 + 1 \quad \text{or} \quad y = 2(x + 2)^2 + 1.$$

To write this equation in the form $y = ax^2 + bx + c$, do the following.

$$\begin{aligned} y &= 2(x^2 + 4x + 4) + 1 & \text{Multiply } (x + 2)^2. \\ &= (2x^2 + 8x + 8) + 1 & \text{Distributive property} \\ &= 2x^2 + 8x + 9 & \text{Add.} \end{aligned}$$

The equivalent equation is $y = 2x^2 + 8x + 9$.

COMPLETING THE SQUARE TO FIND THE VERTEX If we are given the equation $y = x^2 + 4x + 2$, can we write it in an equivalent vertex form? The answer is *yes*, and in this process we use the **completing the square method**. In Chapter 5 we learned that $(a + b)^2 = a^2 + 2ab + b^2$. If we apply this result to $\left(x + \frac{b}{2}\right)^2$, we obtain the following.

$$\left(x + \frac{b}{2}\right)^2 = x^2 + 2x\left(\frac{b}{2}\right) + \left(\frac{b}{2}\right)^2$$

$$= x^2 + bx + \left(\frac{b}{2}\right)^2$$

CRITICAL THINKING

Use the figure to *complete the square* for $x^2 + 8x$.

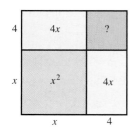

Thus we can always complete the square for $x^2 + bx$ by adding $\left(\frac{b}{2}\right)^2$. For example, if $y = x^2 + 8x + 2$, then $b = 8$ and so $\left(\frac{b}{2}\right)^2 = \left(\frac{8}{2}\right)^2 = 16$. If we simply add 16 to the right side of this equation to complete the square, the resulting equation will *not be equivalent* to the given equation. To avoid this situation, we *add and subtract* 16 on the right side of the equation. Adding 16 and then subtracting 16 is equivalent to adding 0, which does not change the equation.

$y = x^2 + 8x + 2$ Given equation

$ = (x^2 + 8x + 16) - 16 + 2$ Add and subtract 16.

$ = (x + 4)^2 - 14$ Perfect square trinomial

Thus $y = x^2 + 8x + 2$ and $y = (x + 4)^2 - 14$ are equivalent equations. The vertex for this parabola is $(-4, -14)$. Note that we added *and* subtracted 16, so the right side of the equation did not change in value.

EXAMPLE 4   Writing vertex form

Write each equation in vertex form. Identify the vertex.
(a) $y = x^2 - 6x - 1$ (b) $y = x^2 + 3x + 4$ (c) $y = 2x^2 + 4x - 1$

Solution
(a) Because $\left(\frac{b}{2}\right)^2 = \left(\frac{-6}{2}\right)^2 = 9$, add and subtract 9 on the right side.

$y = x^2 - 6x - 1$ Given equation

$ = (x^2 - 6x + 9) - 9 - 1$ Add and subtract 9.

$ = (x - 3)^2 - 10$ Perfect square trinomial

The vertex is $(3, -10)$.

(b) Because $\left(\frac{b}{2}\right)^2 = \left(\frac{3}{2}\right)^2 = \frac{9}{4}$, add and subtract $\frac{9}{4}$ on the right side.

$y = x^2 + 3x + 4$ Given equation

$ = \left(x^2 + 3x + \frac{9}{4}\right) - \frac{9}{4} + 4$ Add and subtract $\frac{9}{4}$.

$ = \left(x + \frac{3}{2}\right)^2 + \frac{7}{4}$ Perfect square trinomial

The vertex is $\left(-\frac{3}{2}, \frac{7}{4}\right)$.

(c) This equation is slightly different because the leading coefficient is 2 rather than 1. Start by factoring 2 from the first two terms on the right side.

$y = 2x^2 + 4x - 1$ Given equation

$ = 2(x^2 + 2x) - 1$ Factor out 2.

$ = 2(x^2 + 2x + 1 - 1) - 1$ $\left(\frac{b}{2}\right)^2 = \left(\frac{2}{2}\right)^2 = 1$

$ = 2(x^2 + 2x + 1) - 2 - 1$ Distributive property $(2 \cdot -1)$

$ = 2(x + 1)^2 - 3$ Perfect square trinomial; add.

The vertex is $(-1, -3)$.

Modeling with Quadratic Functions (Optional)

▶ **REAL-WORLD CONNECTION** In the introduction to this section we discussed airport taxiway curves designed for airplanes. The data previously shown in Figure 8.17 are listed in Table 8.9.

TABLE 8.9

x (mph)	10	20	30	40	50	60
R (ft)	50	200	450	800	1250	1800

Source: Federal Aviation Administration.

A second scatterplot of the data is shown in Figure 8.23. The data may be modeled by $R(x) = ax^2$ for some value a. To illustrate this relation, graph R for different values of a. In Figures 8.24–8.26, R has been graphed for $a = 2, -1$, and $\frac{1}{2}$, respectively. When $a > 0$ the parabola opens upward and when $a < 0$ the parabola opens downward. Larger values of $|a|$ make a parabola narrower, whereas smaller values of $|a|$ make the parabola wider. Through trial and error, $a = \frac{1}{2}$ gives a good fit to the data, so $R(x) = \frac{1}{2}x^2$ models the data.

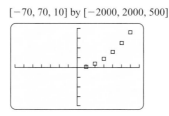

Figure 8.23

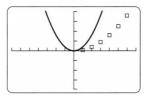

Figure 8.24 $a = 2$

CALCULATOR HELP
To make a scatterplot, see the Appendix (page AP-4).

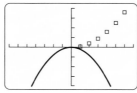

Figure 8.25 $a = -1$

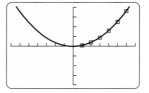

Figure 8.26 $a = \frac{1}{2}$

This value of a can also be found *symbolically*, as demonstrated in the next example.

EXAMPLE 5 Modeling safe taxiway speed

Find a value for the constant a symbolically so that $R(x) = ax^2$ models the data in Table 8.9 above. Check your result by making a table of values for $R(x)$.

Solution
When $x = 10$ miles per hour, the curve radius is $R(x) = 50$ feet. Therefore
$$R(10) = 50 \quad \text{or} \quad a(10)^2 = 50.$$
Solving for a gives
$$a = \frac{50}{10^2} = \frac{1}{2}.$$

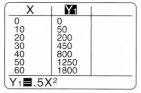

Figure 8.27

To be sure that $R(x) = \frac{1}{2}x^2$ is correct, construct a table, as shown in Figure 8.27. Its values agree with those in Table 8.9.

8.2 PARABOLAS AND MODELING

▶ **REAL-WORLD CONNECTION** In 1981, the first cases of AIDS were reported in the United States. Since then, AIDS has become one of the most devastating diseases of recent times. Table 8.10 lists the *cumulative* number of AIDS cases in the United States for various years. For example, between 1981 and 1990, a total of 199,608 AIDS cases were reported.

A scatterplot of these data is shown in Figure 8.28. To model these nonlinear data, we want to find (the right half of) a parabola with the shape illustrated in Figure 8.29. We do so in the next example.

TABLE 8.10

Year	AIDS Cases
1981	425
1984	11,106
1987	71,414
1990	199,608
1993	417,835
1996	609,933

Source: Department of Health and Human Services.

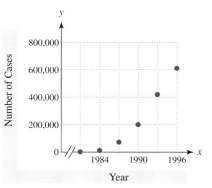

Figure 8.28 AIDS Cases

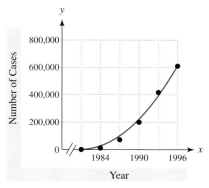

Figure 8.29 Modeling AIDS Cases

EXAMPLE 6 Modeling AIDS cases

Use the data in Table 8.10 to complete the following.
(a) Make a scatterplot of the data in [1980, 1997, 2] by [−10000, 800000, 100000].
(b) The lowest data point in Table 8.10 is (1981, 425). Let this point be the vertex of a parabola that opens upward. Graph $y = a(x - 1981)^2 + 425$ together with the data by first letting $a = 1000$.
(c) Use trial and error to adjust the value of a until the graph models the data.
(d) Use your final equation to estimate the number of AIDS cases in 1992. Compare it to the known value of 338,786.

Solution
(a) A scatterplot of the data is shown in Figure 8.30.

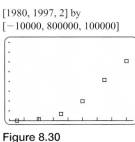

Figure 8.30

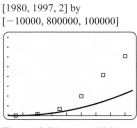

Figure 8.31 $a = 1000$

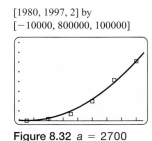

Figure 8.32 $a = 2700$

(b) A graph of $y = 1000(x - 1981)^2 + 425$ is shown in Figure 8.31. To have a better fit of the data, a larger value for a is needed.
(c) Figure 8.32 shows the effect of adjusting the value of a to 2700. This value provides a reasonably good fit. (Note that you may decide on a slightly different value for a.)
(d) If $a = 2700$, the modeling equation becomes

$$y = 2700(x - 1981)^2 + 425.$$

To estimate the number of AIDS cases in 1992, substitute $x = \mathbf{1992}$ to obtain

$$y = 2700(\mathbf{1992} - 1981)^2 + 425 = 327{,}125.$$

This number is about 12,000 less than the known value of 338,786.

8.2 PUTTING IT ALL TOGETHER

The following table summarizes some of the important topics in this section.

Concept	Explanation	Examples
Translations of Parabolas	Compared to the graph $y = x^2$, the graph of $y = x^2 + k$ is shifted vertically k units and the graph of $y = (x - h)^2$ is shifted horizontally h units.	Compared to the graph of $y = x^2$, the graph of $y = x^2 - 4$ is shifted downward 4 units. Compared to the graph of $y = x^2$, the graph of $y = (x - 4)^2$ is shifted right 4 units and the graph of $y = (x + 4)^2$ is shifted left 4 units.
Vertex Form	The vertex form of the equation of a parabola with vertex (h, k) is $$y = a(x - h)^2 + k,$$ where $a \neq 0$ is a constant. If $a > 0$, the parabola opens upward; if $a < 0$, the parabola opens downward.	The graph of $y = 3(x + 2)^2 - 7$ has a vertex of $(-2, -7)$ and opens upward because $3 > 0$.
Completing the Square Method	To complete the square to obtain the vertex form, add and subtract $\left(\frac{b}{2}\right)^2$ on the right side of $y = x^2 + bx + c$. Then factor the perfect square trinomial.	If $y = x^2 + 10x - 3$, then add and subtract $\left(\frac{b}{2}\right)^2 = \left(\frac{10}{2}\right)^2 = 25$ on the right side of this equation. $y = (x^2 + 10x + 25) - 25 - 3$ $= (x + 5)^2 - 28$ The vertex is $(-5, -28)$.

8.3 QUADRATIC EQUATIONS

Basics of Quadratic Equations ▪ The Square Root Property ▪ Completing the Square ▪ Solving an Equation for a Variable ▪ Applications of Quadratic Equations

A LOOK INTO MATH ▷ In Section 8.2 we modeled curves on airport taxiways by using $R(x) = \frac{1}{2}x^2$. In this formula x represented the airplane's speed in miles per hour, and R represented the radius of the curve in feet. This formula may be used to determine the speed limit for a curve with a radius of 650 feet by solving the *quadratic equation*

$$\frac{1}{2}x^2 = 650.$$

In this section we demonstrate techniques for solving this and other quadratic equations.

Basics of Quadratic Equations

Any quadratic function f can be represented by $f(x) = ax^2 + bx + c$ with $a \neq 0$. Examples of quadratic functions include

$$f(x) = 2x^2 - 1, \quad g(x) = -\frac{1}{3}x^2 + 2x, \quad \text{and} \quad h(x) = x^2 + 2x - 1.$$

Quadratic functions can be used to write quadratic equations. Examples of quadratic equations include

$$2x^2 - 1 = 0, \quad -\frac{1}{3}x^2 + 2x = 0, \quad \text{and} \quad x^2 + 2x - 1 = 0.$$

QUADRATIC EQUATION

A **quadratic equation** is an equation that can be written as

$$ax^2 + bx + c = 0,$$

where a, b, and c are constants with $a \neq 0$.

Solutions to the quadratic equation $ax^2 + bx + c = 0$ correspond to x-intercepts of the graph of $y = ax^2 + bx + c$. Because the graph of a quadratic function is either ∪-shaped or ∩-shaped, it can intersect the x-axis zero, one, or two times, as illustrated in Figure 8.33. Hence a quadratic equation can have zero, one, or two real solutions.

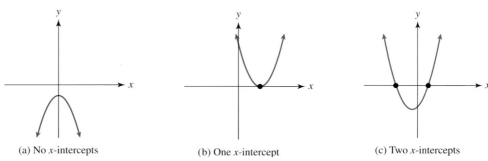

(a) No x-intercepts (b) One x-intercept (c) Two x-intercepts

Figure 8.33

We have already solved quadratic equations by factoring, graphing, and constructing tables. In the next example we apply these three techniques to quadratic equations that have no real solutions, one real solution, and two real solutions.

EXAMPLE 1 Solving quadratic equations

Solve each quadratic equation. Support your results numerically and graphically.
(a) $2x^2 + 1 = 0$ (No real solutions)
(b) $x^2 + 4 = 4x$ (One real solution)
(c) $x^2 - 6x + 8 = 0$ (Two real solutions)

Solution
(a) *Symbolic Solution*

$$\begin{aligned} 2x^2 + 1 &= 0 & &\text{Given equation} \\ 2x^2 &= -1 & &\text{Subtract 1.} \\ x^2 &= -\frac{1}{2} & &\text{Divide by 2.} \end{aligned}$$

This equation has no real-number solutions because $x^2 \geq 0$ for all real numbers x.

Numerical and Graphical Solution   The points in Table 8.11 for $y = 2x^2 + 1$ are plotted in Figure 8.34 and connected with a parabolic graph. The graph of $y = 2x^2 + 1$ has no x-intercepts, indicating that there are no real solutions.

TABLE 8.11

x	y
−2	9
−1	3
0	1
1	3
2	9

For all x, $y \neq 0$.

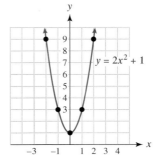

Figure 8.34 No Solutions

(b) Symbolic Solution

$$x^2 + 4 = 4x \qquad \text{Given equation}$$
$$x^2 - 4x + 4 = 0 \qquad \text{Subtract } 4x \text{ from each side.}$$
$$(x - 2)(x - 2) = 0 \qquad \text{Factor.}$$
$$x - 2 = 0 \quad \text{or} \quad x - 2 = 0 \qquad \text{Zero-product property}$$
$$x = 2 \qquad \text{There is one solution.}$$

Numerical and Graphical Solution Because the given quadratic equation is equivalent to $x^2 - 4x + 4 = 0$, we let $y = x^2 - 4x + 4$. The points in Table 8.12 are plotted in Figure 8.35 and connected with a parabolic graph. The graph of $y = x^2 - 4x + 4$ has one x-intercept, 2. Note that in Table 8.12, $y = 0$ when $x = 2$, indicating that the equation has one solution.

TABLE 8.12

x	y
0	4
1	1
2	**0**
3	1
4	4

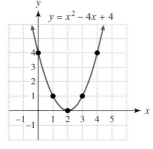

Figure 8.35 One Solution

(c) Symbolic Solution

$$x^2 - 6x + 8 = 0 \qquad \text{Given equation}$$
$$(x - 2)(x - 4) = 0 \qquad \text{Factor.}$$
$$x - 2 = 0 \quad \text{or} \quad x - 4 = 0 \qquad \text{Zero-product property}$$
$$x = 2 \quad \text{or} \quad x = 4 \qquad \text{There are two solutions.}$$

Numerical and Graphical Solution The points in Table 8.13 for $y = x^2 - 6x + 8$ are plotted in Figure 8.36 and connected with a parabolic graph. The graph of $y = x^2 - 6x + 8$ has two *x*-intercepts, 2 and 4, indicating two solutions. Note that in Table 8.13 $y = 0$ when $x = 2$ or $x = 4$.

TABLE 8.13

x	y
0	8
1	3
2	**0**
3	−1
4	**0**
5	3
6	8

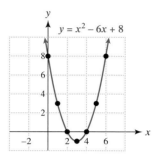

Figure 8.36 Two Solutions

The Square Root Property

The **square root property** is used to solve quadratic equations that have no *x*-terms. The following is an example of the square root property.

$$x^2 = 25 \quad \text{is equivalent to} \quad x = \pm 5.$$

The equation $x = \pm 5$ (read "*x* equals plus or minus 5") indicates that either $x = 5$ or $x = -5$. Each value is a solution because $(5)^2 = 25$ and $(-5)^2 = 25$.

We can derive this result in general for $k \geq 0$.

$x^2 = k$	Given quadratic equation
$\sqrt{x^2} = \sqrt{k}$	Take the square root of each side.
$\lvert x \rvert = \sqrt{k}$	$\sqrt{x^2} = \lvert x \rvert$ for all *x*.
$x = \pm\sqrt{k}$	$\lvert x \rvert = b$ implies $x = \pm b, b \geq 0$.

This result is summarized by the *square root property*.

> **SQUARE ROOT PROPERTY**
>
> Let *k* be a nonnegative number. Then the solutions to the equation
>
> $$x^2 = k$$
>
> are given by $x = \pm\sqrt{k}$. If $k < 0$, then this equation has no real solutions.

Before applying the square root property in the next two examples, we review a quotient property of square roots. If *a* and *b* are positive numbers, then

$$\sqrt{\frac{a}{b}} = \frac{\sqrt{a}}{\sqrt{b}}.$$

For example, $\sqrt{\frac{25}{36}} = \frac{\sqrt{25}}{\sqrt{36}} = \frac{5}{6}$.

EXAMPLE 2 Using the square root property

Solve each equation.
(a) $x^2 = 7$ (b) $16x^2 - 9 = 0$ (c) $(x - 4)^2 = 25$

Solution
(a) $x^2 = 7$ is equivalent to $x = \pm\sqrt{7}$ by the square root property. The solutions are $\sqrt{7}$ and $-\sqrt{7}$.

(b)
$$16x^2 - 9 = 0 \quad \text{Given equation}$$
$$16x^2 = 9 \quad \text{Add 9.}$$
$$x^2 = \frac{9}{16} \quad \text{Divide by 16.}$$
$$x = \pm\sqrt{\frac{9}{16}} \quad \text{Square root property}$$
$$x = \pm\frac{3}{4} \quad \text{Simplify.}$$

The solutions are $\frac{3}{4}$ and $-\frac{3}{4}$.

(c)
$$(x - 4)^2 = 25 \quad \text{Given equation}$$
$$(x - 4) = \pm\sqrt{25} \quad \text{Square root property}$$
$$x - 4 = \pm 5 \quad \text{Simplify.}$$
$$x = 4 \pm 5 \quad \text{Add 4.}$$
$$x = 9 \quad \text{or} \quad x = -1 \quad \text{Evaluate } 4 + 5 \text{ and } 4 - 5.$$

The solutions are 9 and -1.

▶ **REAL-WORLD CONNECTION** If an object is dropped from a height of h feet, its distance d above the ground after t seconds is given by
$$d(t) = h - 16t^2.$$
This formula can be used to estimate the time it takes for a falling object to hit the ground.

EXAMPLE 3 Modeling a falling object

A toy falls 30 feet from a window. How long does the toy take to hit the ground?

Solution
The height of the window above the ground is 30 feet so let $d(t) = 30 - 16t^2$. The toy strikes the ground when the distance d above the ground equals 0.

$$30 - 16t^2 = 0 \quad \text{Equation to solve for } t$$
$$-16t^2 = -30 \quad \text{Subtract 30.}$$
$$t^2 = \frac{30}{16} \quad \text{Divide by } -16.$$
$$t = \pm\sqrt{\frac{30}{16}} \quad \text{Square root property}$$
$$t = \pm\frac{\sqrt{30}}{4} \quad \text{Simplify.}$$

Time cannot be negative in this problem, so the appropriate solution is $t = \frac{\sqrt{30}}{4} \approx 1.4$. The toy hits the ground after about 1.4 seconds.

Completing the Square

In Section 8.2 we used the *method of completing the square* to find the vertex of a parabola. This method can also be used to solve quadratic equations. Because

$$x^2 + bx + \left(\frac{b}{2}\right)^2 = \left(x + \frac{b}{2}\right)^2,$$

we can solve a quadratic equation in the form $x^2 + bx = d$, where b and d are constants, by adding $\left(\frac{b}{2}\right)^2$ to each side and then factoring the resulting perfect square trinomial.

In the equation $x^2 + 6x = 7$ we have $b = 6$, so we add $\left(\frac{6}{2}\right)^2 = 9$ to each side.

$x^2 + 6x = 7$	Given equation
$x^2 + 6x + 9 = 7 + 9$	Add 9 to each side.
$(x + 3)^2 = 16$	Perfect square trinomial
$x + 3 = \pm 4$	Square root property
$x = -3 \pm 4$	Add -3 to each side.
$x = 1 \quad \text{or} \quad x = -7$	Simplify $-3 + 4$ and $-3 - 4$.

The solutions are 1 and -7. Note that the left side of the equation becomes a perfect square trinomial. We show how to create one in the next example.

CRITICAL THINKING

Use the figure to *complete the square* for $x^2 + 6x$.

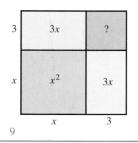

EXAMPLE 4   Creating a perfect square trinomial

Find the term that should be added to $x^2 - 10x$ to form a perfect square trinomial.

Solution

The coefficient of the x-term is -10, so we let $b = -10$. To complete the square we divide b by 2 and then square the result.

$$\left(\frac{b}{2}\right)^2 = \left(\frac{-10}{2}\right)^2 = 25$$

If we add 25 to $x^2 - 10x$, a perfect square trinomial is formed.

$$x^2 - 10x + 25 = (x - 5)^2$$

Completing the square can be used to solve quadratic equations when a trinomial does not factor easily, as illustrated in the next two examples.

EXAMPLE 5   Completing the square when the leading coefficient is 1

Solve the equation $x^2 - 4x + 2 = 0$.

Solution

Start by writing the equation in the form $x^2 + bx = d$.

$x^2 - 4x + 2 = 0$	Given equation
$x^2 - 4x = -2$	Subtract 2.
$x^2 - 4x + 4 = -2 + 4$	Add $\left(\frac{b}{2}\right)^2 = \left(\frac{-4}{2}\right)^2 = 4.$
$(x - 2)^2 = 2$	Perfect square trinomial; add.
$x - 2 = \pm \sqrt{2}$	Square root property
$x = 2 \pm \sqrt{2}$	Add 2.

The solutions are $2 + \sqrt{2} \approx 3.41$ and $2 - \sqrt{2} \approx 0.59$.

EXAMPLE 6 Completing the square when the leading coefficient is not 1

Solve the equation $2x^2 + 7x - 5 = 0$.

Solution
Start by writing the equation in the form $x^2 + bx = d$. That is, add 5 to each side and then divide the equation by 2 so that the leading coefficient of the x^2-term becomes 1.

$$2x^2 + 7x - 5 = 0 \quad \text{Given equation}$$
$$2x^2 + 7x = 5 \quad \text{Add 5.}$$
$$x^2 + \frac{7}{2}x = \frac{5}{2} \quad \text{Divide by 2.}$$
$$x^2 + \frac{7}{2}x + \frac{49}{16} = \frac{5}{2} + \frac{49}{16} \quad \text{Add } \left(\frac{b}{2}\right)^2 = \left(\frac{7}{4}\right)^2 = \frac{49}{16}.$$
$$\left(x + \frac{7}{4}\right)^2 = \frac{89}{16} \quad \text{Perfect square trinomial; add.}$$
$$x + \frac{7}{4} = \pm \frac{\sqrt{89}}{4} \quad \text{Square root property}$$
$$x = -\frac{7}{4} \pm \frac{\sqrt{89}}{4} \quad \text{Add } -\frac{7}{4}.$$
$$x = \frac{-7 \pm \sqrt{89}}{4} \quad \text{Combine fractions.}$$

The solutions are $\frac{-7 + \sqrt{89}}{4} \approx 0.61$ and $\frac{-7 - \sqrt{89}}{4} \approx -4.1$.

> **CRITICAL THINKING**
>
> What happens if you try to solve
>
> $2x^2 - 13 = 1$
>
> by completing the square? What method could you use to solve this problem?

Solving an Equation for a Variable

We often need to solve an equation or formula for a variable. For example, the formula $V = \frac{1}{3}\pi r^2 h$ calculates the volume of the cone shown in Figure 8.37. Let's say that we know the volume V is 120 cubic inches and the height h is 15 inches. We can then find the radius of the cone by solving the equation for r.

$$V = \frac{1}{3}\pi r^2 h \quad \text{Solve the equation for } r.$$
$$3V = \pi r^2 h \quad \text{Multiply by 3.}$$
$$\frac{3V}{\pi} = r^2 h \quad \text{Divide by } \pi.$$
$$\frac{3V}{\pi h} = r^2 \quad \text{Divide by } h.$$
$$r = \pm\sqrt{\frac{3V}{\pi h}} \quad \text{Square root property; rewrite.}$$

Because $r \geq 0$, we use the positive or *principal square root*. Thus for $V = 120$ cubic inches and $h = 15$ inches,

$$r = \sqrt{\frac{3(120)}{\pi(15)}} = \sqrt{\frac{24}{\pi}} \approx 2.8 \text{ inches.}$$

Figure 8.37

EXAMPLE 7　Solving equations for variables

Solve each equation for the specified variable.
(a) $s = -\frac{1}{2}gt^2 + h$, for t (b) $d^2 = x^2 + y^2$, for y

Solution
(a) Begin by subtracting h from each side of the equation.

$$s = -\tfrac{1}{2}gt^2 + h \qquad \text{Solve the equation for } t.$$
$$s - h = -\tfrac{1}{2}gt^2 \qquad \text{Subtract } h.$$
$$-2(s - h) = gt^2 \qquad \text{Multiply by } -2.$$
$$\frac{2h - 2s}{g} = t^2 \qquad \text{Divide by } g; \text{ simplify.}$$
$$t = \pm\sqrt{\frac{2h - 2s}{g}} \qquad \text{Square root property; rewrite.}$$

(b) Begin by subtracting x^2 from each side of the equation.

$$d^2 = x^2 + y^2 \qquad \text{Solve the equation for } y.$$
$$d^2 - x^2 = y^2 \qquad \text{Subtract } x^2.$$
$$y = \pm\sqrt{d^2 - x^2} \qquad \text{Square root property; rewrite.}$$

Applications of Quadratic Equations

▶ **REAL-WORLD CONNECTION**　In the introduction to this section we discussed how the solution to $\frac{1}{2}x^2 = 650$ would give a safe speed limit for a curve with a radius of 650 feet on an airport taxiway. We solve this problem in the next example.　(*Source:* FAA.)

EXAMPLE 8　Finding a safe speed limit

Solve $\frac{1}{2}x^2 = 650$ and interpret any solutions.

Solution
Use the square root property to solve this problem.

$$\tfrac{1}{2}x^2 = 650 \qquad \text{Given equation}$$
$$x^2 = 1300 \qquad \text{Multiply by 2.}$$
$$x = \pm\sqrt{1300} \qquad \text{Square root property}$$

The solutions are $\sqrt{1300} \approx 36$ and $-\sqrt{1300} \approx -36$. The solution of $x \approx 36$ indicates that a safe speed limit for a curve with a radius of 650 feet should be 36 miles per hour. (The negative solution has no physical meaning in this problem.)

▶ **REAL-WORLD CONNECTION**　In applications, solving a quadratic equation either graphically or numerically is often easier than solving it symbolically. We do so in the next example.

EXAMPLE 9   Modeling numbers of Internet users

Use of the Internet in Western Europe has increased dramatically. Figure 8.38 shows a scatterplot of online users in Western Europe, together with a graph of a function f that models the data. The function f is given by

$$f(x) = 0.976x^2 - 4.643x + 0.238,$$

where the output is in millions of users. In this formula $x = 6$ corresponds to 1996, $x = 7$ to 1997, and so on, until $x = 12$ represents 2002.   (**Source:** Nortel Networks.)

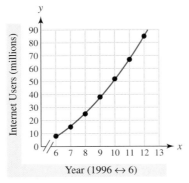

Figure 8.38 Internet Usage in Western Europe

(a) Evaluate $f(10)$ and interpret the result.
(b) Graph f and estimate the year when the number of Internet users reached 85 million. Compare your result with Figure 8.38.
(c) Solve part (b) numerically.

Solution
(a) Substituting $x = 10$ in the formula gives

$$f(10) = 0.976(10)^2 - 4.643(10) + 0.238 \approx 51.4.$$

Because $x = 10$ corresponds to 2000, there were about 51.4 million Internet users in Western Europe in 2000.
(b) Graph $Y_1 = .976X^2 - 4.643X + .238$ and $Y_2 = 85$, as shown in Figure 8.39. Their graphs intersect near $x = 12$, which corresponds to 2002 and agrees with Figure 8.38.
(c) The table for y_1, shown in Figure 8.40, reveals $y_1 \approx 85$ when $x = 12$.

CALCULATOR HELP
To find a point of intersection, see the Appendix (page AP-7).

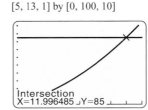

Figure 8.39   Figure 8.40

8.3 PUTTING IT ALL TOGETHER

Quadratic equations can be expressed in the form $ax^2 + bx + c = 0$, $a \neq 0$. They can have no real solutions, one real solution, or two real solutions and may be solved symbolically, graphically, and numerically. Symbolic techniques for solving quadratic equations include factoring, the square root property, and completing the square. We discussed factoring extensively in Chapter 5, so the following table summarizes only the square root property and the method of completing the square.

Technique	Description	Examples
Square Root Property	If $k \geq 0$, the solutions to the equation $x^2 = k$ are $\pm\sqrt{k}$.	$x^2 = 100$ is equivalent to $x = \pm 10$ and $x^2 = 13$ is equivalent to $x = \pm\sqrt{13}$. $x^2 = -2$ has no real solutions.
Method of Completing the Square	To solve an equation in the form $x^2 + bx = d$, add $\left(\frac{b}{2}\right)^2$ to each side of the equation. Factor the resulting perfect square trinomial and solve for x by applying the square root property.	To solve $x^2 + 8x = 3$, add $\left(\frac{8}{2}\right)^2 = 16$ to each side. $x^2 + 8x + 16 = 3 + 16$ Add 16 to each side. $(x + 4)^2 = 19$ Perfect square trinomial $x + 4 = \pm\sqrt{19}$ Square root property $x = -4 \pm \sqrt{19}$ Add -4.

8.4 THE QUADRATIC FORMULA

Solving Quadratic Equations ▪ The Discriminant ▪ Quadratic Equations Having Complex Solutions

A LOOK INTO MATH ▷ To model the stopping distance of a car, highway engineers compute two quantities. The first quantity is the *reaction distance*, which is the distance a car travels from the time a driver first recognizes a hazard until the brakes are applied. The second quantity is *braking distance*, which is the distance a car travels after a driver applies the brakes. *Stopping distance* equals the sum of the reaction distance and the braking distance. If a car is traveling x miles per hour, highway engineers estimate the reaction distance in feet as $\frac{11}{3}x$ and the braking distance in feet as $\frac{1}{9}x^2$. See Figure 8.41. (**Source:** L. Haefner, *Introduction to Transportation Systems*.)

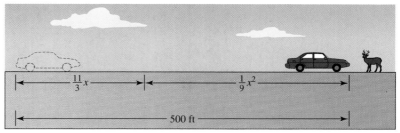

Figure 8.41 Stopping Distance

To estimate the total stopping distance d in feet, add the two expressions to obtain

$$d(x) = \frac{1}{9}x^2 + \frac{11}{3}x.$$

If a car's headlights don't illuminate the road beyond 500 feet, a safe nighttime speed limit x for the car can be determined by solving the quadratic equation

$$\frac{1}{9}x^2 + \frac{11}{3}x = 500.$$

Braking Distance + Reaction Distance = Stopping Distance

In this section we learn how to solve this equation with the quadratic formula.

Solving Quadratic Equations

Recall that any quadratic equation can be written in the form

$$ax^2 + bx + c = 0.$$

If we solve this equation for x in terms of a, b, and c by completing the square, we obtain the **quadratic formula**. We assume that $a > 0$ and derive it as follows.

$ax^2 + bx + c = 0$	Quadratic equation
$ax^2 + bx = -c$	Subtract c.
$x^2 + \dfrac{b}{a}x = -\dfrac{c}{a}$	Divide by a.
$x^2 + \dfrac{b}{a}x + \dfrac{b^2}{4a^2} = -\dfrac{c}{a} + \dfrac{b^2}{4a^2}$	Add $\left(\dfrac{b/a}{2}\right)^2 = \dfrac{b^2}{4a^2}$.
$\left(x + \dfrac{b}{2a}\right)^2 = -\dfrac{c}{a} + \dfrac{b^2}{4a^2}$	Perfect square trinomial
$\left(x + \dfrac{b}{2a}\right)^2 = -\dfrac{c \cdot 4a}{a \cdot 4a} + \dfrac{b^2}{4a^2}$	Multiply $-\dfrac{c}{a}$ by $\dfrac{4a}{4a}$.
$\left(x + \dfrac{b}{2a}\right)^2 = -\dfrac{4ac}{4a^2} + \dfrac{b^2}{4a^2}$	Simplify.
$\left(x + \dfrac{b}{2a}\right)^2 = \dfrac{-4ac + b^2}{4a^2}$	Add fractions.
$\left(x + \dfrac{b}{2a}\right)^2 = \dfrac{b^2 - 4ac}{4a^2}$	Rewrite.
$x + \dfrac{b}{2a} = \pm\sqrt{\dfrac{b^2 - 4ac}{4a^2}}$	Square root property
$x = -\dfrac{b}{2a} \pm \sqrt{\dfrac{b^2 - 4ac}{4a^2}}$	Add $-\dfrac{b}{2a}$.
$x = -\dfrac{b}{2a} \pm \dfrac{\sqrt{b^2 - 4ac}}{2a}$	Property of square roots
$x = \dfrac{-b \pm \sqrt{b^2 - 4ac}}{2a}$	Combine fractions.

8.4 THE QUADRATIC FORMULA

QUADRATIC FORMULA

The solutions to $ax^2 + bx + c = 0$ with $a \neq 0$ are given by

$$x = \frac{-b \pm \sqrt{b^2 - 4ac}}{2a}.$$

NOTE: The quadratic formula can be used to solve *any* quadratic equation. It always "works."

MAKING CONNECTIONS

Completing the Square and the Quadratic Formula

The quadratic formula results from completing the square for the equation $ax^2 + bx + c = 0$. When you use the quadratic formula, the work of completing the square has already been done for you. However, you can always complete the square rather than use the quadratic formula. See Example 10.

The next three examples show symbolic and graphical solutions to quadratic equations.

EXAMPLE 1 Solving a quadratic equation having two solutions

Solve the equation $2x^2 - 3x - 1 = 0$. Support your results graphically.

Solution
Symbolic Solution Let $a = 2$, $b = -3$, and $c = -1$.

$x = \dfrac{-b \pm \sqrt{b^2 - 4ac}}{2a}$ Quadratic formula

$x = \dfrac{-(-3) \pm \sqrt{(-3)^2 - 4(2)(-1)}}{2(2)}$ Substitute for a, b, and c.

$x = \dfrac{3 \pm \sqrt{17}}{4}$ Simplify.

The solutions are $\dfrac{3 + \sqrt{17}}{4} \approx 1.78$ and $\dfrac{3 - \sqrt{17}}{4} \approx -0.28$.

Graphical Solution The graph of $y = 2x^2 - 3x - 1$ is shown in Figure 8.42. Note that the two x-intercepts correspond to the two solutions to $2x^2 - 3x - 1 = 0$. Estimating from this graph, we see that the solutions are approximately -0.25 and 1.75, which supports our symbolic solution. (You could also use a graphing calculator to find the x-intercepts.)

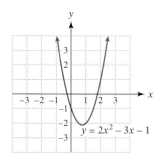

Figure 8.42 Two x-intercepts

CRITICAL THINKING
Use the results of Example 1 to evaluate each expression mentally.

$$2\left(\frac{3 + \sqrt{17}}{4}\right)^2 - 3\left(\frac{3 + \sqrt{17}}{4}\right) - 1 \quad \text{and} \quad 2\left(\frac{3 - \sqrt{17}}{4}\right)^2 - 3\left(\frac{3 - \sqrt{17}}{4}\right) - 1$$

EXAMPLE 2 Solving a quadratic equation having one solution

Solve the equation $25x^2 + 20x + 4 = 0$. Support your result graphically.

Solution
Symbolic Solution Let $a = 25$, $b = 20$, and $c = 4$.

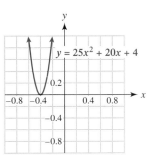

Figure 8.43 One x-intercept

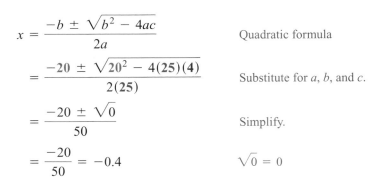

$$x = \frac{-b \pm \sqrt{b^2 - 4ac}}{2a} \quad \text{Quadratic formula}$$

$$= \frac{-20 \pm \sqrt{20^2 - 4(25)(4)}}{2(25)} \quad \text{Substitute for } a, b, \text{ and } c.$$

$$= \frac{-20 \pm \sqrt{0}}{50} \quad \text{Simplify.}$$

$$= \frac{-20}{50} = -0.4 \quad \sqrt{0} = 0$$

There is one solution, -0.4.

Graphical Solution The graph of $y = 25x^2 + 20x + 4$ is shown in Figure 8.43. Note that the one x-intercept, -0.4, corresponds to the solution to $25x^2 + 20x + 4 = 0$.

EXAMPLE 3 Recognizing a quadratic equation having no real solutions

Solve the equation $5x^2 - x + 3 = 0$. Support your result graphically.

Solution
Symbolic Solution Let $a = 5$, $b = -1$, and $c = 3$.

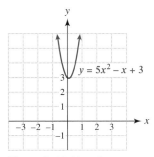

Figure 8.44 No x-intercepts

$$x = \frac{-b \pm \sqrt{b^2 - 4ac}}{2a} \quad \text{Quadratic formula}$$

$$= \frac{-(-1) \pm \sqrt{(-1)^2 - 4(5)(3)}}{2(5)} \quad \text{Substitute for } a, b, \text{ and } c.$$

$$= \frac{1 \pm \sqrt{-59}}{10} \quad \text{Simplify.}$$

There are no real solutions to this equation because $\sqrt{-59}$ is not a real number. (Later in this section we discuss how to find complex solutions to quadratic equations like this one.)

Graphical Solution The graph of $y = 5x^2 - x + 3$ is shown in Figure 8.44. There are no x-intercepts, indicating that the equation $5x^2 - x + 3 = 0$ has no real solutions.

▶ **REAL-WORLD CONNECTION** Earlier in this section we discussed how engineers estimate safe stopping distances for automobiles. In the next example we solve the equation presented in the introduction.

EXAMPLE 4 Modeling stopping distance

If a car's headlights do not illuminate the road beyond 500 feet, estimate a safe nighttime speed limit x for the car by solving $\frac{1}{9}x^2 + \frac{11}{3}x = 500$.

Solution
Begin by subtracting 500 from each side of the given equation.

$$\frac{1}{9}x^2 + \frac{11}{3}x - 500 = 0. \qquad \text{Subtract 500.}$$

To eliminate fractions multiply each side by the LCD, which is 9. (This step is not necessary, but it makes the problem easier to work.)

$$x^2 + 33x - 4500 = 0. \qquad \text{Multiply by 9.}$$

Now let $a = 1$, $b = 33$, and $c = -4500$ in the quadratic formula.

$$\begin{aligned} x &= \frac{-b \pm \sqrt{b^2 - 4ac}}{2a} & \text{Quadratic formula} \\ &= \frac{-33 \pm \sqrt{33^2 - 4(1)(-4500)}}{2(1)} & \text{Substitute for } a, b, \text{ and } c. \\ &= \frac{-33 \pm \sqrt{19{,}089}}{2} & \text{Simplify.} \end{aligned}$$

The solutions are

$$\frac{-33 + \sqrt{19{,}089}}{2} \approx 52.6 \quad \text{and} \quad \frac{-33 - \sqrt{19{,}089}}{2} \approx -85.6.$$

The negative solution has no physical meaning because negative speeds are not possible. The other solution is 52.6, so an appropriate speed limit might be 50 miles per hour.

The Discriminant

The expression $b^2 - 4ac$ in the quadratic formula is called the **discriminant**. It provides information about the number of solutions to a quadratic equation.

> **THE DISCRIMINANT AND QUADRATIC EQUATIONS**
>
> To determine the number of solutions to the quadratic equation $ax^2 + bx + c = 0$, evaluate the discriminant $b^2 - 4ac$.
>
> 1. If $b^2 - 4ac > 0$, there are two real solutions.
> 2. If $b^2 - 4ac = 0$, there is one real solution.
> 3. If $b^2 - 4ac < 0$, there are no real solutions; there are two complex solutions.

EXAMPLE 5 Using the discriminant

Use the discriminant to determine the number of solutions to $4x^2 + 25 = 20x$. Then solve the equation, using the quadratic formula.

Solution
Write the equation as $4x^2 - 20x + 25 = 0$ so that $a = 4$, $b = -20$, and $c = 25$. The discriminant evaluates to
$$b^2 - 4ac = (-20)^2 - 4(4)(25) = 0.$$
Thus there is one real solution.

$$\begin{aligned} x &= \frac{-b \pm \sqrt{b^2 - 4ac}}{2a} &&\text{Quadratic formula} \\ &= \frac{-(-20) \pm \sqrt{0}}{2(4)} &&\text{Substitute.} \\ &= \frac{20}{8} = 2.5 &&\text{Simplify.} \end{aligned}$$

The only solution is 2.5.

We also need to be able to analyze graphs of quadratic functions, which we demonstrate in the next example.

EXAMPLE 6 Analyzing graphs of quadratic functions

A graph of $f(x) = ax^2 + bx + c$ is shown in Figure 8.45.
(a) State whether $a > 0$ or $a < 0$.
(b) Solve the equation $ax^2 + bx + c = 0$.
(c) Determine whether the discriminant is positive, negative, or zero.

Solution
(a) The parabola opens downward, so $a < 0$.
(b) The graph of $f(x) = ax^2 + bx + c$ intersects the x-axis at -3 and 2. Therefore $f(-3) = 0$ and $f(2) = 0$. The solutions to $ax^2 + bx + c = 0$ are -3 and 2.
(c) There are two solutions, so the discriminant is positive.

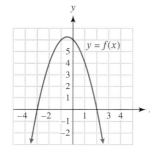
Figure 8.45

Quadratic Equations Having Complex Solutions

The quadratic equation written as $ax^2 + bx + c = 0$ has no real solutions if the discriminant, $b^2 - 4ac$, is negative. For example, the quadratic equation $x^2 + 4 = 0$ has $a = 1$, $b = 0$, and $c = 4$. Its discriminant is
$$b^2 - 4ac = 0^2 - 4(1)(4) = -16 < 0,$$
so this equation has no real solutions. However, if we use complex numbers, we can solve this equation as follows.

$$\begin{aligned} x^2 + 4 &= 0 &&\text{Given equation} \\ x^2 &= -4 &&\text{Subtract 4.} \\ x &= \pm\sqrt{-4} &&\text{Square root property} \\ x = \sqrt{-4} \quad &\text{or} \quad x = -\sqrt{-4} &&\text{Meaning of } \pm \\ x = 2i \quad &\text{or} \quad x = -2i &&\text{The expression } \sqrt{-a} \end{aligned}$$

The solutions are $\pm 2i$. We check each solution to $x^2 + 4 = 0$ as follows.

$$\begin{aligned} (2i)^2 + 4 &= (2)^2 i^2 + 4 = 4(-1) + 4 = 0 &&\text{It checks.} \\ (-2i)^2 + 4 &= (-2)^2 i^2 + 4 = 4(-1) + 4 = 0 &&\text{It checks.} \end{aligned}$$

The fact that the equation $x^2 + 4 = 0$ has only imaginary solutions can be seen visually from the graph of $y = x^2 + 4$, shown in Figure 8.46. This parabola does not intersect the x-axis, and so the equation $x^2 + 4 = 0$ has no real solutions.

These results can be generalized as follows.

THE EQUATION $x^2 + k = 0$

If $k > 0$, the solutions to $x^2 + k = 0$ are given by $x = \pm i\sqrt{k}$.

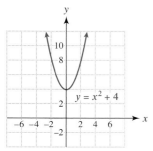

Figure 8.46 No x-intercepts

NOTE: This result is a form of the *square root property* that includes complex solutions.

EXAMPLE 7 Solving a quadratic equation having complex solutions

Solve $x^2 + 5 = 0$.

Solution
The solutions are $\pm i\sqrt{5}$. That is, $x = i\sqrt{5}$ or $x = -i\sqrt{5}$.

When $b \neq 0$, the preceding method cannot be used. Consider the quadratic equation $2x^2 + x + 3 = 0$, which has $a = 2$, $b = 1$, and $c = 3$. Its discriminant is

$$b^2 - 4ac = 1^2 - 4(2)(3) = -23 < 0.$$

This equation has two complex solutions as demonstrated in the next example.

EXAMPLE 8 Solving a quadratic equation having complex solutions

Solve $2x^2 + x + 3 = 0$. Write your answer in standard form: $a + bi$.

CALCULATOR HELP
To set your calculator in $a + bi$ mode or to access the imaginary unit i, see the Appendix (page AP-11).

Solution
Let $a = 2$, $b = 1$, and $c = 3$.

$$x = \frac{-b \pm \sqrt{b^2 - 4ac}}{2a} \quad \text{Quadratic formula}$$

$$= \frac{-1 \pm \sqrt{1^2 - 4(2)(3)}}{2(2)} \quad \text{Substitute for } a, b, \text{ and } c.$$

$$= \frac{-1 \pm \sqrt{-23}}{4} \quad \text{Simplify.}$$

$$= \frac{-1 \pm i\sqrt{23}}{4} \quad \sqrt{-23} = i\sqrt{23}$$

$$= -\frac{1}{4} \pm i\frac{\sqrt{23}}{4} \quad \text{Divide each term by 4.}$$

The solutions are $-\frac{1}{4} + i\frac{\sqrt{23}}{4}$ and $-\frac{1}{4} - i\frac{\sqrt{23}}{4}$.

CRITICAL THINKING

Use the results of Example 8 to evaluate each expression mentally.

$$2\left(-\tfrac{1}{4} + i\tfrac{\sqrt{23}}{4}\right)^2 + \left(-\tfrac{1}{4} + i\tfrac{\sqrt{23}}{4}\right) + 3 \quad \text{and} \quad 2\left(-\tfrac{1}{4} - i\tfrac{\sqrt{23}}{4}\right)^2 + \left(-\tfrac{1}{4} - i\tfrac{\sqrt{23}}{4}\right) + 3$$

Sometimes we can use properties of radicals to simplify a solution to a quadratic equation, as demonstrated in the next example.

EXAMPLE 9 Solving a quadratic equation having complex solutions

Solve $\tfrac{3}{4}x^2 + 1 = x$. Write your answer in standard form: $a + bi$.

Solution
Begin by subtracting x from each side of the equation and then multiply by 4 to clear fractions. The resulting equation is $3x^2 - 4x + 4 = 0$. Substitute $a = 3$, $b = -4$, and $c = 4$ in the quadratic formula.

$$x = \frac{-b \pm \sqrt{b^2 - 4ac}}{2a} \qquad \text{Quadratic formula}$$

$$= \frac{-(-4) \pm \sqrt{(-4)^2 - 4(3)(4)}}{2(3)} \qquad \text{Substitute.}$$

$$= \frac{4 \pm \sqrt{-32}}{6} \qquad \text{Simplify.}$$

$$= \frac{4 \pm 4i\sqrt{2}}{6} \qquad \sqrt{-32} = i\sqrt{32} = i\sqrt{16}\sqrt{2} = 4i\sqrt{2}$$

$$= \frac{2}{3} \pm \frac{2}{3}i\sqrt{2} \qquad \text{Divide 6 into each term and simplify.}$$

In the next example, we use completing the square to obtain complex solutions to a quadratic equation.

EXAMPLE 10 Completing the square to find complex solutions

Solve $x(x + 2) = -2$ by completing the square.

Solution
After applying the distributive property, the equation becomes $x^2 + 2x = -2$. Because $b = 2$, add $\left(\tfrac{b}{2}\right)^2 = \left(\tfrac{2}{2}\right)^2 = 1$ to each side of the equation.

$$x^2 + 2x = -2 \qquad \text{Equation to be solved}$$
$$x^2 + 2x + 1 = -2 + 1 \qquad \text{Add 1 to each side.}$$
$$(x + 1)^2 = -1 \qquad \text{Perfect square trinomial; add.}$$
$$x + 1 = \pm\sqrt{-1} \qquad \text{Square root property}$$
$$x + 1 = \pm i \qquad \sqrt{-1} = i, \text{ the imaginary unit}$$
$$x = -1 \pm i \qquad \text{Add } -1 \text{ to each side.}$$

The solutions are $-1 + i$ and $-1 - i$.

8.4 PUTTING IT ALL TOGETHER

Quadratic equations can be solved symbolically by using factoring, the square root property, completing the square, and the quadratic formula. Graphical and numerical methods can also be used to solve quadratic equations. In this section we discussed the quadratic formula and its discriminant, which we summarize in the following table.

Concept	Explanation	Examples
Quadratic Formula	The quadratic formula can be used to solve *any* quadratic equation written as $ax^2 + bx + c = 0$. The solutions are given by $$x = \frac{-b \pm \sqrt{b^2 - 4ac}}{2a}.$$	For the equation $$2x^2 - 3x + 1 = 0$$ with $a = 2$, $b = -3$, and $c = 1$, the solutions are $$\frac{-(-3) \pm \sqrt{(-3)^2 - 4(2)(1)}}{2(2)} = \frac{3 \pm \sqrt{1}}{4} = 1, \frac{1}{2}.$$
The Discriminant	The expression $b^2 - 4ac$ is called the discriminant. 1. $b^2 - 4ac > 0$ indicates two real solutions. 2. $b^2 - 4ac = 0$ indicates one real solution. 3. $b^2 - 4ac < 0$ indicates no real solutions; rather, there are two complex solutions.	For the equation $$x^2 + 4x - 1 = 0$$ with $a = 1$, $b = 4$, and $c = -1$, the discriminant is $$b^2 - 4ac = 4^2 - 4(1)(-1) = 20 > 0,$$ indicating two real solutions.
Quadratic Formula and Complex Solutions	If the discriminant is negative ($b^2 - 4ac < 0$), the solutions to a quadratic equation are complex numbers. If $k > 0$, the solutions to $x^2 + k = 0$ are given by $x = \pm i\sqrt{k}$.	$2x^2 - x + 3 = 0$ $$x = \frac{1 \pm \sqrt{(-1)^2 - 4(2)(3)}}{2(2)}$$ $$= \frac{1 \pm \sqrt{-23}}{4} = \frac{1}{4} \pm i\frac{\sqrt{23}}{4}$$ $x^2 + 7 = 0$ is equivalent to $x = \pm i\sqrt{7}$.

Like other equations that we have studied, quadratic equations can be solved symbolically, numerically, and graphically. The equation $x(x - 1) = 2$ is solved by these three methods.

NOTE: Symbolic solutions are *exact*. Numerical and graphical solutions are often *approximate*, particularly when solutions contain fractions or square roots.

Symbolic Solution

$x(x - 1) = 2$
$x^2 - x = 2$
$x^2 - x - 2 = 0$
$(x + 1)(x - 2) = 0$
$x = -1$ or $x = 2$

Solutions are -1 and 2.

Numerical Solution

X	Y1	Y2
-3	12	2
-2	6	2
-1	2	2
0	0	2
1	0	2
2	2	2
3	6	2

Y1≡X(X−1)

$x(x - 1) = 2$ when $x = -1$ or 2.

Graphical Solution

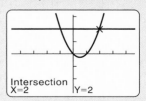

Intersection X=2 Y=2

Points of intersectios are $(-1, 2)$ and $(2, 2)$, where $y_1 = x(x - 1)$ and $y_2 = 2$.

8.5 QUADRATIC INEQUALITIES

Basic Concepts ▪ Graphical and Numerical Solutions ▪ Symbolic Solutions

A LOOK INTO MATH ▷

Parabolas are frequently used in highway design. Sometimes it is necessary for engineers to determine where a highway is below a particular elevation. To do this, they may need to solve a *quadratic inequality*. Quadratic inequalities are nonlinear inequalities. These inequalities are often simple enough to be solved by hand. In this section we discuss how to solve them graphically, numerically, and symbolically.

Basic Concepts

If the equals sign in a quadratic equation is replaced with $>, \geq, <,$ or $\leq$, a **quadratic inequality** results. Examples of quadratic inequalities include

$$x^2 + 4x - 3 < 0, \quad 5x^2 \geq 5, \quad \text{and} \quad 1 - z \leq z^2.$$

Any quadratic equation can be written as

$$ax^2 + bx + c = 0, \quad a \neq 0,$$

so any quadratic inequality can be written as

$$ax^2 + bx + c > 0, \quad a \neq 0,$$

where $>$ may be replaced with $\geq, <,$ or $\leq$.

EXAMPLE 1 Identifying a quadratic inequality

Determine whether the inequality is quadratic.
(a) $5x + x^2 - x^3 \leq 0$ **(b)** $4 + 5x^2 > 4x^2 + x$

Solution
(a) The inequality $5x + x^2 - x^3 \leq 0$ is not quadratic because it has an x^3-term.
(b) Write the inequality as follows.

$$\begin{aligned}
4 + 5x^2 &> 4x^2 + x &&\text{Given inequality} \\
4 + 5x^2 - 4x^2 - x &> 0 &&\text{Subtract } 4x^2 \text{ and } x. \\
4 + x^2 - x &> 0 &&\text{Combine like terms.} \\
x^2 - x + 4 &> 0 &&\text{Rewrite.}
\end{aligned}$$

Because the inequality can be written in the form $ax^2 + bx + c > 0$ with $a = 1$, $b = -1$, and $c = 4$, it is a quadratic inequality.

Graphical and Numerical Solutions

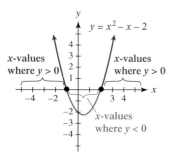

Figure 8.47

Equality often is the boundary between *greater than* and *less than*, so a first step in solving an inequality is to determine the x-values where equality occurs. We begin by using this concept with graphical and numerical techniques.

A graph of $y = x^2 - x - 2$ with x-intercepts -1 and 2 is shown in Figure 8.47. The solutions to $x^2 - x - 2 = 0$ are given by $x = -1$ or $x = 2$. Between the x-intercepts the graph dips below the x-axis and the y-values are negative. Thus the solutions to $x^2 - x - 2 < 0$ satisfy $-1 < x < 2$. To check this result we select a **test value**. For example, 0 lies between -1 and 2. If we substitute $x = 0$ in $x^2 - x - 2 < 0$, it results in a true statement.

$$0^2 - 0 - 2 < 0 \qquad \text{True}$$

When $x < -1$ or $x > 2$, the graph lies above the x-axis and the y-values are positive. Thus the solutions to $x^2 - x - 2 > 0$ satisfy $x < -1$ or $x > 2$. For example, 3 is greater than 2 and -3 is less than -1. Therefore both 3 and -3 are solutions. We can verify this result by substituting 3 and -3 as test values in $x^2 - x - 2 > 0$.

$$3^2 - 3 - 2 > 0 \qquad \text{True}$$
$$(-3)^2 - (-3) - 2 > 0 \qquad \text{True}$$

In the next three examples, we use these concepts to solve quadratic inequalities.

EXAMPLE 2 Solving a quadratic inequality

Make a table of values for $y = x^2 - 3x - 4$ and then sketch the graph. Use the table and graph to solve $x^2 - 3x - 4 \leq 0$. Write your answer in interval notation.

Solution
The points calculated for Table 8.14 are plotted in Figure 8.48 and connected with a smooth ∪-shaped graph.

Numerical Solution Table 8.14 shows that $x^2 - 3x - 4$ equals 0 when $x = -1$ or $x = 4$. Between these values, $x^2 - 3x - 4$ is negative so the solution set to $x^2 - 3x - 4 \leq 0$ is given by $-1 \leq x \leq 4$ or in interval notation, $[-1, 4]$.

Graphical Solution In Figure 8.48 the graph of $y = x^2 - 3x - 4$ shows that the x-intercepts are -1 and 4. Between these values, the graph dips *below* the x-axis. Thus the solution set is $[-1, 4]$.

TABLE 8.14

x	$y = x^2 - 3x - 4$
-2	6
-1	0
0	-4
1	-6
2	-6
3	-4
4	0
5	6

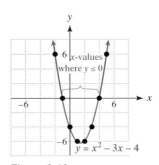

Figure 8.48

EXAMPLE 3 Solving a quadratic inequality

Solve $x^2 > 1$. Write your answer in interval notation.

Solution
First, rewrite $x^2 > 1$ as $x^2 - 1 > 0$. The graph $y = x^2 - 1$ is shown in Figure 8.49 with x-intercepts -1 and 1. The graph lies *above* the x-axis and is shaded green to the left of $x = -1$ and to the right of $x = 1$. Thus the solution set is given by $x < -1$ or $x > 1$, which can be written in interval notation as $(-\infty, -1) \cup (1, \infty)$.

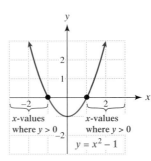

Figure 8.49

EXAMPLE 4 Solving some special cases

Solve each of the inequalities graphically.
(a) $x^2 + 1 > 0$ (b) $x^2 + 1 < 0$ (c) $(x - 1)^2 \leq 0$

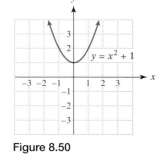

Figure 8.50

Solution
(a) Because the graph of $y = x^2 + 1$, shown in Figure 8.50, is always above the x-axis, $x^2 + 1$ is always greater than 0. The solution set includes all real numbers, or $(-\infty, \infty)$.
(b) Because the graph of $y = x^2 + 1$, shown in Figure 8.50, never goes below the x-axis, $x^2 + 1$ is never less than 0. Thus there are no real solutions.
(c) Because the graph of $y = (x - 1)^2$, shown in Figure 8.51, never goes below the x-axis, $(x - 1)^2$ is never less than 0. When $x = 1$, $y = 0$, so 1 is the only solution to the inequality $(x - 1)^2 \leq 0$.

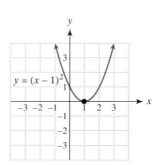

Figure 8.51

▶ **REAL-WORLD CONNECTION**   In the next example we show how quadratic inequalities are used in highway design.

EXAMPLE 5   Determining elevations on a sag curve

Parabolas are frequently used in highway design to model hills and sags (valleys) along a proposed route. Suppose that the elevation E in feet of a sag, or *sag curve*, is given by

$$E(x) = 0.00004x^2 - 0.4x + 2000,$$

where x is the horizontal distance in feet along the sag curve and $0 \leq x \leq 10{,}000$. See Figure 8.52. Estimate graphically the x-values where the elevation is 1500 feet or less.

(*Source:* F. Mannering and W. Kilareski, *Principles of Highway Engineering and Traffic Analysis.*)

Figure 8.52

Solution

Graphical Solution   We must solve the quadratic inequality

$$0.00004x^2 - 0.4x + 2000 \leq 1500.$$

To do so, we let $Y_1 = .00004X^2 - .4X + 2000$ represent the sag or valley in the road and $Y_2 = 1500$ represent a horizontal line with an elevation of 1500 feet. Their graphs intersect at $x \approx 1464$ and $x \approx 8536$, as shown in Figure 8.53. The elevation of the proposed route is less than 1500 feet between these x-values. Therefore the elevation of the road is 1500 feet or less when $1464 \leq x \leq 8536$ (approximately).

CALCULATOR HELP

To find a point of intersection, see the Appendix (page AP-7).

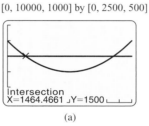
(a)

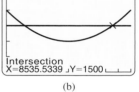
(b)

Figure 8.53

Symbolic Solutions

To solve a quadratic inequality symbolically we first solve the corresponding equality. We can then write the solution to the inequality, using the following method.

> ### SOLUTIONS TO QUADRATIC INEQUALITIES
>
> Let $ax^2 + bx + c = 0$, $a > 0$, have two real solutions p and q, where $p < q$.
>
> $ax^2 + bx + c < 0$ is equivalent to $p < x < q$ (see left-hand figure).
>
> $ax^2 + bx + c > 0$ is equivalent to $x < p$ or $x > q$ (see right-hand figure).
>
> Quadratic inequalities involving $\leq$ or $\geq$ can be solved similarly.
>
>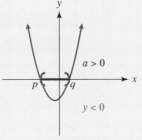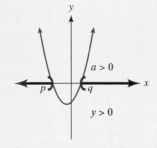
>
> Solutions lie between p and q.   Solutions lie "outside" p and q.

One way to handle the situation, where $a < 0$, is to multiply each side of the inequality by -1, in which case we must be sure to *reverse* the inequality symbol. For example, the inequality $-2x^2 + 8 \leq 0$ has $a = -2$, which is negative. If we multiply each side of this inequality by -1, we obtain $2x^2 - 8 \geq 0$ and now $a = 2$, which is positive. Another way to solve this inequality is to graph $y = -2x^2 + 8$, which is a parabola that opens downward. The following Critical Thinking addresses this situation. However, when $a < 0$ we will usually rewrite the quadratic inequality so that $a > 0$.

CRITICAL THINKING

The graph of $y = -x^2 + x + 12$ is a parabola opening downward with x-intercepts -3 and 4. Solve each inequality.

i. $-x^2 + x + 12 > 0$ ii. $-x^2 + x + 12 < 0$

EXAMPLE 6 Solving quadratic inequalities

Solve each inequality symbolically. Write your answer in interval notation.
(a) $6x^2 - 7x - 5 \geq 0$ (b) $x(3 - x) > -18$

Solution
(a) Begin by solving $6x^2 - 7x - 5 = 0$.

$$6x^2 - 7x - 5 = 0 \quad \text{Quadratic equation}$$
$$(2x + 1)(3x - 5) = 0 \quad \text{Factor.}$$
$$2x + 1 = 0 \quad \text{or} \quad 3x - 5 = 0 \quad \text{Zero-product property}$$
$$x = -\frac{1}{2} \quad \text{or} \quad x = \frac{5}{3} \quad \text{Solve.}$$

Therefore the solutions to $6x^2 - 7x - 5 \geq 0$ lie "outside" these two values and satisfy $x \leq -\frac{1}{2}$ or $x \geq \frac{5}{3}$. In interval notation the solution set is $\left(-\infty, -\frac{1}{2}\right] \cup \left[\frac{5}{3}, \infty\right)$.

(b) First, rewrite the inequality as follows.

$$x(3 - x) > -18 \quad \text{Given inequality}$$
$$3x - x^2 > -18 \quad \text{Distributive property}$$
$$3x - x^2 + 18 > 0 \quad \text{Add 18.}$$
$$-x^2 + 3x + 18 > 0 \quad \text{Rewrite.}$$
$$x^2 - 3x - 18 < 0 \quad \text{Multiply by } -1 \text{ because } a < 0;$$
$$\text{reverse the inequality symbol.}$$

Next, solve $x^2 - 3x - 18 = 0$.

$$(x + 3)(x - 6) = 0 \quad \text{Factor.}$$
$$x = -3 \quad \text{or} \quad x = 6 \quad \text{Solve.}$$

Solutions to $x^2 - 3x - 18 < 0$ lie between these two values and satisfy $-3 < x < 6$. In interval notation the solution set is $(-3, 6)$.

EXAMPLE 7 Finding the dimensions of a building

A rectangular building needs to be 7 feet longer than it is wide, as illustrated in Figure 8.54. The area of the building must be at least 450 square feet. What widths x are possible for this building? Support your results with a table of values.

Figure 8.54

Solution

Symbolic Solution   If x is the width of the building, $x + 7$ is the length of the building and its area is $x(x + 7)$. The area must be at least 450 square feet, so the inequality $x(x + 7) \geq 450$ must be satisfied. First solve the following quadratic equation.

$$x(x + 7) = 450 \quad \text{Quadratic equation}$$
$$x^2 + 7x = 450 \quad \text{Distributive property}$$
$$x^2 + 7x - 450 = 0 \quad \text{Subtract 450.}$$
$$x = \frac{-7 \pm \sqrt{7^2 - 4(1)(-450)}}{2(1)} \quad \text{Quadratic formula; } a = 1, b = 7, \text{ and } c = -450$$
$$= \frac{-7 \pm \sqrt{1849}}{2} \quad \text{Simplify.}$$
$$= \frac{-7 \pm 43}{2} \quad \sqrt{1849} = 43$$
$$= 18, -25 \quad \text{Evaluate.}$$

Thus the solutions to $x(x + 7) \geq 450$ are $x \leq -25$ or $x \geq 18$. The width is positive, so the building width must be 18 feet or more.

Numerical Solution   A table of values is shown in Figure 8.55, where $Y_1 = X(X + 7)$ equals 450 when $x = 18$. For $x \geq 18$ the area is *at least* 450 square feet.

X	Y1
15	330
16	368
17	408
18	450
19	494
20	540
21	588

Y1=X(X+7)

Figure 8.55

8.5 PUTTING IT ALL TOGETHER

The following table summarizes solutions of quadratic inequalities containing the symbols $<$ or $>$. Cases involving $\leq$ or $\geq$ are solved similarly.

Method	Explanation
Solving a Quadratic Inequality Symbolically	Let $ax^2 + bx + c = 0$, $a > 0$, have two real solutions p and q, where $p < q$. $ax^2 + bx + c < 0$ is equivalent to $p < x < q$. $ax^2 + bx + c > 0$ is equivalent to $x < p$ or $x > q$. *Examples:* The solutions to $x^2 - 3x + 2 = 0$ are given by $x = 1$ or $x = 2$. The solutions to $x^2 - 3x + 2 < 0$ are given by $1 < x < 2$. The solutions to $x^2 - 3x + 2 > 0$ are given by $x < 1$ or $x > 2$.
Solving a Quadratic Inequality Graphically	Given $ax^2 + bx + c < 0$ with $a > 0$, graph $y = ax^2 + bx + c$ and locate any x-intercepts. If there are two x-intercepts, then solutions correspond to x-values between the x-intercepts. Solutions to $ax^2 + bx + c > 0$ correspond to x-values "outside" the x-intercepts. See the box on page 401.
Solving a Quadratic Inequality Numerically	If a quadratic inequality is expressed as $ax^2 + bx + c < 0$ with $a > 0$, then we can solve $y = ax^2 + bx + c = 0$ with a table. If there are two solutions, then the solutions to the given inequality lie between these values. Solutions to $ax^2 + bx + c > 0$ lie "outside" these values.

8.6 EQUATIONS IN QUADRATIC FORM

Higher Degree Polynomial Equations ▪ Equations Having Rational Exponents ▪ Equations Having Complex Solutions

A LOOK INTO MATH ▷ Although many equations are *not* quadratic equations, they can sometimes be written in quadratic form. These equations are *reducible to quadratic form*. To express such an equation in quadratic form we often use substitution. In this section we discuss this process.

Higher Degree Polynomial Equations

Sometimes a fourth degree polynomial can be factored like a quadratic trinomial, provided it does not have an *x*-term or an x^3-term. Let's consider the equation $x^4 - 5x^2 + 4 = 0$.

$$x^4 - 5x^2 + 4 = 0 \qquad \text{Given equation}$$
$$(x^2)^2 - 5(x^2) + 4 = 0 \qquad \text{Properties of exponents}$$

We use the substitution $u = x^2$.

$$u^2 - 5u + 4 = 0 \qquad \text{Let } u = x^2.$$
$$(u - 4)(u - 1) = 0 \qquad \text{Factor.}$$
$$u - 4 = 0 \quad \text{or} \quad u - 1 = 0 \qquad \text{Zero-product property}$$
$$u = 4 \quad \text{or} \quad u = 1 \qquad \text{Solve each equation.}$$

Because the given equation uses the variable *x*, we must give the solutions in terms of *x*. We substitute x^2 for *u* and then solve to obtain the following four solutions.

$$x^2 = 4 \quad \text{or} \quad x^2 = 1 \qquad \text{Substitute } x^2 \text{ for } u.$$
$$x = \pm 2 \quad \text{or} \quad x = \pm 1 \qquad \text{Square root property}$$

The solutions are $-2, -1, 1,$ and 2.

In the next example we solve a sixth degree polynomial equation.

EXAMPLE 1 Solving equations by substitution

Solve $2x^6 + x^3 = 1$.

Solution
Start by subtracting 1 from each side.

$$2x^6 + x^3 - 1 = 0 \qquad \text{Subtract 1.}$$
$$2(x^3)^2 + (x^3) - 1 = 0 \qquad \text{Properties of exponents}$$
$$2u^2 + u - 1 = 0 \qquad \text{Let } u = x^3.$$
$$(2u - 1)(u + 1) = 0 \qquad \text{Factor.}$$
$$2u - 1 = 0 \quad \text{or} \quad u + 1 = 0 \qquad \text{Zero-product property}$$
$$u = \tfrac{1}{2} \quad \text{or} \quad u = -1 \qquad \text{Solve.}$$

Now substitute x^3 for *u*, and solve for *x* to obtain the following two solutions.

$$x^3 = \tfrac{1}{2} \quad \text{or} \quad x^3 = -1 \qquad \text{Substitute } x^3 \text{ for } u.$$
$$x = \sqrt[3]{\tfrac{1}{2}} \quad \text{or} \quad x = -1 \qquad \text{Take cube root of each side.}$$

Equations Having Rational Exponents

Equations that have negative exponents are sometimes reducible to quadratic form. Consider the following example, in which two methods are presented.

EXAMPLE 2 Solving an equation having negative exponents

Solve $-6m^{-2} + 13m^{-1} + 5 = 0$.

Solution
Method I Use the substitution $u = m^{-1} = \frac{1}{m}$ and $u^2 = m^{-2} = \frac{1}{m^2}$.

$-6m^{-2} + 13m^{-1} + 5 = 0$	Given equation
$-6u^2 + 13u + 5 = 0$	Let $u = m^{-1}$ and $u^2 = m^{-2}$.
$6u^2 - 13u - 5 = 0$	Multiply by -1.
$(2u - 5)(3u + 1) = 0$	Factor.
$2u - 5 = 0$ or $3u + 1 = 0$	Zero-product property
$u = \frac{5}{2}$ or $u = -\frac{1}{3}$	Solve for u.

Because $u = \frac{1}{m}$, $m = \frac{1}{u}$. Thus the solutions are given by $m = \frac{2}{5}$ or $m = -3$.

Method II Another way to solve this equation is to multiply each side by the LCD, m^2.

$-6m^{-2} + 13m^{-1} + 5 = 0$	Given equation
$m^2(-6m^{-2} + 13m^{-1} + 5) = m^2 \cdot 0$	Multiply by m^2.
$-6m^2m^{-2} + 13m^2m^{-1} + 5m^2 = 0$	Distributive property
$-6 + 13m + 5m^2 = 0$	Add exponents.
$5m^2 + 13m - 6 = 0$	Rewrite the equation.
$(5m - 2)(m + 3) = 0$	Factor.
$5m - 2 = 0$ or $m + 3 = 0$	Zero-product property
$m = \frac{2}{5}$ or $m = -3$	Solve.

In the next example we solve an equation having fractional exponents.

EXAMPLE 3 Solving an equation having fractional exponents

Solve $x^{2/3} - 2x^{1/3} - 8 = 0$.

Solution
Use the substitution $u = x^{1/3}$.

$x^{2/3} - 2x^{1/3} - 8 = 0$	Given equation
$(x^{1/3})^2 - 2(x^{1/3}) - 8 = 0$	Properties of exponents
$u^2 - 2u - 8 = 0$	Let $u = x^{1/3}$.
$(u - 4)(u + 2) = 0$	Factor.
$u - 4 = 0$ or $u + 2 = 0$	Zero-product property
$u = 4$ or $u = -2$	Solve.

Because $u = x^{1/3}$, $u^3 = (x^{1/3})^3 = x$. Thus $x = 4^3 = 64$ or $x = (-2)^3 = -8$. The solutions are -8 and 64.

Equations Having Complex Solutions

Sometimes an equation that is reducible to quadratic form also has complex solutions. This situation is discussed in the next two examples.

EXAMPLE 4 Solving the fourth degree equation

Find all complex solutions to $x^4 - 1 = 0$.

Solution

$$x^4 - 1 = 0 \quad \text{Given equation}$$
$$(x^2)^2 - 1 = 0 \quad \text{Properties of exponents}$$
$$u^2 - 1 = 0 \quad \text{Let } u = x^2.$$
$$(u - 1)(u + 1) = 0 \quad \text{Factor difference of squares.}$$
$$u - 1 = 0 \quad \text{or} \quad u + 1 = 0 \quad \text{Zero-product property}$$
$$u = 1 \quad \text{or} \quad u = -1 \quad \text{Solve for } u.$$

Now substitute x^2 for u, and solve for x.

$$x^2 = 1 \quad \text{or} \quad x^2 = -1 \quad \text{Let } x^2 = u.$$
$$x = \pm 1 \quad \text{or} \quad x = \pm i \quad \text{Square root property}$$

There are four complex solutions: -1, 1, $-i$, and i.

EXAMPLE 5 Solving a rational equation

Find all complex solutions to $\frac{1}{x} + \frac{1}{x^2} = -1$.

Solution
This equation is a rational equation. However, if we multiply through by the LCD, x^2, we clear fractions and obtain a quadratic equation with complex solutions.

$$\frac{1}{x} + \frac{1}{x^2} = -1 \quad \text{Given equation}$$
$$\frac{x^2}{x} + \frac{x^2}{x^2} = -1x^2 \quad \text{Multiply each term by } x^2.$$
$$x + 1 = -x^2 \quad \text{Simplify.}$$
$$x^2 + x + 1 = 0 \quad \text{Add } x^2.$$
$$x = \frac{-1 \pm \sqrt{1^2 - 4(1)(1)}}{2(1)} \quad \text{Quadratic formula}$$
$$x = \frac{-1 \pm i\sqrt{3}}{2} \quad \sqrt{-3} = i\sqrt{3}$$
$$x = -\frac{1}{2} \pm \frac{i\sqrt{3}}{2} \quad \frac{a \pm b}{c} = \frac{a}{c} \pm \frac{b}{c}$$

8.6 PUTTING IT ALL TOGETHER

The following table demonstrates how to reduce some types of equations to quadratic form.

Equation	Substitution	Examples
Higher Degree Polynomial	Let $u = x^n$ for some integer n.	To solve $x^4 - 3x^2 - 4 = 0$, let $u = x^2$. This equation becomes $$u^2 - 3u - 4 = 0.$$
Rational Exponents	Pick a substitution that reduces the equation to quadratic form.	To solve $n^{-2} + 6n^{-1} + 9 = 0$, let $u = n^{-1}$. This equation becomes $$u^2 + 6u + 9 = 0.$$ To solve $6x^{2/5} - 5x^{1/5} - 4 = 0$, let $u = x^{1/5}$. This equation becomes $$6u^2 - 5u - 4 = 0.$$
Equations Having Complex Solutions	Both polynomial and rational equations can have complex solutions. Use the fact that if $a > 0$, then $\sqrt{-a} = i\sqrt{a}$.	$$1 + \frac{1}{x^2} = 0$$ $$x^2 \cdot \left(1 + \frac{1}{x^2}\right) = 0 \cdot x^2$$ $$x^2 + 1 = 0$$ $$x^2 = -1$$ $$x = \pm i$$

CHAPTER 8 SUMMARY

SECTION 8.1 ■ QUADRATIC FUNCTIONS AND THEIR GRAPHS

Quadratic Function Any quadratic function f can be written as
$$f(x) = ax^2 + bx + c \quad (a \neq 0).$$

Graph of a Quadratic Function Its graph is a parabola that is wider than the graph of $y = x^2$, if $0 < |a| < 1$ and narrower, if $|a| > 1$. The y-intercept is c.

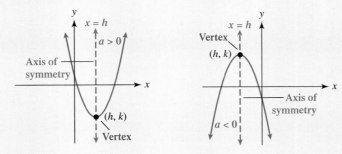

Axis of Symmetry The parabola is symmetric with respect to this vertical line. The axis of symmetry passes through the vertex.

Vertex Formula The x-coordinate of the vertex is $-\frac{b}{2a}$.

Example: Let $y = x^2 - 4x + 1$.
$$x = -\frac{-4}{2(1)} = 2 \quad \text{and} \quad y = 2^2 - 4(2) + 1 = -3. \text{ The vertex is } (2, -3).$$

SECTION 8.2 ■ PARABOLAS AND MODELING

Vertical and Horizontal Translations   Let h and k be positive numbers.

To graph	shift the graph of $y = x^2$ by k units
$y = x^2 + k$	upward.
$y = x^2 - k$	downward.

To graph	shift the graph of $y = x^2$ by h units
$y = (x - h)^2$	right.
$y = (x + h)^2$	left.

Example: Compared to $y = x^2$, the graph of $y = (x - 1)^2 + 2$ is translated right 1 unit and upward 2 units.

Vertex Form   Any quadratic function can be expressed as $f(x) = a(x - h)^2 + k$. In this form the point (h, k) is the vertex. A quadratic function can be put in this form by completing the square.

Example:
$$y = x^2 + 10x - 4 \quad \text{Given equation}$$
$$= (x^2 + 10x + 25) - 25 - 4 \quad \left(\tfrac{b}{2}\right)^2 = \left(\tfrac{10}{2}\right)^2 = 25$$
$$= (x + 5)^2 - 29 \quad \text{Perfect square trinomial; add.}$$

The vertex is $(-5, -29)$.

SECTION 8.3 ■ QUADRATIC EQUATIONS

Quadratic Equations   Any quadratic equation can be written as $ax^2 + bx + c = 0$ and can have no real solutions, one real solution, or two real solutions. These solutions correspond to the x-intercepts on the graph of $y = ax^2 + bx + c$.

Example:
$$x^2 + x - 2 = 0$$
$$(x + 2)(x - 1) = 0$$
$$x = -2 \quad \text{or} \quad x = 1$$

The x-intercepts for $y = x^2 + x - 2$ are -2 and 1.

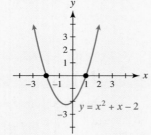

Completing the Square   Write the equation in the form $x^2 + bx = d$. Complete the square by adding $\left(\tfrac{b}{2}\right)^2$ to each side of the equation.

SECTION 8.4 ■ THE QUADRATIC FORMULA

The Quadratic Formula   The solutions to $ax^2 + bx + c = 0$ ($a \neq 0$) are given by

$$x = \frac{-b \pm \sqrt{b^2 - 4ac}}{2a}.$$

Example: Solve $2x^2 + 3x - 1 = 0$ by letting $a = 2$, $b = 3$, and $c = -1$.

$$x = \frac{-3 \pm \sqrt{3^2 - 4(2)(-1)}}{2(2)} = \frac{-3 \pm \sqrt{17}}{4} \approx 0.28, -1.78$$

The Discriminant   The expression $b^2 - 4ac$ is called the discriminant. If $b^2 - 4ac > 0$, there are two real solutions; if $b^2 - 4ac = 0$, there is one real solution; and if $b^2 - 4ac < 0$, there are no real solutions, rather there are two complex solutions.

Example: For $2x^2 + 3x - 1 = 0$, the discriminant is

$$b^2 - 4ac = 3^2 - 4(2)(-1) = 17 > 0.$$

There are two real solutions to this quadratic equation.

SECTION 8.5 ■ QUADRATIC INEQUALITIES

Quadratic Inequalities When the equals sign in a quadratic equation is replaced with $<, >, \leq,$ or $\geq$, a quadratic inequality results. For example,

$$3x^2 - x + 1 = 0$$

is a quadratic equation and

$$3x^2 - x + 1 > 0$$

is a quadratic inequality. Like quadratic equations, quadratic inequalities can be solved symbolically, graphically, and numerically. An important first step in solving a quadratic inequality is to solve the corresponding quadratic equation.

Examples: The solutions to $x^2 - 5x - 6 = 0$ are given by $x = -1, 6$.

The solutions to $x^2 - 5x - 6 < 0$ satisfy $-1 < x < 6$.

The solutions to $x^2 - 5x - 6 > 0$ satisfy $x < -1$ or $x > 6$.

SECTION 8.6 ■ EQUATIONS IN QUADRATIC FORM

Equations Reducible to Quadratic Form An equation that is not quadratic, but can be put into quadratic form by using a substitution is reducible to quadratic form.

Example: To solve $x^{2/3} - 2x^{1/3} - 15 = 0$ let $u = x^{1/3}$. This equation becomes

$$u^2 - 2u - 15 = 0.$$

Factoring results in $(u + 3)(u - 5) = 0$, or $u = -3$ or 5. Because $u = x^{1/3}$, $x = u^3$ and $x = (-3)^3 = -27$ or $x = (5)^3 = 125$.

CHAPTER 8 QUADRATIC FUNCTIONS AND EQUATIONS

Assignment Name _____ Name _____ Date _____

Show all work for these items:_____

# _____	# _____
# _____	# _____
# _____	# _____
# _____	# _____

Assignment Name _____ Name _____ Date _____
Show all work for these items: _____

# _____	# _____
# _____	# _____
# _____	# _____
# _____	# _____

CHAPTER 9
Exponential and Logarithmic Functions

9.1 Composite and Inverse Functions
9.2 Exponential Functions
9.3 Logarithmic Functions
9.4 Properties of Logarithms
9.5 Exponential and Logarithmic Equations

The Swedish scientist Svante Arrhenius first predicted in 1900 a greenhouse effect resulting from emissions of carbon dioxide by industrialized countries. His classic calculation made use of logarithms and predicted that a doubling of the carbon dioxide concentration in the atmosphere would raise the average global temperature by 7°F to 11°F. An increase in world population has resulted in higher emissions of greenhouse gases, such as carbon dioxide, and these emissions have the potential to alter Earth's climate and destroy portions of the ozone layer.

Today there is renewed interest in global warming. Since 1958 the amount of carbon dioxide in the atmosphere has increased by 21%. According to the United Nation's Intergovernmental Panel on Climate Change, the world faces an average temperature rise of 3°C, or 5.4°F, this century, if greenhouse gas emissions continue to rise at their current pace.

In this chapter we use exponential and logarithmic functions to model a wide variety of phenomena, such as greenhouse gases, acid rain, the decline of the bluefin tuna, the demand for liver transplants, diversity of bird species, hurricanes, and earthquakes. Mathematics plays a key role in understanding, controlling, and predicting natural phenomena and people's effect on them.

Human history becomes more and more a race between education and catastrophe.
—H. G. WELLS

Source: M. Kraljic, *The Greenhouse Effect*.

9.1 COMPOSITE AND INVERSE FUNCTIONS

Composition of Functions ■ One-to-One Functions ■ Inverse Functions ■ Tables and Graphs of Inverse Functions

A LOOK INTO MATH ▷

Suppose that you walk into a classroom, turn on the lights, and sit down at your desk. How could you undo or reverse these actions? You might stand up from the desk, turn off the lights, and walk out of the classroom. Note that you must not only perform the "inverse" of each action, but you also must do them in the *reverse order*. In mathematics, we undo an arithmetic operation by performing its inverse operation. For example, the inverse operation of addition is subtraction, and the inverse operation of multiplication is division. In this section, we explore these concepts further by discussing inverse functions.

Composition of Functions

▶ **REAL-WORLD CONNECTION** Many tasks in life are performed in *sequence*, such as putting on your socks and then your shoes. These types of situations also occur in mathematics. For example, suppose that we want to calculate the number of ounces in 3 tons. Because there are 2000 pounds in one ton, we might first multiply **3** by 2000 to obtain **6000** pounds. There are 16 ounces in a pound, so we could multiply 6000 by 16 to obtain **96,000** ounces. This particular calculation involves a *sequence* of calculations that can be represented by the diagram shown in Figure 9.1.

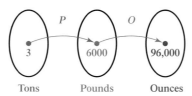

Figure 9.1

The results shown in Figure 9.1 can be calculated by using functions. Suppose that we let $P(x) = 2000x$ convert x tons to P pounds and also let $O(x) = 16x$ convert x pounds to O ounces. We can calculate the number of ounces in 3 tons by performing the *composition of O and P*. This method can be expressed symbolically as

$$(O \circ P)(3) = O(P(3)) \qquad \text{First compute } P(3).$$
$$= O(2000 \cdot 3) \qquad \text{Let } x = 3 \text{ in } P(x) = 2000x.$$
$$= O(6000) \qquad \text{Simplify.}$$
$$= 16(6000) \qquad \text{Let } x = 6000 \text{ in } O(x) = 16x.$$
$$= 96{,}000. \qquad \text{Multiply.}$$

Note that the output for function P—namely, $P(3)$—becomes the input for function O. That is, we evaluate $O(P(3))$ by calculating $P(3)$ first and then substituting the result of 6000 in function O.

COMPOSITION OF FUNCTIONS

If f and g are functions, then the **composite function** $g \circ f$, or **composition** of g and f, is defined by

$$(g \circ f)(x) = g(f(x)).$$

NOTE: We read $g(f(x))$ as "g of f of x."

9.1 COMPOSITE AND INVERSE FUNCTIONS

The compositions $g \circ f$ and $f \circ g$ represent evaluating functions f and g in two different ways. When evaluating $g \circ f$, function f is performed first followed by function g, whereas for $f \circ g$ function g is performed first followed by function f. Note that in general $(g \circ f)(x) \neq (f \circ g)(x)$. That is, *the order in which functions are applied makes a difference,* the same way that putting on your socks and then your shoes is quite different from putting on your shoes and then your socks.

EXAMPLE 1 Finding composite functions

Evaluate $(g \circ f)(2)$ and then find a formula for $(g \circ f)(x)$.
(a) $f(x) = x^3, g(x) = 3x - 2$ (b) $f(x) = 5x, g(x) = x^2 - 3x + 1$
(c) $f(x) = \sqrt{2x}, g(x) = \dfrac{1}{x - 1}$

Solution

(a) $(g \circ f)(2) = g(f(2))$ Composition of functions
$\quad\quad\quad\quad\quad\; = g(8)$ $f(2) = 2^3 = 8$
$\quad\quad\quad\quad\quad\; = 22$ $g(8) = 3(8) - 2 = 22$

Note that the output from f—namely, $f(2)$—becomes the input for g.

$(g \circ f)(x) = g(f(x))$ Composition of functions
$\quad\quad\quad\quad\; = g(x^3)$ $f(x) = x^3$
$\quad\quad\quad\quad\; = 3x^3 - 2$ $g(x) = 3x - 2$

(b) $(g \circ f)(2) = g(f(2))$ Composition of functions
$\quad\quad\quad\quad\quad\; = g(10)$ $f(2) = 5(2) = 10$
$\quad\quad\quad\quad\quad\; = 71$ $g(10) = 10^2 - 3(10) + 1 = 71$

$(g \circ f)(x) = g(f(x))$ Composition of functions
$\quad\quad\quad\quad\; = g(5x)$ $f(x) = 5x$
$\quad\quad\quad\quad\; = (5x)^2 - 3(5x) + 1$ $g(x) = x^2 - 3x + 1$
$\quad\quad\quad\quad\; = 25x^2 - 15x + 1$ Simplify.

(c) $(g \circ f)(2) = g(f(2))$ Composition of functions
$\quad\quad\quad\quad\quad\; = g(2)$ $f(2) = \sqrt{2(2)} = 2$
$\quad\quad\quad\quad\quad\; = 1$ $g(2) = \dfrac{1}{2 - 1} = 1$

$(g \circ f)(x) = g(f(x))$ Composition of functions
$\quad\quad\quad\quad\; = g(\sqrt{2x})$ $f(x) = \sqrt{2x}$
$\quad\quad\quad\quad\; = \dfrac{1}{\sqrt{2x} - 1}$ $g(x) = \dfrac{1}{x - 1}$

Composite functions can also be evaluated numerically and graphically, as demonstrated in the next two examples.

EXAMPLE 2 Evaluating composite functions with tables

Use Tables 9.1 and 9.2 to evaluate each expression.
(a) $(f \circ g)(2)$ **(b)** $(g \circ f)(3)$ **(c)** $(f \circ f)(0)$

TABLE 9.1

x	0	1	2	3
$f(x)$	3	2	0	1

TABLE 9.2

x	0	1	2	3
$g(x)$	1	3	2	0

Solution

(a) $(f \circ g)(2) = f(g(2))$ Composition of functions
$= f(2)$ $g(2) = 2$
$= 0$ $f(2) = 0$

(b) $(g \circ f)(3) = g(f(3))$ Composition of functions
$= g(1)$ $f(3) = 1$
$= 3$ $g(1) = 3$

(c) $(f \circ f)(0) = f(f(0))$ Composition of functions
$= f(3)$ $f(0) = 3$
$= 1$ $f(3) = 1$

EXAMPLE 3 Evaluating composite functions graphically

Use Figure 9.2 to evaluate $(g \circ f)(2)$.

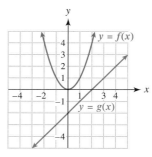

Figure 9.2

Solution
Because $(g \circ f)(2) = g(f(2))$, start by using Figure 9.2 to evaluate $f(2)$. Figure 9.3(a) shows that $f(2) = 4$, which becomes the input for g. Figure 9.3(b) reveals that $g(4) = 2$.

$$(g \circ f)(2) = g(f(2)) = g(4) = 2$$

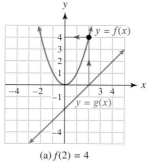 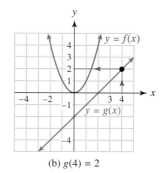

(a) $f(2) = 4$ (b) $g(4) = 2$

Figure 9.3

One-to-One Functions

If we change the input for a function, does the output also change? Do *different inputs* always result in *different outputs* for every function? The answer is no. For example, if $f(x) = x^2 + 1$, then the inputs -2 and 2 result in the *same* output, 5. That is, $f(-2) = 5$ and $f(2) = 5$. However, for $g(x) = 2x$, *different inputs* always result in *different outputs*. Thus we say that g is a *one-to-one function*, whereas f is not.

ONE-TO-ONE FUNCTION

A function f is **one-to-one** if, for any c and d in the domain of f,

$$c \neq d \quad \text{implies that} \quad f(c) \neq f(d).$$

That is, different inputs always result in different outputs.

One way to determine whether a function f is one-to-one is to look at its graph. Suppose that a function has two different inputs that result in the same output. Then there must be two points on its graph that have the same y-value but different x-values. For example, if $f(x) = 2x^2$, then $f(-1) = 2$ and $f(1) = 2$. Thus the points $(-1, 2)$ and $(1, 2)$ both lie on the graph of f, as shown in Figure 9.4(a). Two points with different x-values and the same y-value determine a horizontal line, as shown in Figure 9.4(b). This horizontal line intersects the graph of f more than once, indicating that different inputs do *not* always have different outputs. Thus $f(x) = 2x^2$ is *not* one-to-one.

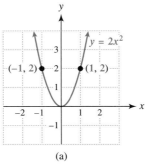
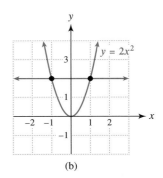

Figure 9.4

This discussion motivates the **horizontal line test**.

HORIZONTAL LINE TEST

If every horizontal line intersects the graph of a function f at most once, then f is a one-to-one function.

We apply the horizontal line test in the next example.

EXAMPLE 4 Using the horizontal line test

Determine whether each graph in Figure 9.5 represents a one-to-one function.

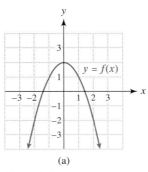
(a)

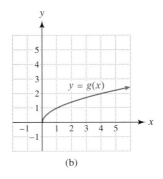
(b)

Figure 9.5

Solution
Figure 9.6(a) shows one of many horizontal lines that intersect the graph of $y = f(x)$ twice. Therefore function f is *not* one-to-one.

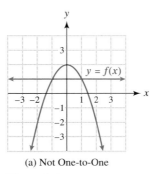
(a) Not One-to-One

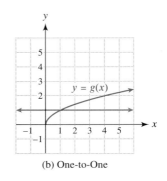
(b) One-to-One

Figure 9.6

Figure 9.6(b) suggests that every horizontal line will intersect the graph of $y = g(x)$ at *most* once. Therefore function g is one-to-one.

MAKING CONNECTIONS

Vertical and Horizontal Line Tests

The *vertical line test* is used to identify functions, whereas the *horizontal line test* is used to identify one-to-one functions. For example, consider the graph of $f(x) = x^2$. A vertical line never intersects the graph more than once, so f is a function. A horizontal line can intersect the graph twice, so f is *not* a one-to-one function.

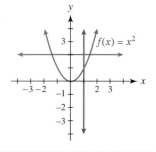

Inverse Functions

▶ **REAL-WORLD CONNECTION**   Turning on a light and turning off a light are inverse operations from ordinary life. Inverse operations undo each other. In mathematics, adding 5 to x, and subtracting 5 from x are inverse operations because

$$x + 5 - 5 = x.$$

Similarly, multiplying x by 5 and dividing x by 5 are inverse operations because

$$\frac{5x}{5} = x.$$

In general, addition and subtraction are inverse operations and multiplication and division are inverse operations.

EXAMPLE 5 Finding inverse operations

State the inverse operations for each statement. Then write a function f for the given statement and a function g for its inverse operations.
(a) Divide x by 3.
(b) Cube x and then add 1 to the result.

Solution

(a) The inverse of dividing x by 3 is to *multiply x by 3*. Thus

$$f(x) = \frac{x}{3} \quad \text{and} \quad g(x) = 3x.$$

(b) The inverse of cubing a number is taking a cube root, and the inverse of adding 1 is subtracting 1. The inverse operations of "cubing a number and then adding 1" are "subtracting 1 and then taking a cube root." For example, 2 cubed plus 1 is $2^3 + 1 = 9$. For the inverse operations, we *first* subtract 1 from 9 and then take the cube root to obtain 2. That is, $\sqrt[3]{9 - 1} = 2$. When there is more than one operation, we must perform the inverse operations in *reverse order*. Thus

$$f(x) = x^3 + 1 \quad \text{and} \quad g(x) = \sqrt[3]{x - 1}.$$

Functions f and g in each part of Example 5 are examples of *inverse functions*. Note that in part (a), if $f(x) = \frac{x}{3}$ and $g(x) = 3x$, then

$$f(15) = 5 \quad \text{and} \quad g(5) = 15.$$

In general, *if f and g are inverse functions, then $f(a) = b$ implies $g(b) = a$*. Thus

$$(g \circ f)(a) = g(f(a)) = g(b) = a$$

for any a in the domain of f, whenever g and f are inverse functions. The composition of a function with its inverse leaves the input unchanged.

INVERSE FUNCTIONS

Let f be a one-to-one function. Then f^{-1} is the **inverse function** of f, if

$(f^{-1} \circ f)(x) = f^{-1}(f(x)) = x,$ for every x in the domain of f, and
$(f \circ f^{-1})(x) = f(f^{-1}(x)) = x,$ for every x in the domain of f^{-1}.

NOTE: In the expression $f^{-1}(x)$, the -1 is *not* an exponent. That is, $f^{-1}(x) \neq \frac{1}{f(x)}$. Rather, if $f(x) = \frac{x}{3}$, then $f^{-1}(x) = 3x$ and, if $f(x) = x^3 + 1$, then $f^{-1}(x) = \sqrt[3]{x - 1}$.

EXAMPLE 6 Verifying inverses

Verify that $f^{-1}(x) = 3x$ if $f(x) = \frac{x}{3}$.

Solution

We must show that $(f^{-1} \circ f)(x) = x$ and that $(f \circ f^{-1})(x) = x$.

$$\begin{aligned}
(f^{-1} \circ f)(x) &= f^{-1}(f(x)) && \text{Composition of functions} \\
&= f^{-1}\left(\frac{x}{3}\right) && f(x) = \frac{x}{3} \\
&= 3\left(\frac{x}{3}\right) && f^{-1}(x) = 3x \\
&= x && \text{Simplify.} \\
(f \circ f^{-1})(x) &= f(f^{-1}(x)) && \text{Composition of functions} \\
&= f(3x) && f^{-1}(x) = 3x \\
&= \frac{3x}{3} && f(x) = \frac{x}{3} \\
&= x && \text{Simplify.}
\end{aligned}$$

The definition of inverse functions states that f must be a one-to-one function. To understand why, consider Figure 9.7. In Figure 9.7(a) a one-to-one function f is represented by a diagram. To find f^{-1} the arrows are reversed. For example, $f(1) = 3$ implies that $f^{-1}(3) = 1$, so the arrow from 1 to 3 for f must be redrawn from 3 to 1 for f^{-1}.

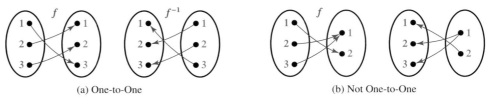

(a) One-to-One   (b) Not One-to-One

Figure 9.7

To be a *function* each input must correspond to exactly one output, which is the case in Figure 9.7(a). However, a different function f that is *not* a one-to-one function because inputs 2 and 3 both result in output 1 is shown in Figure 9.7(b). If the arrows for f are reversed to represent its inverse, then input 1 has two outputs, 2 and 3. Because no inverse *function* can satisfy both $f^{-1}(1) = 2$ and $f^{-1}(1) = 3$ at once, f^{-1} does not exist here.

The following steps can be used to find the inverse of a function symbolically.

FINDING AN INVERSE FUNCTION

To find f^{-1} for a one-to-one function f perform the following steps.

STEP 1: Let $y = f(x)$.

STEP 2: Interchange x and y.

STEP 3: Solve the formula for y. The resulting formula is $y = f^{-1}(x)$.

We apply these steps in the next example.

EXAMPLE 7 Finding an inverse function

Find the inverse of each one-to-one function.
(a) $f(x) = 3x - 7$ (b) $g(x) = (x + 2)^3$

Solution

(a) **STEP 1:** Let $y = 3x - 7$.

STEP 2: Write the formula as $x = 3y - 7$.

STEP 3: To solve for y start by adding 7 to each side.

$$x + 7 = 3y \quad \text{Add 7 to each side.}$$
$$\frac{x + 7}{3} = y \quad \text{Divide each side by 3.}$$

Thus $f^{-1}(x) = \frac{x+7}{3}$ or $f^{-1}(x) = \frac{1}{3}x + \frac{7}{3}$.

(b) **STEP 1:** Let $y = (x + 2)^3$.

STEP 2: Write the formula as $x = (y + 2)^3$.

STEP 3: To solve for y start by taking the cube root of each side.

$$\sqrt[3]{x} = y + 2 \quad \text{Take cube root of each side.}$$
$$\sqrt[3]{x} - 2 = y \quad \text{Subtract 2 from each side.}$$

Thus $g^{-1}(x) = \sqrt[3]{x} - 2$.

Tables and Graphs of Inverse Functions

TABLES   Inverse functions can be represented with tables and graphs. Table 9.3 shows a table of values for a function f.

TABLE 9.3

x	1	2	3	4	5
$f(x)$	3	6	9	12	15

Because $f(1) = 3$, $f^{-1}(3) = 1$. Similarly, $f(2) = 6$ implies that $f^{-1}(6) = 2$ and so on. Table 9.4 lists values for $f^{-1}(x)$.

TABLE 9.4

x	3	6	9	12	15
$f^{-1}(x)$	1	2	3	4	5

Note that the domain of f is {1, 2, 3, 4, 5} and that the range of f is {3, 6, 9, 12, 15}, whereas the domain of f^{-1} is {3, 6, 9, 12, 15} and the range of f^{-1} is {1, 2, 3, 4, 5}. *The domain of f is the range of f^{-1}, and the range of f is the domain of f^{-1}.* This statement is true in general for a function and its inverse.

GRAPHS   If $f(a) = b$, then the point (a, b) lies on the graph of f. This statement also means that $f^{-1}(b) = a$ and that the point (b, a) lies on the graph of f^{-1}. These points are shown in Figure 9.8(a) with a blue line segment connecting them. The line $y = x$ is a perpendicular bisector of this line segment. As a result, the graph of f^{-1} can be obtained from the graph of f by reflecting the graph of f across the line $y = x$. For example, the graphs of a function f and its inverse are shown in Figure 9.8(b).

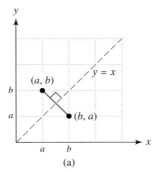
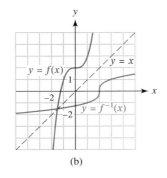

Figure 9.8

The relationship between the graph of a function and the graph of its inverse is summarized as follows.

GRAPHS OF FUNCTIONS AND THEIR INVERSES

The graph of f^{-1} is a reflection of the graph of f across the line $y = x$.

EXAMPLE 8 Graphing an inverse function

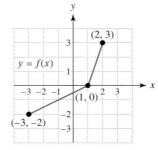

Figure 9.9

The line graph shown in Figure 9.9 represents a function f.
(a) Is f a one-to-one function?
(b) Sketch a graph of $y = f^{-1}(x)$.

Solution
(a) Every horizontal line intersects the graph of f at most once. By the horizontal line test, the graph represents a one-to-one function.
(b) The points $(-3, -2)$, $(1, 0)$, and $(2, 3)$ lie on the graph of f. It follows that the points $(-2, -3)$, $(0, 1)$, and $(3, 2)$ lie on the graph of f^{-1}. Plot these three points and then connect them with line segments, as shown in Figure 9.10(a). Note that the graph of $y = f^{-1}(x)$ is a reflection of the graph of $y = f(x)$ across the line $y = x$, as shown in Figure 9.10(b).

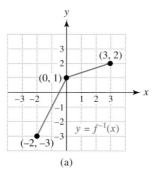
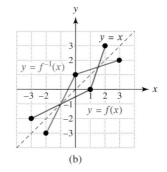

Figure 9.10

CRITICAL THINKING

The graph of a linear function f passes through the points (1, 2) and (2, 1). What two points does the graph of f^{-1} pass through? Find $f(x)$ and $f^{-1}(x)$.

EXAMPLE 9 Graphing an inverse function

The graph of $y = f(x)$ is shown in Figure 9.11. Sketch a graph of $y = f^{-1}(x)$.

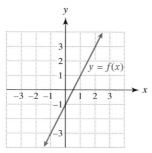

Figure 9.11

Solution
The graph of $y = f^{-1}(x)$ is the reflection of the graph of $y = f(x)$ across the line $y = x$ and is shown in Figure 9.12.

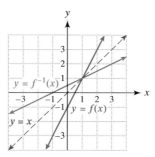

Figure 9.12

9.1 PUTTING IT ALL TOGETHER

The following table summarizes some concepts about composite and inverse functions.

Concept	Explanation	Examples
Composite Functions	The composite of g and f is given by $(g \circ f)(x) = g(f(x))$, and represents a *new* function whose name is $g \circ f$.	If $f(x) = 1 - 4x$ and $g(x) = x^3$, then $(g \circ f)(x) = g(f(x))$ $= g(1 - 4x)$ $= (1 - 4x)^3$.
One-to-One Functions	Function f is one-to-one if different inputs always give different outputs.	$f(x) = x^2$ is not one-to-one because $f(-4) = f(4) = 16$, whereas $g(x) = x + 1$ is one-to-one because, if two inputs differ, then adding 1 to each does not change this difference.

continued on next page

Concept	Explanation	Examples
Horizontal Line Test	This test is used to determine whether a function is one-to-one from its graph.	$f(x) = x^2$ is not one-to-one because a horizontal line can intersect its graph more than once.
Inverse Functions	f^{-1} will undo the operations performed by f. That is, $(f^{-1} \circ f)(x) = x$ and $(f \circ f^{-1})(x) = x.$	If $f(x) = x^3$, then $f^{-1}(x) = \sqrt[3]{x}$ because cubing a number x and then taking its cube root results in the number x.

9.2 EXPONENTIAL FUNCTIONS

Basic Concepts ▪ Graphs of Exponential Functions ▪ Models Involving Exponential Functions ▪ The Natural Exponential Function

A LOOK INTO MATH ▷ Many times the growth of a quantity depends on the amount or number present. The more money deposited in an account, the more interest the account earns; that is, the interest earned is proportional to the amount of money in an account. For example, if a person has $100 in an account and receives 10% annual interest, the interest accrued the first year will be $10, and the balance in the account will be $110 at the end of 1 year. Similarly, if a person begins with $1000 in a similar account, the balance in the account will be $1100 at the end of 1 year. This type of growth is called *exponential growth* and can be modeled by an *exponential function*.

Basic Concepts

▶ **REAL-WORLD CONNECTION** Suppose that an insect population doubles each week. Table 9.5 shows the populations after x weeks. Note that, as the population of insects becomes larger, the *increase* in population each week becomes greater. The population is increasing by 100%, or doubling numerically, each week. When a quantity increases by a constant percentage (or constant factor) at regular intervals, its growth is exponential.

TABLE 9.5

Week	0	1	2	3	4	5
Population	100	200	400	800	1600	3200

We can model the data in Table 9.5 by using the exponential function

$$f(x) = 100(2)^x.$$

For example,

$$f(0) = 100(2)^0 = 100 \cdot 1 = 100,$$
$$f(1) = 100(2)^1 = 100 \cdot 2 = 200,$$
$$f(2) = 100(2)^2 = 100 \cdot 4 = 400,$$

and so on. Note that the exponential function f has a *variable as an exponent*.

EXPONENTIAL FUNCTION

A function represented by

$$f(x) = Ca^x, \quad a > 0 \quad \text{and} \quad a \neq 1,$$

is an **exponential function with base a and coefficient C**. (Unless stated otherwise, we assume that $C > 0$.)

In the formula $f(x) = Ca^x$, a is called the **growth factor** when $a > 1$ and the **decay factor** when $0 < a < 1$. For an exponential function, each time x increases by 1 unit $f(x)$ increases by a *factor* of a when $a > 1$ and decreases by a factor of a when $0 < a < 1$. Moreover, as

$$f(0) = Ca^0 = C(1) = C,$$

the value of C equals the value of $f(x)$ when $x = 0$. If x represents time, C represents the initial value of f when time equals 0. Figure 9.13 illustrates **exponential growth** and **exponential decay** for $x > 0$.

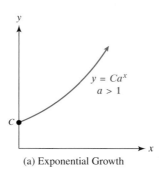

(a) Exponential Growth (b) Exponential Decay

Figure 9.13

The set of valid inputs (domain) for an exponential function includes all real numbers. The set of corresponding outputs (range) includes all positive real numbers.

In the next example we evaluate some exponential functions. When evaluating an exponential function, we evaluate a^x before multiplying by C. This standard order of precedence is much like doing multiplication before addition. For example, $2(3)^2$ should be evaluated as

$$2(9) = 18 \quad \text{not as} \quad (6)^2 = 36.$$

EXAMPLE 1 Evaluating exponential functions

Evaluate $f(x)$ for the given value of x.

(a) $f(x) = 10(3)^x \quad x = 2$ (b) $f(x) = 5\left(\frac{1}{2}\right)^x \quad x = 3$
(c) $f(x) = \frac{1}{3}(2)^x \quad x = -1$

Solution
(a) $f(2) = 10(3)^2 = 10 \cdot 9 = 90$
(b) $f(3) = 5\left(\frac{1}{2}\right)^3 = 5 \cdot \frac{1}{8} = \frac{5}{8}$
(c) $f(-1) = \frac{1}{3}(2)^{-1} = \frac{1}{3} \cdot \frac{1}{2} = \frac{1}{6}$

MAKING CONNECTIONS

The Expressions a^{-x} and $\left(\frac{1}{a}\right)^x$

Using properties of exponents, we can write 2^{-x} as

$$2^{-x} = \frac{1}{2^x} = \left(\frac{1}{2}\right)^x.$$

In general, the expressions a^{-x} and $\left(\frac{1}{a}\right)^x$ are equal for positive a.

In the next example, we determine whether a function is linear or exponential.

EXAMPLE 2 Finding linear and exponential functions

For each table, determine whether f is a linear function or an exponential function. Find a formula for f.

(a)

x	0	1	2	3	4
$f(x)$	16	8	4	2	1

(b)

x	0	1	2	3	4
$f(x)$	5	7	9	11	13

(c)

x	0	1	2	3	4
$f(x)$	1	3	9	27	81

Solution

(a) Each time x increases by 1 unit, $f(x)$ decreases by a factor of $\frac{1}{2}$. Therefore f is an exponential function with a decay factor of $\frac{1}{2}$. Because $f(0) = 16$, $C = 16$ and so $f(x) = 16\left(\frac{1}{2}\right)^x$. This formula can also be written as $f(x) = 16(2)^{-x}$.

(b) Each time x increases by 1 unit, $f(x)$ increases by 2 units. Therefore f is a linear function, and the slope of its graph equals 2. The y-intercept is 5, so $f(x) = 2x + 5$.

(c) Each time x increases by 1 unit, $f(x)$ increases by a factor of 3. Therefore f is an exponential function with a growth factor of 3. Because $f(0) = 1$, $C = 1$ and so $f(x) = 1(3)^x$, or $f(x) = 3^x$.

MAKING CONNECTIONS

Linear and Exponential Functions

For a *linear function*, given by $f(x) = ax + b$, each time x increases by 1 unit y increases (or decreases) by a units, where a equals the slope of the graph of f.

For an *exponential function*, given by $f(x) = Ca^x$, each time x increases by 1 unit y increases *by a factor of a* when $a > 1$ and decreases by a factor of a when $0 < a < 1$. The constant a equals either the growth factor or the decay factor.

▶ **REAL-WORLD CONNECTION** If $100 are deposited in a savings account paying 10% annual interest, the interest earned after 1 year equals $100 \times 0.10 = \$10$. The total amount of money in the account after 1 year is $100(1 + 0.10) = \$110$. Each year the money in the account increases by a factor of 1.10, so after x years there will be $100(1.10)^x$ dollars in the account. Thus compound interest is an example of exponential growth.

9.2 EXPONENTIAL FUNCTIONS

> **COMPOUND INTEREST**
>
> If C dollars are deposited in an account and if interest is paid at the end of each year with an annual rate of interest r, expressed in decimal form, then after x years the account will contain A dollars, where
>
> $$A = C(1 + r)^x.$$
>
> The growth factor is $(1 + r)$.

NOTE: The compound interest formula takes the form of an exponential function with

$$a = 1 + r.$$

EXAMPLE 3 Calculating compound interest

A 20-year-old worker deposits $2000 in a retirement account that pays 13% annual interest at the end of each year. How much money will be in the account when the worker is 65 years old? What is the growth factor?

Solution
Here, $C = 2000$, $r = 0.13$, and $x = 45$. The amount in the account after 45 years is

$$A = 2000(1 + 0.13)^{45} \approx \$489{,}282.80,$$

which is supported by Figure 9.14. In this dramatic example of exponential growth, $2000 grows to nearly half a million dollars in 45 years. Each year the amount of money on deposit is multiplied by a factor of $(1 + 0.13)$, so the growth factor is 1.13.

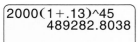

Figure 9.14

Graphs of Exponential Functions

We can graph $f(x) = 2^x$ by first plotting some points, as in Table 9.6.

TABLE 9.6

x	-2	-1	0	1	2
2^x	$\frac{1}{4}$	$\frac{1}{2}$	1	2	4

If we plot these points and sketch the graph, we obtain Figure 9.15. Note that, for negative values of x, $0 < 2^x < 1$ and that, for positive values of x, $2^x > 1$. The graph of $y = 2^x$ passes through the point $(0, 1)$, never intersects the x-axis, and always lies above the x-axis.

We can investigate the graphs of exponential functions further by graphing $y = 1.3^x$, $y = 1.7^x$, and $y = 2.5^x$ (see Figure 9.16). For $a > 1$ the graph of $y = a^x$ *increases* at a faster rate for larger values of a. We now graph $y = 0.7^x$, $y = 0.5^x$, and $y = 0.15^x$ (see Figure 9.17). Note that, if $0 < a < 1$, the graph of $y = a^x$ decreases more rapidly for smaller values of a. The graph of $y = a^x$ is *increasing* when $a > 1$ and *decreasing* when $0 < a < 1$ (from left to right).

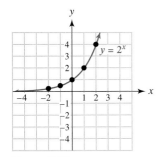

Figure 9.15

CRITICAL THINKING

Every graph of $y = a^x$ passes through what point? Why?

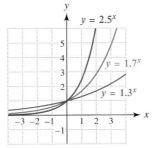

Figure 9.16

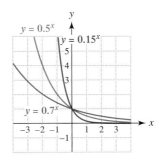

Figure 9.17

EXAMPLE 4 Comparing exponential and linear functions

Compare $f(x) = 3^x$ and $g(x) = 3x$ graphically and numerically for $x \geq 0$.

Solution

Graphical Comparison The graphs of $Y_1 = 3\wedge X$ and $Y_2 = 3X$ are shown in Figure 9.18. The graph of the exponential function y_1 increases much faster than the graph of the linear function y_2.

Numerical Comparison The tables of $Y_1 = 3\wedge X$ and $Y_2 = 3X$ are shown in Figure 9.19. The values for y_1 increase much faster than the values for y_2.

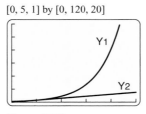

Figure 9.18   Figure 9.19

The results of Example 4 are true in general: For large enough inputs, exponential functions with $a > 1$ grow far faster than any linear function.

MAKING CONNECTIONS

Exponential and Polynomial Functions

The function $f(x) = 2^x$ is an exponential function. The base 2 is a constant and the exponent x is a variable, so $f(3) = 2^3 = 8$.

The function $g(x) = x^2$ is a polynomial function. The base x is a variable and the exponent 2 is a constant, so $g(3) = 3^2 = 9$.

The table clearly shows that the exponential function grows much faster than the polynomial function for larger values of x.

x	0	2	4	6	8	10	12
2^x	1	4	16	64	256	1024	4096
x^2	0	4	16	36	64	100	144

Models Involving Exponential Functions

▶ **REAL-WORLD CONNECTION** Traffic flow on highways can be modeled by exponential functions whenever traffic patterns occur randomly. In the next example we model traffic at an intersection by using an exponential function.

9.2 EXPONENTIAL FUNCTIONS

EXAMPLE 5   Modeling traffic flow

On average, a particular intersection has 360 vehicles arriving randomly each hour. Highway engineers use $f(x) = (0.905)^x$ to estimate the likelihood, or probability, that *no* vehicle will enter the intersection within an interval of x seconds. (*Source:* F. Mannering and W. Kilareski, *Principles of Highway Engineering and Traffic Analysis*.)

(a) Compute $f(5)$ and interpret the results.
(b) A graph of $y = f(x)$ is shown in Figure 9.20. Discuss this graph.
(c) Is this function an example of exponential growth or decay?

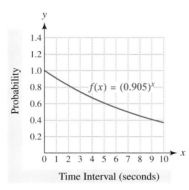

Figure 9.20

Solution

(a) The result $f(5) = (0.905)^5 \approx 0.61$ indicates that there is a 61% chance that no vehicle will enter the intersection during any particular 5-second interval.
(b) The graph decreases, which means that as the interval of time increases there is less chance (probability) that a car will not enter the intersection.
(c) Because the graph is decreasing and $a = 0.905 < 1$, this function is an example of exponential decay.

In the next example, we use an exponential function to model how trees grow in a forest.

EXAMPLE 6   Modeling tree density in a forest

Ecologists studied the spacing of individual trees in a British Columbia forest. This pine forest was 40 to 50 years old and contained approximately 1600 randomly spaced trees per acre. The probability or likelihood that *no* tree is located within a circle of radius x feet can be estimated by $P(x) = (0.892)^x$. For example, $P(4) \approx 0.63$ means that, if a person picks a point at random in the forest, there is a 63% chance that no tree will be located within 4 feet of the person. (*Source:* E. Pielou, *Populations and Community Ecology*.)

(a) Evaluate $P(8)$ and interpret the result.
(b) Graph P in $[0, 20, 5]$ by $[0, 1, 0.1]$ and discuss the graph.

[0, 20, 5] by [0, 1, 0.1]

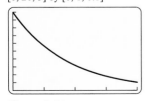
Figure 9.21

Solution

(a) The probability $P(8) = (0.892)^8 \approx 0.40$ means that there is a 40% chance that no tree is growing within any particular circle of radius 8 feet.
(b) The graph of $Y_1 = 0.892\wedge X$, as shown in Figure 9.21, indicates that the larger the circle, the less the likelihood is of no tree being inside the circle.

The Natural Exponential Function

▶ **REAL-WORLD CONNECTION** A special type of exponential function is called the *natural exponential function*, expressed as $f(x) = e^x$. The base e is a special number in mathematics similar to π. The number π is approximately 3.14, whereas the number e is approximately 2.72. The number e is named for the great Swiss mathematician, Leonhard Euler (1707–1783). Most calculators have a special key that can be used to compute the natural exponential function.

> **NATURAL EXPONENTIAL FUNCTION**
>
> The function represented by
> $$f(x) = e^x$$
> is the **natural exponential function**, where $e \approx 2.71828$.

▶ **REAL-WORLD CONNECTION** The natural exponential function is frequently used to model **continuous growth**. For example, the fact that births and deaths occur throughout the year, not just at one time during the year, must be recognized when population growth is being modeled. If a population P is growing continuously at r percent per year, expressed as a decimal, we can model this population after x years by

$$P = Ce^{rx},$$

where C is the initial population. To evaluate natural exponential functions, we use a calculator, as in the next example.

EXAMPLE 7 Modeling population

CALCULATOR HELP
To evaluate the natural exponential function, see the Appendix (page AP-1).

In 2000 Florida's population was 16 million people and was growing at a continuous rate of 2% per year. This population in millions x years after 2000 can be modeled by

$$f(x) = 16e^{0.02x}.$$

Estimate the population in 2010.

```
16e^(.02*10)
         19.54244413
```

Figure 9.22

Solution
As 2010 is 10 years after 2000, we evaluate $f(10)$ to obtain

$$f(10) = 16e^{0.02(10)} \approx 19.5,$$

which is supported by Figure 9.22. (Be sure to include parentheses around the exponent of e.) This model estimates the population of Florida to be about 19.5 million in 2010.

CRITICAL THINKING

Sketch a graph of $y = 2^x$ and $y = 3^x$ in the same xy-plane. Then use these two graphs to sketch a graph of $y = e^x$. How do these graphs compare?

9.2 PUTTING IT ALL TOGETHER

The following table summarizes some important concepts of exponential functions and compound interest.

Topic	Explanation	Example
Exponential Function	An exponential function can be written as $f(x) = Ca^x$, where $a > 0$ and $a \neq 1$. If $a > 1$, the function models exponential growth, and if $0 < a < 1$, the function models exponential decay. The natural exponential function has $C = 1$ and $a = e \approx 2.71828$; that is, $f(x) = e^x$.	$f(x) = 3(2)^x$ models exponential growth and $g(x) = 2\left(\frac{1}{3}\right)^x$ models exponential decay.
Compound Interest	If C dollars are deposited in an account and if interest is paid at the end of each year with an annual rate of interest r, expressed as a decimal, then after x years the account will contain A dollars, where $$A = C(1 + r)^x.$$ The growth factor is $(1 + r)$.	If $1000 are deposited in an account paying 5% annual interest, then after 6 years the amount A in the account is $$A = 1000(1 + 0.05)^6 \approx \$1340.10.$$

9.3 LOGARITHMIC FUNCTIONS

The Common Logarithmic Function ▪ The Inverse of the Common Logarithmic Function ▪ Logarithms with Other Bases

A LOOK INTO MATH ▷ Logarithmic functions are used in many applications. For example, if one airplane weighs twice as much as another, does the heavier airplane typically need a runway that is twice as long? Using a logarithmic function, we can answer this question. (See Example 9.) Logarithmic functions are also used to measure the intensity of sound. In this section we discuss logarithmic functions and several of their applications.

The Common Logarithmic Function

In applications, measurements can vary greatly in size. Table 9.7 lists some examples of objects, with the approximate distances in meters across each.

TABLE 9.7

Object	Distance (meters)
Atom	10^{-9}
Protozoan	10^{-4}
Small Asteroid	10^2
Earth	10^7
Universe	10^{26}

Source: C. Ronan, *The Natural History of the Universe.*

Each distance is listed in the form 10^k for some k. The value of k distinguishes one measurement from another. The *common logarithmic function* or *base-10 logarithmic function*, denoted *log* or log_{10}, outputs k if the input x can be expressed as 10^k for some real number k. For example, $\log 10^{-9} = -9$, $\log 10^2 = 2$, and $\log 10^{1.43} = 1.43$. For any real number k, $\log 10^k = k$. Some values for $f(x) = \log x$ are given in Table 9.8.

TABLE 9.8

x	10^{-4}	10^{-3}	10^{-2}	10^{-1}	10^0	10^1	10^2	10^3	10^4
$\log x$	-4	-3	-2	-1	0	1	2	3	4

NOTE: A common logarithm is an *exponent* having base 10.

We use this information to define the common logarithm.

COMMON LOGARITHM

The **common logarithm of a positive number** x, denoted $\log x$, is calculated as follows. If x is written as $x = 10^k$, then

$$\log x = k,$$

where k is a real number. That is, $\log 10^k = k$.
The function given by

$$f(x) = \log x$$

is called the **common logarithmic function**.

The common logarithmic function outputs an exponent k, which may be positive, negative, or zero. However, a valid input must be positive because 10^k is always positive. *The expression log x equals the exponent k on base* 10 *that gives the number x.* For example, $\log 1000 = 3$ because $1000 = 10^3$.

NOTE: Previously, we have always used one letter, such as f or g, to name a function. The common logarithm is the *first* function for which we use *three* letters, *log*, to name it. Thus $f(x), g(x)$, and $\log(x)$ all represent functions. Generally, $\log(x)$ is written without parentheses as $\log x$. We can also define a function f to be the common logarithmic function by writing $f(x) = \log x$.

EXAMPLE 1 Evaluating common logarithms

Simplify each common logarithm.
(a) $\log 100$ (b) $\log \frac{1}{10}$ (c) $\log \sqrt{1000}$ (d) $\log 45$

Solution
(a) $100 = 10^2$, so $\log 100 = \log 10^2 = 2$
(b) $\log \frac{1}{10} = \log 10^{-1} = -1$
(c) $\log \sqrt{1000} = \log (1000)^{1/2} = \log (10^3)^{1/2} = \log 10^{3/2} = \frac{3}{2}$
(d) How to write 45 as a power of 10 is not obvious. However, we can use a calculator to determine that $\log 45 \approx 1.6532$. Thus $10^{\wedge}(1.6532) \approx 45$. Figure 9.23 supports these answers.

CALCULATOR HELP

To evaluate the common logarithmic function, see the Appendix (page AP-1).

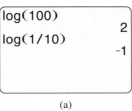

(a)

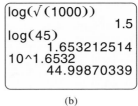

(b)

Figure 9.23

The points $(10^{-1}, -1)$, $(10^0, 0)$, $(10^{0.5}, 0.5)$, and $(10^1, 1)$ are on the graph of $y = \log x$. Plotting these points, as shown in Figure 9.24(a), and sketching the graph of $y = \log x$ results in Figure 9.24(b). Note some important features of this graph.

- The graph of the common logarithm increases very slowly for large values of x. For example, x must be 100 for $\log x$ to reach 2 and x must be 1000 for $\log x$ to reach 3.
- The graph passes through the point $(1, 0)$. Thus $\log 1 = 0$.
- The graph does not exist for negative values of x. The domain of $\log x$ includes only positive numbers. The range of $\log x$ includes all real numbers.
- When $0 < x < 1$, $\log x$ outputs negative values. The y-axis is a vertical asymptote, so as x approaches 0, $\log x$ approaches $-\infty$.

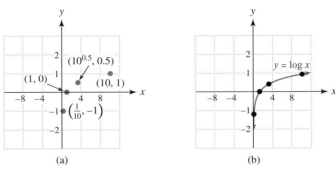

Figure 9.24

MAKING CONNECTIONS

The Common Logarithmic Function and the Square Root Function

Much like the square root function, the common logarithmic function does not have an easy-to-evaluate formula. For example, we can calculate $\sqrt{4} = 2$ and $\sqrt{100} = 10$ mentally, but for $\sqrt{2}$ we usually rely on a calculator. Similarly, we can mentally calculate $\log 1000 = \log 10^3 = 3$, whereas we can use a calculator to approximate $\log 45$. The notation $\log x$ is implied to be $\log_{10} x$ much as $\sqrt{x}$ equals $\sqrt[2]{x}$. Another similarity between the square root function and the common logarithmic function is that their domains do not include negative numbers. If only real numbers are allowed as outputs, both $\sqrt{-3}$ and $\log(-3)$ are undefined expressions.

The Inverse of the Common Logarithmic Function

The graph of $y = \log x$ shown in Figure 9.24(b) is a one-to-one function because it passes the horizontal line test. Different inputs always result in different outputs. Thus the common logarithmic function has an inverse function. To determine this inverse function for $\log x$, consider Tables 9.9 and 9.10.

TABLE 9.9

x	−2	−1	0	1	2
10^x	10^{-2}	10^{-1}	10^0	10^1	10^2

TABLE 9.10

x	10^{-2}	10^{-1}	10^0	10^1	10^2
$\log x$	−2	−1	0	1	2

If we start with the number 2, compute 10^2, and then calculate $\log 10^2$, the result is 2.
$$\log(10^2) = 2$$
In general, $\log 10^x = x$ for any real number x. Now suppose that we perform the calculations in reverse order by taking the common logarithm and then computing a power of 10. For example, suppose that we start with the number 100. The result is
$$10^{\log 100} = 10^2 = 100.$$
In general, $10^{\log x} = x$ for any positive number x.

The *inverse function* of $f(x) = \log x$ is $f^{-1}(x) = 10^x$. That is, if $\log x = y$, then $10^y = x$. Note that composition of these two functions satisfies the definition of an inverse function.

$$(f \circ f^{-1})(x) = f(f^{-1}(x)) \quad \text{and} \quad (f^{-1} \circ f)(x) = f^{-1}(f(x))$$
$$= f(10^x) \qquad\qquad\qquad\qquad = f^{-1}(\log x)$$
$$= \log 10^x \qquad\qquad\qquad\qquad = 10^{\log x}$$
$$= x \qquad\qquad\qquad\qquad\qquad = x$$

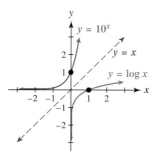

Figure 9.25

In general, the graph of $y = f^{-1}(x)$ is a reflection of the graph of $y = f(x)$ across the line $y = x$. The graphs of $y = \log x$ and $y = 10^x$ are shown in Figure 9.25. Note that the graph of $y = 10^x$ is a reflection of the graph of $y = \log x$ across the line $y = x$.

These inverse properties of $\log x$ and 10^x are summarized as follows.

INVERSE PROPERTIES OF THE COMMON LOGARITHM

The following properties hold for common logarithms.
$$\log 10^x = x, \quad \text{for any real number } x$$
$$10^{\log x} = x, \quad \text{for any positive real number } x$$

EXAMPLE 2 Applying inverse properties

Use inverse properties to simplify each expression.
(a) $\log 10^\pi$ **(b)** $\log 10^{x^2+1}$ **(c)** $10^{\log 7}$ **(d)** $10^{\log 3x}, x > 0$

Solution
(a) Because $\log 10^x = x$ for any real number x, $\log 10^\pi = \pi$.
(b) $\log 10^{x^2+1} = x^2 + 1$
(c) Because $10^{\log x} = x$ for any positive real number x, $10^{\log 7} = 7$.
(d) $10^{\log 3x} = 3x$, provided x is a positive number.

We can also graph logarithmic functions, as demonstrated in the next example.

EXAMPLE 3 Graphing logarithmic functions

Graph each function f and compare its graph to $y = \log x$.
(a) $f(x) = \log(x - 2)$ **(b)** $f(x) = \log(x) + 1$

Solution

(a) We can use our knowledge of translations to sketch the graph of $y = \log(x - 2)$. For example, the graph of $y = (x - 2)^2$ is similar to the graph of $y = x^2$, except that it is translated 2 units to the *right*. Thus the graph of $y = \log(x - 2)$ is similar to the graph of $y = \log x$ (see Figure 9.25) except that it is translated 2 units to the right, as shown in Figure 9.26. The graph of $y = \log x$ passes through $(1, 0)$, so the graph of $y = \log(x - 2)$ passes through $(3, 0)$. Also, instead of the y-axis being a vertical asymptote, the line $x = 2$ is the vertical asymptote.

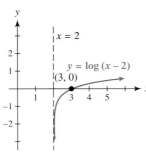

Figure 9.26   Figure 9.27

(b) The graph of $y = \log(x) + 1$ is similar to the graph of $y = \log x$, except that it is translated 1 unit *upward*. This graph is shown in Figure 9.27. Note that the graph of $y = \log(x) + 1$ passes through the point $(1, 1)$.

NOTE: The graph of $y = \log x$ has a vertical asympotote when $x = 0$ and is undefined when $x < 0$. Thus the graph of $\log(x - 2)$ has a vertical asymptote when $x - 2 = 0$, or $x = 2$. It is undefined when $x - 2 < 0$ or $x < 2$.

▶ **REAL-WORLD CONNECTION** Logarithms are used to model quantities that vary greatly in intensity. For example, the human ear is extremely sensitive and able to detect intensities on the eardrum ranging from 10^{-16} watts per square centimeter (w/cm^2) to 10^{-4} w/cm^2, which usually is painful. The next example illustrates modeling sound with logarithms.

EXAMPLE 4 Modeling sound levels

Sound levels in decibels (dB) can be computed by $f(x) = 160 + 10 \log x$, where x is the intensity of the sound in watts per square centimeter. Ordinary conversation has an intensity of 10^{-10} w/cm^2. What decibel level is this? (*Source:* R. Weidner and R. Sells, *Elementary Classical Physics, Vol. 2.*)

CRITICAL THINKING

If the sound level increases by 10 dB, by what factor does the intensity x increase?

Solution
To find the decibel level for ordinary conversation, evaluate $f(10^{-10})$.

$f(10^{-10}) = 160 + 10 \log(10^{-10})$ Substitute $x = 10^{-10}$.
$\phantom{f(10^{-10})} = 160 + 10(-10)$ Evaluate $\log(10^{-10})$.
$\phantom{f(10^{-10})} = 60$ Simplify.

Ordinary conversation corresponds to 60 dB.

Logarithms with Other Bases

Common logarithms are base-10 logarithms, but we can define logarithms having other bases. For example, base-2 logarithms are frequently used in computer science. Some values for the base-2 logarithmic function, denoted $f(x) = \log_2 x$, are shown in Table 9.11. If x can be expressed as $x = 2^k$ for some real number k, then $\log_2 x = \log_2 2^k = k$.

TABLE 9.11

x	2^{-3}	2^{-2}	2^{-1}	2^0	2^1	2^2	2^3
$\log_2 x$	-3	-2	-1	0	1	2	3

NOTE: A base-2 logarithm is an *exponent* having base 2.

Logarithms with other bases are evaluated in the next three examples.

EXAMPLE 5 Evaluating base-2 logarithms

Simplify each logarithm.
(a) $\log_2 8$ (b) $\log_2 \frac{1}{4}$

Solution
(a) The logarithmic expression $\log_2 8$ represents the exponent on base 2 that gives 8. Because $8 = 2^3$, $\log_2 8 = \log_2 2^3 = 3$.
(b) Because $\frac{1}{4} = \frac{1}{2^2} = 2^{-2}$, $\log_2 \frac{1}{4} = \log_2 2^{-2} = -2$.

Some values of base-e logarithms are shown in Table 9.12. A base-e logarithm is referred to as a **natural logarithm** and denoted either $\log_e x$ or $\ln x$. Natural logarithms are used in mathematics, science, economics, electronics, and communications.

CALCULATOR HELP

To evaluate the natural logarithmic function, see the Appendix (page AP-1).

TABLE 9.12

x	e^{-3}	e^{-2}	e^{-1}	e^0	e^1	e^2	e^3
$\ln x$	-3	-2	-1	0	1	2	3

NOTE: A natural logarithm is an *exponent* having base e.

To evaluate natural logarithms we usually use a calculator.

EXAMPLE 6 Evaluating natural logarithms

Approximate to the nearest hundredth.
(a) $\ln 10$ (b) $\ln \frac{1}{2}$

Solution
(a) Figure 9.28 shows that $\ln 10 \approx 2.30$.
(b) Figure 9.28 shows that $\ln \frac{1}{2} \approx -0.69$.

We now define base-a logarithms.

```
ln(10)
         2.302585093
ln(1/2)
         -.6931471806
```

Figure 9.28

BASE-a LOGARITHMS

The **logarithm with base a of a positive number x**, denoted $\log_a x$, is calculated as follows. If x is written as $x = a^k$, then

$$\log_a x = k,$$

where $a > 0$, $a \neq 1$, and k is a real number. That is, $\log_a a^k = k$.
The function given by

$$f(x) = \log_a x$$

is called the **logarithmic function with base a**.

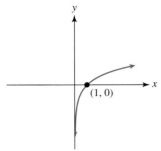

Figure 9.29 $y = \log_a x$, $a > 1$

Remember that *a logarithm is an exponent*. The expression $\log_a x$ equals the exponent k such that $a^k = x$. The graph of $y = \log_a x$ with $a > 1$ is shown in Figure 9.29. Note that the graph passes through the point $(1, 0)$. Thus $\log_a 1 = 0$.

CRITICAL THINKING
Explain why $\log_a 1 = 0$ for any positive base a, $a \neq 1$.

NOTE: The natural logarithm, $\ln x$, is a base-a logarithm with $a = e$. That is, $\ln x = \log_e x$.

EXAMPLE 7 Evaluating base-a logarithms

Simplify each logarithm.
(a) $\log_5 25$ (b) $\log_4 \frac{1}{64}$ (c) $\log_7 1$ (d) $\log_3 9^{-1}$

Solution
(a) $25 = 5^2$, so $\log_5 25 = \log_5 5^2 = 2$.
(b) $\frac{1}{64} = \frac{1}{4^3} = 4^{-3}$, so $\log_4 \frac{1}{64} = \log_4 4^{-3} = -3$.
(c) $1 = 7^0$, so $\log_7 1 = \log_7 7^0 = 0$. (The logarithm of 1 is 0, regardless of the base.)
(d) $9^{-1} = (3^2)^{-1} = 3^{-2}$, so $\log_3 9^{-1} = \log_3 3^{-2} = -2$.

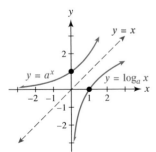

Figure 9.30 $a > 1$

The graph of $y = \log_a x$ in Figure 9.29 passes the horizontal line test, so it is a one-to-one function and it has an inverse. If we let $f(x) = \log_a x$, then $f^{-1}(x) = a^x$. This statement is a generalization of the fact that $\log x$ and 10^x represent inverse functions. The graphs of $y = \log_a x$ and $y = a^x$ with $a > 1$ are shown in Figure 9.30. Note that the graph of $y = a^x$ is a reflection of the graph of $y = \log_a x$ across the line $y = x$.

These inverse properties for logarithmic and exponential functions are summarized by the following.

INVERSE PROPERTIES OF BASE-a LOGARITHMS

The following properties hold for logarithms with base a.

$$\log_a a^x = x, \quad \text{for any real number } x$$
$$a^{\log_a x} = x, \quad \text{for any positive real number } x$$

EXAMPLE 8 Applying inverse properties

Simplify each expression.
(a) $\ln e^{0.5x}$ (b) $e^{\ln 4}$ (c) $2^{\log_2 7x}$ (d) $10^{\log(9x-3)}$

Solution
(a) $\ln e^{0.5x} = 0.5x$ because $\ln e^k = k$ for all k.
(b) $e^{\ln 4} = 4$ because $e^{\ln k} = k$ for all positive k.
(c) $2^{\log_2 7x} = 7x$ for $x > 0$ because $a^{\log_a k} = k$ for all positive k.
(d) $10^{\log(9x-3)} = 9x - 3$ for $x > \frac{1}{3}$ because $10^{\log k} = k$ for all positive k.

▶ **REAL-WORLD CONNECTION** Logarithms occur in many applications. One application is runway length for airplanes, which we discuss in the next example.

EXAMPLE 9 Calculating runway length

There is a mathematical relationship between an airplane's weight x and the runway length required at takeoff. For certain types of airplanes, the minimum runway length L in thousands of feet may be modeled by $L(x) = 1.3 \ln x$, where x is in thousands of pounds.
(*Source:* L. Haefner, *Introduction to Transportation Systems*.)
(a) Estimate the runway length needed for an airplane weighing 10,000 pounds.
(b) Does a 20,000-pound airplane need twice the runway length that a 10,000-pound airplane needs? Explain.

Solution
(a) Because $L(x) = 1.3 \ln x$, it follows that $L(10) = 1.3 \ln(10) \approx 3$. An airplane weighing 10,000 pounds requires a runway (at least) 3000 feet long.
(b) Because $L(20) = 1.3 \ln(20) \approx 3.9$, a 20,000-pound airplane does not need twice the runway length needed by a 10,000-pound airplane. Rather the heavier airplane needs roughly 3900 feet of runway, or only an extra 900 feet.

9.3 PUTTING IT ALL TOGETHER

Common logarithms are base-10 logarithms. If a positive number x is written as $x = 10^k$, then $\log x = k$. The value of $\log x$ represents the exponent on the base 10 that gives x. We can define logarithms that have other bases. For example, the natural logarithm is a base-e logarithm that is usually evaluated using a calculator. The following table summarizes some important concepts related to base-a logarithms.

Concept	Description	Examples
Base-a Logarithms	*Definition:* $\log_a x = k$ means $x = a^k$, where $a > 0$ and $a \neq 1$. *Domain:* all *positive* real numbers *Range:* all real numbers *Graph:* $a > 1$ (shown to the right) Passes through $(1, 0)$; vertical asymptote: y-axis *Common Logarithm:* Base-10 logarithm and denoted $\log x$ *Natural Logarithm:* Base-e logarithm, where $e \approx 2.718$, and denoted $\ln x$	$\log 1000 = \log 10^3 = 3,$ $\log_2 16 = \log_2 2^4 = 4,$ and $\log_3 \dfrac{1}{81} = \log_3 3^{-4} = -4$ $y = \log_a x$, $a > 1$, passes through $(1, 0)$

continued on next page

continued from previous page

Concept	Description	Examples
Inverse Properties	The following properties hold for base-a logarithms. $\log_a a^x = x,$ for any real number x $a^{\log_a x} = x,$ for any positive number x	$\log 10^{7.48} = 7.48$ and $2^{\log_2 63} = 63$

9.4 PROPERTIES OF LOGARITHMS

Basic Properties ■ Change of Base Formula

A LOOK INTO MATH ▷ The discovery of logarithms by John Napier (1550–1617) played an important role in the history of science. Logarithms were instrumental in allowing Johannes Kepler (1571–1630) to calculate the positions of the planet Mars, which led to his discovery of the laws of planetary motion. Kepler's laws were used by Isaac Newton (1642–1727) to discover the universal laws of gravity. Although calculators and computers have made tables of logarithms obsolete, applications involving logarithms still play an important role in modern-day computation. One reason for their continued importance is that logarithms possess several important properties.

Basic Properties

In this subsection we discuss three important properties of logarithms. The first property is the product rule for logarithms.

> **PRODUCT RULE FOR LOGARITHMS**
> For positive numbers m, n, and $a \neq 1$,
> $$\log_a mn = \log_a m + \log_a n.$$

This property may be verified by using properties of exponents and the fact that $\log_a a^k = k$ for any real number k. Here, we verify the product property for logarithms. Other properties presented can be verified in a similar manner.

If m and n are positive numbers, we can write $m = a^c$ and $n = a^d$ for some real numbers c and d.

$$\log_a mn = \log_a(a^c a^d) = \log_a(a^{c+d}) = c + d \quad \text{and}$$
$$\log_a m + \log_a n = \log_a a^c + \log_a a^d = c + d$$

Thus $\log_a mn = \log_a m + \log_a n$.

This property is illustrated in Figure 9.31, which shows that
$$\log 10 = \log(2 \cdot 5) = \log 2 + \log 5.$$

In the next two examples, we demonstrate various operations involving logarithms.

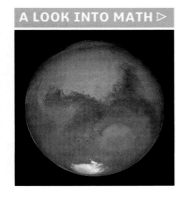

Figure 9.31

EXAMPLE 1 Writing logarithms as sums

Write each expression as a sum of logarithms. Assume that x is positive.
(a) $\log 21$ (b) $\ln 5x$ (c) $\log_2 x^3$

Solution
(a) $\log 21 = \log(3 \cdot 7) = \log 3 + \log 7$
(b) $\ln 5x = \ln(5 \cdot x) = \ln 5 + \ln x$
(c) $\log_2 x^3 = \log_2(x \cdot x \cdot x) = \log_2 x + \log_2 x + \log_2 x$

EXAMPLE 2 Combining logarithms

Write each expression as one logarithm. Assume that x and y are positive.
(a) $\log 5 + \log 6$ (b) $\ln x + \ln xy$ (c) $\log_3 2x + \log_3 5x$

Solution
(a) $\log 5 + \log 6 = \log(5 \cdot 6) = \log 30$
(b) $\ln x + \ln xy = \ln(x \cdot xy) = \ln x^2 y$
(c) $\log_3 2x + \log_3 5x = \log_3(2x \cdot 5x) = \log_3 10x^2$

The second property is the quotient rule for logarithms.

QUOTIENT RULE FOR LOGARITHMS

For positive numbers m, n, and $a \neq 1$,

$$\log_a \frac{m}{n} = \log_a m - \log_a n.$$

```
log(10)
              1
log(20)-log(2)
              1
```
Figure 9.32

This property is illustrated in Figure 9.32, which shows that

$$\log 10 = \log \frac{20}{2} = \log 20 - \log 2.$$

EXAMPLE 3 Writing logarithms as differences

Write each expression as a difference of two logarithms. Assume that variables are positive.
(a) $\log \frac{3}{2}$ (b) $\ln \frac{3x}{y}$ (c) $\log_5 \frac{x}{z^4}$

Solution
(a) $\log \frac{3}{2} = \log 3 - \log 2$ (b) $\ln \frac{3x}{y} = \ln 3x - \ln y$
(c) $\log_5 \frac{x}{z^4} = \log_5 x - \log_5 z^4$

NOTE: $\log_a(m + n) \neq \log_a m + \log_a n$; $\log_a(m - n) \neq \log_a m - \log_a n$;

$$\log_a(mn) \neq \log_a m \cdot \log_a n; \quad \log_a\left(\frac{m}{n}\right) \neq \frac{\log_a m}{\log_a n}$$

EXAMPLE 4 Combining logarithms

Write each expression as one term. Assume that x is positive.
(a) $\log 50 - \log 25$ (b) $\ln x^3 - \ln x$ (c) $\log_4 15x - \log_4 5x$

Solution

(a) $\log 50 - \log 25 = \log \dfrac{50}{25} = \log 2$

(b) $\ln x^3 - \ln x = \ln \dfrac{x^3}{x} = \ln x^2$

(c) $\log_4 15x - \log_4 5x = \log_4 \dfrac{15x}{5x} = \log_4 3$

The third property is the power rule for logarithms. To illustrate this rule we use
$$\log x^3 = \log(x \cdot x \cdot x) = \log x + \log x + \log x = 3 \log x.$$
Thus $\log x^3 = 3 \log x$. This example is generalized in the following rule.

POWER RULE FOR LOGARITHMS

For positive numbers m and $a \neq 1$ and any real number r,
$$\log_a(m^r) = r \log_a m.$$

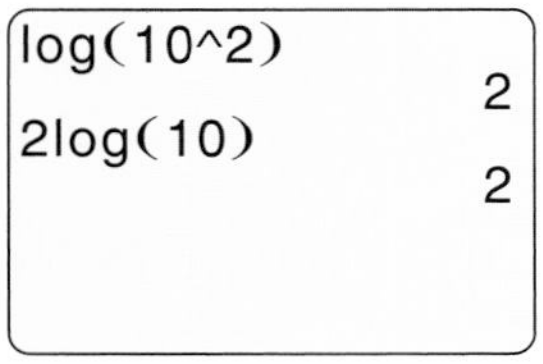

```
log(10^2)
           2
2log(10)
           2
```

Figure 9.33

We use a calculator and the equation
$$\log 10^2 = 2 \log 10$$
to illustrate this property. Figure 9.33 shows the result.

We apply the power rule in the next example.

EXAMPLE 5 Applying the power rule

Rewrite each expression, using the power rule.
(a) $\log 5^6$ (b) $\ln(0.55)^{x-1}$ (c) $\log_5 8^{kx}$

Solution
(a) $\log 5^6 = 6 \log 5$
(b) $\ln(0.55)^{x-1} = (x-1)\ln(0.55)$
(c) $\log_5 8^{kx} = kx \log_5 8$

Sometimes we use more than one property to simplify an expression. We assume that all variables are positive in the next two examples.

EXAMPLE 6 Combining logarithms

Write each expression as the logarithm of a single expression.
(a) $3 \log x + \log x^2$ (b) $2 \ln x - \ln \sqrt{x}$

Solution
(a) $3 \log x + \log x^2 = \log x^3 + \log x^2$ Power rule
$ = \log(x^3 \cdot x^2)$ Product rule
$ = \log x^5$ Properties of exponents

(b) $2 \ln x - \ln \sqrt{x} = 2 \ln x - \ln x^{1/2}$ $\sqrt{x} = x^{1/2}$
$\phantom{2 \ln x - \ln \sqrt{x}} = \ln x^2 - \ln x^{1/2}$ Power rule
$\phantom{2 \ln x - \ln \sqrt{x}} = \ln \dfrac{x^2}{x^{1/2}}$ Quotient rule
$\phantom{2 \ln x - \ln \sqrt{x}} = \ln x^{3/2}$ Properties of exponents

EXAMPLE 7 Expanding logarithms

Write each expression in terms of logarithms of x, y, and z.

(a) $\log \dfrac{x^2 y^3}{\sqrt{z}}$ 　　(b) $\ln \sqrt[3]{\dfrac{xy}{z}}$

Solution

(a) $\log \dfrac{x^2 y^3}{\sqrt{z}} = \log x^2 y^3 - \log \sqrt{z}$ 　　Quotient rule

$\quad\quad\quad\quad\quad = \log x^2 + \log y^3 - \log z^{1/2}$ 　　Product rule; $\sqrt{z} = z^{1/2}$

$\quad\quad\quad\quad\quad = 2 \log x + 3 \log y - \tfrac{1}{2} \log z$ 　　Power rule

(b) $\ln \sqrt[3]{\dfrac{xy}{z}} = \ln \left(\dfrac{xy}{z} \right)^{1/3}$ 　　$\sqrt[3]{m} = m^{1/3}$

$\quad\quad\quad\quad = \tfrac{1}{3} \ln \dfrac{xy}{z}$ 　　Power rule

$\quad\quad\quad\quad = \tfrac{1}{3} (\ln xy - \ln z)$ 　　Quotient rule

$\quad\quad\quad\quad = \tfrac{1}{3} (\ln x + \ln y - \ln z)$ 　　Product rule

$\quad\quad\quad\quad = \tfrac{1}{3} \ln x + \tfrac{1}{3} \ln y - \tfrac{1}{3} \ln z$ 　　Distributive property

EXAMPLE 8 Applying properties of logarithms

Using only properties of logarithms and the approximations $\ln 2 \approx 0.7$, $\ln 3 \approx 1.1$, and $\ln 5 \approx 1.6$, find an approximation for each expression.

(a) $\ln 8$ 　　(b) $\ln 15$ 　　(c) $\ln \dfrac{10}{3}$

Solution

(a) $\ln 8 = \ln 2^3 = 3 \ln 2 \approx 3(0.7) = 2.1$

(b) $\ln 15 = \ln(3 \cdot 5) = \ln 3 + \ln 5 \approx 1.1 + 1.6 = 2.7$

(c) $\ln \dfrac{10}{3} = \ln \left(\dfrac{2 \cdot 5}{3} \right) = \ln 2 + \ln 5 - \ln 3 \approx 0.7 + 1.6 - 1.1 = 1.2$

Change of Base Formula

▶ **REAL-WORLD CONNECTION** Most calculators only have keys to evaluate common and natural logarithms. Occasionally, it is necessary to evaluate a logarithmic function with a base other than 10 or e. In these situations we use the following change of base formula, which we illustrate in the next example.

CHANGE OF BASE FORMULA

Let x and $a \neq 1$ be positive real numbers. Then

$$\log_a x = \dfrac{\log x}{\log a} \quad \text{or} \quad \log_a x = \dfrac{\ln x}{\ln a}.$$

EXAMPLE 9 Change of base formula

Approximate $\log_2 14$ to the nearest thousandth.

Solution
Using the change of base formula,

$$\log_2 14 = \frac{\log 14}{\log 2} \approx 3.807 \quad \text{or} \quad \log_2 14 = \frac{\ln 14}{\ln 2} \approx 3.807.$$

Figure 9.34 supports these results.

```
log(14)/log(2)
      3.807354922
ln(14)/ln(2)
      3.807354922
```
Figure 9.34

9.4 PUTTING IT ALL TOGETHER

The following table summarizes some important properties for base-a logarithms. Common and natural logarithms satisfy the same properties.

Concepts	Description	Examples
Properties of Logarithms 1. Product Rule 2. Quotient Rule 3. Power Rule	The following properties hold for positive numbers m, n, and $a \neq 1$ and for any real number r. 1. $\log_a mn = \log_a m + \log_a n$ 2. $\log_a \frac{m}{n} = \log_a m - \log_a n$ 3. $\log_a (m^r) = r \log_a m$	1. $\log 20 = \log 10 + \log 2$ 2. $\log \frac{45}{6} = \log 45 - \log 6$ 3. $\ln x^6 = 6 \ln x$
Change of Base Formula	Let x and $a \neq 1$ be positive numbers. Then $$\log_a x = \frac{\log x}{\log a} \quad \text{and} \quad \log_a x = \frac{\ln x}{\ln a}.$$	The expression $\log_3 6$ is equivalent to either $$\frac{\log 6}{\log 3} \quad \text{or} \quad \frac{\ln 6}{\ln 3}.$$

9.5 EXPONENTIAL AND LOGARITHMIC EQUATIONS

Exponential Equations and Models ■ Logarithmic Equations and Models

A LOOK INTO MATH ▷ Although we have solved many equations throughout this course, one equation that we have not solved *symbolically* is $a^x = k$. This exponential equation occurs frequently in applications and is used to model either exponential growth or decay. For example, one bluefin tuna can be worth over $30,000. As a result, the numbers of bluefin tuna declined exponentially from 1974 to 1991. Logarithmic equations contain logarithms and are also used in modeling real-world data. To solve exponential equations we use logarithms, and to solve logarithmic equations we use exponential expressions.

Exponential Equations and Models

To solve the equation $10 + x = 100$, we subtract 10 from each side because addition and subtraction are inverse operations.

$$10 + x - 10 = 100 - 10$$
$$x = 90$$

To solve the equation $10x = 100$, we divide each side by 10 because multiplication and division are inverse operations.

$$\frac{10x}{10} = \frac{100}{10}$$
$$x = 10$$

Now suppose that we want to solve the exponential equation

$$10^x = 100.$$

What is new about this type of equation is that the variable x is an *exponent*. The inverse operation of 10^x is $\log x$. Rather than subtracting 10 from each side or dividing each side by 10, we take the base-10 logarithm of each side. Doing so results in

$$\log 10^x = \log 100.$$

Because $\log 10^x = x$ for all real numbers x, the equation becomes

$$x = \log 100, \quad \text{or equivalently,} \quad x = 2.$$

These concepts are applied in the next example.

EXAMPLE 1   Solving exponential equations

Solve.
(a) $10^x = 150$ **(b)** $e^x = 40$ **(c)** $2^x = 50$ **(d)** $0.9^x = 0.5$

Solution

(a)
$$10^x = 150 \quad \text{Given equation}$$
$$\log 10^x = \log 150 \quad \text{Take the common logarithm of each side.}$$
$$x = \log 150 \approx 2.18 \quad \text{Inverse property: } \log 10^k = k \text{ for all } k$$

(b) The inverse operation of e^x is $\ln x$, so we take the natural logarithm of each side.
$$e^x = 40 \quad \text{Given equation}$$
$$\ln e^x = \ln 40 \quad \text{Take the natural logarithm of each side.}$$
$$x = \ln 40 \approx 3.69 \quad \text{Inverse property: } \ln e^k = k \text{ for all } k$$

(c) The inverse operation of 2^x is $\log_2 x$. Calculators do not usually have a base-2 logarithm key, so we take the common logarithm of each side and then apply the power rule.
$$2^x = 50 \quad \text{Given equation}$$
$$\log 2^x = \log 50 \quad \text{Take the common logarithm of each side.}$$
$$x \log 2 = \log 50 \quad \text{Power rule: } \log(m^r) = r \log m$$
$$x = \frac{\log 50}{\log 2} \approx 5.64 \quad \text{Divide by } \log 2 \text{ and approximate.}$$

(d) This time we begin by taking the natural logarithm of each side.
$$0.9^x = 0.5 \quad \text{Given equation}$$
$$\ln 0.9^x = \ln 0.5 \quad \text{Take the natural logarithm of each side.}$$
$$x \ln 0.9 = \ln 0.5 \quad \text{Power rule: } \ln(m^r) = r \ln m$$
$$x = \frac{\ln 0.5}{\ln 0.9} \approx 6.58 \quad \text{Divide by } \ln 0.9 \text{ and approximate.}$$

MAKING CONNECTIONS

Logarithms of Quotients and Quotients of Logarithms

The solution in Example 1(c) is $\frac{\log 50}{\log 2}$. Note that

$$\frac{\log 50}{\log 2} \neq \log 50 - \log 2.$$

However, $\log 50 - \log 2 = \log \frac{50}{2} = \log 25$ by the quotient rule for logarithms, as shown in the figure.

```
log(50)/log(2)
         5.64385619
log(50)-log(2)
         1.397940009
log(25)
         1.397940009
```

The next two examples illustrate methods for solving exponential equations.

EXAMPLE 2 Solving exponential equations

Solve each equation.
(a) $2e^x - 1 = 5$ (b) $3^{x-5} = 15$ (c) $e^{2x} = e^{x+5}$ (d) $3^{2x} = 2^{x+3}$

Solution

(a) Begin by solving for e^x.

$2e^x - 1 = 5$ Given equation
$2e^x = 6$ Add 1 to each side.
$e^x = 3$ Divide each side by 2.
$\ln e^x = \ln 3$ Take the natural logarithm.
$x = \ln 3 \approx 1.10$ Inverse property: $\ln e^k = k$

(b) Start by taking the common logarithm of each side. (We could also take the natural logarithm of each side.)

$3^{x-5} = 15$ Given equation
$\log 3^{x-5} = \log 15$ Take the common logarithm of each side.
$(x - 5) \log 3 = \log 15$ Power rule for logarithms
$x - 5 = \frac{\log 15}{\log 3}$ Divide by log 3.
$x = \frac{\log 15}{\log 3} + 5 \approx 7.46$ Add 5 to each side and approximate.

(c) In $e^{2x} = e^{x+5}$, the bases are equal, so the exponents must also be equal. To verify this assertion, take the natural logarithm of each side.

$e^{2x} = e^{x+5}$ Given equation
$\ln e^{2x} = \ln e^{x+5}$ Take the natural logarithm.
$2x = x + 5$ Inverse property: $\ln e^k = k$
$x = 5$ Subtract x.

(d) In $3^{2x} = 2^{x+3}$, the bases are not equal. However, we can still solve the equation by taking the common logarithm of each side. A logarithm of any base could be used.

$$3^{2x} = 2^{x+3} \qquad \text{Given equation}$$
$$\log 3^{2x} = \log 2^{x+3} \qquad \text{Take the common logarithm.}$$
$$2x \log 3 = (x + 3) \log 2 \qquad \text{Power rule for logarithms}$$
$$2x \log 3 = x \log 2 + 3 \log 2 \qquad \text{Distributive property}$$
$$2x \log 3 - x \log 2 = 3 \log 2 \qquad \text{Subtract } x \log 2.$$
$$x(2 \log 3 - \log 2) = 3 \log 2 \qquad \text{Factor out } x.$$
$$x = \frac{3 \log 2}{2 \log 3 - \log 2} \qquad \text{Divide by } 2 \log 3 - \log 2.$$
$$x \approx 1.38 \qquad \text{Approximate.}$$

EXAMPLE 3 Solving an exponential equation

Graphs for $f(x) = 0.2e^x$ and $g(x) = 4$ are shown in Figure 9.35.
(a) Use the graphs to estimate the solution to the equation $f(x) = g(x)$.
(b) Check your estimate by solving the equation symbolically.

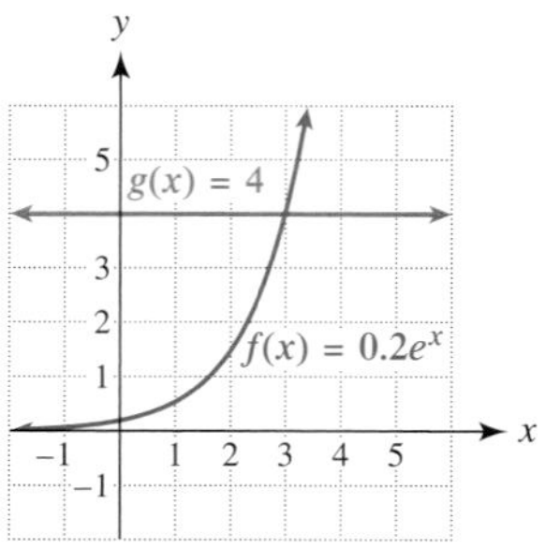

Figure 9.35

Solution
(a) The graphs intersect near the point (3, 4). Therefore the solution is given by $x \approx 3$.
(b) We must solve the equation $0.2e^x = 4$.

$$0.2e^x = 4 \qquad \text{Given equation}$$
$$e^x = 20 \qquad \text{Divide each side by 0.2.}$$
$$\ln e^x = \ln 20 \qquad \text{Take the natural logarithm of each side.}$$
$$x = \ln 20 \qquad \text{Inverse property: } \ln e^k = k$$
$$x \approx 2.996 \qquad \text{Approximate.}$$

NOTE: The graphical estimate did not give the *exact* solution of ln 20.

▶ **REAL-WORLD CONNECTION** In Section 9.2 we showed that if $1000 are deposited in a savings account paying 10% annual interest at the end of each year, the amount A in the account after x years is given by

$$A(x) = 1000(1.1)^x.$$

After 10 years there will be

$$A(10) = 1000(1.1)^{10} \approx \$2593.74$$

in the account. To calculate how long it will take for $4000 to accrue in the account, we need to solve the exponential equation

$$1000(1.1)^x = 4000.$$

We do so in the next example.

EXAMPLE 4 Solving exponential equations

Solve $1000(1.1)^x = 4000$ symbolically. Give graphical support for your answer.

Solution
Symbolic Solution Begin by dividing each side of the equation by 1000.

$1000(1.1)^x = 4000$	Given equation
$1.1^x = 4$	Divide by 1000.
$\log 1.1^x = \log 4$	Take the common logarithm of each side.
$x \log 1.1 = \log 4$	Power rule for logarithms
$x = \dfrac{\log 4}{\log 1.1} \approx 14.5$	Divide by log 1.1 and approximate.

Interest is paid at the end of the year, so it will take 15 years for $1000 earning 10% annual interest to grow to (at least) $4000.

[0, 20, 5] by [0, 6000, 1000]

Figure 9.36

Graphical Solution Graphical support is shown in Figure 9.36, where the graphs of $Y_1 = 1000 * 1.1 \wedge X$ and $Y_2 = 4000$ intersect when $x \approx 14.5$.

▶ **REAL-WORLD CONNECTION** In the next example, we model the life span of a robin with an exponential function.

EXAMPLE 5 Modeling the life spans of robins

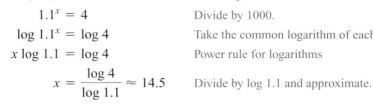

The life spans of 129 robins were monitored over a 4-year period in one study. The formula $f(x) = 10^{-0.42x}$ can be used to calculate the percentage of robins remaining after x years. For example, $f(1) \approx 0.38$ means that after 1 year 38% of the robins were still alive. (*Source: D. Lack, The Life Span of a Robin.*)

(a) Evaluate $f(2)$ and interpret the result.
(b) Determine when 5% of the robins remained.

Solution
(a) $f(2) = 10^{-0.42(2)} \approx 0.145$. After 2 years about 14.5% of the robins were still alive.
(b) Use 5% = 0.05 and solve the following equation.

$10^{-0.42x} = 0.05$	Equation to solve
$\log 10^{-0.42x} = \log 0.05$	Take the common logarithm of each side.
$-0.42x = \log 0.05$	Inverse property: $\log 10^k = k$
$x = \dfrac{\log 0.05}{-0.42} \approx 3.1$	Divide by -0.42.

After about 3 years only 5% of the robins were still alive.

Logarithmic Equations and Models

To solve an exponential equation we use logarithms. To solve a logarithmic equation we *exponentiate* each side of the equation. To do so we use the fact that if $x = y$, then $a^x = a^y$ for any positive base a. For example, to solve

$$\log x = 3$$

we exponentiate each side of the equation, using base 10.

$$10^{\log x} = 10^3$$

Because $10^{\log x} = x$ for all positive x,
$$x = 10^3 = 1000.$$

To solve logarithmic equations, we frequently use the inverse property
$$a^{\log_a x} = x.$$

Examples of this inverse property include
$$e^{\ln 2k} = 2k, \quad 2^{\log_2 x} = x, \quad \text{and} \quad 10^{\log(x+5)} = x + 5.$$

The next two examples show how to solve logarithmic equations, followed by two applications of these methods.

EXAMPLE 6 Solving logarithmic equations

Solve and approximate solutions to the nearest hundredth when appropriate.
(a) $2 \log x = 4$ **(b)** $\ln 3x = 5.5$ **(c)** $\log_2 (x + 4) = 7$

Solution
(a)
$2 \log x = 4$	Given equation
$\log x = 2$	Divide each side by 2.
$10^{\log x} = 10^2$	Exponentiate each side, using base 10.
$x = 100$	Inverse property: $10^{\log k} = k$

(b)
$\ln 3x = 5.5$	Given equation
$e^{\ln 3x} = e^{5.5}$	Exponentiate each side, using base e.
$3x = e^{5.5}$	Inverse property: $e^{\ln k} = k$
$x = \dfrac{e^{5.5}}{3} \approx 81.56$	Divide each side by 3 and approximate.

(c)
$\log_2 (x + 4) = 7$	Given equation
$2^{\log_2 (x+4)} = 2^7$	Exponentiate each side, using base 2.
$x + 4 = 2^7$	Inverse property: $2^{\log_2 k} = k$
$x = 2^7 - 4$	Subtract 4 from each side.
$x = 124$	Simplify.

Because the domain of any logarithmic function includes only positive numbers, it is important to check answers, as emphasized in the next example.

EXAMPLE 7 Solving a logarithmic equation

Solve $\log(x + 2) + \log(x - 2) = \log 5$. Check any answers.

Solution
Start by applying the product rule for logarithms.

$\log(x + 2) + \log(x - 2) = \log 5$	Given equation
$\log((x + 2)(x - 2)) = \log 5$	Product rule
$\log(x^2 - 4) = \log 5$	Multiply.
$10^{\log(x^2-4)} = 10^{\log 5}$	Exponentiate using base 10.
$x^2 - 4 = 5$	Inverse properties
$x^2 = 9$	Add 4.
$x = \pm 3$	Square root property

Check each answer.

$$\log(3+2) + \log(3-2) \stackrel{?}{=} \log 5 \qquad \log(-3+2) + \log(-3-2) \stackrel{?}{=} \log 5$$
$$\log 5 + \log 1 \stackrel{?}{=} \log 5 \qquad \log(-1) + \log(-5) \neq \log 5$$
$$\log 5 + 0 \stackrel{?}{=} \log 5 \qquad \text{Undefined}$$
$$\log 5 = \log 5$$

Although 3 is a solution, -3 is not, because both $\log(-1)$ and $\log(-5)$ are undefined expressions. Be sure to check your answers.

EXAMPLE 8 Modeling runway length

For some types of airplanes with weight x, the minimum runway length L required at takeoff is modeled by

$$L(x) = 3 \log x.$$

In this equation L is measured in thousands of feet and x is measured in thousands of pounds. Estimate the weight of the heaviest airplane that can take off from a runway 5100 feet long. (**Source:** L. Haefner, *Introduction to Transportation Systems*.)

Solution
Runway length is measured in thousands of feet, so we must solve the equation $L(x) = 5.1$.

$$3 \log x = 5.1 \qquad L(x) = 5.1$$
$$\log x = 1.7 \qquad \text{Divide each side by 3.}$$
$$10^{\log x} = 10^{1.7} \qquad \text{Exponentiate each side, using base 10.}$$
$$x = 10^{1.7} \qquad \text{Inverse property: } 10^{\log k} = k$$
$$x \approx 50.1 \qquad \text{Approximate.}$$

The largest airplane that can take off from this runway weighs about 50,000 pounds.

CRITICAL THINKING

In Example 9, Section 9.3, we used the formula $L(x) = 1.3 \ln x$ to model runway length. Are $L(x) = 1.3 \ln x$ and $L(x) = 3 \log x$ equivalent formulas? Explain.

EXAMPLE 9 Modeling bird populations

Near New Guinea there is a relationship between the number of different species of birds and the size of an island. Larger islands tend to have a greater variety of birds. Table 9.13 lists the number of species of birds y found on islands with an area of x square kilometers.

TABLE 9.13

x (km²)	0.1	1	10	100	1000
y (species)	10	15	20	25	30

Source: B. Freedman, *Environmental Ecology*.

(a) Find values for the constants a and b so that $y = a + b \log x$ models the data.
(b) Predict the number of bird species on an island of 4000 square kilometers.

Solution
(a) Because $\log 1 = 0$, substitute $x = 1$ and $y = 15$ in the equation to find a.

$$15 = a + b \log 1$$
$$15 = a + b \cdot 0$$
$$15 = a$$

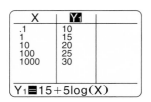

Figure 9.37

Thus $y = 15 + b \log x$. To find b substitute $x = 10$ and $y = 20$.

$$20 = 15 + b \log 10$$
$$20 = 15 + b \cdot 1$$
$$5 = b$$

The data in Table 9.13 are modeled by $y = 15 + 5 \log x$. This result is supported by Figure 9.37.

(b) To predict the number of species on an island of 4000 square kilometers, let $x = 4000$ and find y.

$$y = 15 + 5 \log 4000 \approx 33$$

The model estimates about 33 different species of birds on this island.

9.5 PUTTING IT ALL TOGETHER

Basic steps for solving exponential and logarithmic equations are summarized in the following table.

Type of Equation	Procedure	Example	
Exponential	Begin by solving for the exponential expression a^x. Then take a logarithm of each side.	$4e^x + 1 = 9$	Given equation
		$e^x = 2$	Solve for e^x.
		$\ln e^x = \ln 2$	Take the natural logarithm.
		$x = \ln 2$	Inverse property: $\ln e^k = k$
Logarithmic	Begin by solving for the logarithm in the equation. Then exponentiate each side of the equation, using the same base as the logarithm. That is, if $x = y$, then $a^x = a^y$ for any positive base a.	$\frac{1}{3} \log 2x = 1$	Given equation
		$\log 2x = 3$	Multiply by 3.
		$10^{\log 2x} = 10^3$	Exponentiate using base 10.
		$2x = 1000$	Inverse property: $10^{\log k} = k$
		$x = 500$	Divide by 2.

CHAPTER 9 SUMMARY

SECTION 9.1 ■ COMPOSITE AND INVERSE FUNCTIONS

Composition of Functions If f and g are functions, then the composite function $g \circ f$, or composition of g and f, is defined by $(g \circ f)(x) = g(f(x))$.

Example: If $f(x) = x - 5$ and $g(x) = 2x^2 + 4x - 6$, then $(g \circ f)(x)$ is

$$g(f(x)) = g(x - 5)$$
$$= 2(x - 5)^2 + 4(x - 5) - 6.$$

One-to-One Function A function f is one-to-one if, for any c and d in the domain of f,

$$c \neq d \quad \text{implies that} \quad f(c) \neq f(d).$$

That is, different inputs always result in different outputs.

Example: $f(x) = x^2 + 4$ is *not* one-to-one; $-3 \neq 3$, but $f(-3) = f(3) = 13$.

Horizontal Line Test If every horizontal line intersects the graph of a function f at most once, then f is a one-to-one function.

Examples:

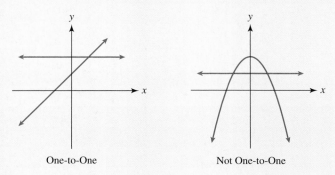

One-to-One   Not One-to-One

Inverse Functions If f is one-to-one, then f has an inverse function, denoted f^{-1}, that satisfies $(f^{-1} \circ f)(x) = x$ and $(f \circ f^{-1})(x) = x$.

Example: $f(x) = 7x$ and $f^{-1}(x) = \frac{x}{7}$ are inverse functions.

SECTION 9.2 ■ EXPONENTIAL FUNCTIONS

Exponential Function An exponential function is defined by $f(x) = Ca^x$, where $a > 0, C > 0$, and $a \neq 1$. Its domain (set of valid inputs) is all real numbers and its range (outputs) is all positive real numbers.

Example: $f(x) = e^x$ is the natural exponential function and $e \approx 2.71828$.

Exponential Growth and Decay When $a > 1$, the graph of $f(x) = Ca^x$ models exponential growth, and when $0 < a < 1$, it models exponential decay. The base a either represents the growth factor or the decay factor. The constant C equals $f(0)$.

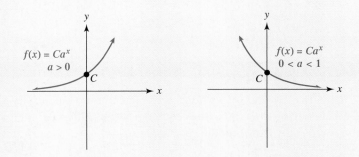

Example: $f(x) = 1.5(2)^x$ is an exponential function with $a = 2$ and $C = 1.5$. It models exponential growth because $a > 1$. The growth factor is 2 because for each unit increase in x, the output from $f(x)$ increases by a *factor* of 2.

SECTION 9.3 ■ LOGARITHMIC FUNCTIONS

Base-a Logarithms The logarithm with base a of a positive number x is denoted $\log_a x$. If $\log_a x = b$, then $x = a^b$. That is, $\log_a x$ represents the exponent on base a that results in x.

Example: $\log_2 16 = 4$ because $16 = 2^4$.

Domain and Range of Logarithmic Functions The domain (set of valid inputs) of a logarithmic function is the set of all positive real numbers and the range (outputs) is the set of real numbers.

Graph of a Logarithmic Function The graph of a logarithmic function passes through the point (1, 0), as illustrated in the following graph. As x becomes large, $\log_a x$ with $a > 1$ grows very slowly.

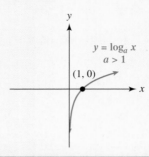

SECTION 9.4 ■ PROPERTIES OF LOGARITHMS

Basic Properties Logarithms have several important properties. For positive numbers m, n, and $a \neq 1$ and any real number r,

1. $\log_a mn = \log_a m + \log_a n$.
2. $\log_a \frac{m}{n} = \log_a m - \log_a n$.
3. $\log_a (m^r) = r \log_a m$.

Examples:
1. $\log 5 + \log 20 = \log (5 \cdot 20) = \log 100 = 2$
2. $\log 100 - \log 5 = \log \frac{100}{5} = \log 20 \approx 1.301$
3. $\ln 2^6 = 6 \ln 2 \approx 4.159$

NOTE: $\log_a 1 = 0$ for any valid base a. Thus $\log 1 = 0$ and $\log_2 1 = 0$.

Inverse Properties The following inverse properties are important for solving exponential and logarithmic equations.

1. $\log_a a^x = x$, for any real number x
2. $a^{\log_a x} = x$, for any positive number x

Examples:
1. $\log_2 2^\pi = \pi$
2. $10^{\log 2.5} = 2.5$

SECTION 9.5 ■ EXPONENTIAL AND LOGARITHMIC EQUATIONS

Solving Equations The calculations a^x and $\log_a x$ are inverse operations, much like addition and subtraction or multiplication and division. When solving an exponential equation, we usually take a logarithm of each side. When solving a logarithmic equation, we usually exponentiate each side.

Examples:

$2(5)^x = 22$	Exponential equation
$5^x = 11$	Divide by 2.
$\log 5^x = \log 11$	Take the common logarithm.
$x \log 5 = \log 11$	Power rule
$x = \dfrac{\log 11}{\log 5}$	Divide by log 5.

$\log 2x = 2$	Logarithmic equation
$10^{\log 2x} = 10^2$	Exponentiate each side.
$2x = 100$	Inverse properties
$x = 50$	Divide by 2.

CHAPTER 9 SHOW YOUR WORK   453

Assignment Name _____ Name _____ Date _____
Show all work for these items: _____

# _____	# _____
# _____	# _____
# _____	# _____
# _____	# _____

454　CHAPTER 9　EXPONENTIAL AND LOGARITHMIC FUNCTIONS

Assignment Name _____ Name _____ Date _____

Show all work for these items: _____

# _____	# _____
# _____	# _____
# _____	# _____
# _____	# _____

CHAPTER 10
Conic Sections

10.1 Parabolas and Circles
10.2 Ellipses and Hyperbolas
10.3 Nonlinear Systems of Equations and Inequalities

Throughout history people have been fascinated with the universe around them and compelled to understand its mysteries. Conic sections, which include parabolas, circles, ellipses, and hyperbolas, have played an important role in gaining this understanding. Although conic sections were described and named by the Greek astronomer Apollonius in 200 B.C., not until much later were they used to model motion in the universe. In the sixteenth century Tycho Brahe, the greatest observational astronomer of the age, recorded precise data on planetary movement in the sky. Using Brahe's data, in 1619 Johannes Kepler determined that planets move in elliptical orbits around the sun. In 1686 Newton used Kepler's work to show that elliptical orbits are the result of his famous theory of gravitation. We now know that all celestial objects—including planets, comets, asteroids, and satellites—travel in paths described by conic sections.

Today scientists search the sky for information about the universe with enormous radio telescopes in the shape of parabolic dishes. The Hubble telescope also makes use of a parabolic mirror. As a result, our understanding of the universe has changed dramatically in recent years.

Conic sections have had a profound influence on people's understanding of their world and the cosmos. In this chapter we introduce you to these age-old curves.

> *The art of asking the right questions in mathematics is more important than the art of solving them.*
> —GEORG CANTOR

Source: Historical Topics for the Mathematics Classroom, Thirty-first Yearbook, NCTM.

10.1 PARABOLAS AND CIRCLES

Types of Conic Sections ▪ Parabolas with Horizontal Axes of Symmetry ▪ Equations of Circles

A LOOK INTO MATH ▷

In this section we discuss two types of conic sections: parabolas and circles. Recall that we discussed parabolas with vertical axes of symmetry in Chapter 8. In this section we discuss parabolas with horizontal axes of symmetry, but first we introduce the three basic types of conic sections. The Hubble telescope travels in a path described by a conic section called an ellipse.

Types of Conic Sections

Conic sections are named after the different ways that a plane can intersect a cone. The three basic curves are parabolas, ellipses, and hyperbolas. A circle is a special case of an ellipse. Figure 10.1 shows the three types of conic sections along with an example of the graph associated with each.

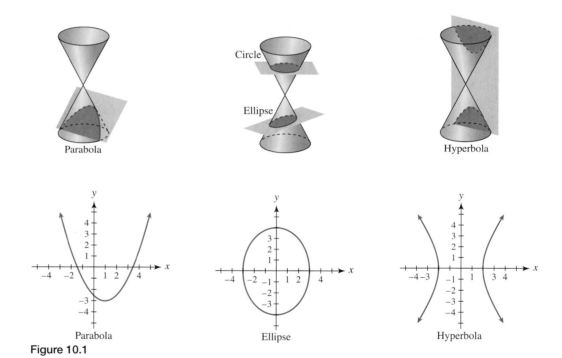

Figure 10.1

Parabolas with Horizontal Axes of Symmetry

Recall that the *vertex form of a parabola* with a vertical axis of symmetry is

$$y = a(x - h)^2 + k,$$

where (h, k) is the vertex. If $a > 0$, the parabola opens upward; if $a < 0$, the parabola opens downward, as shown in Figure 10.2. The preceding equation can also be expressed in the form

$$y = ax^2 + bx + c.$$

In this form the x-coordinate of the vertex is $x = -\frac{b}{2a}$.

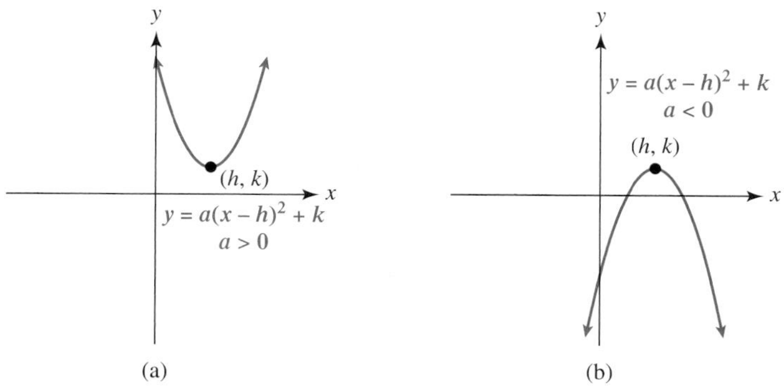

10.1 PARABOLAS AND CIRCLES   457

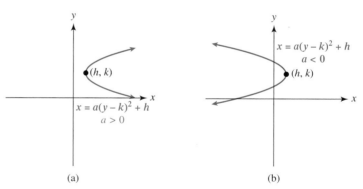

Figure 10.2 Vertical Axes of Symmetry

Interchanging the roles of x and y (and also h and k) gives equations for parabolas that open to the right or the left. In this case, their axes of symmetry are horizontal.

PARABOLAS WITH HORIZONTAL AXES OF SYMMETRY

The graph of $x = a(y - k)^2 + h$ is a parabola that opens to the right if $a > 0$ and to the left if $a < 0$. The vertex of the parabola is located at (h, k).

The graph of $x = ay^2 + by + c$ is a parabola opening to the right if $a > 0$ and to the left if $a < 0$. The y-coordinate of its vertex is given by $y = -\frac{b}{2a}$.

These parabolas are illustrated in Figure 10.3.

Figure 10.3 Horizontal Axes of Symmetry

EXAMPLE 1 Graphing a parabola

Graph $x = -\frac{1}{2}y^2$. Find its vertex and axis of symmetry.

Solution
The equation can be written in vertex form because $x = -\frac{1}{2}(y - 0)^2 + 0$. The vertex is $(0, 0)$, and because $a = -\frac{1}{2} < 0$, the parabola opens to the left. We can make a table of values, as shown in Table 10.1, and plot a few points to help determine the location and shape of the graph. To obtain Table 10.1, we first choose a y-value and then calculate an x-value. The resulting graph is shown in Figure 10.4. Its axis of symmetry is the x-axis, or $y = 0$.

TABLE 10.1

y	x
−2	−2
−1	−$\frac{1}{2}$
0	0
1	−$\frac{1}{2}$
2	−2

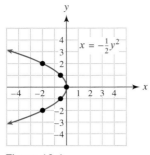

Figure 10.4

EXAMPLE 2 Graphing a parabola

Graph $x = (y - 3)^2 + 2$. Find its vertex and axis of symmetry.

Solution
As $h = 2$ and $k = 3$ in the equation $x = a(y - k)^2 + h$, the vertex is $(2, 3)$, and because $a = 1 > 0$, the parabola opens to the right. This parabola has the same shape as $y = x^2$, except that it opens to the right rather than upward. To graph this parabola we can make a table of values and plot a few points. Table 10.2 can be obtained by first choosing y-values and then calculating corresponding x-values using $x = (y - 3)^2 + 2$.

Sometimes, finding the x- and y-intercepts of the parabola is helpful when you are graphing. To find the x-intercept let $y = 0$ in $x = (y - 3)^2 + 2$. The x-intercept is $x = (0 - 3)^2 + 2 = 11$. To find any y-intercepts let $x = 0$ in $x = (y - 3)^2 + 2$. Here $0 = (y - 3)^2 + 2$ means that $(y - 3)^2 = -2$, which has no real solutions, and that this parabola has no y-intercepts.

Both the graph of the parabola and the points from Table 10.2 are shown in Figure 10.5. Note that there are no y-intercepts and that the x-intercept is 11. The axis of symmetry is $y = 3$ because, if we fold the graph on the horizontal line $y = 3$, the two sides match.

TABLE 10.2

y	x
1	6
2	3
3	2
4	3
5	6

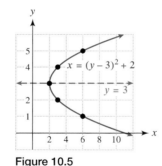

Figure 10.5

EXAMPLE 3 Graphing a parabola and finding its vertex

Identify the vertex and then graph each parabola.
(a) $x = -y^2 + 1$ **(b)** $x = y^2 - 2y - 1$

Solution

(a) If we rewrite $x = -y^2 + 1$ as $x = -(y - 0)^2 + 1$, then $h = 1$ and $k = 0$, so the vertex is $(1, 0)$. By letting $y = 0$ in $x = -y^2 + 1$ we find that the x-intercept is $x = -0^2 + 1 = 1$. Similarly, we let $x = 0$ in $x = -y^2 + 1$ to find the y-intercepts. The equation $0 = -y^2 + 1$ has solutions -1 and 1.

The parabola opens to the left because $a = -1 < 0$. Additional points given in Table 10.3 will help in graphing the parabola shown in Figure 10.6.

TABLE 10.3

y	x
−2	−3
−1	0
0	1
1	0
2	−3

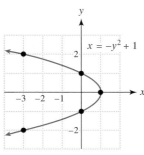

Figure 10.6

(b) The y-coordinate of the vertex for the graph of $x = y^2 - 2y - 1$ is given by

$$y = -\frac{b}{2a} = -\frac{-2}{2(1)} = 1.$$

To find the x-coordinate of the vertex, substitute $y = 1$ into the given equation.

$$x = (1)^2 - 2(1) - 1 = -2$$

The vertex is $(-2, 1)$. The parabola opens to the right because $a = 1 > 0$. The additional points given in Table 10.4 help in graphing the parabola shown in Figure 10.7. Note that the y-intercepts do not have integer values and that the quadratic formula could be used to find approximations for these values. The x-intercept is -1.

TABLE 10.4

y	x
−1	2
0	−1
1	−2
2	−1
3	2

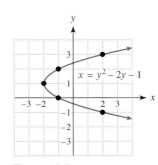

Figure 10.7

Equations of Circles

A **circle** consists of the set of points in a plane that are the same distance from a fixed point. The fixed distance is called the **radius**, and the fixed point is called the **center**. In Figure 10.8 all points lying on the circle are a distance of 2 units from the center $(2, 1)$. Therefore the radius of the circle equals 2.

We can find the equation of the circle shown in Figure 10.8 by using the distance formula. If a point (x, y) lies on the graph of a circle, its distance from the center $(2, 1)$ is 2 and

$$\sqrt{(x - 2)^2 + (y - 1)^2} = 2.$$

Squaring each side gives

$$(x - 2)^2 + (y - 1)^2 = 2^2.$$

This equation represents the standard equation for a circle with center $(2, 1)$ and radius 2.

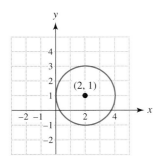

Figure 10.8

STANDARD EQUATION OF A CIRCLE

The **standard equation of a circle** with center (h, k) and radius r is

$$(x - h)^2 + (y - k)^2 = r^2.$$

EXAMPLE 4 Graphing a circle

Graph $x^2 + y^2 = 9$. Find the radius and center.

Solution
The equation $x^2 + y^2 = 9$ can be written in standard form as
$$(x - 0)^2 + (y - 0)^2 = 3^2.$$
Therefore the center is $(0, 0)$ and the radius is **3**. Its graph is shown in Figure 10.9.

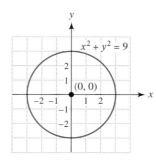

Figure 10.9

EXAMPLE 5 Graphing a circle

Graph $(x + 1)^2 + (y - 3)^2 = 4$. Find the radius and center.

Solution
Write the equation as
$$(x - (-1))^2 + (y - 3)^2 = 2^2.$$

The center is $(-1, 3)$, and the radius is **2**. Its graph is shown in Figure 10.10.

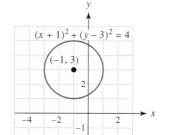

Figure 10.10

TECHNOLOGY NOTE: Graphing Circles

The graph of a circle does not represent a function. One way to graph a circle with a graphing calculator is to solve the equation for y and obtain two equations. One equation gives the upper half of the circle, and the other equation gives the lower half.

For example, to graph $x^2 + y^2 = 4$ begin by solving for y.

$$y^2 = 4 - x^2 \quad \text{Subtract } x^2.$$
$$y = \pm\sqrt{4 - x^2} \quad \text{Square root property}$$

Then graph $Y_1 = \sqrt{(4 - X^2)}$ and $Y_2 = -\sqrt{(4 - X^2)}$. The graph of y_1 is the upper half of the circle, and the graph of y_2 is the lower half of the circle, as shown in Figure 10.11.

[−4.7, 4.7, 1] by [−3.1, 3.1, 1]   [−4.7, 4.7, 1] by [−3.1, 3.1, 1]   [−4.7, 4.7, 1] by [−3.1, 3.1, 1]

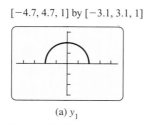 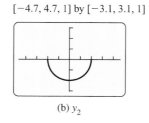 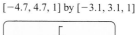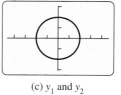

(a) y_1 (b) y_2 (c) y_1 and y_2

Figure 10.11

In the next example we use the *method of completing the square* to find the center and radius of a circle. (To review completing the square, refer to Sections 8.2 and 8.3.)

EXAMPLE 6 Finding the center of a circle

Find the center and radius of the circle given by $x^2 + 4x + y^2 - 6y = 5$.

Solution
Begin by writing the equation as

$$(x^2 + 4x + \underline{\quad}) + (y^2 - 6y + \underline{\quad}) = 5.$$

To complete the square, add $\left(\frac{4}{2}\right)^2 = 4$ and $\left(\frac{-6}{2}\right)^2 = 9$ to each side of the equation.

$$(x^2 + 4x + 4) + (y^2 - 6y + 9) = 5 + 4 + 9$$

Factoring each perfect square trinomial yields

$$(x + 2)^2 + (y - 3)^2 = 18.$$

The center is $(-2, 3)$, and because $18 = \left(\sqrt{18}\right)^2$, the radius is $\sqrt{18}$, or $3\sqrt{2}$.

CRITICAL THINKING

Does the following equation represent a circle? If so, give its center and radius.

$$x^2 + y^2 + 10y = -32$$

$[-4, 4, 1]$ by $[-5, 5, 1]$

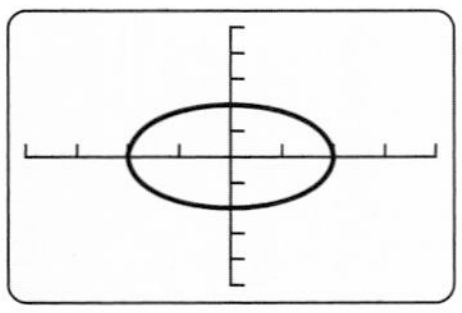

Figure 10.12

NOTE: If a circle is not graphed in a *square viewing rectangle*, it will appear to be an oval rather than a circle. In a square viewing rectangle a circle will appear circular. Figure 10.12 shows the circle graphed in a viewing rectangle that is not square.

10.1 PUTTING IT ALL TOGETHER

The following table summarizes some basic concepts about parabolas and circles.

Concept	Explanation	Example
Parabola with Horizontal Axis	Vertex form: $x = a(y - k)^2 + h$. If $a > 0$, it opens to the right; if $a < 0$, it opens to the left. The vertex is (h, k). See Figure 10.3. These parabolas may also be expressed as $x = ay^2 + by + c$, where the y-coordinate of the vertex is $y = -\frac{b}{2a}$.	$x = 2(y - 1)^2 + 4$ opens to the right and its vertex is $(4, 1)$. 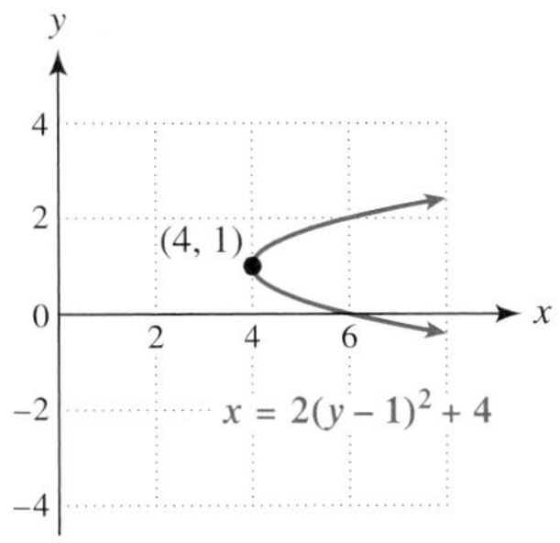
Standard Equation of a Circle	Standard equation: $(x - h)^2 + (y - k)^2 = r^2$. The radius is r and the center is (h, k).	$(x + 2)^2 + (y - 1)^2 = 16$ has center $(-2, 1)$ and radius 4.

10.2 ELLIPSES AND HYPERBOLAS

Equations of Ellipses ▪ Equations of Hyperbolas

A LOOK INTO MATH ▷ Celestial objects travel in paths or trajectories determined by conic sections. For this reason, conic sections have been studied for centuries. In modern times physicists have learned that subatomic particles can also travel in trajectories determined by conic sections. Recall that the three main types of conic sections are parabolas, ellipses, and hyperbolas and that circles are a special type of ellipse. In Section 8.2 and Section 10.1, we discussed parabolas and circles. In this section we focus on ellipses and hyperbolas and some of their applications.

Equations of Ellipses

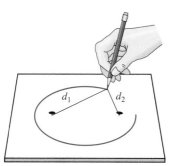

Figure 10.13

One method used to sketch an ellipse is to tie the ends of a string to two nails driven into a flat board. If a pencil is placed against the string anywhere between the nails, as shown in Figure 10.13, and is used to draw a curve, the resulting curve is an ellipse. The sum of the distances d_1 and d_2 between the pencil and each of the nails is always fixed by the length of the string. The location of the nails corresponds to the foci of the ellipse. An **ellipse** is the set of points in a plane, the sum of whose distances from two fixed points is constant. Each fixed point is called a **focus** (plural *foci*) of the ellipse.

CRITICAL THINKING

What happens to the shape of the ellipse shown in Figure 10.13 as the nails are moved farther apart? What happens to its shape as the nails are moved closer together? When would a circle be formed?

In Figure 10.14 the **major axis** and the **minor axis** are labeled for each ellipse. The major axis is the longer of the two axes. Figure 10.14(a) shows an ellipse with a *horizontal* major axis, and Figure 10.14(b) shows an ellipse with a *vertical* major axis. The **vertices**, V_1 and V_2, of each ellipse are located at the endpoints of the major axis, and the **center** of the ellipse is the midpoint of the major axis (or the intersection of the major and minor axes).

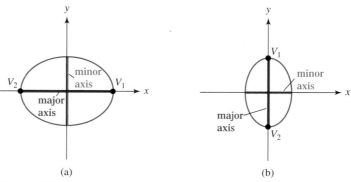

Figure 10.14

A vertical line can intersect the graph of an ellipse twice, so an ellipse cannot be represented by a function. However, some ellipses can be represented by the following equations.

STANDARD EQUATIONS FOR ELLIPSES CENTERED AT (0, 0)

The ellipse with center at the origin, *horizontal* major axis, and equation

$$\frac{x^2}{a^2} + \frac{y^2}{b^2} = 1, \quad a > b > 0,$$

has vertices $(\pm a, 0)$ and endpoints of the minor axis $(0, \pm b)$.

The ellipse with center at the origin, *vertical* major axis, and equation

$$\frac{x^2}{b^2} + \frac{y^2}{a^2} = 1, \quad a > b > 0,$$

has vertices $(0, \pm a)$ and endpoints of the minor axis $(\pm b, 0)$.

Figure 10.15(a) shows an ellipse having a horizontal major axis; Figure 10.15(b) shows one having a vertical major axis. The coordinates of the vertices V_1 and V_2 and endpoints of the minor axis U_1 and U_2 are labeled.

CRITICAL THINKING

Suppose that $a = b$ for an ellipse centered at (0, 0). What can be said about the ellipse? Explain.

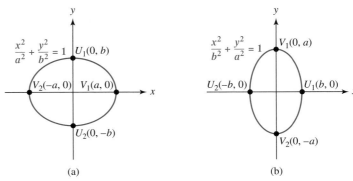

Figure 10.15

In the next example we show how to sketch graphs of ellipses.

EXAMPLE 1 Sketching ellipses

Sketch a graph of each ellipse. Label the vertices and endpoints of the minor axes.

(a) $\dfrac{x^2}{25} + \dfrac{y^2}{4} = 1$ (b) $9x^2 + 4y^2 = 36$

Solution

(a) The equation $\frac{x^2}{25} + \frac{y^2}{4} = 1$ describes an ellipse with $a^2 = 25$ and $b^2 = 4$. (When you are deciding whether 25 or 4 represents a^2, let a^2 be the larger of the two numbers.) Thus $a = 5$ and $b = 2$, so the ellipse has a horizontal major axis with vertices $(\pm 5, 0)$ and the endpoints of the minor axis are $(0, \pm 2)$. Plot these four points and then sketch the ellipse, as shown in Figure 10.16(a).

(b) To put $9x^2 + 4y^2 = 36$ in standard form, divide each term by 36.

$$9x^2 + 4y^2 = 36 \quad \text{Given equation}$$

$$\frac{9x^2}{36} + \frac{4y^2}{36} = \frac{36}{36} \quad \text{Divide each side by 36.}$$

$$\frac{x^2}{4} + \frac{y^2}{9} = 1 \quad \text{Simplify.}$$

This ellipse has a vertical major axis with $a = 3$ and $b = 2$. The vertices are $(0, \pm 3)$, and the endpoints of the minor axis are $(\pm 2, 0)$, as shown in Figure 10.16(b).

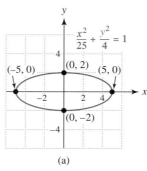
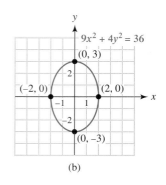

Figure 10.16

EXAMPLE 2 Finding the equation of an ellipse

Use the graph in Figure 10.17 to determine an equation of the ellipse.

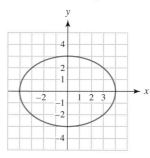

Figure 10.17

Solution
The ellipse is centered at (0, 0) with a horizontal major axis. The length of the major axis is 8, so $a = 4$. The length of the minor axis is 6, so $b = 3$. Thus $a^2 = 16$ and $b^2 = 9$, and the equation of the ellipse is

$$\frac{x^2}{16} + \frac{y^2}{9} = 1.$$

▶ **REAL-WORLD CONNECTION** Planets travel around the sun in elliptical orbits. Astronomers have measured the values of a and b for each planet. Using this information, we can find the equation of a planet's orbit, as illustrated in the next example.

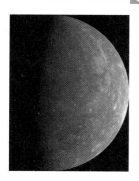

EXAMPLE 3 Modeling the orbit of Mercury

The planet Mercury has the least circular orbit of the eight major planets. For Mercury $a = 0.387$ and $b = 0.379$. The units are astronomical units (A.U.), where 1 A.U. equals 93 million miles—the distance between Earth and the sun. Graph $\frac{x^2}{a^2} + \frac{y^2}{b^2} = 1$ to model the orbit of Mercury in $[-0.6, 0.6, 0.1]$ by $[-0.4, 0.4, 0.1]$. Then plot the sun at the point $(0.08, 0)$. (*Source:* M. Zeilik, *Introductory Astronomy and Astrophysics*.)

Solution
The orbit of Mercury is given by

$$\frac{x^2}{0.387^2} + \frac{y^2}{0.379^2} = 1.$$

To graph an ellipse with some graphing calculators, we must solve the equation for y. Doing so results in two equations.

$$\frac{x^2}{0.387^2} + \frac{y^2}{0.379^2} = 1$$

$$\frac{y^2}{0.379^2} = 1 - \frac{x^2}{0.387^2}$$

$$\frac{y}{0.379} = \pm\sqrt{1 - \frac{x^2}{0.387^2}}$$

$$y = \pm 0.379\sqrt{1 - \frac{x^2}{0.387^2}}$$

The orbit of Mercury results from graphing these two equations. See Figures 10.18(a) and (b). The point (0.08, 0) represents the position of the sun in Figure 10.18(b).

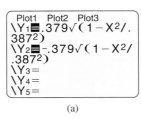
(a)

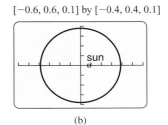
(b)

Figure 10.18

CRITICAL THINKING

Use Figure 10.18 and the information in Example 3 to estimate the minimum and maximum distances that Mercury is from the sun.

Equations of Hyperbolas

The third type of conic section is the **hyperbola**, which is the set of points in a plane, the difference of whose distances from two fixed points is constant. Each fixed point is called a **focus** of the hyperbola. Figure 10.19 shows a hyperbola whose equation is

$$\frac{x^2}{4} - \frac{y^2}{9} = 1.$$

This hyperbola is centered at the origin and has two **branches**, a *left branch* and a *right branch*. The **vertices** are $(-2, 0)$ and $(2, 0)$, and the line segment connecting the vertices is called the **transverse axis**. (The transverse axis is not part of the hyperbola.)

By the vertical line test, a hyperbola cannot be represented by a function, but many hyperbolas can be described by the following equations.

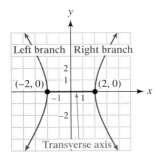

Figure 10.19

STANDARD EQUATIONS FOR HYPERBOLAS CENTERED AT (0, 0)

The hyperbola with center at the origin, *horizontal* transverse axis, and equation

$$\frac{x^2}{a^2} - \frac{y^2}{b^2} = 1$$

has vertices $(\pm a, 0)$.

The hyperbola with center at the origin, *vertical* transverse axis, and equation

$$\frac{y^2}{a^2} - \frac{x^2}{b^2} = 1$$

has vertices $(0, \pm a)$.

Hyperbolas, along with the coordinates of their vertices, are shown in Figure 10.20. The two parts of the hyperbola in Figure 10.20(a) are the *left branch* and *right branch*, whereas in Figure 10.20(b) the hyperbola has an *upper branch* and a *lower branch*. The dashed rectangle in each figure is called the **fundamental rectangle**, and its four vertices are determined by either $(\pm a, \pm b)$ or $(\pm b, \pm a)$. If its diagonals are extended, they correspond to the asymptotes of the hyperbola. The lines $y = \pm \frac{b}{a}x$ and $y = \pm \frac{a}{b}x$ are **asymptotes** for the hyperbolas, respectively, and may be used as an aid to graphing them.

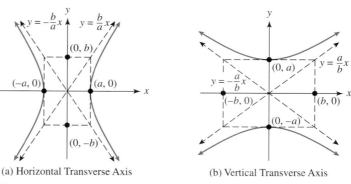

(a) Horizontal Transverse Axis   (b) Vertical Transverse Axis

Figure 10.20

NOTE: A hyperbola consists of two solid curves, or branches. The dashed lines and rectangles are not part of the actual graph but are used as an aid for sketching the hyperbola.

▶ **REAL-WORLD CONNECTION** One interpretation of an asymptote of a hyperbola can be based on trajectories of comets as they approach the sun. Comets travel in parabolic, elliptic, or hyperbolic trajectories. If the speed of a comet is too slow, the gravitational pull of the sun captures the comet in an elliptic orbit (see Figure 10.21(a)). If the speed of the comet is too fast, the sun's gravity is too weak to capture the comet and the comet passes by it in a hyperbolic trajectory. Near the sun the gravitational pull is stronger, and the comet's trajectory is curved. Farther from the sun, the gravitational pull becomes weaker, and the comet eventually returns to a straight-line trajectory determined by the *asymptote* of the hyperbola (see Figure 10.21(b)). Finally, if the speed is neither too slow nor too fast, the comet will travel in a parabolic path (see Figure 10.21(c)).

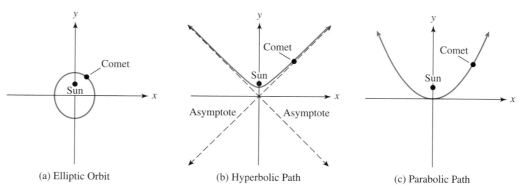

(a) Elliptic Orbit   (b) Hyperbolic Path   (c) Parabolic Path

Figure 10.21

EXAMPLE 4   Sketching a hyperbola

Sketch a graph of $\frac{y^2}{4} - \frac{x^2}{9} = 1$. Label the vertices and show the asymptotes.

Solution

The equation is in standard form with $a^2 = 4$ and $b^2 = 9$, so $a = 2$ and $b = 3$. It has a vertical transverse axis with vertices $(0, -2)$ and $(0, 2)$. The vertices of the fundamental rectangle are $(\pm 3, \pm 2)$, that is, $(3, 2)$, $(3, -2)$, $(-3, 2)$, and $(-3, -2)$. The asymptotes are the diagonals of this rectangle and are given by $y = \pm \frac{a}{b}x$, or $y = \pm \frac{2}{3}x$. Figure 10.22 shows the hyperbola and these features.

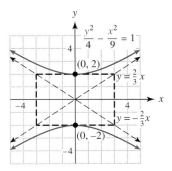

Figure 10.22

TECHNOLOGY NOTE: Graphing a Hyperbola

The graph of a hyperbola does not represent a function. One way to graph a hyperbola with a graphing calculator is to solve the equation for y and obtain two equations. One equation gives the upper (or right) half of the hyperbola and the other equation gives the lower (or left) half. For example, to graph $\frac{y^2}{4} - \frac{x^2}{8} = 1$, begin by solving for y.

$$\frac{y^2}{4} = 1 + \frac{x^2}{8} \qquad \text{Add } \frac{x^2}{8}.$$

$$y^2 = 4\left(1 + \frac{x^2}{8}\right) \qquad \text{Multiply by 4.}$$

$$y = \pm 2\sqrt{1 + \frac{x^2}{8}} \qquad \text{Square root property}$$

Graph $Y_1 = 2\sqrt{(1 + X^2/8)}$ and $Y_2 = -2\sqrt{(1 + X^2/8)}$. See Figure 10.23.

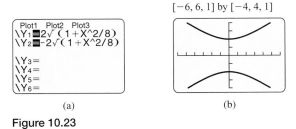

(a)   (b)

Figure 10.23

EXAMPLE 5   Determining the equation of a hyperbola from its graph

Use the graph shown in Figure 10.24 to determine an equation of the hyperbola.

Solution
The hyperbola has a horizontal transverse axis, so the x^2-term must come first in the equation. The vertices of the hyperbola are $(\pm 3, 0)$, which indicates that $a = 3$ and so $a^2 = 9$. The value of b can be found by noting that one of the asymptotes passes through the point $(3, 4)$. This asymptote has the equation $y = \frac{b}{a}x$ or $y = \frac{4}{3}x$, so let $b = 4$ and $b^2 = 16$. The equation of the hyperbola is $\frac{x^2}{9} - \frac{y^2}{16} = 1$.

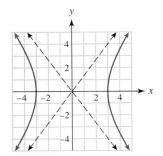

Figure 10.24

10.2 PUTTING IT ALL TOGETHER

The following table summarizes some basic concepts about ellipses and hyperbolas.

Concept	Description	
Ellipses Centered at (0, 0) with $a > b > 0$	**Horizontal Major Axis** Vertices: $(a, 0)$ and $(-a, 0)$ Endpoints of minor axis: $(0, b)$ and $(0, -b)$ $$\frac{x^2}{a^2} + \frac{y^2}{b^2} = 1$$	**Vertical Major Axis** Vertices: $(0, a)$ and $(0, -a)$ Endpoints of minor axis: $(-b, 0)$ and $(b, 0)$ $$\frac{x^2}{b^2} + \frac{y^2}{a^2} = 1$$
Hyperbolas Centered at (0, 0) with $a > 0$ and $b > 0$	**Horizontal Transverse Axis** Vertices: $(a, 0)$ and $(-a, 0)$ Asymptotes: $y = \pm \frac{b}{a} x$ $$\frac{x^2}{a^2} - \frac{y^2}{b^2} = 1$$	**Vertical Transverse Axis** Vertices: $(0, a)$ and $(0, -a)$ Asymptotes: $y = \pm \frac{a}{b} x$ $$\frac{y^2}{a^2} - \frac{x^2}{b^2} = 1$$

10.3 NONLINEAR SYSTEMS OF EQUATIONS AND INEQUALITIES

Basic Concepts ■ Solving Nonlinear Systems of Equations ■ Solving Nonlinear Systems of Inequalities

A LOOK INTO MATH ▷

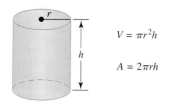

Figure 10.25 Cylindrical Container

To describe characteristics of curved objects we often need *nonlinear equations*. The equations of the conic sections discussed in this chapter are but a few examples of nonlinear equations. For instance, cylinders have a curved shape, as illustrated in Figure 10.25. If the radius of a cylinder is denoted r and its height h, then its volume V is given by the nonlinear equation $V = \pi r^2 h$ and its side area A is given by the nonlinear equation $A = 2\pi rh$.

If we want to manufacture a cylindrical container that holds 35 cubic inches and whose side area is 50 square inches, we need to solve the following **nonlinear system of equations**. (This system is solved in Example 4.)

$$\pi r^2 h = 35$$
$$2\pi rh = 50$$

Basic Concepts

One way to locate the points at which the line $y = 2x$ intersects the circle $x^2 + y^2 = 5$, is to graph both equations (see Figure 10.26).

The equation describing the circle is nonlinear. Another way to locate the points of intersection is symbolically, by solving the nonlinear system of equations.

$$y = 2x$$
$$x^2 + y^2 = 5$$

Linear systems of equations can have no solutions, one solution, or infinitely many solutions. It is possible for a nonlinear system of equations to have *any number* of solutions. Figure 10.26 shows that this nonlinear system of equations has two solutions: $(-1, -2)$ and $(1, 2)$.

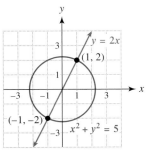

Figure 10.26

Solving Nonlinear Systems of Equations

Nonlinear systems of equations in two variables can sometimes be solved graphically, numerically, and symbolically. One symbolic technique is the **method of substitution**, which we demonstrate in the next example.

EXAMPLE 1 Solving a nonlinear system of equations symbolically

Solve the system of equations symbolically. Check any solutions.

$$y = 2x$$
$$x^2 + y^2 = 5$$

Solution
Substitute $2x$ for y in the second equation and solve for x.

$x^2 + (2x)^2 = 5$	Let $y = 2x$ in the second equation.
$x^2 + 4x^2 = 5$	Properties of exponents
$5x^2 = 5$	Combine like terms.
$x^2 = 1$	Divide by 5.
$x = \pm 1$	Square root property

To determine corresponding y-values, substitute $x = \pm 1$ in $y = 2x$; the solutions are $(1, 2)$ and $(-1, -2)$. To check $(1, 2)$, substitute $x = 1$ and $y = 2$ in the given equations.

$2 \stackrel{?}{=} 2(1)$	True
$(1)^2 + (2)^2 \stackrel{?}{=} 5$	True

To check $(-1, -2)$, substitute $x = -1$ and $y = -2$ in the given equations.

$-2 \stackrel{?}{=} 2(-1)$	True
$(-1)^2 + (-2)^2 \stackrel{?}{=} 5$	True

The solutions check.

In the next example we solve a nonlinear system of equations graphically and symbolically. The symbolic solution gives the exact solutions.

EXAMPLE 2 Solving a nonlinear system of equations

Solve the nonlinear system of equations graphically and symbolically.

$$x^2 - y = 2$$
$$x^2 + y = 4$$

Solution

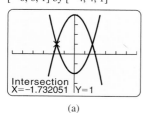

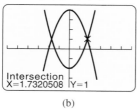

Figure 10.27

Graphical Solution Begin by solving each equation for y.

$$y = x^2 - 2$$
$$y = 4 - x^2$$

Graph $Y_1 = X^\wedge 2 - 2$ and $Y_2 = 4 - X^\wedge 2$. The solutions are approximately $(-1.73, 1)$ and $(1.73, 1)$, as shown in Figure 10.27. The graphs consist of two parabolas intersecting at two points.

Symbolic Solution Solving the first equation for y gives $y = x^2 - 2$. Substitute this expression for y in the second equation and solve for x.

$$
\begin{aligned}
x^2 + y &= 4 && \text{Second equation} \\
x^2 + (x^2 - 2) &= 4 && \text{Substitute } y = x^2 - 2. \\
2x^2 &= 6 && \text{Combine like terms; add 2.} \\
x^2 &= 3 && \text{Divide by 2.} \\
x &= \pm\sqrt{3} && \text{Square root property}
\end{aligned}
$$

To determine y, substitute $x = \pm\sqrt{3}$ in $y = x^2 - 2$.

$$y = (\sqrt{3})^2 - 2 = 3 - 2 = 1$$
$$y = (-\sqrt{3})^2 - 2 = 3 - 2 = 1$$

The *exact* solutions are $(\sqrt{3}, 1)$ and $(-\sqrt{3}, 1)$.

EXAMPLE 3 Solving a nonlinear system of equations

Solve the nonlinear system of equations symbolically and graphically.

$$x^2 - y^2 = 3$$
$$x^2 + y^2 = 5$$

Solution

Symbolic Solution Instead of using substitution on this nonlinear system of equations, we use elimination. Note that, if we add the two equations, the y-variable will be eliminated.

$$
\begin{aligned}
x^2 - y^2 &= 3 \\
\underline{x^2 + y^2} &= \underline{5} \\
2x^2 &= 8 \quad \text{Add equations.}
\end{aligned}
$$

Solving gives $x^2 = 4$, or $x = \pm 2$. To determine y, substitute 4 for x^2 in $x^2 + y^2 = 5$.

$$4 + y^2 = 5 \quad \text{or} \quad y^2 = 1$$

Because $y^2 = 1$, $y = \pm 1$. There are four solutions: $(2, 1)$, $(2, -1)$, $(-2, 1)$, and $(-2, -1)$.

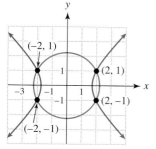

Figure 10.28

Graphical Solution The graph of the first equation is a hyperbola, and the graph of the second is a circle with radius $\sqrt{5}$. The four points of intersection (solutions) are $(\pm 2, \pm 1)$, as shown in Figure 10.28.

In the next example we solve the system of equations presented in the introduction.

EXAMPLE 4 Modeling the dimensions of a can

Find the dimensions of a can having a volume V of 35 cubic inches and a side area A of 50 square inches by solving the following nonlinear system of equations symbolically.

$$\pi r^2 h = 35$$
$$2\pi r h = 50$$

Solution
We can find r by solving each equation for h and then setting them equal. This eliminates the variable h.

$$\frac{50}{2\pi r} = \frac{35}{\pi r^2} \qquad h = \frac{50}{2\pi r} \text{ and } h = \frac{35}{\pi r^2}$$
$$50\pi r^2 = 70\pi r \qquad \text{Clear fractions (cross multiply).}$$
$$50\pi r^2 - 70\pi r = 0 \qquad \text{Subtract } 70\pi r.$$
$$10\pi r(5r - 7) = 0 \qquad \text{Factor out } 10\pi r.$$
$$10\pi r = 0 \quad \text{or} \quad 5r - 7 = 0 \qquad \text{Zero-product property}$$
$$r = 0 \quad \text{or} \quad r = \tfrac{7}{5} = 1.4 \qquad \text{Solve.}$$

Because $h = \frac{50}{2\pi r}$, $r = 0$ is not possible, but we can find h by substituting 1.4 for r in the formula.

$$h = \frac{50}{2\pi(1.4)} \approx 5.68$$

A can having a volume of 35 cubic inches and a side area of 50 square inches has a radius of 1.4 inches and a height of about 5.68 inches.

Solving Nonlinear Systems of Inequalities

In Section 4.3 we solved systems of linear inequalities. A **nonlinear system of inequalities** in two variables can be solved similarly by using graphical techniques. For example, consider the nonlinear system of inequalities

$$y \geq x^2 - 2$$
$$y \leq 4 - x^2.$$

The graph of $y = x^2 - 2$ is a parabola opening upward. The solution set to $y \geq x^2 - 2$ includes all points lying on or above this parabola. See Figure 10.29(a). Similarly, the graph of $y = 4 - x^2$ is a parabola opening downward. The solution set to $y \leq 4 - x^2$ includes all points lying on or below this parabola. See Figure 10.29(b).

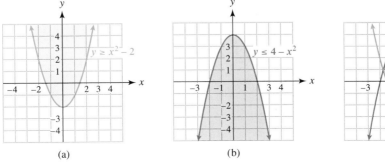

Figure 10.29

The solution set for this nonlinear *system* of inequalities includes all points (x, y) in *both* shaded regions. The *intersection* of the shaded regions is shown in Figure 10.29(c).

EXAMPLE 5 Solving a nonlinear system of inequalities graphically

Shade the solution set for the system of inequalities.

$$\frac{x^2}{4} + \frac{y^2}{9} < 1$$
$$y > 1$$

Solution
The solutions to $\frac{x^2}{4} + \frac{y^2}{9} < 1$ lie *inside* the ellipse $\frac{x^2}{4} + \frac{y^2}{9} = 1$. See Figure 10.30(a). Solutions to $y > 1$ lie above the line $y = 1$, as shown in Figure 10.30(b). The intersection of these two regions is shown in Figure 10.30(c). Any point in this region is a solution. For example, the point (0, 2) lies in the shaded region and is a solution to the system. Note that a dashed curve and line are used when equality is not included.

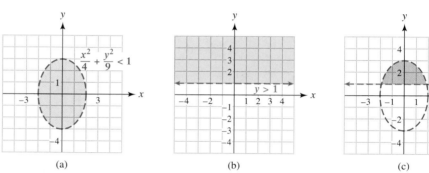

(a)   (b)   (c)

Figure 10.30

EXAMPLE 6 Solving a nonlinear system of inequalities graphically

Shade the solution set for the following system of inequalities.

$$x^2 + y \leq 4$$
$$-x + y \geq 2$$

Solution
The solutions to $x^2 + y \leq 4$ lie on or below the parabola $y = -x^2 + 4$, and the solutions to $-x + y \geq 2$ lie on or above the line $y = x + 2$. The appropriate shaded region is shown in Figure 10.31. Both the parabola and the line are solid because equality is included in both inequalities.

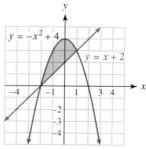

Figure 10.31

EXAMPLE 7 Solving a system of inequalities with a graphing calculator

Shade the solution set for the following system of inequalities.

$$y \geq x^2 - 2$$
$$y \geq -1 - x$$

Solution
Enter $Y_1 = X^2 - 2$ and $Y_2 = -1 - X$, as shown in Figure 10.32(a). Note that the option to shade above the graphs of Y_1 and Y_2 was selected to the left of Y_1 and Y_2. Then the two inequalities were graphed in Figure 10.32(b). The solution set corresponds to the region where there are both vertical and horizontal lines.

CALCULATOR HELP
To shade the solution set to a system of inequalities, see the Appendix (page AP-7).

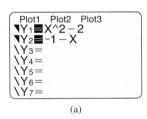
(a)

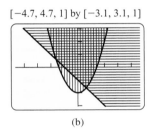
[−4.7, 4.7, 1] by [−3.1, 3.1, 1]
(b)

Figure 10.32

10.3 PUTTING IT ALL TOGETHER

In this section we discussed nonlinear systems of equations in two variables. Unlike a linear system of equations, a nonlinear system of equations can have *any number of solutions*. Nonlinear systems of equations can be solved symbolically and graphically. Nonlinear systems of inequalities involving two variables usually have infinitely many solutions, which can be represented by a shaded region in the *xy*-plane. The following table summarizes these concepts.

Concept	Explanation
Nonlinear Systems of Equations in Two Variables	To solve the following system of equations symbolically, using *substitution*, solve the first equation for *y*. $$x + y = 5 \quad \text{or} \quad y = 5 - x$$ $$x^2 - y = 1$$ Substitute $5 - x$ for *y* in the second equation and solve the resulting quadratic equation. (*Elimination* can also be used on this system.) $$x^2 - (5 - x) = 1 \quad \text{or} \quad x^2 + x - 6 = 0$$ implies that $$x = -3 \quad \text{or} \quad x = 2.$$

continued on next page

continued from previous page

Concept	Explanation
Nonlinear Systems of Equations in Two Variables *(continued)*	Then $y = 5 - (-3) = 8$ or $y = 5 - 2 = 3$. The solutions are $(-3, 8)$ and $(2, 3)$. Graphical support is shown in the accompanying figure.
Nonlinear Systems of Inequalities in Two Variables	To solve the following system of inequalities graphically, solve each inequality for y. $x + y \leq 5 \quad \text{or} \quad y \leq 5 - x$ $x^2 - y \leq 1 \quad \text{or} \quad y \geq x^2 - 1$ The solutions lie on or above the parabola and on or below the line, as shown in the figure.

CHAPTER 10 SUMMARY

SECTION 10.1 ■ PARABOLAS AND CIRCLES

Parabolas There are three basic types of conic sections: parabolas, ellipses, and hyperbolas. A parabola can have a vertical or a horizontal axis. Two forms of an equation for a parabola with a *vertical axis* are

$$y = ax^2 + bx + c \quad \text{and} \quad y = a(x - h)^2 + k.$$

If $a > 0$ the parabola opens upward, and if $a < 0$ it opens downward (see the figure on the left). The vertex is located at (h, k). Two forms of an equation for a parabola with a *horizontal axis* are

$$x = ay^2 + by + c \quad \text{and} \quad x = a(y - k)^2 + h.$$

If $a > 0$ the parabola opens to the right, and if $a < 0$ it opens to the left (see the figure on the right). The vertex is located at (h, k).

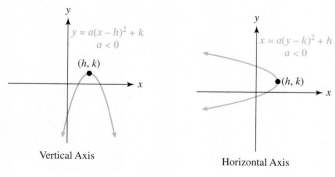

Vertical Axis   Horizontal Axis

Circles The standard equation for a circle with center (h, k) and radius r is

$$(x - h)^2 + (y - k)^2 = r^2.$$

SECTION 10.2 ■ ELLIPSES AND HYPERBOLAS

Ellipses The standard equation for an ellipse centered at the origin with a *horizontal major axis* is $\frac{x^2}{a^2} + \frac{y^2}{b^2} = 1$, $a > b > 0$, and the vertices are $(\pm a, 0)$, as shown in the figure on the left. The standard equation for an ellipse centered at the origin with a *vertical major axis* is $\frac{x^2}{b^2} + \frac{y^2}{a^2} = 1$, $a > b > 0$, and the vertices are $(0, \pm a)$, as shown in the figure on the right. Circles are a special type of ellipse, with the major and minor axes having equal lengths.

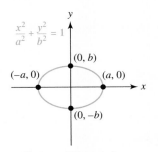
Horizontal Major Axis

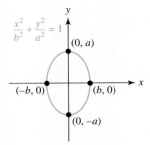
Vertical Major Axis

Hyperbolas The standard equation for a hyperbola centered at the origin with a *horizontal transverse axis* is $\frac{x^2}{a^2} - \frac{y^2}{b^2} = 1$, the asymptotes are given by $y = \pm \frac{b}{a}x$, and the vertices are $(\pm a, 0)$, as shown in the figure on the left. The standard equation for a hyperbola centered at the origin with a *vertical transverse axis* is $\frac{y^2}{a^2} - \frac{x^2}{b^2} = 1$, the asymptotes are $y = \pm \frac{a}{b}x$, and the vertices are $(0, \pm a)$, as shown in the figure on the right.

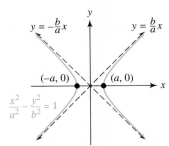
Horizontal Transverse Axis

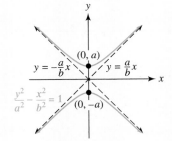
Vertical Transverse Axis

SECTION 10.3 ■ NONLINEAR SYSTEMS OF EQUATIONS AND INEQUALITIES

Nonlinear Systems Nonlinear systems of equations can have any number of solutions. The methods of substitution or elimination can often be used to solve a nonlinear system of equations symbolically. Nonlinear systems can also be solved graphically. The solution set for a nonlinear system of two inequalities in two variables is typically a region in the xy-plane. A solution is an ordered pair (x, y) that satisfies both inequalities.

Example: Solve $y \geq x^2 - 2$
$y \leq 4 - \frac{1}{2}x^2$.

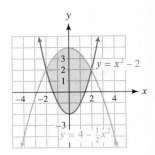

476 CHAPTER 10 CONIC SECTIONS

Assignment Name _____ Name _____ Date _____
Show all work for these items: _____

# _____	# _____
# _____	# _____
# _____	# _____
# _____	# _____

CHAPTER 10 SHOW YOUR WORK 477

Assignment Name _____ Name _____ Date _____
Show all work for these items: _____

# _____	# _____
# _____	# _____
# _____	# _____
# _____	# _____

CHAPTER 11

Sequences and Series

11.1 Sequences
11.2 Arithmetic and Geometric Sequences
11.3 Series
11.4 The Binomial Theorem

In this final chapter we present sequences and series, which are essential topics because they are used to model and approximate important quantities. Complicated population growth can be modeled with sequences, and accurate approximations for numbers such as π and e are made with series. For example, in about 2000 B.C. the Babylonians thought that π equaled 3.125, whereas at the same time the Chinese thought π equaled 3. By 1700, mathematicians were able to calculate π to 100 decimal places due to the invention of series. Series also are essential to the solution of many modern applied mathematics problems.

Although you may not always recognize the impact of mathematics on everyday life, its influence is nonetheless profound. Mathematics is the *language of technology*—it allows experiences to be quantified. In the preceding chapters we showed numerous examples of mathematics being used to model the real world. Computers, DVD players, cars, highway design, weather, hurricanes, electricity, government data, cellular phones, medicine, ecology, business, sports, and psychology represent only some of the applications of mathematics. In fact, if a subject is studied in enough detail, mathematics usually appears in one form or another. Although predicting what the future may bring is difficult, one thing *is* certain—mathematics will continue to play an important role in both theoretical research and new technology.

> *The essence of mathematics is not to make simple things complicated, but to make complicated things simple.*
>
> —STANLEY GUDDER

Source: Mathforum.org, Drexel University 1994–2007.

11.1 SEQUENCES

Basic Concepts ▪ Representations of Sequences ▪ Models and Applications

A LOOK INTO MATH ▷ Sequences are *ordered lists*. For example, names listed alphabetically represent a sequence. Figure 11.1 shows an insect population in thousands per acre over a 6-year period. Listing populations by year is another example of a sequence. In mathematics a sequence is a function, for which valid inputs must be natural numbers. For example, we can use a function f to define this sequence by letting $f(1)$ represent the insect population after 1 year, $f(2)$ represent the insect population after 2 years, and in general let $f(n)$ represent the population after n years. In this section we discuss sequences.

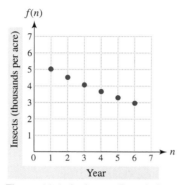

Figure 11.1 An Insect Population

Basic Concepts

▶ **REAL-WORLD CONNECTION** Suppose that an individual's starting salary is $40,000 per year and that the person's salary is increased by 10% each year. This situation is modeled by the formula

$$f(n) = 40{,}000(1.10)^n.$$

We do not allow the input n to be any real number, but rather limit n to a *natural number* because the individual's salary is constant throughout a particular year. The first 5 terms of the sequence are

$$f(1), f(2), f(3), f(4), f(5).$$

They can be computed as follows.

$$f(1) = 40{,}000(1.10)^1 = 44{,}000$$
$$f(2) = 40{,}000(1.10)^2 = 48{,}400$$
$$f(3) = 40{,}000(1.10)^3 = 53{,}240$$
$$f(4) = 40{,}000(1.10)^4 = 58{,}564$$
$$f(5) = 40{,}000(1.10)^5 \approx 64{,}420$$

This sequence is represented *numerically* in Table 11.1.

TABLE 11.1 Numerical Representation

n	1	2	3	4	5
$f(n)$	44,000	48,400	53,240	58,564	64,420

A *graphical* representation results when each data point in Table 11.1 is plotted, as illustrated in Figure 11.2.

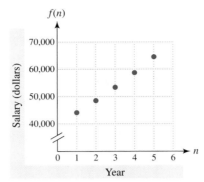

Figure 11.2 Graphical Representation

NOTE: Graphs of sequences are scatterplots.

The preceding sequence is an example of a *finite sequence* of numbers. The even natural numbers,

$$2, 4, 6, 8, 10, 12, 14, \ldots$$

are an example of an *infinite sequence* represented by $f(n) = 2n$, where n is a natural number. The three dots, or periods (called an *ellipsis*), indicate that the pattern continues indefinitely.

> **SEQUENCES**
>
> A **finite sequence** is a function whose domain is $D = \{1, 2, 3, \ldots, n\}$ for some fixed natural number n.
>
> An **infinite sequence** is a function whose domain is the set of natural numbers.

Because sequences are functions, many of the concepts discussed in previous chapters apply to sequences. Instead of letting y represent the output, however, the convention is to write $a_n = f(n)$, where n is a natural number in the domain of the sequence. The *terms* of a sequence are

$$a_1, a_2, a_3, \ldots, a_n, \ldots.$$

The first term is $a_1 = f(1)$, the second term is $a_2 = f(2)$, and so on. The **nth term**, or **general term**, of a sequence is $a_n = f(n)$.

EXAMPLE 1 Computing terms of a sequence

Write the first four terms of each sequence for $n = 1, 2, 3$, and 4.

(a) $f(n) = 2n - 1$ **(b)** $f(n) = 3(-2)^n$ **(c)** $f(n) = \dfrac{n}{n+1}$

Solution

(a) For $a_n = f(n) = 2n - 1$, we write

$$a_1 = f(1) = 2(1) - 1 = 1;$$
$$a_2 = f(2) = 2(2) - 1 = 3;$$
$$a_3 = f(3) = 2(3) - 1 = 5;$$
$$a_4 = f(4) = 2(4) - 1 = 7.$$

The first four terms are **1, 3, 5,** and **7**.

(b) For $a_n = f(n) = 3(-2)^n$, we write

$$a_1 = f(1) = 3(-2)^1 = -6;$$
$$a_2 = f(2) = 3(-2)^2 = 12;$$
$$a_3 = f(3) = 3(-2)^3 = -24;$$
$$a_4 = f(4) = 3(-2)^4 = 48.$$

The first four terms are $-6, 12, -24,$ and 48.

(c) For $a_n = f(n) = \frac{n}{n+1}$, we write

$$a_1 = f(1) = \frac{1}{1+1} = \frac{1}{2};$$
$$a_2 = f(2) = \frac{2}{2+1} = \frac{2}{3};$$
$$a_3 = f(3) = \frac{3}{3+1} = \frac{3}{4};$$
$$a_4 = f(4) = \frac{4}{4+1} = \frac{4}{5}.$$

The first four terms are $\frac{1}{2}, \frac{2}{3}, \frac{3}{4}$, and $\frac{4}{5}$. Note that, although the input to a sequence is a natural number, the output need not be a natural number.

TECHNOLOGY NOTE: Generating Sequences

Many graphing calculators can generate sequences if you change the MODE from function (Func) to sequence (Seq). In Figures 11.3 and 11.4 the sequences from Example 1 are generated. On some calculators the sequence utility is found in the LIST OPS menus. The expression

$$\text{seq}(2n - 1, n, 1, 4)$$

represents terms 1 through 4 of the sequence $a_n = 2n - 1$ with the variable n.

```
seq(2n-1,n,1,4)
       {1 3 5 7}
seq(3(-2)^n,n,1,
4)
     {-6 12 -24 48}
```

```
seq(n/(n+1),n,1,
4)▶Frac
     {1/2 2/3 3/4 4/...
```

Figure 11.3   Figure 11.4

Representations of Sequences

Because sequences are functions, they can be represented symbolically, graphically, and numerically. The next two examples illustrate such representations.

EXAMPLE 2   Using a graphical representation

Use Figure 11.5 to write the terms of the sequence.

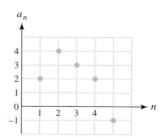

Figure 11.5 Finite Sequence

Solution

The points $(1, 2), (2, 4), (3, 3), (4, 2)$, and $(5, -1)$ are shown in the graph. The terms of the sequence are $2, 4, 3, 2,$ and -1.

EXAMPLE 3 Representing a sequence

In 2002 the average person in the United States used 100 gallons of water each day. Give symbolic, numerical, and graphical representations for a sequence that models the total amount of water used over a 7-day period.

Solution
Symbolic Representation Let $a_n = 100n$ for $n = 1, 2, 3, \ldots, 7$.

Numerical Representation Table 11.2 contains the sequence.

Graphical Representation Plot the points $(1, 100)$, $(2, 200)$, $(3, 300)$, $(4, 400)$, $(5, 500)$, $(6, 600)$, and $(7, 700)$, as shown in Figure 11.6.

TABLE 11.2

n	a_n
1	100
2	200
3	300
4	400
5	500
6	600
7	700

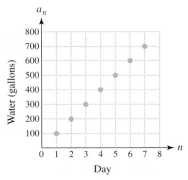

Figure 11.6 Water Usage

Models and Applications

▶ REAL-WORLD CONNECTION A population model for a species of insect with a life span of 1 year can be described with a sequence. Suppose that each adult female insect produces, on average, r female offspring that survive to reproduce the following year. Let a_n represent the female insect population at the beginning of year n. Then the number of female insects is given by

$$a_n = Cr^{n-1},$$

where C is the initial population of female insects. (*Source:* D. Brown and P. Rothery, *Models in Biology*.)

EXAMPLE 4 Modeling numbers of insects

Suppose that the initial population of adult female insects is 500 per acre and that $r = 1.04$. Then the average number of female insects per acre at the beginning of year n is described by

$$a_n = 500(1.04)^{n-1}.$$

Represent the female insect population numerically and graphically for 7 years. Discuss the results. By what percent is the population increasing each year?

Solution
Numerical Representation Table 11.3 contains *approximations* for the first 7 terms of the sequence. The insect population increases from 500 to about 633 insects per acre during this time period.

TABLE 11.3

n	1	2	3	4	5	6	7
a_n	500	520	540.8	562.43	584.93	608.33	632.66

Graphical Representation Plot the points (1, 500), (2, 520), (3, 540.8), (4, 562.43), (5, 584.93), (6, 608.33), and (7, 632.66), as shown in Figure 11.7. These results indicate that the insect population gradually increases. Because the growth factor is 1.04, the population is increasing by 4% each year.

CRITICAL THINKING

Explain how the value of *r* in Example 4 affects the population of female insects over time. Assume that $r > 0$.

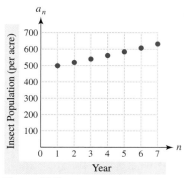

Figure 11.7

TECHNOLOGY NOTE: Graphs and Tables of Sequences

In the sequence mode, many graphing calculators are capable of representing sequences graphically and numerically. Figure 11.8(a) shows how to enter the sequence from Example 4 to produce the table of values shown in Figure 11.8(b).

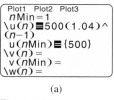   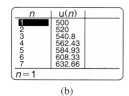

(a)   (b)

Figure 11.8

Figures 11.9(a) and (b) show the set-up for graphing the sequence from Example 4 to produce the graph shown in Figure 11.9(c).

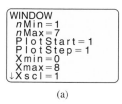   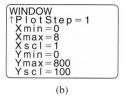   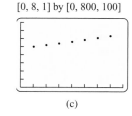

(a)   (b)   (c)

Figure 11.9

11.1 PUTTING IT ALL TOGETHER

An infinite sequence is a function whose domain is the set of natural numbers. A finite sequence has the domain $D = \{1, 2, 3, \ldots, n\}$ for some fixed natural number n. Graphs of sequences are scatterplots, *not* continuous lines and curves.

Sequences are functions that may be represented symbolically, numerically, and graphically. Examples of representations of sequences are shown in the following table.

Representation	Example
Symbolic	$a_n = n - 3$ represents a sequence. The first four terms of the sequence are $-2, -1, 0,$ and 1: $$a_1 = 1 - 3 = -2,\ a_2 = 2 - 3 = -1,$$ $$a_3 = 3 - 3 = 0,\ \text{and}\ a_4 = 4 - 3 = 1.$$
Numerical	A numerical representation for $a_n = n - 3$ with $n = 1, 2, 3,$ and 4 is shown in the table. | n | 1 | 2 | 3 | 4 | |---|---|---|---|---| | a_n | -2 | -1 | 0 | 1 |
Graphical	For a graphical representation of the first four terms of $a_n = n - 3$, the points $(1, -2), (2, -1), (3, 0),$ and $(4, 1)$ are plotted.

11.2 ARITHMETIC AND GEOMETRIC SEQUENCES

Representations of Arithmetic Sequences ■ Representations of Geometric Sequences ■ Applications and Models

A LOOK INTO MATH ▷ Indoor air pollution has become more hazardous as people spend 80% to 90% of their time in tightly sealed, energy-efficient buildings, which often lack proper ventilation. Many contaminants such as tobacco smoke, formaldehyde, radon, lead, and carbon monoxide are often allowed to increase to unsafe levels. One way to alleviate this problem is to use efficient ventilation systems. Mathematics plays an important role in determining the proper amount of ventilation. In this section we use sequences to model ventilation in classrooms. Before implementing this model, however, we discuss the basic concepts relating to two special types of sequences.

Representations of Arithmetic Sequences

If a sequence is defined by a linear function, it is an *arithmetic sequence*. For example,

$$f(n) = 2n - 3$$

represents an arithmetic sequence because $f(x) = 2x - 3$ defines a linear function. The first five terms of this sequence are shown in Table 11.4.

Each time n increases by 1, the next term is 2 more than the previous term. We say that the *common difference* of this arithmetic sequence is $d = 2$. That is, the difference between successive terms equals 2. When the points associated with these terms are graphed, they lie on the line $y = 2x - 3$, as illustrated in Figure 11.10. Arithmetic sequences are represented by linear functions and so their graphical representations consist of collinear points (points that lie on a line). The slope m of the line equals the common difference d.

TABLE 11.4

n	$f(n)$
1	-1
2	1
3	3
4	5
5	7

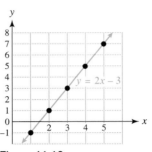

Figure 11.10

ARITHMETIC SEQUENCE

An **arithmetic sequence** is a linear function given by $a_n = dn + c$ whose domain is the set of natural numbers. The value of d is called the **common difference**.

NOTE: If each term after the first term is obtained by adding a fixed number to the previous term, then the sequence is an arithmetic sequence. The fixed number is the *common difference*. For example, 1, 6, 11, 16, 21, ... is an arithmetic sequence because each term (after the first) is found by adding the common difference of 5 to the previous term.

EXAMPLE 1 Recognizing arithmetic sequences

Determine whether f is an arithmetic sequence. If it is, identify the common difference d.
(a) $f(n) = 2 - 3n$

(b)

n	$f(n)$
1	10
2	5
3	0
4	-5
5	-10

(c) A graph of f is shown in Figure 11.11.

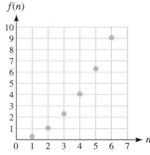

Figure 11.11

Solution
(a) This sequence is arithmetic because $f(x) = -3x + 2$ defines a linear function. The common difference is $d = -3$.
(b) The table reveals that each term is found by adding -5 to the previous term. This represents an arithmetic sequence with common difference $d = -5$.
(c) The sequence shown in Figure 11.11 is not an arithmetic sequence because the points are not collinear. That is, there is no common difference.

MAKING CONNECTIONS

Common Difference and Slope

The common difference d of an arithmetic sequence equals the slope of the line passing through the collinear points. For example, if $a_n = -2n + 4$, the common difference is -2, and the slope of the line passing through the points on the graph of a_n is also -2 (see Figure 11.12).

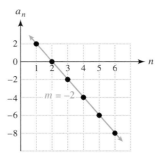

Figure 11.12

EXAMPLE 2 Finding symbolic representations

Find the general term a_n for each arithmetic sequence.
(a) $a_1 = 3$ and $d = 4$ (b) $a_1 = 3$ and $a_4 = 12$

Solution
(a) Let $a_n = dn + c$. For $d = 4$, we write $a_n = 4n + c$, and to find c we use $a_1 = 3$.

$$a_1 = 4(1) + c = 3 \quad \text{or} \quad c = -1$$

Thus $a_n = 4n - 1$.

(b) Because $a_1 = 3$ and $a_4 = 12$, the common difference d equals the slope of the line passing through the points $(1, 3)$ and $(4, 12)$, or

$$d = \frac{12 - 3}{4 - 1} = 3.$$

Therefore $a_n = 3n + c$. To find c we use $a_1 = 3$ and obtain

$$a_1 = 3(1) + c = 3 \quad \text{or} \quad c = 0.$$

Thus $a_n = 3n$.

Consider the arithmetic sequence

$$1, 5, 9, 13, 17, 21, 25, 29, \ldots.$$

The common difference is $d = 4$, and the first term is $a_1 = 1$. To find the second term we add d to the first term. To find the third term we add $2d$ to the first term, and to find the fourth term we add $3d$ to the first term a_1. That is,

$$a_1 = 1,$$
$$a_2 = a_1 + 1d = 1 + 1 \cdot 4 = 5,$$
$$a_3 = a_1 + 2d = 1 + 2 \cdot 4 = 9,$$
$$a_4 = a_1 + 3d = 1 + 3 \cdot 4 = 13,$$

and, in general, a_n is determined by

$$a_n = a_1 + (n - 1)d = 1 + (n - 1)4.$$

This result suggests the following formula.

GENERAL TERM OF AN ARITHMETIC SEQUENCE

The nth term a_n of an arithmetic sequence is given by

$$a_n = a_1 + (n-1)d,$$

where a_1 is the first term and d is the common difference.

EXAMPLE 3 Finding terms of an arithmetic sequence

If $a_1 = 5$ and $d = 3$, find a_{54}.

Solution
To find a_{54}, apply the formula $a_n = a_1 + (n-1)d$.

$$a_{54} = 5 + (54-1)3 = 164$$

Representations of Geometric Sequences

If a sequence is defined by an exponential function, it is a *geometric sequence*. For example,

$$f(n) = 3(2)^{n-1}$$

represents a geometric sequence because $f(x) = 3(2)^{x-1}$ defines an exponential function. The first five terms of this sequence are shown in Table 11.5.

TABLE 11.5

n	1	2	3	4	5
$f(n)$	3	6	12	24	48

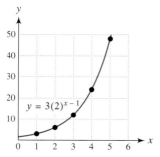

Figure 11.13

Successive terms are found by multiplying the previous term by 2. We say that the *common ratio* of this geometric sequence equals 2. Note that the ratios of successive terms are $\frac{6}{3}, \frac{12}{6}, \frac{24}{12}$, and $\frac{48}{24}$ and that they all equal the common ratio 2. When the points associated with the terms in Table 11.5 are graphed, they do *not* lie on a line. Rather they lie on the exponential curve $y = 3(2)^{x-1}$, as shown in Figure 11.13. A geometric sequence with a positive common ratio is an exponential function whose domain is the set of natural numbers. Its terms reflect either *exponential growth* or *exponential decay*.

GEOMETRIC SEQUENCE

A **geometric sequence** is given by $a_n = a_1(r)^{n-1}$, where n is a natural number and $r \neq 0$ or 1. The value of r is called the **common ratio**, and a_1 is the first term of the sequence.

NOTE: If each term after the first term is obtained by multiplying the previous term by a fixed number, the sequence is a geometric sequence. The fixed number is the *common ratio* and cannot be 0 or 1. For example, 3, 6, 12, 24, 48, ... is a geometric sequence. Each term (after the first) is found by multiplying the previous term by the common ratio of 2.

EXAMPLE 4 Recognizing geometric sequences

Determine whether f is a geometric sequence. If it is, identify the common ratio.
(a) $f(n) = 2(0.9)^{n-1}$

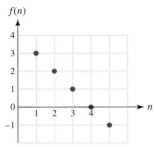

Figure 11.14

(b)

n	1	2	3	4	5
f(n)	8	4	2	1	$\frac{1}{2}$

(c) A graph of f is shown in Figure 11.14.

Solution
(a) This sequence is geometric because $f(x) = 2(0.9)^{x-1}$ defines an exponential function. The common ratio is $r = 0.9$.
(b) The table shows that each successive term is half the previous term. This sequence represents a geometric sequence with a common ratio of $r = \frac{1}{2}$.
(c) The sequence shown in Figure 11.14 is not a geometric sequence because the points are collinear. There is no common ratio.

MAKING CONNECTIONS
Common Ratios and Growth or Decay Factors

If the common ratio r of a geometric sequence is positive, then r equals either the growth factor or the decay factor for an exponential function.

EXAMPLE 5 Finding symbolic representations

Find a general term a_n for each geometric sequence.
(a) $a_1 = \frac{1}{2}$ and $r = 5$ (b) $a_1 = 2$, $a_3 = 18$, and $r < 0$.

CRITICAL THINKING

If we are given a_1 and a_5, can we determine the common ratio of a geometric series? Explain.

Solution
(a) Let $a_n = a_1(r)^{n-1}$. Because $a_1 = \frac{1}{2}$ and $r = 5$, we can write $a_n = \frac{1}{2}(5)^{n-1}$.
(b) $a_1 = 2$ and $a_3 = 18$, so

$$a_3 = a_1(r)^{3-1}$$
$$18 = 2(r)^2.$$

This equation simplifies to

$$r^2 = 9 \quad \text{or} \quad r = \pm 3.$$

It is specified that $r < 0$, so $r = -3$ and $a_n = 2(-3)^{n-1}$.

EXAMPLE 6 Finding a term of a geometric sequence

If $a_1 = 5$ and $r = 3$, find a_{10}.

Solution
To find a_{10}, apply the formula $a_n = a_1(r)^{n-1}$ with $a_1 = 5$, $r = 3$, and $n = 10$.

$$a_{10} = 5(3)^{10-1} = 5(3)^9 = 98{,}415$$

Applications and Models

▶ **REAL-WORLD CONNECTION** Sequences are frequently used to describe a variety of situations. In the next example, we use a sequence to model classroom ventilation.

EXAMPLE 7 Modeling classroom ventilation

Ventilation is an effective means for removing indoor air pollutants. According to the American Society of Heating, Refrigerating, and Air-Conditioning Engineers (ASHRAE), a classroom should have a ventilation rate of 900 cubic feet per hour per person.

(a) Write a sequence that gives the hourly ventilation necessary for 1, 2, 3, 4, and 5 people in a classroom. Is this sequence arithmetic, geometric, or neither?

(b) Write the general term for this sequence. Why is it reasonable to limit the domain to natural numbers?

(c) Find a_{30} and interpret the result.

Solution

(a) One person requires 900 cubic feet of air circulated per hour, two people require 1800, three people 2700, and so on. The first five terms of this sequence are

$$900, 1800, 2700, 3600, 4500.$$

This sequence is arithmetic, with a common difference of 900.

(b) The nth term equals $900n$, so we let $a_n = 900n$. Because we cannot have a fraction of a person, limiting the domain to the natural numbers is reasonable.

(c) The result $a_{30} = 900(30) = 27{,}000$ indicates that a classroom with 30 people should have a ventilation rate of 27,000 cubic feet per hour.

▶ **REAL-WORLD CONNECTION** Chlorine is frequently added to the water to disinfect swimming pools. The chlorine concentration should remain between 1.5 and 2.5 parts per million (ppm). On a warm, sunny day 30% of the chlorine may dissipate from the water. In the next example we use a sequence to model the amount of chlorine in a pool at the beginning of each day. (*Source*: D. Thomas, *Swimming Pool Operator's Handbook*.)

EXAMPLE 8 Modeling chlorine in a swimming pool

A swimming pool on a warm, sunny day begins with a high chlorine content of 4 parts per million. (Assume that each day 30% of the chlorine dissipates.)

(a) Write a sequence that models the amount of chlorine in the pool at the beginning of the first 3 days, assuming that no additional chlorine is added and that the days are warm and sunny. Is this sequence arithmetic, geometric, or neither?

(b) Write the general term for this sequence.

(c) At the beginning of what day does the chlorine first drop below 1.5 parts per million?

Solution

(a) Because 30% of the chlorine dissipates, 70% remains in the water at the beginning of the next day. If the concentration at the beginning of the first day is 4 parts per million, then at the beginning of the second day it is

$$4 \cdot 0.70 = 2.8 \text{ parts per million,}$$

and at the start of the third day it is

$$2.8 \cdot 0.70 = 1.96 \text{ parts per million.}$$

The first three terms are 4, 2.8, 1.96. Successive terms are found by multiplying the previous term by 0.7. Thus the sequence is geometric, with common ratio 0.7.

(b) The initial amount is $a_1 = 4$ and the common ratio is $r = 0.7$, so the sequence can be represented by $a_n = 4(0.7)^{n-1}$.

(c) The table shown in Figure 11.15 reveals that $a_4 = 4(0.7)^{4-1} \approx 1.372 < 1.5$. Thus, at the beginning of the fourth day, the chlorine level in the swimming pool drops below the recommended minimum of 1.5 parts per million.

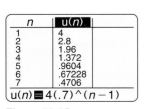

Figure 11.15

11.2 PUTTING IT ALL TOGETHER

In this section we discussed two types of sequences: arithmetic and geometric. Arithmetic sequences are linear functions, and geometric sequences with positive r are exponential functions. The inputs for both are limited to the natural numbers. Each successive term in an arithmetic sequence is found by adding the common difference d to the previous term. For a geometric sequence each successive term is found by multiplying the previous term by the common ratio r. The graph of an arithmetic sequence consists of points that lie on a line, whereas the graph of a geometric sequence (with a positive r) consists of points that lie on an exponential curve. Examples are shown in the following table.

Sequence	Formula	Example
Arithmetic	$a_n = dn + c$ or $a_n = a_1 + (n-1)d$, where d is the common difference and a_1 is the first term.	If $a_n = 5n + 2$, then the common difference is $d = 5$ and the terms of the sequence are $$7, 12, 17, 22, 27, 32, 37, \ldots.$$ Each term after the first is found by adding 5 to the previous term. The general term can be written as $$a_n = 7 + 5(n-1).$$
Geometric	$a_n = a_1(r)^{n-1}$, where r is the common ratio ($r \neq 0$, $r \neq 1$) and a_1 is the first term.	If $a_n = 4(-2)^{n-1}$, then the common ratio is $r = -2$ and the first term is $a_1 = 4$. The terms of the sequence are $$4, -8, 16, -32, 64, -128, 256, \ldots.$$ Each term after the first is found by multiplying the previous term by -2.

11.3 SERIES

Basic Concepts ■ Arithmetic Series ■ Geometric Series ■ Summation Notation

A LOOK INTO MATH ▷ Although the terms *sequence* and *series* are sometimes used interchangeably in everyday life, they represent different mathematical concepts. In mathematics a sequence is a function whose domain is the set of natural numbers, whereas a series is a summation of the terms in a sequence. Series have played a central role in the development of modern mathematics. Today series are often used to approximate functions that are too complicated to have formulas. Series are also used to calculate accurate approximations for π and e.

Basic Concepts

Suppose that a person has a starting salary of $30,000 per year and receives a $2000 raise each year. Then the *sequence*

$$30{,}000, 32{,}000, 34{,}000, 36{,}000, 38{,}000$$

lists these salaries over a 5-year period. The total amount earned is given by the *series*

$$30{,}000 + 32{,}000 + 34{,}000 + 36{,}000 + 38{,}000,$$

whose sum is $170,000. We now define the concept of a series.

FINITE SERIES

A **finite series** is an expression of the form

$$a_1 + a_2 + a_3 + \cdots + a_n.$$

EXAMPLE 1 Computing total reported AIDS cases

Table 11.6 presents a sequence a_n that computes the number of AIDS cases diagnosed each year from 1997 through 2003, where $n = 1$ corresponds to 1997.

TABLE 11.6 Annual U.S. AIDS Cases

n	1	2	3	4	5	6	7
a_n	50,167	43,225	41,314	41,239	41,227	42,136	43,171

Source: Department of Health and Human Services.

(a) Write a series whose sum represents the total number of AIDS cases diagnosed from 1997 to 2003. Find its sum.
(b) Interpret the series $a_1 + a_2 + a_3 + \cdots + a_{10}$.

Solution
(a) The required series and sum are given by

$$50{,}167 + 43{,}225 + 41{,}314 + 41{,}239 + 41{,}227 + 42{,}136 + 43{,}171 = 302{,}479.$$

(b) The series $a_1 + a_2 + a_3 + \cdots + a_{10}$ represents the total number of AIDS cases diagnosed over 10 years from 1997 through 2006.

MAKING CONNECTIONS

Sequences and Series

A *sequence* is an *ordered list*; a *series* is the *sum of the terms of a sequence*. For example, the even integers from 2 to 20 are represented by the sequence

$$2, 4, 6, 8, 10, 12, 14, 16, 18, 20. \quad \text{Sequence}$$

The corresponding series is

$$2 + 4 + 6 + 8 + 10 + 12 + 14 + 16 + 18 + 20, \quad \text{Series}$$

which sums to 110.

Arithmetic Series

Summing the terms of an arithmetic sequence results in an **arithmetic series**. For example, $a_n = 2n - 1$ for $n = 1, 2, 3, \ldots, 7$ defines the arithmetic sequence

$$1, 3, 5, 7, 9, 11, 13.$$

The corresponding arithmetic *series* is

$$1 + 3 + 5 + 7 + 9 + 11 + 13,$$

whose sum is 49. The following formula gives the sum of the first n terms of an arithmetic sequence.

SUM OF THE FIRST n TERMS OF AN ARITHMETIC SEQUENCE

The **sum of the first n terms of an arithmetic sequence**, denoted S_n, is found by averaging the first and nth terms and then multiplying by n. That is,

$$S_n = a_1 + a_2 + a_3 + \cdots + a_n = n\left(\frac{a_1 + a_n}{2}\right).$$

The series $1 + 3 + 5 + 7 + 9 + 11 + 13$ consists of 7 terms, where the first term is 1 and the last term is 13. Substituting in the formula gives

$$S_7 = 7\left(\frac{1 + 13}{2}\right) = 49,$$

which agrees with the sum obtained by adding the 7 terms.

Because $a_n = a_1 + (n - 1)d$ for an arithmetic sequence, S_n can also be written

$$S_n = n\left(\frac{a_1 + a_n}{2}\right)$$
$$= \frac{n}{2}(a_1 + a_n)$$
$$= \frac{n}{2}(a_1 + a_1 + (n - 1)d)$$
$$= \frac{n}{2}(2a_1 + (n - 1)d).$$

EXAMPLE 2 Finding the sum of a finite arithmetic series

Suppose that a person has a starting annual salary of $30,000 and receives a $1500 raise each year. Calculate the total amount earned after 10 years.

Solution
The sequence that gives the salary during year n is given by

$$a_n = 30{,}000 + 1500(n - 1).$$

One way to calculate the sum of the first 10 terms, denoted S_{10}, is to find a_1 and a_{10}, or

$$a_1 = 30{,}000 + 1500(1 - 1) = 30{,}000$$
$$a_{10} = 30{,}000 + 1500(10 - 1) = 43{,}500.$$

Thus the total amount earned during this 10-year period is

$$S_{10} = 10\left(\frac{a_1 + a_{10}}{2}\right)$$
$$= 10\left(\frac{30{,}000 + 43{,}500}{2}\right)$$
$$= \$367{,}500.$$

This sum can also be found with the second formula by letting $d = 1500$.

$$S_n = \frac{n}{2}(2a_1 + (n - 1)d)$$
$$= \frac{10}{2}(2 \cdot 30{,}000 + (10 - 1)1500)$$
$$= 5(60{,}000 + 9 \cdot 1500)$$
$$= \$367{,}500.$$

EXAMPLE 3 Finding the sum of an arithmetic series

Find the sum of the series $2 + 4 + 6 + \cdots + 100$.

Solution

The first term of this series is $a_1 = 2$, and the common difference is $d = 2$. This series represents the even numbers from 2 to 100, so the number of terms is $n = 50$. Using the formula

$$S_n = n\left(\frac{a_1 + a_n}{2}\right)$$

for the sum of an arithmetic series, we obtain

$$S_{50} = 50\left(\frac{2 + 100}{2}\right) = 2550.$$

TECHNOLOGY NOTE: Sum of a Series

The "seq(" utility, found on some calculators under the LIST OPS menus, generates a sequence. The "sum(" utility found on some calculators under the LIST MATH menus, calculates the sum of the sequence inside the parentheses. To verify the result in Example 2, let $a_n = 30{,}000 + 1500(n - 1)$. The value 367,500 for S_{10} is shown in Figure 11.16(a). The result found in Example 3 is shown in Figure 11.16(b).

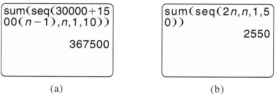

(a)   (b)

Figure 11.16

Geometric Series

A **geometric series** is the sum of the terms of a geometric sequence. For example,

$$1, 2, 4, 8, 16, 32$$

is a geometric sequence with $a_1 = 1$ and $r = 2$. Then

$$1 + 2 + 4 + 8 + 16 + 32$$

is a geometric series. We can use the following formula to sum a finite geometric series.

SUM OF THE FIRST n TERMS OF A GEOMETRIC SEQUENCE

If its first term is a_1 and its common ratio is r, then the **sum of the first n terms of a geometric sequence** is given by

$$S_n = a_1\left(\frac{1 - r^n}{1 - r}\right),$$

provided $r \neq 1$.

EXAMPLE 4 Finding the sum of finite geometric series

Find the sum of each series.
(a) $1 + 2 + 4 + 8 + 16 + 32 + 64 + 128 + 256$
(b) $\frac{1}{2} - \frac{1}{4} + \frac{1}{8} - \frac{1}{16} + \frac{1}{32}$

Solution
(a) This series is geometric, with $n = 9$, $a_1 = 1$, and $r = 2$, so
$$S_9 = 1\left(\frac{1 - 2^9}{1 - 2}\right) = 511.$$

(b) This series is geometric, with $n = 5$, $a_1 = \frac{1}{2} = 0.5$, and $r = -\frac{1}{2} = -0.5$, so
$$S_5 = 0.5\left(\frac{1 - (-0.5)^5}{1 - (-0.5)}\right) = \frac{11}{32} = 0.34375.$$

▶ **REAL-WORLD CONNECTION** A sum of money from which regular payments are made is called an **annuity**. An annuity may be purchased with a lump sum deposit or by deposits made at various intervals. Suppose that $1000 are deposited at the end of each year in an annuity account that pays an annual interest rate I expressed as a decimal. At the end of the first year the account contains $1000. At the end of the second year $1000 are deposited again. In addition, the first deposit of $1000 would have received interest during the second year. Therefore the value of the annuity after 2 years is
$$1000 + 1000(1 + I).$$

After 3 years the balance is
$$1000 + 1000(1 + I) + 1000(1 + I)^2,$$

and after n years this amount is given by
$$1000 + 1000(1 + I) + 1000(1 + I)^2 + \cdots + 1000(1 + I)^{n-1}.$$

This series is a geometric series with its first term $a_1 = 1000$ and the common ratio $r = (1 + I)$. The sum of the first n terms is given by
$$S_n = a_1\left(\frac{1 - (1 + I)^n}{1 - (1 + I)}\right) = a_1\left(\frac{(1 + I)^n - 1}{I}\right).$$

EXAMPLE 5 Finding the future value of an annuity

Suppose that a 20-year-old worker deposits $1000 into an annuity account at the end of each year. If the interest rate is 12%, find the future value of the annuity when the worker is 65 years old.

Solution
Let $a_1 = 1000$, $I = 0.12$, and $n = 45$. The future value of the annuity is
$$S_n = a_1\left(\frac{(1 + I)^n - 1}{I}\right)$$
$$= 1000\left(\frac{(1 + 0.12)^{45} - 1}{0.12}\right)$$
$$\approx \$1{,}358{,}230.$$

Summation Notation

Summation notation is used to write series efficiently. The symbol Σ, the uppercase Greek letter *sigma*, is used to indicate a sum.

SUMMATION NOTATION

$$\sum_{k=1}^{n} a_k = a_1 + a_2 + a_3 + \cdots + a_n$$

The letter k is called the **index of summation**. The numbers 1 and n represent the subscripts of the first and last terms in the series. They are called the **lower limit** and **upper limit** of the summation, respectively.

EXAMPLE 6 Using summation notation

Evaluate each series.

(a) $\sum_{k=1}^{5} k^2$ (b) $\sum_{k=1}^{4} 5$ (c) $\sum_{k=3}^{6} (2k - 5)$

Solution

(a) $\sum_{k=1}^{5} k^2 = 1^2 + 2^2 + 3^2 + 4^2 + 5^2 = 55$

(b) $\sum_{k=1}^{4} 5 = 5 + 5 + 5 + 5 = 20$

(c) $\sum_{k=3}^{6} (2k - 5) = \underbrace{(2(3) - 5)}_{k=3} + \underbrace{(2(4) - 5)}_{k=4} + \underbrace{(2(5) - 5)}_{k=5} + \underbrace{(2(6) - 5)}_{k=6}$

$= 1 + 3 + 5 + 7 = 16$

Summation notation is used frequently in statistics. The next example demonstrates how averages can be expressed in summation notation.

EXAMPLE 7 Applying summation notation

Express the average of the n numbers $x_1, x_2, x_3, \ldots, x_n$ in summation notation.

Solution
The average of n numbers can be written as

$$\frac{x_1 + x_2 + x_3 + \cdots + x_n}{n}.$$

This expression is equivalent to $\frac{1}{n}\left(\sum_{k=1}^{n} x_k\right)$.

NOTE: $\sum_{k=1}^{n} x_k$ is equivalent to $\sum_{k=1}^{n} x_k$.

▶ **REAL-WORLD CONNECTION** Series play an essential role in various applications.

EXAMPLE 8 Modeling air filtration

Suppose that an air filter removes 90% of the impurities entering it.
(a) Find a series that represents the amount of impurities removed by a sequence of n air filters. Express this answer in summation notation.
(b) How many air filters would be necessary to remove 99.99% of the impurities?

Solution

(a) The first filter removes 90% of the impurities, so 10%, or 0.1, passes through it. Of the 0.1 that passes through the first filter, 90% is removed by the second filter, while 10% of 10%, or 0.01, passes through. Then, 10% of 0.01, or 0.001, passes through the third filter. Figure 11.17 depicts these results, from which we can establish a pattern. If we let 100%, or 1, represent the amount of impurities entering the first air filter, the amount removed by n filters equals

$$(0.9)(1) + (0.9)(0.1) + (0.9)(0.01) + (0.9)(0.001) + \cdots + (0.9)(0.1)^{n-1}.$$

In summation notation we write this series as $\sum_{k=1}^{n} 0.9(0.1)^{k-1}$.

(b) To remove 99.99%, or 0.9999, of the impurities requires 4 air filters, because

$$\sum_{k=1}^{4} 0.9(0.1)^{k-1} = (0.9)(1) + (0.9)(0.1) + (0.9)(0.01) + (0.9)(0.001)$$
$$= 0.9 + 0.09 + 0.009 + 0.0009$$
$$= 0.9999.$$

Figure 11.17 Impurities Passing Through Air Filters

11.3 PUTTING IT ALL TOGETHER

A finite sequence is an ordered list such as

$$a_1, a_2, a_3, a_4, a_5, \ldots, a_n.$$

A finite series is the summation of the terms of a sequence and can be expressed as

$$a_1 + a_2 + a_3 + a_4 + a_5 + \cdots + a_n.$$

The following table summarizes concepts related to arithmetic and geometric series.

Series	Description	Example
Finite Arithmetic	$a_1 + a_2 + a_3 + \cdots + a_n$, where $a_n = dn + c$ or $a_n = a_1 + (n-1)d$. The sum of the first n terms is $$S_n = n\left(\frac{a_1 + a_n}{2}\right)$$ or $$S_n = \frac{n}{2}(2a_1 + (n-1)d),$$ where a_1 is the first term and d is the common difference.	The series $$4 + 7 + 10 + 13 + 16 + 19 + 22$$ is obtained from the sequence $$a_n = 3n + 1 \quad \text{or} \quad a_n = 4 + 3(n-1).$$ Its sum is $$S_7 = 7\left(\frac{4 + 22}{2}\right) = 91 \quad \text{or}$$ $$S_7 = \frac{7}{2}(2 \cdot 4 + (7-1)3) = 91.$$
Finite Geometric	$a_1 + a_2 + a_3 + \cdots + a_n$, where $a_n = a_1(r)^{n-1}$ for nonzero constants a_1 and r. The sum of the first n terms is $$S_n = a_1\left(\frac{1 - r^n}{1 - r}\right),$$ where a_1 is the first term and r is the common ratio ($r \neq 1$).	The series $$3 + 6 + 12 + 24 + 48 + 96$$ has $n = 6$, $a_1 = 3$, and $r = 2$. Its sum is $$S_6 = 3\left(\frac{1 - 2^6}{1 - 2}\right) = 189.$$

11.4 THE BINOMIAL THEOREM

Pascal's Triangle ▪ Factorial Notation and Binomial Coefficients ▪ Using the Binomial Theorem

A LOOK INTO MATH ▷

In this section we demonstrate how to expand expressions of the form $(a + b)^n$, where n is a natural number. These expressions occur in statistics, finite mathematics, computer science, and calculus. The two methods that we discuss are Pascal's triangle and the binomial theorem.

Pascal's Triangle

Expanding $(a + b)^n$ for increasing values of n gives the following results.

$$(a + b)^0 = 1$$
$$(a + b)^1 = 1a + 1b$$
$$(a + b)^2 = 1a^2 + 2ab + 1b^2$$
$$(a + b)^3 = 1a^3 + 3a^2b + 3ab^2 + 1b^3$$
$$(a + b)^4 = 1a^4 + 4a^3b + 6a^2b^2 + 4ab^3 + 1b^4$$
$$(a + b)^5 = 1a^5 + 5a^4b + 10a^3b^2 + 10a^2b^3 + 5ab^4 + 1b^5$$

Note that $(a + b)^1$ has two terms, starting with a and ending with b; $(a + b)^2$ has three terms, starting with a^2 and ending with b^2; and in general, $(a + b)^n$ has $n + 1$ terms, starting with a^n and ending with b^n. The exponent on a decreases by 1 each successive term, and the exponent on b increases by 1 each successive term.

```
        1
       1 1
      1 2 1
     1 3 3 1
    1 4 6 4 1
   1 5 10 10 5 1
```
Figure 11.18 Pascal's Triangle

The triangle formed by the highlighted numbers is called **Pascal's triangle**. This triangle consists of 1s along the sides, and each element inside the triangle is the sum of the two numbers above it, as shown in Figure 11.18. Pascal's triangle is usually written without variables and can be extended to include as many rows as needed.

We can use this triangle to expand $(a + b)^n$, where n is a natural number. For example, the expression $(m + n)^4$ consists of five terms written as

$$(m + n)^4 = \_m^4 + \_m^3n^1 + \_m^2n^2 + \_m^1n^3 + \_n^4.$$

Because there are five terms, the coefficients can be found in the fifth row of Pascal's triangle, which is

$$1 \quad 4 \quad 6 \quad 4 \quad 1.$$

Thus

$$(m + n)^4 = \underline{1}\,m^4 + \underline{4}\,m^3n^1 + \underline{6}\,m^2n^2 + \underline{4}\,m^1n^3 + \underline{1}\,n^4$$
$$= m^4 + 4m^3n + 6m^2n^2 + 4mn^3 + n^4.$$

EXAMPLE 1 Expanding a binomial

Expand each binomial, using Pascal's triangle.
(a) $(x + 2)^5$ **(b)** $(2m - n)^3$

Solution
(a) To find the coefficients, use the sixth row in Pascal's triangle.

$$(x + 2)^5 = \underline{1}x^5 + \underline{5}x^4 \cdot 2^1 + \underline{10}x^3 \cdot 2^2 + \underline{10}x^2 \cdot 2^3 + \underline{5}x^1 \cdot 2^4 + \underline{1}(2^5)$$
$$= x^5 + 10x^4 + 40x^3 + 80x^2 + 80x + 32$$

(b) To find the coefficients, use the fourth row in Pascal's triangle.

$$(2m - n)^3 = \underline{1}\,(2m)^3 + \underline{3}\,(2m)^2(-n)^1 + \underline{3}\,(2m)^1(-n)^2 + \underline{1}\,(-n)^3$$
$$= 8m^3 - 12m^2n + 6mn^2 - n^3$$

Factorial Notation and Binomial Coefficients

An alternative to Pascal's triangle is the binomial theorem, which uses **factorial notation**.

n FACTORIAL (n!)

For any positive integer n,

$$n! = 1 \cdot 2 \cdot 3 \cdot \cdots \cdot n.$$

We also define $0! = 1$.

NOTE: Because multiplication is commutative, n factorial can also be defined as

$$n! = n \cdot (n-1) \cdot (n-2) \cdot \cdots \cdot 2 \cdot 1.$$

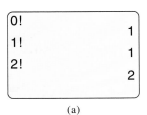

Figure 11.19

Examples include the following.

$$0! = 1$$
$$1! = 1$$
$$2! = 1 \cdot 2 = 2$$
$$3! = 1 \cdot 2 \cdot 3 = 6$$
$$4! = 1 \cdot 2 \cdot 3 \cdot 4 = 24$$
$$5! = 1 \cdot 2 \cdot 3 \cdot 4 \cdot 5 = 120$$

Figure 11.19 supports these results. On some calculators, factorial (!) can be accessed in the MATH PRB menus.

EXAMPLE 2 Evaluating factorial expressions

Simplify the expression.

(a) $\dfrac{5!}{3!2!}$ **(b)** $\dfrac{4!}{4!0!}$

Solution

(a) $\dfrac{5!}{3!2!} = \dfrac{1 \cdot 2 \cdot 3 \cdot 4 \cdot 5}{(1 \cdot 2 \cdot 3)(1 \cdot 2)} = \dfrac{120}{6 \cdot 2} = 10$

(b) $0! = 1$, so $\dfrac{4!}{4!0!} = \dfrac{4!}{4!(1)} = \dfrac{4!}{4!} = 1$

The expression $_nC_r$ represents a *binomial coefficient* that can be used to calculate the numbers in Pascal's triangle.

BINOMIAL COEFFICIENT $_nC_r$

For n and r nonnegative integers, $n \geq r$,

$$_nC_r = \dfrac{n!}{(n-r)!r!}$$

is a **binomial coefficient**.

Values of $_nC_r$ for $r = 0, 1, 2, \ldots, n$ correspond to the $n + 1$ numbers in row $n + 1$ of Pascal's triangle.

EXAMPLE 3 Calculating $_nC_r$

Calculate $_3C_r$ for $r = 0, 1, 2, 3$ by hand. Check your results on a calculator. Compare these numbers with the fourth row in Pascal's triangle.

Solution

$$_3C_0 = \frac{3!}{(3-0)!0!} = \frac{6}{6 \cdot 1} = 1 \qquad _3C_1 = \frac{3!}{(3-1)!1!} = \frac{6}{2 \cdot 1} = 3$$

$$_3C_2 = \frac{3!}{(3-2)!2!} = \frac{6}{1 \cdot 2} = 3 \qquad _3C_3 = \frac{3!}{(3-3)!3!} = \frac{6}{1 \cdot 6} = 1$$

These results are supported in Figure 11.20. The fourth row of Pascal's triangle is

$$1 \quad 3 \quad 3 \quad 1,$$

which agrees with the calculated values for $_3C_r$. On some calculators, the MATH PRB menus are used to calculate $_nC_r$.

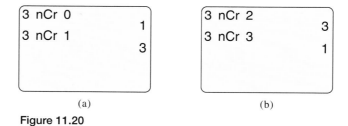

(a)   (b)

Figure 11.20

Using the Binomial Theorem

The binomial coefficients can be used to expand expressions of the form $(a + b)^n$. To do so, we use the **binomial theorem**.

BINOMIAL THEOREM

For any positive integer n and any numbers a and b,
$$(a + b)^n = {_nC_0}a^n + {_nC_1}a^{n-1}b^1 + \cdots + {_nC_{n-1}}a^1b^{n-1} + {_nC_n}b^n.$$

Using the results of Example 3, we write
$$(a + b)^3 = {_3C_0}a^3 + {_3C_1}a^2b^1 + {_3C_2}a^1b^2 + {_3C_3}b^3$$
$$= 1a^3 + 3a^2b + 3ab^2 + 1b^3$$
$$= a^3 + 3a^2b + 3ab^2 + b^3.$$

EXAMPLE 4 Expanding a binomial

Use the binomial theorem to expand each expression.
(a) $(x + y)^5$ **(b)** $(3 - 2x)^4$

Solution

(a) The coefficients are calculated as follows.

$$_5C_0 = \frac{5!}{(5-0)!0!} = 1, \quad _5C_1 = \frac{5!}{(5-1)!1!} = 5, \quad _5C_2 = \frac{5!}{(5-2)!2!} = 10$$

$$_5C_3 = \frac{5!}{(5-3)!3!} = 10, \quad _5C_4 = \frac{5!}{(5-4)!4!} = 5, \quad _5C_5 = \frac{5!}{(5-5)!5!} = 1$$

Using the binomial theorem, we arrive at the following result.

$$(x + y)^5 = {_5C_0}x^5 + {_5C_1}x^4y^1 + {_5C_2}x^3y^2 + {_5C_3}x^2y^3 + {_5C_4}x^1y^4 + {_5C_5}y^5$$
$$= 1x^5 + 5x^4y + 10x^3y^2 + 10x^2y^3 + 5xy^4 + 1y^5$$
$$= x^5 + 5x^4y + 10x^3y^2 + 10x^2y^3 + 5xy^4 + y^5$$

(b) The coefficients are calculated as follows.

$$_4C_0 = \frac{4!}{(4-0)!0!} = 1, \quad _4C_1 = \frac{4!}{(4-1)!1!} = 4, \quad _4C_2 = \frac{4!}{(4-2)!2!} = 6,$$

$$_4C_3 = \frac{4!}{(4-3)!3!} = 4, \quad _4C_4 = \frac{4!}{(4-4)!4!} = 1$$

Using the binomial theorem with $a = 3$ and $b = (-2x)$, we arrive at the following result.

$$(3 - 2x)^4 = {_4C_0}(3)^4 + {_4C_1}(3)^3(-2x) + {_4C_2}(3)^2(-2x)^2$$
$$+ {_4C_3}(3)(-2x)^3 + {_4C_4}(-2x)^4$$
$$= 1(81) + 4(27)(-2x) + 6(9)(4x^2) + 4(3)(-8x^3) + 1(16x^4)$$
$$= 81 - 216x + 216x^2 - 96x^3 + 16x^4$$

The binomial theorem gives *all* of the terms of $(a + b)^n$. However, we can find any individual term by noting that the $(r + 1)$st term in the binomial expansion for $(a + b)^n$ is given by the formula $_nC_r a^{n-r} b^r$, for $0 \leq r \leq n$. The next example shows how to use this formula to find the $(r + 1)$st term of $(a + b)^n$.

EXAMPLE 5 Finding the kth term in a binomial expansion

Find the third term of $(x - y)^5$.

Solution

In this example the $(r + 1)$st term is the *third* term in the expansion of $(x - y)^5$. That is, $r + 1 = 3$, or $r = 2$. Also, the exponent in the expression is $n = 5$. To get this binomial into the form $(a + b)^n$, we note that the first term in the binomial is $a = x$ and that the second term in the binomial is $b = -y$. Substituting the values for r, n, a, and b in the formula $_nC_r a^{n-r} b^r$ for the $(r + 1)$st term yields

$$_5C_2(x)^{5-2}(-y)^2 = 10x^3y^2.$$

The third term in the binomial expansion of $(x - y)^5$ is $10x^3y^2$.

11.4 PUTTING IT ALL TOGETHER

In this section we showed how to expand the expression $(a + b)^n$ by using Pascal's triangle and the binomial theorem. The following table outlines important topics from this section.

Topic	Explanation	Example
Pascal's Triangle	1 1 1 1 2 1 1 3 3 1 1 4 6 4 1 1 5 10 10 5 1	$(a + b)^3 = 1a^3 + 3a^2b + 3ab^2 + 1b^3$ (Row 4) To expand $(a + b)^n$, use row $n + 1$ in the triangle.
Factorial Notation	The expression $n!$ equals $1 \cdot 2 \cdot 3 \cdot \cdots \cdot n.$	$5! = 1 \cdot 2 \cdot 3 \cdot 4 \cdot 5 = 120$
Binomial Coefficient $_nC_r$	$_nC_r = \dfrac{n!}{(n-r)!\,r!}$	$_6C_4 = \dfrac{6!}{(6-4)!\,4!} = \dfrac{6!}{2!\,4!} = \dfrac{720}{2 \cdot 24} = 15$
Binomial Theorem	$(a + b)^n = {}_nC_0 a^n + {}_nC_1 a^{n-1}b^1 + \cdots$ $+ {}_nC_{n-1} a^1 b^{n-1} + {}_nC_n b^n$	$(a + b)^4 = {}_4C_0 a^4 + {}_4C_1 a^3 b + {}_4C_2 a^2 b^2$ $+ {}_4C_3 ab^3 + {}_4C_4 b^4$ $= 1a^4 + 4a^3b + 6a^2b^2 + 4ab^3 + 1b^4$ $= a^4 + 4a^3b + 6a^2b^2 + 4ab^3 + b^4$

CHAPTER 11 SUMMARY

SECTION 11.1 ■ SEQUENCES

Sequences An *infinite sequence* is a function whose domain is the natural numbers. A *finite sequence* is a function whose domain is $D = \{1, 2, 3, \ldots, n\}$ for some natural number n.

Example: $a_n = 2n$ is a symbolic representation of the even natural numbers. The first six terms of this sequence are represented numerically and graphically in the table and figure.

n	a_n
1	2
2	4
3	6
4	8
5	10
6	12

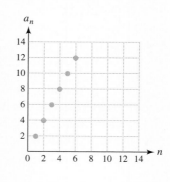

SECTION 11.2 ■ ARITHMETIC AND GEOMETRIC SEQUENCES

Two common types of sequences are arithmetic and geometric.

Arithmetic Sequence   An arithmetic sequence is determined by a linear function of the form $f(n) = dn + c$ or $f(n) = a_1 + (n-1)d$. Successive terms in an arithmetic sequence are found by adding the common difference d to the previous term. The sequence $1, 3, 5, 7, 9, 11, \ldots$ is an arithmetic sequence with its first term $a_1 = 1$, common difference $d = 2$, and general term $a_n = 2n - 1$.

Geometric Sequence   The general term for a geometric sequence is given by $f(n) = a_1 r^{n-1}$. Successive terms in a geometric sequence are found by multiplying the previous term by the common ratio r. The sequence $3, 6, 12, 24, 48, \ldots$ is a geometric sequence with its first term $a_1 = 3$, common ratio $r = 2$, and general term $a_n = 3(2)^{n-1}$.

SECTION 11.3 ■ SERIES

Series   A series results when the terms of a sequence are summed. The series associated with the sequence $2, 4, 6, 8, 10$ is

$$2 + 4 + 6 + 8 + 10,$$

and its sum equals 30. An arithmetic series results when the terms of an arithmetic sequence are summed, and a geometric series results when the terms of a geometric sequence are summed. In this chapter, we discussed formulas for finding sums of arithmetic and geometric series. See Putting It All Together for Section 11.3.

Summation Notation   Summation notation can be used to write series efficiently.

Example: $1^2 + 2^2 + 3^2 + 4^2 + 5^2 = \sum_{k=1}^{5} k^2.$

SECTION 11.4 ■ THE BINOMIAL THEOREM

Pascal's triangle may be used to find the coefficients for the expansion of $(a + b)^n$, where n is a natural number.

$$\begin{array}{ccccccccccc}
 & & & & & 1 & & & & & \\
 & & & & 1 & & 1 & & & & \\
 & & & 1 & & 2 & & 1 & & & \\
 & & 1 & & 3 & & 3 & & 1 & & \\
 & 1 & & 4 & & 6 & & 4 & & 1 & \\
1 & & 5 & & 10 & & 10 & & 5 & & 1
\end{array}$$

Example: To expand $(x + y)^4$, use the fifth row of Pascal's triangle.

$$(x + y)^4 = 1x^4 + 4x^3y + 6x^2y^2 + 4xy^3 + 1y^4$$
$$= x^4 + 4x^3y + 6x^2y^2 + 4xy^3 + y^4$$

The binomial theorem can also be used to expand powers of binomials.

504 CHAPTER 11 SEQUENCES AND SERIES

Assignment Name _____ Name _____ Date _____

Show all work for these items: _____

# _____	# _____
# _____	# _____
# _____	# _____
# _____	# _____

Assignment Name _____ Name _____ Date _____

Show all work for these items: _____

_____

_____

_____

_____

_____

_____

_____

_____

Appendix Using the Graphing Calculator

Overview of the Appendix

This appendix provides instruction for the TI-83, TI-83 Plus, and TI-84 Plus graphing calculators that may be used in conjunction with this textbook. It includes specific keystrokes needed to work several examples from the text. Students are advised to consult the *Graphing Calculator Guidebook* provided by the manufacturer.

Entering Mathematical Expressions

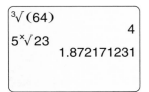

Figure A.1

EVALUATING π To evaluate π, use the following keystrokes, as shown in the first and second lines of Figure A.1. (Do *not* use 3.14 or $\frac{22}{7}$ for π.)

$$\boxed{\text{2nd}}\,\boxed{\wedge[\pi]}\,\boxed{\text{ENTER}}$$

EVALUATING A SQUARE ROOT   To evaluate a square root, such as $\sqrt{200}$, use the following keystrokes, as shown in the third and fourth lines of Figure A.1.

$$\boxed{\text{2nd}}\,\boxed{x^2[\sqrt{\ }]}\,\boxed{2}\,\boxed{0}\,\boxed{0}\,\boxed{)}\,\boxed{\text{ENTER}}$$

EVALUATING AN EXPONENTIAL EXPRESSION   To evaluate an exponential expression, such as 10^4, use the following keystrokes, as shown in the last two lines of Figure A.1.

$$\boxed{1}\,\boxed{0}\,\boxed{\wedge}\,\boxed{4}\,\boxed{\text{ENTER}}$$

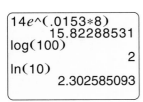

Figure A.2

EVALUATING A CUBE ROOT   To evaluate a cube root, such as $\sqrt[3]{64}$, use the following keystrokes, as shown in the first and second lines of Figure A.2.

$$\boxed{\text{MATH}}\,\boxed{4}\,\boxed{6}\,\boxed{4}\,\boxed{)}\,\boxed{\text{ENTER}}$$

EVALUATING OTHER ROOTS   To evaluate a fifth root, such as $\sqrt[5]{23}$, use the following keystrokes, as shown in the third and fourth lines of Figure A.2.

$$\boxed{5}\,\boxed{\text{MATH}}\,\boxed{5}\,\boxed{2}\,\boxed{3}\,\boxed{\text{ENTER}}$$

```
14e^(.0153*8)
            15.82288531
log(100)
                       2
ln(10)
            2.302585093
```

Figure A.3

EVALUATING THE NATURAL EXPONENTIAL FUNCTION   To evaluate $14e^{0.0153(8)}$, use the following keystrokes, as shown in the first and second lines of Figure A.3.

$$\boxed{1}\,\boxed{4}\,\boxed{\text{2nd}}\,\boxed{\text{LN}[e^x]}\,\boxed{.}\,\boxed{0}\,\boxed{1}\,\boxed{5}\,\boxed{3}\,\boxed{\times}\,\boxed{8}\,\boxed{)}\,\boxed{\text{ENTER}}$$

EVALUATING THE COMMON LOGARITHMIC FUNCTION   To evaluate $\log(100)$, use the following keystrokes, as shown in the third and fourth lines of Figure A.3.

$$\boxed{\text{LOG}}\,\boxed{1}\,\boxed{0}\,\boxed{0}\,\boxed{)}\,\boxed{\text{ENTER}}$$

EVALUATING THE NATURAL LOGARITHMIC FUNCTION   To evaluate $\ln(10)$, use the following keystrokes, as shown in the last two lines of Figure A.3.

$$\boxed{\text{LN}}\,\boxed{1}\,\boxed{0}\,\boxed{)}\,\boxed{\text{ENTER}}$$

To access the *number* π, use (2nd)(^[π]).

To evaluate a *square root*, use (2nd)(x^2[√]).

To evaluate an *exponential expression*, use the (^) key. To square a number, the (x^2) key can also be used.

To evaluate a *cube root*, use (MATH)(4).

To evaluate a *kth root*, use (k)(MATH)(5).

To access the *natural exponential function*, use (2nd)(LN [e^x]).

To access the *common logarithmic function*, use (LOG).

To access the *natural logarithmic function*, use (LN).

Expressing Answers as Fractions

To evaluate $\frac{1}{3} + \frac{2}{5} - \frac{4}{9}$ in fraction form, use the following keystrokes, as shown in the last three lines of Figure 1.8 on page 13.

Enter the arithmetic expression. To access the "Frac" feature, use the keystrokes (MATH)(1). Then press (ENTER).

Displaying Numbers in Scientific Notation

To display numbers in scientific notation, set the graphing calculator in scientific mode (Sci), by using the following keystrokes. See Figure A.4. (These keystrokes assume that the calculator is in normal mode.)

(MODE)(▷)(ENTER)(2nd)(MODE [QUIT])

In scientific mode we can display the numbers 5432 and 0.00001234 in scientific notation, as shown in Figure A.5.

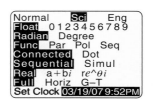

Figure A.4

If your calculator is in normal mode, it can be set in scientific mode by pressing

(MODE)(▷)(ENTER)(2nd)(MODE [QUIT]).

These keystrokes return the graphing calculator to the home screen.

Figure A.5

Entering Numbers in Scientific Notation

Numbers can be entered in scientific notation. For example, to enter 4.2×10^{-3} in scientific notation, use the following keystrokes. (Be sure to use the negation key $(-)$ rather than the subtraction key.)

This number can also be entered using the following keystrokes. See Figure A.6.

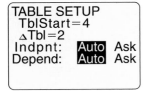

Figure A.6

One way to enter a number in scientific notation is to use the keystrokes

$$\boxed{2\text{nd}}\;\boxed{,[\text{EE}]}$$

to access an exponent (EE) of 10.

Making a Table

To make a table of values for $y = 3x + 1$ starting at $x = 4$ and incrementing by 2, begin by pressing $\boxed{Y=}$ and then entering the formula $Y_1 = 3X + 1$, as shown in Figure A.7. To set the table parameters, press the following keys. See Figure A.8.

$$\boxed{2\text{nd}}\;\boxed{\text{WINDOW [TBLSET]}}\;\boxed{4}\;\boxed{\text{ENTER}}\;\boxed{2}$$

These keystrokes specify a table that starts at $x = 4$ and increments the x-values by 2. Therefore, the values of Y_1 at $x = 4, 6, 8, \ldots$ appear in the table. To create this table, press the following keys.

$$\boxed{2\text{nd}}\;\boxed{\text{GRAPH [TABLE]}}$$

We can scroll through x- and y-values by using the arrow keys. See Figure A.9. Note that there is no first or last x-value in the table.

```
Plot1  Plot2  Plot3
\Y₁■3X+1
\Y₂=
\Y₃=
\Y₄=
\Y₅=
\Y₆=
\Y₇=
```

Figure A.7

```
TABLE SETUP
 TblStart=4
 ΔTbl=2
 Indpnt: Auto Ask
 Depend: Auto Ask
```

Figure A.8

X	Y₁
4	13
6	19
8	25
10	31
12	37
14	43
16	49

Y₁■3X+1

Figure A.9

1. Enter the formula for the equation using $\boxed{Y=}$.
2. Press $\boxed{2\text{nd}}\;\boxed{\text{WINDOW [TBLSET]}}$ to set the starting x-value and the increment between x-values appearing in the table.
3. Create the table by pressing $\boxed{2\text{nd}}\;\boxed{\text{GRAPH [TABLE]}}$.

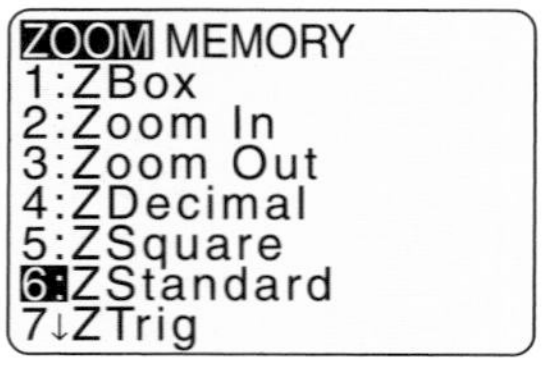

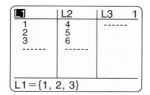

Figure A.10

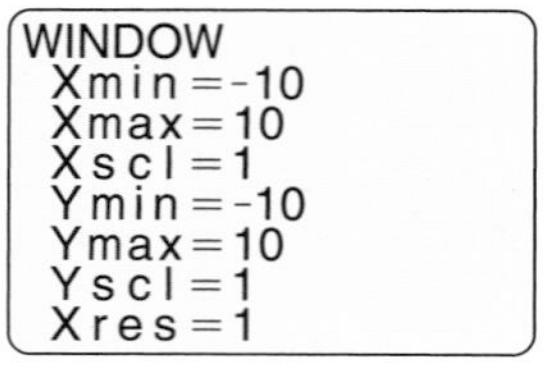

Figure A.11

Setting the Viewing Rectangle (Window)

There are at least two ways to set the standard viewing rectangle of $[-10, 10, 1]$ by $[-10, 10, 1]$. The first involves pressing (ZOOM) followed by (6). See Figure A.10. The second method for setting the standard viewing rectangle is to press (WINDOW) and enter the following keystrokes. See Figure A.11.

(−)(1)(0)(ENTER)(1)(0)(ENTER)(1)(ENTER)
(−)(1)(0)(ENTER)(1)(0)(ENTER)(1)(ENTER)

(Be sure to use the negation key (−) rather than the subtraction key.) Other viewing rectangles can be set in a similar manner by pressing (WINDOW) and entering the appropriate values. To see the viewing rectangle, press (GRAPH). An example is shown in Figure 1.33 on page 38.

To set the standard viewing rectangle, press (ZOOM)(6). To set any viewing rectangle, press (WINDOW) and enter the necessary values. To see the viewing rectangle, press (GRAPH).

NOTE: You do not need to change "Xres" from 1.

Making a Scatterplot or a Line Graph

Figure A.12

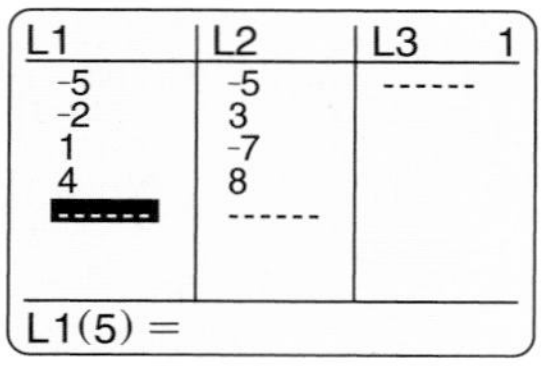

To make a scatterplot with the points $(-5, -5)$, $(-2, 3)$, $(1, -7)$, and $(4, 8)$, begin by following these steps.

1. Press (STAT) followed by (1).
2. If list L1 is not empty, use the arrow keys to place the cursor on L1, as shown in Figure A.12. Then press (CLEAR) followed by (ENTER). This deletes all elements in the list. Similarly, if L2 is not empty, clear the list.
3. Input each x-value into list L1 followed by (ENTER). Input each y-value into list L2 followed by (ENTER). See Figure A.13.

Figure A.13

It is essential that both lists have the same number of values—otherwise an error message appears when a scatterplot is attempted. Before these four points can be plotted, "STAT-PLOT" must be turned on. It is accessed by pressing

(2nd)(Y = [STAT PLOT]),

as shown in Figure A.14.

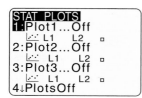

Figure A.14

There are three possible "STATPLOTS," numbered 1, 2, and 3. Any one of the three can be selected. The first plot can be selected by pressing (1). Next, place the cursor over "On" and press (ENTER) to turn "Plot1" on. There are six types of plots that can be selected. The first type is a *scatterplot* and the second type is a *line graph*, so place the cursor over the first type of plot and press (ENTER) to select a scatterplot. (To make the line graph, place the cursor over the second type of plot and press (ENTER).) The x-values are stored in list L1, so select L1 for "Xlist" by pressing (2nd)(1). Similarly, press (2nd)(2) for the "Ylist," since the y-values are stored in list L2. Finally, there are three styles of marks that can be used to show data points in the graph. We will usually use the first because it is largest and shows up the best.

Figure A.15

$[-10, 10, 1]$ by $[-10, 10, 1]$

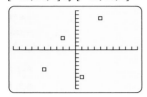

Figure A.16

Make the screen appear as in Figure A.15. Before plotting the four data points, be sure to set an appropriate viewing rectangle. Then press (GRAPH). The data points appear as in Figure A.16.

REMARK 1: A fast way to set the viewing rectangle for any scatterplot is to select the "ZOOMSTAT" feature by pressing (ZOOM)(9). This feature automatically scales the viewing rectangle so that all data points are shown.

REMARK 2: If an equation has been entered into the (Y=) menu and selected, it will be graphed with the data. This feature is used frequently to model data.

The following are basic steps necessary to make either a scatterplot or a line graph.

1. Use (STAT)(1) to access lists L1 and L2.
2. If list L1 is not empty, place the cursor on L1 and press (CLEAR)(ENTER). Repeat for list L2, if it is not empty.
3. Enter the *x*-values into list L1 and the *y*-values into list L2.
4. Use (2nd)(Y = [STAT PLOT]) to select appropriate parameters for the scatterplot or line graph.
5. Set an appropriate viewing rectangle. Press (GRAPH). Otherwise, press (ZOOM)(9). This feature automatically sets the viewing rectangle and plots the data.

NOTE: (ZOOM)(9) *cannot* be used to set a viewing rectangle for the graph of a function.

Entering a Formula

Figure A.17

To enter a formula, press (Y=). For example, use the following keystrokes after "$Y_1 =$" to enter $y = x^2 - 4$. See Figure A.17.

$$(Y=)(CLEAR)(X, T, \theta, n)(\wedge)(2)(-)(4)$$

Note that there is a built-in key to enter the variable X. If "$Y_1 =$" does not appear after pressing (Y=), press (MODE) and make sure the calculator is set in *function mode,* denoted "Func". See Figure A.18.

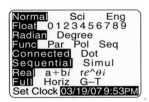

Figure A.18

$[-10, 10, 1]$ by $[-10, 10, 1]$

To enter a formula, press (Y=). To delete formula, press (CLEAR).

Graphing a Function

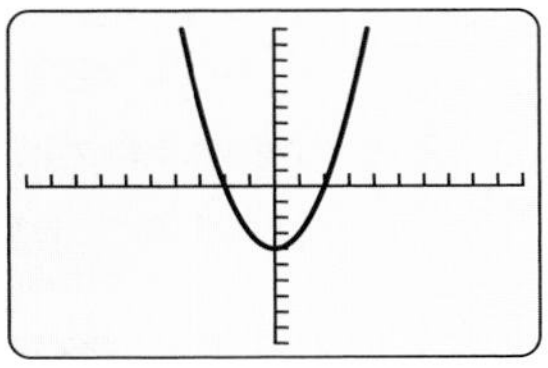

To graph a function, such as $y = x^2 - 4$, start by pressing (Y=) and enter $Y_1 = X^2 - 4$. If there is an equation already entered, remove it by pressing (CLEAR). The equals signs in "$Y_1 =$" should be in reverse video (a dark rectangle surrounding a white equals sign), which indicates that the equation will be graphed. If the equals sign is not in reverse video, place the cursor over it and press (ENTER). Set an appropriate viewing rectangle and then press (GRAPH). The graph of *f* will appear in the specified viewing rectangle. See Figures A.17 and A.19.

Figure A.19

1. Use the (Y=) menu to enter the formula for the function.
2. Use the (WINDOW) menu to set an appropriate viewing rectangle.
3. Press (GRAPH).

Graphing a Vertical Line

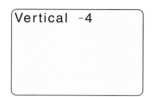

Figure A.20

Set an appropriate window (or viewing rectangle). Then return to the home screen by pressing

$$(2nd)(MODE\ [QUIT]).$$

To graph a vertical line, such as $x = -4$, press

$$(2nd)(PRGM\ [DRAW])(4)(-)(4).$$

See Figure A.20. Pressing (ENTER) will make the vertical line appear, as shown in Figure A.21.

$[-6, 6, 1]$ by $[-6, 6, 1]$

Figure A.21

1. Set an appropriate window by pressing (WINDOW).
2. Return to the home screen by pressing (2nd)(MODE [QUIT]).
3. Draw a vertical line by pressing (2nd)(PRGM [DRAW])(4)(h)(ENTER).

Squaring a Viewing Rectangle

In a square viewing rectangle the graph of $y = x$ is a line that makes a 45° angle with the positive x-axis, a circle appears circular, and all sides of a square have the same length. An approximate square viewing rectangle can be set if the distance along the x-axis is 1.5 times the distance along the y-axis. Examples of viewing rectangles that are (approximately) square include

$$[-6, 6, 1] \text{ by } [-4, 4, 1] \quad \text{and} \quad [-9, 9, 1] \text{ by } [-6, 6, 1].$$

Square viewing rectangles can be set automatically by pressing either

$$(ZOOM)(4) \quad \text{or} \quad (ZOOM)(5).$$

Figure A.22

ZOOM 4 provides a *decimal window*, which is discussed later. See Figure A.22.

Either (ZOOM)(4) or (ZOOM)(5) may be used to produce a square viewing rectangle. An (approximately) square viewing rectangle has the form

$$[-1.5k, 1.5k, 1] \text{ by } [-k, k, 1],$$

where k is a positive number.

Locating a Point of Intersection

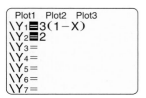

Figure A.23

In Example 8, Section 3.1 on page 96, we find the point of intersection for two lines. To find the point of intersection for the graphs of

$$y_1 = 3(1 - x) \quad \text{and} \quad y_2 = 2,$$

start by entering Y_1 and Y_2, as shown in Figure A.23. Set the window, and graph both equations. Then press the following keys to find the intersection point.

$$\boxed{\text{2nd}}\ \boxed{\text{TRACE [CALC]}}\ \boxed{5}$$

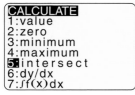

Figure A.24

See Figure A.24, where the "intersect" utility is being selected. The calculator prompts for the first curve, as shown in Figure A.25. Use the arrow keys to locate the cursor near the point of intersection and press $\boxed{\text{ENTER}}$. Repeat these steps for the second curve. Finally we are prompted for a guess. For each of the three prompts, place the free-moving cursor near the point of intersection and press $\boxed{\text{ENTER}}$. The approximate coordinates of the point of intersection are shown in Figure 3.5 on page 96.

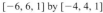

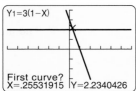

Figure A.25

1. Graph the two equations in an appropriate viewing rectangle.
2. Press $\boxed{\text{2nd}}\ \boxed{\text{TRACE [CALC]}}\ \boxed{5}$.
3. Use the arrow keys to select an approximate location for the point of intersection. Press $\boxed{\text{ENTER}}$ to make the three selections for "First curve?", "Second curve?", and "Guess?". (Note that if the cursor is near the point of intersection, you usually do not need to move the cursor for each selection. Just press $\boxed{\text{ENTER}}$ three times.)

Accessing the Absolute Value

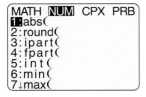

Figure A.26

In Example 9, Section 3.5 on page 129, the absolute value is used to graph $y_1 = |x - 50|$. To graph y_1, begin by entering $Y_1 = \text{abs}(X - 50)$. The absolute value (abs) is accessed by pressing

$$\boxed{\text{MATH}}\ \boxed{\triangleright}\ \boxed{1}.$$

See Figure A.26.

1. Press $\boxed{\text{MATH}}$.
2. Position the cursor over "NUM".
3. Press $\boxed{1}$ to select the absolute value.

Shading a System of Inequalities

In Example 5, Section 4.3 on page 162, we are asked to shade the solution set for the system of linear inequalities $2x + y \leq 5$, $-2x + y \geq 1$. Begin by solving each inequality for y to obtain $y \leq 5 - 2x$ and $y \geq 2x + 1$. Then let $Y_1 = 5 - 2X$ and $Y_2 = 2X + 1$, as shown in

Figure 4.24(b). Position the cursor to the left of Y_1 and press (ENTER) three times. The triangle that appears indicates that the calculator will shade the region below the graph of Y_1. Next locate the cursor to the left of Y_2 and press (ENTER) twice. This triangle indicates that the calculator will shade the region above the graph of Y_2. After setting the viewing rectangle to $[-15, 15, 5]$ by $[-10, 10, 5]$, press (GRAPH). The result is shown in Figure 4.24(c).

1. Solve each inequality for y.
2. Enter each formula as Y_1 and Y_2 in the (Y=) menu.
3. Locate the cursor to the left of Y_1 and press (ENTER) two or three times to shade either above or below the graph of Y_1. Repeat for Y_2.
4. Set an appropriate viewing rectangle.
5. Press (GRAPH).

NOTE: The "Shade" utility in the DRAW menu can also be used to shade the region *between* two graphs.

Entering the Elements of a Matrix

In Example 6(b), Section 4.6 on page 182, the elements of a matrix are entered. The augmented matrix A is given by

$$A = \begin{bmatrix} 1 & 1 & 2 & | & 1 \\ -1 & 0 & 1 & | & -2 \\ 2 & 1 & 5 & | & -1 \end{bmatrix}.$$

Use the following keystrokes on the TI-83 Plus or TI-84 Plus to define a matrix A with dimension 3×4. (*Note:* On the TI-83 the matrix menu is found by pressing (MATRX).)

(2nd) (x^{-1} [MATRIX]) (▷) (▷) (1) (3) (ENTER) (4) (ENTER)

See Figure 4.31(a).

Then input the 12 elements of the matrix A, row by row. Finish each entry by pressing (ENTER). After these elements have been entered, press

(2nd) (MODE [QUIT])

to return to the home screen. To display the matrix A, press

(2nd) (x^{-1} [MATRIX]) (1) (ENTER).

See Figure A.27.

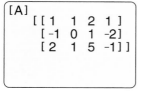

Figure A.27

1. Begin by accessing the matrix A by pressing (2nd) (x^{-1} [MATRIX]) (▷) (▷) (1).
2. Enter the dimension of A by pressing (m) (ENTER) (n) (ENTER), where the dimension of the matrix is $m \times n$.
3. Input each element of the matrix, row by row. Finish each entry by pressing (ENTER). Use (2nd) (MODE [QUIT]) to return to the home screen.

NOTE: On the TI-83, replace the keystrokes (2nd) (x^{-1} [MATRIX]) with (MATRX).

Reduced Row–Echelon Form

In Example 6(b), Section 4.6 on page 182, the reduced row–echelon form of a matrix is found. To find this reduced row–echelon form, use the following keystrokes from the home screen on the TI-83 Plus or TI-84 Plus.

$$\boxed{\text{2nd}}\ \boxed{x^{-1}\text{ [MATRIX]}}\ \boxed{\triangleright}\ \boxed{\text{ALPHA}}\ \boxed{\text{APPS [B]}}\ \boxed{\text{2nd}}\ \boxed{x^{-1}\text{ [MATRIX]}}\ \boxed{1}\ \boxed{)}\ \boxed{\text{ENTER}}$$

The resulting matrix is shown in Figure 4.31(b). On the TI-83 graphing calculator, use the following keystrokes to find the reduced row–echelon form.

$$\boxed{\text{MATRX}}\ \boxed{\triangleright}\ \boxed{\text{ALPHA}}\ \boxed{\text{MATRX [B]}}\ \boxed{\text{MATRX}}\ \boxed{1}\ \boxed{)}\ \boxed{\text{ENTER}}$$

1. To make rref([A]) appear on the home screen, use the following keystrokes for the TI-83 Plus or TI-84 Plus graphing calculator.

$$\boxed{\text{2nd}}\ \boxed{x^{-1}\text{ [MATRIX]}}\ \boxed{\triangleright}\ \boxed{\text{ALPHA}}\ \boxed{\text{APPS [B]}}\ \boxed{\text{2nd}}\ \boxed{x^{-1}\text{ [MATRIX]}}\ \boxed{1}\ \boxed{)}$$

2. Press $\boxed{\text{ENTER}}$ to calculate the reduced row–echelon form.
3. Use arrow keys to access elements that do not appear on the screen.

NOTE: On the TI-83, replace the keystrokes $\boxed{\text{2nd}}\boxed{x^{-1}\text{[MATRIX]}}$ with $\boxed{\text{MATRX}}$ and $\boxed{\text{APPS [B]}}$ with $\boxed{\text{MATRX [B]}}$.

Evaluating a Determinant

In Example 3(a), Section 4.7 on page 186, a graphing calculator is used to evaluate a determinant of a matrix. Start by entering the 9 elements of the 3 × 3 matrix A, as shown in Figure 4.33(a). To compute det A, perform the following keystrokes from the home screen.

$$\boxed{\text{2nd}}\ \boxed{x^{-1}\text{ [MATRIX]}}\ \boxed{\triangleright}\ \boxed{1}\ \boxed{\text{2nd}}\ \boxed{x^{-1}\text{ [MATRIX]}}\ \boxed{1}\ \boxed{)}\ \boxed{\text{ENTER}}$$

The results are shown in the last two lines of Figure 4.33(b).

1. Enter the dimension and elements of the matrix A.
2. Return to the home screen by pressing

$$\boxed{\text{2nd}}\ \boxed{\text{MODE [QUIT]}}.$$

3. On the TI-83 Plus or TI-84 Plus, perform the following keystrokes.

$$\boxed{\text{2nd}}\ \boxed{x^{-1}\text{ [MATRIX]}}\ \boxed{\triangleright}\ \boxed{1}\ \boxed{\text{2nd}}\ \boxed{x^{-1}\text{ [MATRIX]}}\ \boxed{1}\ \boxed{)}\ \boxed{\text{ENTER}}$$

NOTE: On the TI-83, replace the keystrokes $\boxed{\text{2nd}}\boxed{x^{-1}\text{[MATRIX]}}$ with $\boxed{\text{MATRX}}$.

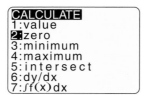

Figure A.28

Locating an x-Intercept or Zero

In Section 5.3 on page 218, we locate an *x*-intercept or *zero* of $f(x) = x^2 - 4$. Start by entering $Y_1 = X^2 - 4$ into the (Y=) menu. Set the viewing rectangle to $[-9, 9, 1]$ by $[-6, 6, 1]$ and graph Y_1. Afterwards, press the following keys to invoke the zero finder. See Figure A.28.

(2nd) (TRACE [CALC]) (2)

The graphing calculator prompts for a left bound. Use the arrow keys to set the cursor to the left of the *x*-intercept and press (ENTER). The graphing calculator then prompts for a right bound. Set the cursor to the right of the *x*-intercept and press (ENTER). Finally, the graphing calculator prompts for a guess. Set the cursor roughly at the *x*-intercept and press (ENTER). See Figures A.29–A.31. The calculator then approximates the *x*-intercept or zero automatically, as shown in the Technology Note on page 219. The zero of -2 can be found similarly.

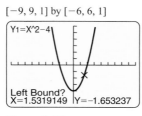

Figure A.29

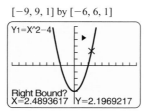

Figure A.30

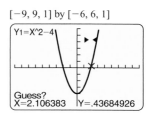

Figure A.31

1. Graph the function in an appropriate viewing rectangle.
2. Press (2nd) (TRACE [CALC]) (2).
3. Select the left and right bounds, followed by a guess. Press (ENTER) after each selection. The calculator then approximates the *x*-intercept or zero.

Setting Connected and Dot Mode

Figure A.32

To set your graphing calculator in dot mode, press (MODE), position the cursor over "Dot", and press (ENTER). See Figure A.32. Graphs will now appear in dot mode rather than connected mode.

1. Press (MODE).
2. Position the cursor over "Connected" or "Dot". Press (ENTER).

Setting a Decimal Window

With a decimal window, the cursor stops on convenient *x*-values. In the decimal window $[-9.4, 9.4, 1]$ by $[-6.2, 6.2, 1]$ the cursor stops on *x*-values that are multiples of 0.2. If we reduce the viewing rectangle to $[-4.7, 4.7, 1]$ by $[-3.1, 3.1, 1]$, the cursor stops on *x*-values

Figure A.33

that are multiples of 0.1. To set this smaller window automatically, press (ZOOM)(4). See Figure A.33. Decimal windows are also useful when graphing rational functions with asymptotes in connected mode.

1. Press (ZOOM)(4) to set the viewing rectangle $[-4.7, 4.7, 1]$ by $[-3.1, 3.1, 1]$.
2. A larger decimal window is $[-9.4, 9.4, 1]$ by $[-6.2, 6.2, 1]$.

Setting $a + bi$ Mode

In Example 1(a), Section 7.6 on page 351, the expression $\sqrt{-25}$ is evaluated. To evaluate expressions containing square roots of negative numbers, set your calculator in $a + bi$ mode by using the following keystrokes.

(MODE)(▽)(▽)(▽)(▽)(▽)(▽)(▷)(ENTER)(2nd)(MODE [QUIT])

Figure A.34

See Figures A.34 and A.35.

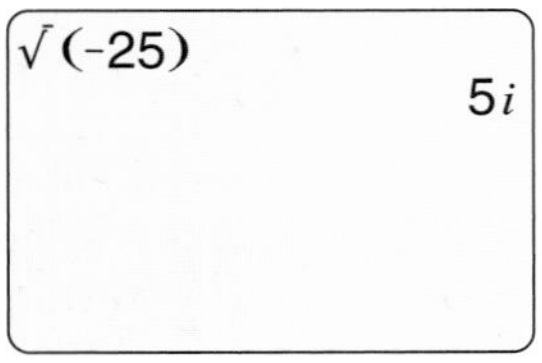

Figure A.35

1. Press (MODE).
2. Move the cursor to the seventh line, highlight $a + bi$, and press (ENTER).
3. Press (2nd)(MODE [QUIT]) and return to the home screen.

Evaluating Complex Arithmetic

Complex arithmetic can be performed much like other arithmetic expressions. This is done by entering

(2nd)(. [i])

to obtain the imaginary unit i from the home screen. For example, to find the sum $(-2 + 3i) + (4 - 6i)$, perform the following keystrokes on the home screen.

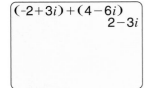

Figure A.36

((−) 2 + 3 (2nd)(. [i])) + (4 − 6 (2nd)(. [i])) (ENTER)

The result is shown in Figure A.36. Other complex arithmetic operations are done similarly.

Enter a complex expression in the same way as you would any arithmetic expression. To obtain the complex number i, use (2nd)(. [i]).

Finding Maximum and Minimum Values

To find a minimum y-value (or vertex) on the graph of $f(x) = 1.5x^2 - 6x + 4$, start by entering $Y_1 = 1.5X^2 - 6X + 4$ from the (Y=) menu. Set the viewing rectangle and then perform the following keystrokes to find the minimum y-value.

(2nd) (TRACE [CALC]) (3)

See Figure A.37.

The calculator prompts for a left bound. Use the arrow keys to position the cursor left of the vertex and press (ENTER). Similarly, position the cursor to the right of the vertex for the right bound and press (ENTER). Finally, the graphing calculator asks for a guess between the left and right bounds. Place the cursor near the vertex and press (ENTER). See Figures A.38–A.40. The minimum value is shown in Figure A.41.

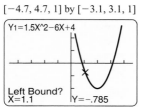

Figure A.37

Figure A.38

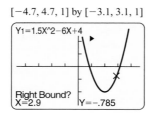

Figure A.39

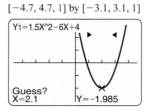

Figure A.40

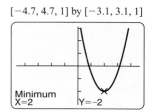

Figure A.41

A maximum of the function f on an interval can be found in a similar manner, except enter

(2nd) (TRACE [CALC]) (4).

The calculator prompts for left and right bounds, followed by a guess. Press (ENTER) after the cursor has been located appropriately for each prompt. The graphing calculator will display the maximum y-value. For example, see the Technology Note on page 371.

1. Graph the function in an appropriate viewing rectangle.
2. Press (2nd) (TRACE [CALC]) (3) to find a minimum y-value.
3. Press (2nd) (TRACE [CALC]) (4) to find a maximum y-value.
4. Use the arrow keys to locate the left and right x-bounds, followed by a guess. Press (ENTER) to select each position of the cursor.

GLOSSARY

absolute value A real number a, written $|a|$, is equal to its distance from the origin on the number line.

absolute value equation An equation that contains an absolute value.

absolute value function The function defined by $f(x) = |x|$.

absolute value inequality An inequality that contains an absolute value.

addends In an addition problem, the two numbers that are added.

addition property of equality If a, b, and c are real numbers, then $a = b$ is equivalent to $a + c = b + c$.

additive identity The number 0.

additive inverse (opposite) The additive inverse or opposite of a real number a is $-a$.

adjacency matrix A matrix used to represent a map showing distances between cities.

algebraic expression An expression consisting of numbers, variables, arithmetic symbols, and grouping symbols, such as parentheses, brackets, and square roots.

annuity A sum of money from which regular payments are made.

approximately equal The symbol $\approx$ indicates that two quantities are nearly equal.

arithmetic sequence A linear function given by $a_n = dn + c$ whose domain is the set of natural numbers.

arithmetic series The sum of the terms of an arithmetic sequence.

associative property for addition For any real numbers a, b, and c, $(a + b) + c = a + (b + c)$.

associative property for multiplication For any real numbers a, b, and c, $(a \cdot b) \cdot c = a \cdot (b \cdot c)$.

asymptotes of a hyperbola The two lines determined by the diagonals of the hyperbola's fundamental rectangle.

augmented matrix A matrix used to represent a system of linear equations; a vertical line is positioned in the matrix where the equals signs occur in the system of equations.

average The result of adding up the numbers of a set and then dividing the sum by the number of elements in the set.

axis of symmetry of a parabola The line passing through the vertex of the parabola that divides the parabola into two symmetric parts.

base The value of a in the expression a^x.

basic principle of fractions When simplifying fractions, the principle which states $\frac{a \cdot c}{b \cdot c} = \frac{a}{b}$.

binomial A polynomial with two terms.

binomial coefficient The expression ${}_nC_r$ that can be used to calculate the numbers in Pascal's triangle.

binomial theorem A theorem that provides a formula to expand expressions of the form $(a + b)^n$.

branches A hyperbola has two branches, a left branch and a right branch, or an upper branch and a lower branch.

break-even point The point at which the cost of producing an item equals the revenue made from selling it.

byte A unit of computer memory, capable of storing one letter of the alphabet.

Cartesian coordinate system (xy-plane) The xy-plane used to plot points and visualize a relation; points are identified by ordered pairs.

center of a circle The point that is a fixed distance from all the points on a circle.

circle The set of points in a plane that are a constant distance from a fixed point called the center.

clearing fractions The process of multiplying both sides of an equation by a common denominator to eliminate the fractions.

coefficient The numeric constant in a monomial.

common denominator A number or expression that each denominator can divide into *evenly* (without a remainder).

common difference The value of d in an arithmetic sequence, $a_n = dn + c$.

common logarithmic function The function given by $f(x) = \log x$.

common logarithm of a positive number x Denoted $\log x$, it may be calculated as follows: if x is expressed as $x = 10^k$, then $\log x = k$, where k is a real number. That is, $\log 10^k = k$.

common ratio The value of r in a geometric sequence, $a_n = a_1(r)^{n-1}$.

commutative property for addition For any real numbers a and b, $a + b = b + a$.

commutative property for multiplication For any real numbers a and b, $a \cdot b = b \cdot a$.

completing the square An important technique in mathematics that involves adding a constant to a binomial so that a perfect square trinomial results.

complex conjugate The complex conjugate of $a + bi$ is $a - bi$.

complex fraction A rational expression that contains fractions in its numerator, denominator, or both.

complex number A complex number can be written in standard form as $a + bi$, where a and b are real numbers and i is the imaginary unit.

compound inequality Two inequalities joined by the words *and* or *or*.

conic section The curve formed by the intersection of a plane and a cone.

conjugate The conjugate of $a + b$ is $a - b$.

consistent system A system of linear equations with at least one solution.

constant function A linear function of the form $f(x) = b$, where b is a constant.

constant of proportionality (constant of variation) In the equation $y = kx$, the number k.

constraint In linear programming, an inequality that limits the objective function.

continuous growth Growth in a quantity that is directly proportional to the amount present.

contradiction An equation that is always false, regardless of the values of any variables.

Cramer's rule A method that uses determinants to solve linear systems of equations.

cubed Process of raising a number or variable to the third power.

cube root The number b is a cube root of a number a if $b^3 = a$.

cube root function The function defined by $g(x) = \sqrt[3]{x}$.

cubic polynomial A polynomial of degree 3 that can be written as $ax^3 + bx^2 + cx + d$, where $a \neq 0$.

decay factor The value of a in an exponential function, $f(x) = Ca^x$, where $0 < a < 1$.

degree of a monomial The sum of the exponents of the variables.

degree of a polynomial The degree of the monomial with highest degree.

dependent equations Equations in a linear system that have the same solution set.

dependent variable The variable that represents the output of a function.

determinant A real number associated with a square matrix.

diagrammatic representation A function represented by a diagram.

difference The answer to a subtraction problem.

difference of two cubes Expression in the form $a^3 - b^3$, which can be factored as $(a - b)(a^2 + ab + b^2)$.

difference of two squares Expression in the form $a^2 - b^2$, which can be factored as $(a - b)(a + b)$.

dimension of a matrix The size expressed in number of rows and columns. For example, if a matrix has m rows and n columns, its dimension is $m \times n$ (m by n).

directly proportional A quantity y is directly proportional to x if there is a nonzero number k such that $y = kx$.

discriminant The expression $b^2 - 4ac$ in the quadratic formula.

distance formula The distance between the points (x_1, y_1) and (x_2, y_2) in the xy-plane is $d = \sqrt{(x_2 - x_1)^2 + (y_2 - y_1)^2}$.

distributive properties For any real numbers a, b, and c, $a(b + c) = ab + ac$ and $a(b - c) = ab - ac$.

dividend In a division problem, the number being divided.

divisor In a division problem, the number being divided *into* another.

domain The set of all x-values of the ordered pairs in a relation.

element Each number in a matrix.

elimination (or addition) method A symbolic method used to solve a system of equations that is based on the property that "equals added to equals are equal."

ellipse The set of points in a plane, the sum of whose distances from two fixed points is constant.

equation A statement that two algebraic expressions are equal.

equivalent equations Two equations that have the same solution set.

even root The nth root, $\sqrt[n]{a}$, where n is even.

expansion of a determinant by minors A method of finding a 3×3 determinant by using determinants of 2×2 matrices.

exponent The value of a in the expression b^a.

exponential decay When $0 < a < 1$, the graph of $f(x) = Ca^x$ models exponential decay.

exponential equation An equation that has a variable as an exponent.

exponential expression An expression that has an exponent.

exponential function with base a and coefficient C A function represented by $f(x) = Ca^x$, where $a > 0$, $C > 0$, and $a \neq 1$.

exponential growth When $a > 1$, the graph of $f(x) = Ca^x$ models exponential growth.

extraneous solution A solution that does not satisfy the given equation.

factors In a multiplication problem, the two numbers multiplied.

factorial notation $n! = 1 \cdot 2 \cdot 3 \cdot \cdots \cdot n$ for any positive integer n.

factoring by grouping A technique that uses the distributive property by grouping four terms of a polynomial in such a way that the polynomial can be factored even though its greatest common factor is 1.

feasible solutions In linear programming, the set of solutions that satisfies the constraints.

finite sequence A function with domain $D = \{1, 2, 3, \ldots, n\}$ for some fixed natural number n.

finite series A series that contains a finite number of terms, and can be expressed in the form $a_1 + a_2 + a_3 + \cdots + a_n$ for some n.

focus (plural foci) A fixed point used to determine the points that form a parabola, an ellipse, or a hyperbola.

FOIL A method for multiplying two binomials $(A + B)$ and $(C + D)$. Multiply First terms AC, Outer terms AD, Inner terms BC, and Last terms BD; then combine like terms.

formula An equation that can be used to calculate one quantity by using a known value of another quantity.

function A relation where each element in the domain corresponds to exactly one element in the range.

function notation $y = f(x)$ is read "y equals f of x". This means that function f with input x produces output y.

fundamental rectangle of a hyperbola The rectangle whose four vertices are determined by either $(\pm a, \pm b)$ or $(\pm b, \pm a)$, where $\frac{x^2}{a^2} - \frac{y^2}{b^2} = 1$ or $\frac{y^2}{a^2} - \frac{x^2}{b^2} = 1$.

Gauss–Jordan elimination A numerical method used to solve a linear system in which matrix row transformations are used.

general term (nth term) of a sequence a_n where n is a natural number in the domain of a sequence $a_n = f(n)$.

geometric sequence A function given by $a_n = a_1(r)^{n-1}$ whose domain is the set of natural numbers.

graphical representation A graph of a function.

graphical solution A solution to an equation obtained by graphing.

greater than If a real number a is located to the right of a real number b on the number line, a is greater than b, or $a > b$.

greatest common factor (GCF) The term with the highest degree and largest coefficient that is a factor of all terms in the polynomial.

growth factor The value of a in the exponential function, $f(x) = Ca^x$, where $a > 1$.

half-life The time it takes for a radioactive sample to decay to half its original amount.

hyperbola The set of points in a plane, the difference of whose distances from two fixed points is constant.

identity An equation that is always true regardless of the values of any variables.

identity property of 1 If any number a is multiplied by 1, the result is a, that is, $a \cdot 1 = 1 \cdot a = a$.

identity property of 0 If 0 is added to any real number, a, the result is a, that is, $a + 0 = 0 + a = a$.

imaginary number A complex number $a + bi$ with $b \neq 0$.

imaginary part The value of b in the complex number $a + bi$.

imaginary unit A number denoted i whose properties are $i = \sqrt{-1}$ and $i^2 = -1$.

inconsistent system A system of linear equations that has no solution.

independent equations Equations in a linear system that have a unique solution.

independent variable The variable that represents the input of a function.

index The value of n in the expression $\sqrt[n]{a}$.

index of summation The variable k in the expression $\sum_{k=1}^{n}$.

inequality When the equals sign in an equation is replaced with any one of the symbols $<, \leq, >,$ or $\geq$.

infinite sequence A function whose domain is the set of natural numbers.

input An element of the domain of a function.

integers A set I of numbers given by $I = \{\ldots, -3, -2, -1, 0, 1, 2, 3, \ldots\}$.

interval notation A notation for number line graphs that eliminates the need to draw the entire line. A **closed interval** is expressed by the endpoints in brackets, such as $[-2, 7]$; an **open interval** is expressed in parentheses, such as $(-2, 7)$, and indicates that the endpoints are not included in the solution. A **half-open interval** has a bracket and a parenthesis, such as $[-2, 7)$.

inversely proportional A quantity y is inversely proportional to x if there is a nonzero number k such that $y = k/x$.

irrational numbers Real numbers that cannot be expressed as fractions, such as π or $\sqrt{2}$.

joint variation A quantity z varies jointly as x and y if there is a nonzero number k such that $z = kxy$.

leading coefficient In a polynomial of one variable, the coefficient of the monomial with highest degree.

least common denominator (LCD) The common denominator with the fewest factors.

least common multiple (LCM) The smallest number that two or more numbers will divide into evenly.

less than If a real number a is located to the left of a real number b on the number line, we say that a is less than b, or $a < b$.

like radicals Radicals that have the same index and the same radicand.

like terms Two terms that contain the same variables raised to the same powers.

linear equation An equation that can be written in the form $ax + b = 0$, where $a \neq 0$.

linear function A function f represented by $f(x) = ax + b$, where a and b are constants.

linear inequality in one variable An inequality that can be written in the form $ax + b > 0$, where $a \neq 0$. (The symbol $>$ may be replaced with $\leq, <,$ or $\geq$.)

linear inequality in two variables When the equals sign in a linear equation of two variables is replaced with $<, \leq, >,$ or $\geq$, a linear inequality in two variables results.

linear polynomial A polynomial of degree 1 that can be written as $ax + b$, where $a \neq 0$.

line graph The graph resulting when consecutive data points in a scatterplot are connected with straight line segments.

linear programming problem A problem consisting of an objective function and a system of linear inequalities called constraints.

logarithm with base a of a positive number x Denoted $\log_a x$, it may be calculated as follows: if x can be expressed as $x = a^k$, then $\log_a x = k$, where $a > 0$, $a \neq 1$, and k is a real number. That is, $\log_a a^k = k$.

logarithmic function with base a The function represented by $f(x) = \log_a x$.

lower limit In summation notation, the number representing the subscript of the first term of the series.

main diagonal The elements $a_{11}, a_{22}, a_{33}, \ldots, a_{nn}$ in a matrix with n rows.

major axis The longest axis of an ellipse, which connects the vertices.

matrix A rectangular array of numbers.

minor axis The shortest axis of an ellipse.

monomial A term whose variables have only nonnegative integer exponents.

multiplication property of equality If a, b, and c are real numbers with $c \neq 0$, then $a = b$ is equivalent to $ac = bc$.

multiplicative identity The number 1.

multiplicative inverse (reciprocal) The multiplicative inverse of a nonzero number a is $1/a$.

natural exponential function The function represented by $f(x) = e^x$, where $e \approx 2.71828$.

natural logarithm The base-e logarithm, denoted either $\log_e x$ or $\ln x$.

natural numbers The set of numbers given by $N = \{1, 2, 3, 4, 5, 6, \ldots\}$.

negative number A number a such that $a < 0$.

negative slope On a graph, the slope of a line that falls from left to right.

negative square root The negative square root is denoted $-\sqrt{a}$.

nonlinear data If data points do not lie on a (straight) line, the data are nonlinear.

nonlinear system of equations Two or more equations, at least one of which is nonlinear.

nonlinear system of inequalities Two or more inequalities, at least one of which is nonlinear.

nth root The number b is an nth root of a if $b^n = a$, where n is a positive integer, and is denoted $\sqrt[n]{a} = b$.

nth term (general term) of a sequence Denoted $a_n = f(n)$.

numerical representation A table of values for a function.

numerical solution A solution often obtained by using a table of values.

objective function The given function in a linear programming problem.

odd root The nth root, $\sqrt[n]{a}$, where n is odd.

opposite (additive inverse) The opposite, or additive inverse, of a real number a is $-a$.

opposite of a polynomial The polynomial obtained by negating each term in a given polynomial.

optimal value In linear programming, the value that often results in maximum revenue or minimum cost.

ordered pair A pair of numbers written in parentheses (x, y), in which the order of the numbers is important.

ordered triple Can be expressed as (x, y, z), where x, y, and z are numbers.

origin On the number line, the point associated with the real number 0; in the xy-plane, the point $(0, 0)$.

output An element of the range of a function.

parabola The U-shaped graph of a quadratic function.

parallel lines Two or more lines in the same plane that never intersect.

Pascal's triangle A triangle made up of numbers in which there are 1s along the sides and each element inside the triangle is the sum of the two numbers above it.

perfect cube An integer with an integer cube root.

perfect nth power The value of a if there exists an integer b such that $b^n = a$.

perfect square An integer with an integer square root.

perfect square trinomial A trinomial that can be factored as the square of a binomial, for example, $a^2 + 2ab + b^2 = (a + b)^2$.

perpendicular lines Two lines in a plane that intersect to form a right (90°) angle.

pixels The tiny units that comprise screens for computer terminals or graphing calculators.

point–slope form The line with slope m passing through the point (x_1, y_1) given by the equation $y = m(x - x_1) + y_1$ or equivalently, $y - y_1 = m(x - x_1)$.

polynomial A monomial or a sum of monomials.

polynomials of one variable Polynomials that contain one variable.

positive number A number a such that $a > 0$.

positive slope On a graph, the slope of a line that rises from left to right.

power function A function that can be represented by $f(x) = x^p$, where p is a rational number.

principle square root The positive square root, denoted $\sqrt{a}$.

product The answer to a multiplication problem.

proportion A statement that two ratios are equal.

Pythagorean theorem If a right triangle has legs a and b with hypotenuse c, then $a^2 + b^2 = c^2$.

quadrants The four regions determined by a Cartesian coordinate system.

quadratic equation An equation that can be written as $ax^2 + bx + c = 0$, where a, b, and c are real numbers, with $a \neq 0$.

quadratic formula The solutions to the quadratic equation, $ax^2 + bx + c = 0$, $a \neq 0$, are $x = (-b \pm \sqrt{b^2 - 4ac})/(2a)$.

quadratic function A function f represented by the equation $f(x) = ax^2 + bx + c$, where a, b, and c are real numbers with $a \neq 0$.

quadratic inequality If the equals sign in a quadratic equation is replaced with $>$, $\geq$, $<$, or $\leq$, a quadratic inequality results.

quadratic polynomial A polynomial of degree 2 that can be written as $ax^2 + bx + c$ with $a \neq 0$.

quotient The answer to a division problem.

radical expression An expression that contains a radical sign.

radical sign The symbol $\sqrt{}$ or $\sqrt[n]{}$ for some positive integer n.

radicand The expression under the radical sign.

radius The fixed distance between the center and any point on the circle.

range The set of all y-values of the ordered pairs in a relation.

rate of change The value of a for the linear function given by $f(x) = ax + b$; slope can be interpreted as a rate of change.

rational equation An equation that involves a rational expression.

rational expression A polynomial divided by a nonzero polynomial.

rational function A function defined by $f(x) = p(x)/q(x)$, where $p(x)$ and $q(x)$ are polynomials and the domain of f includes all x-values such that $q(x) \neq 0$.

rational number Any number that can be expressed as the ratio of two integers p/q, where $q \neq 0$; a fraction.

rationalizing the denominator The process of removing radicals from a denominator so that the denominator contains only rational numbers.

real numbers All rational and irrational numbers; any number that can be written using decimals.

real part The value of a in a complex number $a + bi$.

reciprocal (multiplicative inverse) The reciprocal of a nonzero number a is $1/a$.

reduced row–echelon form A matrix form for representing a system of linear equations in which there are 1s on the main diagonal with 0s above and below each 1.

relation A set of ordered pairs.

rise The vertical change between two points on a line, that is, the change in the y-values.

root function In the power function, $f(x) = x^p$, if $p = 1/n$, where $n \geq 2$ is an integer, then f is also a root function, which is given by $f(x) = \sqrt[n]{x}$.

run The horizontal change between two points on a line, that is, the change in the x-values.

scatterplot A graph of distinct points plotted in the xy-plane.

scientific notation A real number a written as $b \times 10^n$, where $1 \leq |b| < 10$ and n is an integer.

set braces $\{\ \}$, used to enclose the elements of a set.

set-builder notation Notation to describe a set of numbers without having to list all of the elements. For example, $\{x \mid x > 5\}$ is read as "the set of all real numbers x such that x is greater than 5."

similar triangles Triangles that have equal corresponding angles and proportional corresponding sides.

slope The ratio of the change in y (rise) to the change in x (run) along a line. Slope m of a line equals $\frac{y_2 - y_1}{x_2 - x_1}$, where (x_1, y_1) and (x_2, y_2) are points on the line.

slope–intercept form The line with slope m and y-intercept b is given by $y = mx + b$.

solution A value for a variable that makes an equation a true statement.

solution set The set of all solutions to an equation.

squared The process of raising a number or variable to the second power.

square matrix A matrix in which the number of rows and the number of columns are equal.

square root The number b is a square root of a number a if $b^2 = a$.

square root function The function given by $f(x) = \sqrt{x}$, where $x \geq 0$.

square root property If k is a nonnegative number, then the solutions to the equation $x^2 = k$ are $x = \pm\sqrt{k}$. If $k < 0$, then this equation has no real solutions.

standard equation of a circle The standard equation of a circle with center (h, k) and radius r is $(x - h)^2 + (y - k)^2 = r^2$.

standard form of a complex number $a + bi$, where a and b are real numbers.

standard form of an equation for a line The form $ax + by = c$, where a, b, and c are constants with a and b not both 0.

standard viewing rectangle of a graphing calculator Xmin $= -10$, Xmax $= 10$, Xscl $= 1$, Ymin $= -10$, Ymax $= 10$, and Yscl $= 1$, denoted $[-10, 10, 1]$ by $[-10, 10, 1]$.

subscript The expression x_1 has a subscript of 1; it is read "x sub one" or "x one".

substitution method A symbolic method for solving a system of equations in which one equation is solved for one of the variables and then the result is substituted into the other equation.

sum The answer to an addition problem.

summation notation Notation in which the uppercase Greek letter sigma represents the sum, for example, $\sum_{k=1}^{n} k^2$.

sum of two cubes Expression in the form $a^3 + b^3$, which can be factored as $(a + b)(a^2 - ab + b^2)$.

symbolic representation Representing a function with a formula; for example, $f(x) = x^2 - 2x$.

symbolic solution A solution to an equation obtained by using properties of equations, and the resulting solution set is exact.

synthetic division A shortcut that can be used to divide $x - k$, where k is a number, into a polynomial.

system of two linear equations in two variables A system of two equations in which each equation can be written in the form $ax + by = c$; an ordered pair (x, y) is a solution to the system of equations if the values for x and y make *both* equations true.

system of linear inequalities Two or more linear inequalities to be solved at the same time, the solution to which must satisfy both inequalities.

table of values An organized way to display the inputs and outputs of a function; a numerical representation.

term A number, a variable, or a product of numbers and variables raised to powers.

terms of a sequence $a_1, a_2, a_3, \ldots, a_n, \ldots$ where the first term is $a_1 = f(1)$, the second term is $a_2 = f(2)$, and so on.

test value When graphing the solution set of a linear inequality, a point chosen to determine which region of the xy-plane to shade.

three-part inequality A compound inequality written in the form $a < x < b$.

translation The shifting of a graph upward, downward, to the right, or to the left in such a way that the shape of the graph stays the same.

transverse axis In a hyperbola, the line segment that connects the vertices.

trinomial A polynomial with three terms.

upper limit In summation notation, the number representing the subscript of the last term of the series.

variable A symbol, such as x, y, or t, used to represent any unknown number or quantity.

varies directly A quantity y varies directly with x if there is a nonzero number k such that $y = kx$.

varies inversely A quantity y varies inversely with x if there is a nonzero number k such that $y = k/x$.

varies jointly A quantity z varies jointly as x and y if there is a nonzero number k such that $z = kxy$.

verbal representation A description of what a function computes in words.

vertex The lowest point on the graph of a parabola that opens upward or the highest point on the graph of a parabola that opens downward.

vertex form of a parabola The vertex form of a parabola with vertex (h, k) is $y = a(x - h)^2 + k$, where $a \neq 0$ is a constant.

vertical asymptote A vertical asymptote typically occurs in the graph of a rational function when the denominator of the rational expression is 0, but the numerator is not 0; it can be represented by a vertical line in the graph of a rational function.

vertical line test If every vertical line intersects a graph at most once, then the graph represents a function.

vertices of an ellipse The endpoints of the major axis.

vertices of a hyperbola The endpoints of the transverse axis.

viewing rectangle (window) On a graphing calculator, the window that determines the x- and y-values shown in the graph.

whole numbers The set of numbers given by $W = \{0, 1, 2, 3, 4, 5, \ldots\}$.

x-axis The horizontal axis in a Cartesian coordinate system.

x-intercept The x-coordinate of a point where a graph intersects the x-axis.

Xmax Regarding the viewing rectangle of a graphing calculator, Xmax is the maximum x-value along the x-axis.

Xmin Regarding the viewing rectangle of a graphing calculator, Xmin is the minimum x-value along the x-axis.

Xscl The distance represented by consecutive tick marks on the $x$axis.

y-axis The vertical axis in a Cartesian coordinate system.

y-intercept The y-coordinate of a point where a graph intersects the y-axis.

Ymax Regarding the viewing rectangle of a graphing calculator, Ymax is the maximum y-value along the y-axis.

Ymin Regarding the viewing rectangle of a graphing calculator, Ymin is the minimum y-value along the y-axis.

Yscl The distance represented by consecutive tick marks on the y-axis.

zero-product property If the product of two numbers is 0, then at least one of the numbers must equal 0, that is, $ab = 0$ implies $a = 0$ or $b = 0$.

zero of a polynomial An x-value that results in an output of 0 when it is substituted into a polynomial; for example, the zeros of $f(x) = x^2 - 4$ are 2 and -2.

BIBLIOGRAPHY

Baase, S. *Computer Algorithms: Introduction to Design and Analysis.* 2nd ed. Reading, Mass.: Addison-Wesley Publishing Company, 1988.

Baker, S., with B. Leak, "Math Will Rock Your World." *Business Week*, January 23, 2006.

Battan, L. *Weather in Your Life.* San Francisco: W. H. Freeman, 1983.

Beckmann, P. *A History of Pi.* New York: Barnes and Noble, Inc., 1993.

Brown, D., and P. Rothery. *Models in Biology: Mathematics, Statistics and Computing.* West Sussex, England: John Wiley and Sons Ltd, 1993.

Callas, D. *Snapshots of Applications in Mathematics.* Delhi, New York: State University College of Technology, 1994.

Carr, G. *Mechanics of Sport.* Champaign, Ill.: Human Kinetics, 1997.

Conquering the Sciences. Sharp Electronics Corporation, 1986.

Elton, C. S., and M. Nicholson. "The ten year cycle in numbers of lynx in Canada." *J. Anim. Ecol.* 11 (1942): 215–244.

Eves, H. *An Introduction to the History of Mathematics,* 5th ed. Philadelphia: Saunders College Publishing, 1983.

Freedman, B. *Environmental Ecology: The Ecological Effects of Pollution, Disturbance, and Other Stresses.* 2nd ed. San Diego: Academic Press, 1995.

Friedhoff, M., and W. Benzon. *The Second Computer Revolution: Visualization.* New York: W. H. Freeman, 1991.

Garber, N., and L. Hoel. *Traffic and Highway Engineering.* Boston, Mass.: PWS Publishing Co., 1997.

Grigg, D. *The World Food Problem.* Oxford: Blackwell Publishers, 1993.

Haefner, L. *Introduction to Transportation Systems.* New York: Holt, Rinehart and Winston, 1986.

Harrison, F., F. Hills, J. Paterson, and R. Saunders. "The measurement of liver blood flow in conscious calves." *Quarterly Journal of Experimental Physiology* 71: 235–247.

Historical Topics for the Mathematics Classroom, Thirty-first Yearbook. National Council of Teachers of Mathematics, 1969.

Horn, D. *Basic Electronics Theory.* Blue Ridge Summit, Penn.: TAB Books, 1989.

Howells, G. *Acid Rain and Acid Waters.* 2nd ed. New York: Ellis Horwood, 1995.

Karttunen, H., P. Kroger, H. Oja, M. Poutanen, K. Donner, eds. *Fundamental Astronomy.* 2nd ed. New York: Springer-Verlag, 1994.

Kincaid, D., and W. Cheney. *Numerical Analysis.* Pacific Grove, Calif.: Brooks/Cole Publishing Company, 1991.

Kraljic, M. *The Greenhouse Effect.* New York: The H. W. Wilson Company, 1992.

Lack, D. *The Life of a Robin.* London: Collins, 1965.

Lancaster, H. *Quantitative Methods in Biological and Medical Sciences: A Historical Essay.* New York: Springer-Verlag, 1994.

Mannering, F., and W. Kilareski. *Principles of Highway Engineering and Traffic Analysis.* New York: John Wiley and Sons, 1990.

Mar, J., and H. Liebowitz. *Structure Technology for Large Radio and Radar Telescope Systems.* Cambridge, Mass.: The MIT Press, 1969.

Meadows, D. *Beyond the Limits.* Post Mills, Vermont: Chelsea Green Publishing Co., 1992.

Miller, A., and J. Thompson. *Elements of Meteorology.* 2nd ed. Columbus, Ohio: Charles E. Merrill Publishing Company, 1975.

Miller, A., and R. Anthes. *Meteorology.* 5th ed. Columbus, Ohio: Charles E. Merrill Publishing Company, 1985.

Monroe, J. *Steffi Graf.* Mankato, Minn.: Crestwood House, 1988.

Motz, L., and J. Weaver. *The Story of Mathematics.* New York: Plenum Press, 1993.

Nemerow, N., and A. Dasgupta. *Industrial and Hazardous Waste Treatment.* New York: Van Nostrand Reinhold, 1991.

Nicholson, A. J. "An Outline of the dynamics of animal populations." *Austr. J. Zool.* 2 (1935): 9–65.

Nielson, G., and B. Shriver, eds. *Visualization in Scientific Computing.* Los Alamitos, Calif.: IEEE Computer Society Press, 1990.

Nilsson, A. *Greenhouse Earth.* New York: John Wiley and Sons, 1992.

Paetsch, M. *Mobile Communications in the U.S. and Europe: Regulation, Technology, and Markets.* Norwood, Mass.: Artech House, Inc., 1993.

Pearl, R., T. Edwards, and J. Miner. "The growth of *Cucumis melo* seedlings at different temperatures." *J. Gen. Physiol.* 17: 687–700.

Pennycuick, C. *Newton Rules Biology.* New York: Oxford University Press, 1992.

Pielou, E. *Population and Community Ecology: Principles and Methods.* New York: Gordon and Breach Science Publishers, 1974.

Pokorny, C., and C. Gerald. *Computer Graphics: The Principles behind the Art and Science.* Irvine, Calif.: Franklin, Beedle, and Associates, 1989.

Ronan, C. *The Natural History of the Universe.* New York: MacMillan Publishing Company, 1991.

Sharov, A., and I. Novikov. *Edwin Hubble, The Discoverer of the Big Bang Universe.* New York: Cambridge University Press, 1993.

Smith, C. *Practical Cellular and PCS Design.* New York: McGraw-Hill, 1998.

Stent, G. S. *Molecular Biology of Bacterial Viruses.* San Francisco: W. H. Freeman, 1963.

Taylor, J. *DVD Demystified.* New York: McGraw-Hill, 1998.

Taylor, W. *The Geometry of Computer Graphics.* Pacific Grove, Calif.: Wadsworth and Brooks/Cole, 1992.

Thomas, D. *Swimming Pool Operators Handbook.* National Swimming Pool Foundation of Washington, D.C., 1972.

Thomas, V. *Science and Sport.* London: Faber and Faber, 1970.

Thomson, W. *Introduction to Space Dynamics.* New York: John Wiley and Sons, 1961.

Toffler, A., and H. Toffler. *Creating a New Civilization: The Politics of the Third Wave.* Kansas City, Mo.: Turner Publications, 1995.

Triola, M. *Elementary Statistics.* 7th ed. Reading, Mass.: Addison-Wesley Publishing Company, 1998.

Tucker, A., A. Bernat, W. Bradley, R. Cupper, and G. Scragg. *Fundamentals of Computing I: Logic, Problem Solving, Programs, and Computers.* New York: McGraw-Hill, 1995.

Turner, R. K., D. Pierce, and I. Bateman. *Environmental Economics, An Elementary Approach.* Baltimore: The Johns Hopkins University Press, 1993.

Varley, G., and G. Gradwell. "Population models for the winter moth." *Symposium of the Royal Entomological Society of London* 4: 132–142.

Wang, T. *ASHRAE Trans.* 81, Part 1 (1975): 32.

Weidner, R., and R. Sells. *Elementary Classical Physics,* Vol. 2. Boston: Allyn and Bacon, Inc., 1965.

Wigner, E., "The Unreasonable Effectiveness of Mathematics in the Natural Sciences." *Communictions of Pure and Applied Mathematics*, Vol. 13, No. I, February, 1960.

Williams, J. *The Weather Almanac 1995.* New York: Vintage Books, 1994.

Wright, J. *The New York Times Almanac 1999.* New York: Penguin Group, 1998.

Zeilik, M., S. Gregory, and D. Smith. *Introductory Astronomy and Astrophysics.* 3rd ed. Philadelphia: Saunders College Publishers, 1992.

INDEX

Absolute value, 123–124
  of a real number, 9, 14
Absolute value equation, 124–127, 130
Absolute value function, 124
Absolute value inequality, 127–129
Addend, 9
Addition. *See also* Sum
  associative property for, 6, 8
  commutative property for, 5, 8
  of complex numbers, 352, 355
  of fractions, 269
  of polynomials, 199–202, 206
  of radical expressions, 325–327, 333
  of rational expressions, 270–271, 275
  of rational functions, 258–259
  of real numbers, 9–11, 14
Addition method, 149–153, 156
Addition property of equality, 91–92, 99
Additive identity, 5, 8
Additive inverse, 9, 14
Algebraic expression, 26, 31
Alpha Centauri, 15
Annuity, 495
Apollonius, 455
Approximate solution, 96
Approximately equal symbol ($\approx$), 4
Approximating square roots, 313
Area
  of a circle, 27, 31
  of a triangle, 31, 186–187, 189
Arithmetic expression, order of operations for, 21–22
Arithmetic sequence, 485–488, 491
  common diffrence of, 486
  general term of, 488
  sum of first n terms of, 493
  symbolic representation of, 487–488
Arithmetic series, 492–494
Arrhenius, Svante, 413
Associative property
  for addition, 6, 8
  for multiplication, 6, 8
Asymptote
  of a hyperbola, 466
  vertical, 254–255
Atanasoff, John, 137
Augmented matrix, 177–178, 181–182, 184
Average, 4–5
Axis of symmetry, 364, 456–459, 461

Base, 15–16, 25
  of an exponential function, 425
Base-2 logarithm, 436
Base-10 logarithm, 431–433, 438
Base-a logarithm, 437, 438
Base-e logarithm, 436
Basic principle of fractions, 260
Binomial, 208
  expanding, 498–499, 502
  factoring, 236, 237
  multiplication of, 208–209
  special products of, 211–213, 214
  square of, 212–213, 214
Binomial coefficient, 499–500, 502
Binomial theorem, 500–501, 502
Brackets ([]), 118
Brahe, Tyco, 455
Branches of a hyperbola, 465, 466
Break-even point, 110

Canceling, 298
Cardano, Girolamo, 297, 311
Cartesian coordinate system, 34–36, 40
Center
  of a circle, 459, 461
  of an ellipse, 463
Change in x, 67, 73
Change in y, 67, 73
Circle
  area of, 27, 31
  center of, 459, 461
  circumference of, 27, 31
  graphing, 460
  radius of, 459–460
  standard equation of, 459–461
Circumference of a circle, 27, 31
Clearing fractions, 290
Coefficient
  binomial, 499
  of an exponential function, 424
  leading, of a polynomial, 199
  of a monomial, 199
Column of a matrix, 177
Common denominator, 269
Common difference, 486
  slope and, 487
Common factor, 214–215, 235, 237
  greatest, 215, 221
Common logarithm, 431–433, 438
  inverse properties of, 434, 439
Common logarithmic function, 431–43???
  inverse of, 434–434
  square root function compared with, 433

Common ratio, 448–489
Commutative property
  for addition, 6, 8
  for multiplication, 6, 8
Completing the square method, 377, 380, 385–386, 389, 389
Complex conjugate, 353–354, 355
Complex fraction, 283–288
  simplifying, 284–288
Complex number, 311, 350–355
  addition of, 352, 355
  division of, 353–354, 355
  imaginary, 350–351, 355
  imaginary unit of, 350, 355
  multiplication of, 352–353, 355
  real part of, 351, 355
  standard form of, 351
  subtraction of, 352, 355
Complex solution, equations with, 394–396, 397, 406–407
Composite function, 414–416, 423
Compound inequality, 117–123
  graphical solutions to, 121
  numerical solutions to, 121
  solving, 118–121
  symbolic solutions to, 118–121
  three-part, 119–121, 123
Compound interest, 426–427, 431
Conditional equation, 97
Conic section, 455–474. *See also* Ellipse; Hyperbola; Parabola
  types of, 456
Conjugate
  complex, 353–354, 355
  of the denominator, 332, 333
Constant function, 65, 66
Constant of proportionality (variation), 292
  ratios and, 293
Constraint, 164
Continuous growth, 430
Contradiction, 97, 152
Counting number, 2, 8
Cramer, Gabriel, 187
Cramer's rule, 187–188, 189
Cube, 15
  factoring sums and differences of, 232–233
  perfect, 321
  sums of, factoring, 236
Cube (geometric solid), volume of, 27, 31
Cube root, 29–30, 31, 313–314
Cube root function, 337–338, 340
Cubic polynomial, 199

I-1

Cubic polynomial function, 202–203
Curvature, point of, 347
Cylinder, volume of, 296

Data. *See also* Modeling data
    visualization of, 31–32
Data point, 8
Decay, exponential, 425, 488–489
Decay factor, 425
    common ratio and, 489
Decimal
    changing decimals to, 105–106, 109
    real numbers as, 4
    repeating (terminating), 3
    solving linear equations with, 94
Decimal places, 4
Decimal point, moving, 24
Decimal window, 255
Degree
    of a monomial, 199
    of a polynomial, 199
Demand, 211
Denominator
    common, 269
    conjugate of, 331–332, 333
    least common, 269, 275, 286, 288
    like, adding rational expressions with, 270–271, 275
    like, subtracting rational expressions with, 272, 275
    rationalizing, 330–333
    simplifying, 285–286, 288
    unlike, adding rational expressions with, 271, 275
    unlike, subtracting rational expressions with, 273, 275
Dependent variable, 48, 58
Descartes, René, 311
Determinant, 184–189
    of a 2 × 2 matrix, 184–185, 189
    of a 3 × 3 matrix, 185, 189
    area of regions and, 186–187, 189
    calculation of, 184–187
    Cramer's rule and, 187–188, 189
    expansion of, by minors, 183
Diagrammatic representation of a function, 50–51, 58
Difference, 11. *See also* Subtraction
    common, 486
    of polynomial functions, 204–206
    product of a sum and a difference and, 211–212, 214
    of rational expressions, 274, 275
    of two cubes, factoring, 232–233, 234
    of two squares, factoring, 230–231, 233, 235, 237
Dimension of a matrix, 177
Direct variation, 292–293, 297
Directly proportional, 292–293, 297
Discriminant, 393–394, 397
Distance formula, 345–347, 349

Distributive properties, 7, 8
    multiplication of polynomials and, 207–208, 213
Dividend, 12
Division. *See also* Quotient
    of complex numbers, 353–354, 355
    of fractions, 462, 365–366
    of rational expressions, 264–266, 267
    of rational functions, 258–259
    of real numbers, 12, 14
    synthetic, 301–302, 303
Division of polynomials, 297–303
    by monomials, 297–298, 303
    by polynomials, 299–301, 303
    synthetic division and, 301–302, 303
Divisor, 12
Domain
    of a function, 52–55, 58
    of a radical function, 336–337
    of rational functions, 253
    of a relation, 33, 40

Element of a matrix, 177
Elimination method, 149–153, 156, 171–174, 176
Ellipse, 462
    center of, 462
    equations of, 463–465, 468
    foci of, 462
    focus of, 462
    major axis of, 462
    minor axis of, 462
    vertices of, 462
Equality
    addition property of, 91–92, 99
    multiplication property of, 91–92, 99
Equation, 26, 31
    absolute value, 124–127, 136
    of a circle, 459–460
    with complex solutions, 406, 407
    contradictions, 97, 152
    of an ellipse, 463–465, 468
    exponential, 443–447, 450
    expressions compared with, 92
    expressions contrasted with, 312
    of a hyperbola, 465–467, 468
    linear. *See* Linear equation
    logarithmic, 447–450
    nonlinear systems of. *See* Nonlinear systems of equations
    of parallel lines, 81
    of perpendicular lines, 81–83
    polynomial. *See* Polynomial equation
    quadratic. *See* Quadratic equation
    in quadratic form, 404–407
    radical. *See* Radical equation
    rational, 256–258, 259, 275–283
    with rational exponents, 405, 407
Equation of a line, 74
    conditional, 97
    horizontal, 79–80, 83
    identities, 97

    point–slope form of, 74–79, 83
    slope–intercept form of, 69–71, 74
    vertical, 79–80, 83
Equivalent linear equations, 90
Estimation, 13–14
Evaluation
    of a function, 52
    of a polynomial function, 202–203, 206
    of a rational function, 254
Even root, 314
Expansion
    of a binomial, 498–499, 502
    of a determinant, by minors, 185
Exponent, 15, 25. *See also* Logarithm
    integer, 17, 25
    negative, 18–19, 25, 318, 319
    positive, 15
    power rule for, 318, 319
    product rule for, 318, 319
    properties of, 207–208, 213–214
    quotient rule for, 318, 319
    raising powers to powers, 19–20, 25
    raising quotients to powers, 20–21, 25
    rational, 315–319, 405, 407
    zero, 16–17, 25
Exponential decay, 425, 488–489
Exponential equation, 443–447, 450
Exponential expression, 15
    products of, 18, 25
    quotients of, 18–19, 25
Exponential function, 424–431
    base of, 425
    coefficient of, 425
    compound interest and, 427, 431
    decay factor and, 425
    graphs of, 427–428
    growth factor and, 425
    linear functions contrasted with, 426
    models involving, 428–429
    natural, 430
Exponential growth, 425, 488–489
Exponential notation, 16
Expression
    algebraic, 26, 31
    equations compared with, 92
    equations contrasted with, 312
    exponential, 15
    order of operations for, 21–22
    radical, 312–319
    rational, 252
Extraneous solution, 257

Factor
    common, 214–215, 221, 235, 237
    greatest common, 215, 221
    of a number, 11
    in synthetic division, 302
Factorial notation ($n!$), 499–500, 502
Factoring binomials, 236, 237
Factoring polynomials, 214–221, 234–237
    common factors and, 214–215, 221, 235, 237

difference of two cubes, 232–233
difference of two squares, 230–231, 233, 235–236, 237
equations and, 215–219, 221
by grouping, 219–220, 221, 236, 237
with several techniques, 237
sum of two cubes, 232–233, 234
trinomials. *See* Factoring trinomials
with two variables, 236, 237
zero-product property and, 216–217, 221
Factoring trinomials, 222–229, 235, 237
with FOIL, 225–227
graphical, 224, 227–228, 229
with graphs and tables, 227–228, 229
by grouping, 221–224
numerical, 228, 229
perfect square trinomials, 231–232, 234, 235
symbolic, 224, 229
Feasible solution, 164
Finite sequence, 481
Finite series, 492, 497
Focus
of an ellipse, 462
of a hyperbola, 465
FOIL, factoring trinomials using, 225–227
Formula, 26–27, 31, 49
distance, 345–347, 349
quadratic, 390–391, 397
solving for a variable, 100–102, 109
Fractal image, 350
Fraction
addition of, 269
basic principle of, 260
clearing, 290
complex. *See* Complex fraction
decimal form of, 4
division of, 262, 265–266
multiplication of, 262
solving linear equations with, 94
subtraction of, 269
Friendly window, 255
Function, 53, 57
absolute value, 124–125
addition of, 204, 258–259
composite, 414–416, 423
constant, 65, 66
cube root, 337–338, 340
division of, 258
domain of, 52–55, 57–58
identifying, 55–57, 57–58
inverse, 418–423, 424
linear. *See* Linear function
multiplication of, 213, 258
nonlinear, 60
Numerical representation of. *See* Numerical representation of a function
objective, 163, 165
one-to-one, 417–418, 423

polynomial. *See* Polynomial function
power, 338–340
quadratic. *See* Quadratic function
rational. *See* Rational function
representations of. *See* Representations of a function
root, 338, 340
square root, 334–337, 340–341
subtraction of, 204, 258
tables of values for, 49, 58, 421–423. *See also* Numerical representation of a function
vertical line test for, 55–56, 58
writing, 102, 109
Function notation, 48–49, 203–204
Fundamental rectangle, 466
Fundamental theorem of linear programming, 165
$f(x)$ notation, 48, 57

Gauss, Carl Fredrich, 177
Gaussian elimination, 177
Gauss–Jordan elimination, 178–182, 184
General term of a sequence, 481, 488
Geometric sequence, 488–489, 491
sum of first *n* terms of, 494–495
symbolic representations of, 489
Geometric series, 494–495
Graph, 49–50, 58
of exponential functions, 427–428
factoring trinomials with, 227–228, 229
of inverse functions, 421–423
line, 37, 39–40
of linear equations, 46
of a parabola, 218, 219
of quadratic functions, 364–370, 372
of rational functions, 253, 255
Graphical representation
of a function, 49–50, 51, 52, 58
of a linear function, 62, 66
of a radical function, 335, 338
Graphing, 31–40
with graphing calculator, 39–40, 57
Graphing calculator, AP-1–AP-12
$a + bi$ mode, AP-11
absolute value, AP-7
arithmetic operations using, 13
complex arithmetic, AP-11
connected mode, AP-10
creating tables, AP-3
creating tables using, 57
decimal window, AP-10–AP-11
dot mode, AP-10
entering elements of a matrix, AP-8
entering formulas, AP-5
entering mathematical expressions, AP-1–AP-2
evaluating determinants, AP-9
expressing answers as fractions, AP-2
factoring with, 227–228
for finding sum of a series, 494
generating sequences with, 482

graphing functions, AP-5–AP-6
graphing hyperbolas using, 467
graphing radical expressions using, 345
graphing rational functions using, 255
graphing using, 57
graphing vertical lines, AP-6
graphing vertical lines using, 126
graphing with, 39–40
graphs using, 294
line graphs, AP-4–AP-5
linear functions using, 62
Graphing calculator (*continued*)
locating points of intersection, AP-7
locating *x*-intercepts or zero, AP-10
maximum and minimum values, AP-12
rational exponents and, 316–317
reduced row–echelon form, AP-9
scatterplots using, 294, AP-4–AP-5
scientific notation, AP-2–AP-3
setting viewing rectangle (window), AP-4
shading systems of inequalities, AP-7–AP-8
solving systems of inequalities using, 473
solving systems of linear equations using, 182–183
square roots of negative numbers and, 334
squaring viewing rectangles, AP-6
subtraction of rational expressions using, 273
table feature of, 30
viewing rectangle (window) of, 38–39, 80, 255, 461
Greater than, 9
Greater than or equal to, 54
Greatest common factor (GCF), 215, 221
Grouping, factoring by, 219–220, 221
of polynomials, 236, 237
of trinomials, 224–225
Growth
continuous, 430
exponential, 425, 488–489
Growth factor, 425
common ratio and, 489

Higher degree polynomial equation, 240–241, 244, 404, 407
Horizontal axes of symmetry, parabolas with, 456–459, 461
Horizontal line, 79–80, 83
Horizontal line test, 417–418, 424
Hyperbola, 465
asymptotes of, 466
branches of, 465, 466
equations of, 465–467, 468
focus of, 465
fundamental rectangle and, 466
graphing with graphing calculator, 467
transverse axis of, 465
vertices of, 465
Hypotenuse, 346–347, 349

i, powers of, 353
Identity, 97, 153
    additive, 5, 8
    multiplicative, 5, 8
Identity equation, 97
Identity property
    of 5, 8
    of 5, 8
Imaginary number, 311, 350–351
    pure, 350–351
Imaginary part of a complex number, 350–351, 355
Imaginary unit, 350–351, 355
Inconsistent systems of linear equations, 141, 144, 147, 175
Independent variable, 48, 57
Index
    of a radical, 314, 322, 325–326
    of summation (k), 496
Inequality
    absolute value, 127–129
    compound. *See* Compound inequality
    nonlinear systems of, 471–473
    properties of, 110–111
    quadratic, 398–403
Infinite sequences, 481
Infinity, 122
    negative, 122
Infinity symbol (∞), 122
Input, 48
Integer, 4, 7–8
Integer exponents, 17, 25
Intercept, 98, 99
Intersection of solution sets, 117
Interval, 112
Interval notation, 121–122, 123
Inverse
    additive, 9–10, 14
    of common logarithmic function, 433–435
    multiplicative, 11–12
Inverse function, 418–423, 424
    tables and graphs of, 421–423
Inverse properties of logarithms, 438
Inverse variation, 293–295, 297
Inversely proportional, 293–295, 297
Irrational number, 4–5, 8

Joint variation, 295–296, 297

Kepler, Johannes, 439, 455
Kowa, Seki, 184

Leading coefficient of a polynomial, 199
Least common denominator (LCD), 269, 275, 286, 288
Least common multiple (LCM), 267–268, 275
Legs of a right triangle, 395–346, 349
Leibniz, Gottfried, 184, 207
Less than, 9
Less than or equal to, 54

Like radicals, 326–327, 333
Line
    equation of, 75. *See also* Equation of a line
    horizontal, 79–80, 83
    intercepts of, 98, 99
    parallel, 81
    perpendicular, 81–83
    point–slope form of, 74–80, 83
    slope of. *See* Slope
    slope–intercept form of, 70, 74, 76–77, 83
    standard form of, 98, 99
    vertical, 79–80, 83
Line graph, 37, 39, 40
Linear equation, 90–99
    equivalent, 90
    graphical solution of, 96–97
    identities and contradictions and, 97, 152
    intercepts of a line and, 98
    linear functions and linear inequalities compared with, 110–111
    numerical solution of, 95–96
    solution sets to, 90, 99
    solutions to, 90, 99
    symbolic solutions of, 91–95
    systems of. *See* Systems of linear equations; Systems of linear equations in three variables
Linear function, 59–66, 91, 110
    constant, 65, 66
    exponential functions contrasted with, 426
    linear equations and linear inequalities compared with, 110
    modeling data with, 63–65
    rate of change for, 60–61, 63–65, 66
    representations of, 61–63
Linear inequality, 109–116
    graphical solutions to, 113–114, 159
    linear functions and linear equations compared with, 110
    numerical solutions to, 112–113
    in one variable, 110, 116
    properties of, 111, 116
    solution set to, 109, 110
    solutions to, 109–110
    symbolic solutions to, 110–112
    systems of, 159–162
    in two variables, 157–159, 162
Linear polynomial, 199
Linear polynomial function, 202
Linear programming, 163–167
    fundamental theorem of, 164–165
    problem solving using, 164–167
    region of feasible solutions and, 164
Linear programming problems, 164–165
Logarithm
    base-2, 436
    base-*a*, 436–437, 438
    base-*e*, 436

    change of base formula for, 442–443
    common (base-10), 431–433, 438
    inverse properties of, 437
    natural, 436
    power rule for, 441, 443
    product rule for, 439, 443
    quotient rule for, 440, 443
    of quotients, 445
    quotients of, 445
Logarithmic equation, 447–449, 450
Logarithmic function, 431–439
    with base *a*, 436–437
    common, 431–433
Lower limit, of a summation, 496

Main diagonal of a matrix, 178
Major axis of an ellipse, 462
Matrix, 177–178
    augmented, 177–178, 179, 181–182, 184
    columns of, 177
    determinants of, 184–189
    dimension of, 177
    elements of, 177
    Gauss–Jordan elimination and, 178–182, 184
    main diagonal of, 178
    minors of, 185
    row transformations of, 177–181
    rows of, 177
    square, 177
Min–max applications of quadratic functions, 370–372
Minor, 185
Minor axis of an ellipse, 462
Modeling data
    with absolute value inequalities, 129
    with direct variation, 292–293
    with ellipses, 464–465
    with exponential equations, 447
    with exponential functions, 428–429
    with formulas, 27–28
    with linear functions, 63–65, 71–79
    with logarithmic equations, 449–450
    with logarithmic functions, 435
    with nonlinear systems of equations, 471
    with polynomials, 204–205, 206
    with power functions, 339–340
    with quadratic equations, 385, 387–388
    with quadratic functions, 378–379
    with radical expressions, 348–349
    with rational equations, 280
    with sequences, 484, 490
    with series, 496–497
    with systems of linear equations, 153–154, 161, 176, 183
    with systems of linear equations in three variables, 173–174
Monomial, 198–199, 206
    coefficient of, 199
    degree of, 199
    division of a polynomial by, 297–298, 303

Multiplication. *See also* Product
  associative property for, 6, 8
  of binomials, 208–209
  commutative property for, 6, 8
  of complex numbers, 352–353, 355
  of fractions, 262
  of polynomial functions, 213, 214
  of polynomials, 207–214
  of radical expressions, 330, 336
  of rational expressions, 262–263, 267
  of rational functions, 258–259
  of real numbers, 12, 14
Multiplication property of equality,
    91–92, 99
Multiplicative identity, 5, 8
Multiplicative inverse, 11

n factorial, 499, 502
Napier, John, 439
Natural exponential function, 430
Natural logarithm, 436–437
Natural number, 2, 7
Negation, 11
Negative exponent, 18, 25, 317–318, 319
  for quotients, 318, 319
Negative infinity, 122
Negative number, 9
  square roots of, 234, 351
Negative reciprocal, 82
Negative sign, rational expressions and,
    260–261
Negative slope, 68
Newton, Isaac, 439
Nonlinear function, 60
Nonlinear systems of equations, 468–471,
    473
  graphical solution of, 470
  substitution method for solving, 469–471
  symbolic solution of, 470–471
Nonlinear systems of inequalities,
    471–473
  graphical solution to, 473
Not equal to, 12
Notation
  exponential, 16
  function, 203
  interval, 121–122, 123
  radical, 312–315
  scientific, 22–24, 25
  set-builder, 109, 116
  summation, 496–497
  union, 120
nth power, perfect, 321
nth root, 314–315, 319
  principal, 314
  simplifying, 321, 325
nth term of a sequence, 481
Number
  complex. *See* Complex number
  imaginary, 311, 351
  integer, 3, 7
  irrational, 4–5, 7

  natural (counting), 2, 7
  negative, 8–9, 334, 351
  positive, 8–9
  rational, 3, 7
  real. *See* Real number
  sets of, 2–8
  whole, 2, 7
Number line, real, 8–9
Number sense, 13–14
Numerator, simplifying, 285–286, 288
Numerical representation
  factoring trinomials and, 228, 229
  for an inverse function, 421
Numerical representation of a function,
    49, 50, 52, 58
  of a linear function, 62, 66
  of a radical function, 335, 338

Objective function, 163, 165
Odd root, 314
One, identity property of, 5, 8
One-to-one function, 417–418, 423
Opposite, 9, 14. *See also* Inverse
  of a polynomial, 202, 206
Optimal value, 165
Order of operations, 21–22
Ordered pair, 32–33, 40
Ordered triple, 168
Origin
  of a number line, 8
  of the *xy*-plane, 34–35
Output, 48

Pair, ordered, 32–33, 40
Parabola, 217, 218–219, 364–370, 372,
    456
  axis of symmetry of, 364, 456–459, 461
  basic transformations of, 367–368
  with horizontal axes of symmetry,
    456–459, 461
  translations of, 374–375
  vertex form of, 375–377, 380
  vertex of, 364, 367, 371–372, 373,
    456–457, 458–459
Parallel lines, 81–83
Parentheses (( )), 118
Pascal's triangle, 498–499, 502
Percent problem, 106–108, 109
Percentage, changing to a decimal, 106,
    109
Perfect cube, 321
Perfect nth power, 321
Perfect square, 321
Perfect square trinomial, 385–386
  factoring, 231–232, 234, 235
Perimeter of a rectangle, 27, 31
Perpendicular lines, 81–83
Point, 122
Point of curvature, 347
Point–slope form, 74–79, 83
Polya, George, 102
Polynomial, 197–237

  addition of, 199–201, 206
  cubic, 199
  definition of, 199
  degree of, 199
  distributive properties and, 207, 213
  divided by nonzero polynomials, 252
  division of. *See* Division of
    polynomials
  division of a polynomial by, 297–298,
    303
  leading coefficient of, 199
  linear, 199
  multiplication of, 207–214
  with one term, 208
  of one variable, 199
  opposite of, 201, 206
  properties of exponents and, 207–208,
    214
  quadratic, 199
  reciprocal of, 264
  subtraction of, 199–200, 201–202, 206
  with three terms, 208. *See also*
    Factoring trinomials; Trinomial
  with two terms, 208. *See also* Binomial
  zeros of, 215–216, 218, 221
Polynomial equation, 214, 238–244
  graphical solution to, 217, 218, 219,
    241, 242
  higher degree, 240–241, 244, 404–405,
    407
  numerical solution to, 217, 218, 219,
    241, 242
  quadratic, 238–240, 244
  in quadratic form, 241–242, 244
  solutions to, 218, 221
  solving by factoring. *See* Factoring
    polynomials
  symbolic solution to, 217, 218, 219,
    241, 242
Polynomial function, 202, 206
  applications and models using, 204–205
  cubic, 202
  evaluating, 202–203, 206
  linear, 202
  multiplication of, 213–214
  quadratic, 202
  sums and differences of, 204, 206
Positive number, 9
Positive slope, 68
Positive square root, 29, 31, 313
Power
  of i, 353
  raising to powers, 19–20, 25
Power function, 338–340, 341
Power rule, 19–20, 25
  for exponents, 318, 319
  for products, 318, 319
  for quotients, 318, 319
  for solving radical equations, 341–342,
    349
Principal nth root of a, 314
Principal square root, 29, 31, 313

Problem solving, 100–109
  percentage problems, 106–108, 109
  solving a formula for a variable, 100–102, 109
Problem solving steps, 102–106
  for linear programming problems, 165
  for systems of linear equations, 153–156
Product, 11. *See also* Multiplication
  of exponential expressions, 18–19, 20
  power rule for, 318, 319
  special, of binomials, 211–213, 214
  of a sum and a difference, 211–212, 214
Product rule, 18–19, 25
  for exponents, 318, 319
  for logarithms, 439–440, 443
  for radical expressions, 320–322, 325
Proportion, 289–291, 296
Proportionality. *See also* Variation
  constant of, 292, 293
Pure imaginary number, 351
Pythagorean theorem, 345–347, 349

Quadrant, 34, 40
Quadratic equation, 238–240, 241–242, 244, 380–389
  applications of, 238–239, 387–388
  completing the square for solving, 385–386, 389, 391
  with complex solutions, 394–396, 397
  discriminant and, 393–394, 397
  graphical solution to, 238, 381–382, 391, 392
  numerical solution to, 382–383
  solving, 390–392
  solving for a variable, 386–387
  square root property for solving, 383–384, 389
  symbolic solution to, 238, 281–282, 391, 392
Quadratic form, equations in, 404–407
  polynomial equations and, 241–242, 244
Quadratic formula, 390–391, 397
Quadratic function, 364, 372
  graphs of, 364–366, 372. *See also* Parabola
  min–max applications of, 370–372
Quadratic inequality, 398–403
  graphical solution to, 399–401, 403
  numerical solution to, 399, 403
  symbolic solutions to, 401–403
Quadratic polynomial, 199
Quadratic polynomial function, 202
Quotient, 11. *See also* Division
  of exponential expressions, 18–19, 25
  of logarithms, 445
  logarithms of, 445
  negative exponents for, 318, 319
  power rule for, 318, 319
  raising to powers, 19–20, 25
  of rational expressions, 264
Quotient rule, 18–19, 25
  for exponents, 318, 319
  for logarithms, 440, 443
  for radical expressions, 3222–325

Radical
  index of, 314, 322, 325–326
  like, 325–326, 333
Radical equation
  graphical solution to, 343
  power rule for solving, 341–343, 349
  solving, 341–345, 347–350
  symbolic solution to, 343
Radical expression, 312–319
  addition of, 325–327, 333
  graphing with graphing calculator, 345
  multiplication of, 329–330, 333
  product rule for, 320–322, 325
  quotient rule for, 322–325
  rationalizing the denominator and, 330–333
  simplifying, 320–325
  subtraction of, 328–329, 333
Radical function, 334–341
  cube root function, 337–338, 340
  domain of, 336–337
  power functions, 338–341
  square root function, 334–337, 340–341
Radical notation, 312–315
  Radical sign, 312
Radicand, 312
  of like radicals, 325–326
Radius of a circle, 459–460
Range
  of a function, 53–54, 58
  of a relation, 32–33, 40
Rate of change, 60, 63–65, 66
Ratio
  common, 488–489
  constant of proportionality (variation) and, 293
Rational equation, 256–258, 259, 275–283
  graphical solution to, 258, 277
  numerical solution to, 258, 278
  solving for a variable, 282
  symbolic solution to, 258
Rational exponent, 315–319
  equations with, 405–407
  properties of, 315–319
Rational expression, 252
  addition of, 270–271, 275
  division of, 264–266, 267
  multiplication of, 262–264, 267
  negative signs and, 261
  quotients of, 264
  reciprocal of, 264
  simplifying, 260–261, 267
  subtraction of, 272–274, 275
  sums and differences of, 274, 275
Rational function, 252–255, 259
  domain of, 253

evaluating, 254
  graphing, 253
  operations on, 258–259
Rational number, 3, 7
Rationalizing the denominator, 330–333
Real number, 4–8, 8–14
  absolute value of, 9, 14
  addition of, 9–11, 14
  division of, 11–13, 14
  multiplication of, 11–13, 14
  nth roots of, 314–315, 319
  properties of, 5–7
  standard form of, 24–25
  subtraction of, 9–11, 14
Real number line, 8–9
Real part of a complex number, 351–355
Reciprocal, 11
  negative, 82
  of a polynomial, 264
  of a rational expression, 264
Rectangle, perimeter of, 27, 31
Reduced row–echelon form, 178–182, 184
Region of feasible solutions, 164
Relation, 52–53
  domain of, 32–33, 40
  graphing, 31–40
  range of, 32–33, 40
Remainder in synthetic division, 302
Repeating decimal, 3
Representations of a function
  diagrammatic, 50, 58
  graphical, 49, 50, 52, 58, 62, 66, 335, 338
  numerical, 49, 50, 5, 58, 62, 66, 335, 338
  symbolic, 49, 50, 51, 58, 63–64, 66, 335, 338
  verbal, 49, 50, 58, 62, 66
Right triangle, Pythagorean theorem and, 345–347, 349
Rise, 67–68, 73
Root
  cube, 29–30, 31
  even, 314
  nth, 314–315, 319, 321, 325
  odd, 314
  square, 29–30, 31
  square, principal, 29, 31
Root function, 338, 340
Row of a matrix, 177
Row transformation, 178–182
Run, 67–68, 73

Scatterplot, 36–37, 39–40
Schatz, James R., 1
Scientific notation, 22–24, 25
Sequence, 480–485
  applications and models using, 489–490
  arithmetic, 485–448, 491
  finite, 481

general term of, 481, 488
generating with graphing calculator, 482
geometric, 488–489, 491
graphical representation of, 482–483, 485
infinite, 481
models and applications using, 483–484
numerical representation of, 482–483
series compared with, 492
terms of, 481–482
Series, 491–497
   arithmetic, 492–494
   finite, 492, 497
   geometric, 494–495
   sequences compared with, 492
   summation notation and, 496–497
Set braces ({ }), 2
Set-builder notation, 109, 116
Sets of numbers, 2–8
Sigma (Σ), 496–497
Simplifying
   radical expressions, 320–325
   rational expressions, 260–261, 267
Simplifying complex fractions, 284–288
   by multiplying by least common denominator, 286–288
   by simplifying numerator and denominator first, 285–286, 288
Slope, 67–74
   common difference and, 487
   interpreting, 71–73
   negative, 68
   of parallel lines, 81–83
   of perpendicular lines, 81–83
   positive, 68
   rate of change as, 71–73, 74
   undefined, 68
   zero, 68
Slope–intercept form, 69–71, 76–77, 83
Solution
   approximate, 96
   complex, equations with, 394–396, 397, 406, 407
   extraneous, 257
   to a linear equation, 90, 91–95, 99
   to a linear inequality, 110
   to a polynomial equation, 218, 221
   to a quadratic equation, 239
   to systems of two equations in two variables, 138–141, 146
Solution set
   intersection of, 118
   to a linear equation, 90, 99
   to a linear inequality, 109, 110
Square, 15
   of a binomial, 212, 214
   differences of, factoring, 230–231, 233, 235, 237
   perfect, 321
Square matrix, 177

Square root, 29–30, 31
   approximating, 313
   of negative numbers, 334, 351
   positive, 313
   principal, 313
   principal (positive), 29, 31
Square root function, 334–337, 340
   common logarithmic function compared with, 443
Square root property, 383–384, 389
Square viewing rectangle, 461
Standard equation
   of a circle, 459–461
   of an ellipse, 463, 468
   of a hyperbola, 465, 468
Standard form
   of a complex number, 351
   of a line, 98, 99
   of a parabola with a vertical axis, 275–277, 380
   of a real number, 24–25
   of systems of two equations in two variables, 138
Standard viewing rectangle, 38
Subscript, 68
Substitution method, 147–149, 151, 156
   for solving nonlinear systems of equations, 469–471
   for systems of linear equations in three variables, 171, 184
Subtraction. *See also* Difference
   of complex numbers, 352, 355
   of fractions, 270
   negation and, 11
   of polynomials, 199–200, 201–202, 206
   of radical expressions, 328–329, 353
   of rational expressions, 272–274, 275
   of rational functions, 258–259
   of real numbers, 11, 14
Sum, 9. *See also* Addition
   of first n terms of an arithmetic sequence, 493
   of first n terms of a geometric sequence, 494–495
   of polynomial functions, 204, 206
   product of a sum and a difference and, 211–212, 214
   of rational expressions, 274, 275
   of a series, graphing calculator for finding, 494
   of two cubes, factoring, 232–233, 234
Summation
   index of, 496
   limits of, 496
Summation notation (Σ), 496–497
Symbolic representation
   of a function, 49, 50, 52, 58
   of a linear function, 63–64, 66
   of a radical function, 335, 338
Symmetry, axis of, 364
Synthetic division, 301–302, 303

Systems of equations, 138
   nonlinear, 468–471, 473
Systems of linear equations, 177–184
   consistent, 141, 142, 146
   dependent, 142, 143
   elimination method for solving, 149–153, 156
   Gauss–Jordan elimination method for solving, 178–182, 184
   graphing calculator for solving, 182–183
   inconsistent, 141, 144, 146, 175
   independent, 141, 142, 146
   reduced row–echelon form of, 178–182, 184
   representing with matrices, 177–178
   steps for solving, 153–156
   substitution method for solving, 147–149, 151, 156
   in three variables. *See* Systems of linear equations in three variables
   in two variables, 138–146
   in two variables, Cramer's rule for, 187–188, 189
Systems of linear equations in three variables, 168–176
   elimination method for solving, 171–174, 176
   with infinitely many solutions, 175–176
   with no solutions, 175
   substitution method for solving, 171, 176
Systems of linear inequalities, 159–162, 163
   solution set to, 152, 163
Systems of two equations in two variables, 138–146
   graphical solutions to, 139–141, 145
   numerical solutions to, 139–141, 145
   solutions to, 138–141, 146
   standard form of, 138

Table, 30
   graphing calculator for creating, 57
Table of values, 49, 58
   factoring trinomials with, 227, 229
   for an inverse function, 422–423
Term, 198
Term of a sequence, 481–482
   general, 481
   nth, 481
Terminating decimal, 3
Test point, 158
Three-part inequality, 119–121, 123
Translation of a parabola, 374–375
Transverse axis of a hyperbola, 465
Triangle
   area of, 31, 186–187, 189
   right, 345–347, 349
Trinomial, 208
   factoring. *See* Factoring trinomials

perfect square, 231–232, 234, 236, 285–286
Triple, ordered, 168

Undefined slope, 68
Union notation, 120

Variable, 26
    dependent, 48, 58
    independent, 48, 57
    solving problems for, 100–102, 109
Variation
    constant of, 292, 293
    direct, 292–293, 297
    inverse, 293–295, 297
    joint, 295–296, 297
Verbal representation
    of a function, 40, 50, 58
    of a linear function, 62, 66
Vertex, 165
    of an ellipse, 462–463
    of a hyperbola, 465
    of a parabola, 364, 367, 371, 373, 456–457, 458–459
Vertex form of a parabola, 375–377, 380
Vertical asymptote, 254–255
Vertical line, 79–80, 83
Vertical line test, 55–56, 58, 418
Viewing rectangle (window), 38
    decimal (friendly), 255
    square, 82, 461
    standard, 38
Volume
    of a cube, 27, 31
    of a cylinder, 296

Whole number, 2, 7
Wigner, Eugene, 26
Window, 38
    decimal (friendly), 255
    square, 82, 461
    standard, 38

x-axis, 34
x-intercept, 77, 218–219, 221
Xmax, 38
Xmin, 38
Xscl, 38
xy-plane, 34–35, 40

y-axis, 34
y-intercept, 69–71, 74
Ymax, 38
Ymin, 38
Yscl, 38

Zero
    identity property of, 5, 8
    of a polynomial, 216, 218, 221
Zero exponent, 16–17, 25
Zero slope, 68
Zero-product property, 215–216, 221

Formulas and Equations

$m = \dfrac{y_2 - y_1}{x_2 - x_1}$ Slope of a line

$ax + b = 0$ Linear equation

$y = mx + b$ Slope–intercept form

$y = m(x - x_1) + y_1$ Point–slope form

$x = h$ Vertical line

$y = b$ Horizontal line

$d = rt$ Distance, rate, and time

$s = 16t^2$ Distance for a falling object

$a^2 - b^2 = (a - b)(a + b)$ Difference of two squares

$(a + b)^2 = a^2 + 2ab + b^2$
$(a - b)^2 = a^2 - 2ab + b^2$ Square of a binomial

$x = -\dfrac{b}{2a}$ Vertex formula (x-coordinate)

$y = a(x - h)^2 + k$ Vertex form

$ax^2 + bx + c = 0$ Quadratic equation

$x = \dfrac{-b \pm \sqrt{b^2 - 4ac}}{2a}$ Quadratic formula

$(x - h)^2 + (y - k)^2 = r^2$ Equation of a circle

$d = \sqrt{(x_2 - x_1)^2 + (y_2 - y_1)^2}$ Distance between two points

Geometry

Rectangle
$A = LW$
$P = 2L + 2W$

Triangle
$A = \tfrac{1}{2}bh$
$P = a + b + c$

Pythagorean Theorem
$c^2 = a^2 + b^2$

Circle
$C = 2\pi r$
$A = \pi r^2$

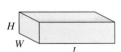

Rectangular (Parallelepiped) Box
$V = LWH$
$S = 2LW + 2LH + 2WH$

Cylinder
$V = \pi r^2 h$
$S = 2\pi rh + 2\pi r^2$

Sphere
$V = \tfrac{4}{3}\pi r^3$
$S = 4\pi r^2$

Cone
$V = \tfrac{1}{3}\pi r^2 h$
$S = \pi r^2 + \pi r\sqrt{r^2 + h^2}$

Library of Functions
Basic Functions

Several important functions are used in algebra. The following provides symbolic, numerical, and graphical representations for several of these basic functions.

Absolute Value Function: $f(x) = |x|$

x	-2	-1	0	1	2
$|x|$	2	1	0	1	2

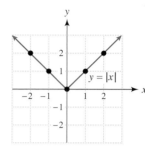

Domain: $(-\infty, \infty)$
Range: $[0, \infty)$

Square Function: $f(x) = x^2$

x	-2	-1	0	1	2
x^2	4	1	0	1	4

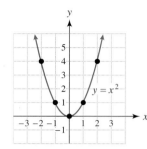

Domain: $(-\infty, \infty)$
Range: $[0, \infty)$

Cube Function: $f(x) = x^3$

x	-2	-1	0	1	2
x^3	-8	-1	0	1	8

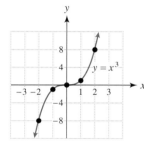

Domain: $(-\infty, \infty)$
Range: $(-\infty, \infty)$

Square Root Function: $f(x) = \sqrt{x}$

x	0	1	4	9
$\sqrt{x}$	0	1	2	3

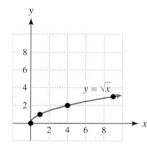

Domain: $[0, \infty)$
Range: $[0, \infty)$

Cube Root Function: $f(x) = \sqrt[3]{x}$

x	-8	-1	0	1	8
$\sqrt[3]{x}$	-2	-1	0	1	2

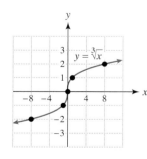

Domain: $(-\infty, \infty)$
Range: $(-\infty, \infty)$

Reciprocal Function: $f(x) = \dfrac{1}{x}$

x	-2	-1	0	1	2
$\dfrac{1}{x}$	$-\dfrac{1}{2}$	-1	—	1	$\dfrac{1}{2}$

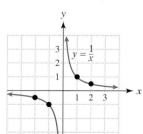

Domain: $(-\infty, 0) \cup (0, \infty)$
Range: $(-\infty, 0) \cup (0, \infty)$